Advanced Textbooks in Control and Signal Processing

Series editors

Michael J. Grimble, Glasgow, UK
Michael A. Johnson, Kidlington, UK

More information about this series at http://www.springer.com/series/4045

Zhiyong Chen · Jie Huang

Stabilization and Regulation of Nonlinear Systems

A Robust and Adaptive Approach

Springer

Zhiyong Chen
School of Electrical Engineering
and Computer Science
University of Newcastle
Callaghan, NSW
Australia

Jie Huang
Department of Mechanical and Automation
Engineering
The Chinese University of Hong Kong
Shatin, Hong Kong
China

ISSN 1439-2232
ISBN 978-3-319-08833-4 ISBN 978-3-319-08834-1 (eBook)
DOI 10.1007/978-3-319-08834-1

Library of Congress Control Number: 2014946592

Springer Cham Heidelberg New York Dordrecht London

Printed on acid-free paper

Springer is part of Springer Science+Business Media (www.springer.com)

To Selina, Ruth, and Andrew

Z. Chen

To Qingwei, Anne, and Jane

J. Huang

Series Editors' Foreword

The topics of control engineering and signal processing continue to flourish and develop. In common with general scientific investigation, new ideas, concepts, and interpretations emerge quite spontaneously and these are then discussed, used, discarded, or subsumed into the prevailing subject paradigm. Sometimes, these innovative concepts coalesce into a new subdiscipline within the broad subject tapestry of control and signal processing. This preliminary battle between old and new usually takes place at conferences, through the Internet and in the journals of the discipline. After a little more maturity has been acquired by the new concepts then archival publication as a scientific or engineering monograph may occur.

A new concept in control and signal processing is known to have arrived when sufficient material has evolved for the topic to be taught as a specialized tutorial workshop or as a course to undergraduate, graduate, or industrial engineers. *Advanced Textbooks in Control and Signal Processing* are designed as a vehicle for the systematic presentation of course material for both popular and innovative topics in the discipline. It is hoped that prospective authors will welcome the opportunity to publish a structured and systematic presentation of some of the newer emerging control and signal processing technologies in the textbook series.

The Editors of *Advanced Textbooks in Control and Signal Processing* are aiming for the series to develop as a library of textbooks of high quality that cover some fundamental topics in the discipline but also include some coherent and systematic presentations of advanced topics that are gaining textbook status. A thorough and systematic presentation of the methods of nonlinear control has been one sought-after entry that has been missing from the series until now. The Editors are therefore understandably pleased to welcome this control textbook *Stabilization and Regulation of Nonlinear Systems: A Robust and Adaptive Approach* by authors Zhiyong Chen and Jie Huang into the series.

Practising control engineers deal with system nonlinearities by using a set of techniques that are extensions from the linear control paradigm. Methods like gain scheduling, controller scheduling, linear controller design across multiple models, and extended Kalman filters are all "constructive" approaches to nonlinear system problems. They have advantages based on simplicity of understanding and implementation. But the field of nonlinear systems theory is now well developed and there is scope for more application of its techniques on industrial processes by control engineers. This course textbook for nonlinear systems methods adopts a structured approach to the theory that should assist engineers to understand the fundamentals, the subtleties, and the application of robust and adaptive approaches to nonlinear systems.

The authors are highly respected for their work in the nonlinear systems field and have interleaved the theory presented in the textbook with instructive demonstration examples. Zhiyong Chen is currently an Associate Professor of the School of Electrical Engineering and Computer Science at the University of Newcastle, NSW, Australia and has been awarded several best paper prizes at prestigious control conferences. Jie Huang is the Chairman and Choh-Ming Li Professor of Mechanical and Automation Engineering at the Chinese University of Hong Kong, Hong Kong, China. He has been a Distinguished Lecturer for the IEEE Control Systems Society, and is an IFAC and IEEE Fellow.

The book of Zhiyong Chen and Jie Huang can be described as a comprehensive and systematic course textbook on nonlinear control systems and will be welcomed by readers seeking to learn about and understand the foundations of the subject. However, the *Advanced Textbooks in Control and Signal Processing* series contains three titles that pursue nonlinear control, robust control, and adaptive control ideas in the context of specific application fields. Readers may also find the titles below of interest:

- *Analysis and Control of Nonlinear Process Systems* by Katalin M. Hangos, József Bokor and Gábor Szederkényi (ISBN 978-1-85233-600-4, 2003);
- *Robust and Adaptive Control with Aerospace Applications* by Eugene Lavretsky and Kevin Wise (ISBN 978-1-4471-4395-6, 2013); and
- *Robust Control Design with MATLAB®* by Da-Wei Gu, Petko Hr. Petrov and Mihail M. Konstantinov (ISBN 978-1-4471-4681-0, Second Edition, 2005).

Glasgow, UK, May 2014

M.J. Grimble
M.A. Johnson

Preface

Stabilization and output regulation are two central design problems in nonlinear control theory and applications. Stabilization refers to the process of designing a feedback control law for a controlled plant such that certain invariant manifold of the closed-loop system is stable in a broad sense. A motion trajectory and an equilibrium point of a dynamic system are special cases of an invariant manifold of the dynamic system. Output regulation, on the other hand, aims to design a feedback control law for a controlled plant subject to some exogenous signal such that certain invariant manifold of the closed-loop system is stable and, in addition, the output of the plant asymptotically approaches a given reference input determined by the exogenous signal. Thus, the output regulation problem is more demanding than the stabilization problem. Both the disturbance and the reference input of a control system can be viewed as the exogenous signal. Thus, the output regulation problem is also referred to as asymptotic tracking and disturbance rejection problem, or servomechanism design problem. In comparison with some other methods dealing with asymptotic tracking and/or disturbance rejection problem, a celebrated feature of the output regulation problem is that the control law is required to be able to handle a class of exogenous signals generated by a dynamic system called exosystem.

The output regulation problem arises from mathematically formulating practical control problems such as vibration suppression of high speed trains, disturbance rejection for flight vehicles, landing of aircraft on carriers under severe weather condition, and coordination and manipulation of robots. Thus this problem has attracted the attention of the control community for several decades and it has also been a driving force for the advancement of modern control theory and applications.

The output regulation problem was first studied for the class of linear time-invariant systems by Davison, Francis, and Wonham, to name just a few. The main tool for dealing with the output regulation problem is the internal model principle. By this principle, the output regulation problem of a given plant can be converted into the stabilization problem of the augmented system composed of the given plant and a well-defined dynamic compensator called internal model. The well-known integral control can be viewed as a special case of the internal model principle when the exogenous signals are constant.

Since the early 1990s, the research on the output regulation problem has been focused on nonlinear systems. The problem has attracted the attention of numerous researchers from the world. By the time of the mid-2000s, the research on the nonlinear output regulation problem had achieved a degree of maturity. In addition to numerous research papers, four monographs on this topic were published.

[1] C. I. Byrnes, F. Delli Priscoli, and A. Isidori, *Output Regulation of Uncertain Nonlinear Systems, Systems and Control: Foundations and Applications*, Birkhauser: Cambridge, 1997.
[2] A. Isidori, L. Marconi, and A. Serrani, *Robust Autonomous Guidance: An Internal Model-Based Approach. Advances in Industrial Control*, Springer Verlag: London, 2003.
[3] J. Huang, *Nonlinear Output Regulation: Theory and Applications*, SIAM: Philadelphia, 2004.
[4] A. Pavlov, N. van de Wouw, and H. Nijmeijer, *Uniform Output Regulation of Nonlinear Systems: A Convergent Dynamics Approach*, Birkhauser: Berlin, 2006.

The book [1] studied the local output regulation problem of nonlinear systems. The book [2] focused on semi-global output regulation for flight vehicles using the internal model approach. The book [3] further studied both local and global output regulation or nonlinear uncertain systems by robust control approach. The book [4] was based on Jacobian analysis of nonlinear systems that is mainly effective for regional output regulation.

It is noted that the main tool for studying the output regulation problem in the first three books [1–3] is the internal model design approach. An internal model is a dynamic compensator which together with the given plant constitutes a so-called augmented system. An internal model is conceived such that the stabilization solution of the augmented system leads to the solution of the output regulation problem of the original plant. Thus, this design framework has made a connection between the stabilization problem and the output regulation problem. In contrast, the book [4] studied the problem using a convergent dynamics approach. In particular, a system is designed to have the property that all its solutions "forget" their initial conditions and converge to each other. The convergent steady-state solution is uniquely defined by an exosystem. Jacobian analysis of nonlinear systems was applied to achieve the convergent property.

Since the mid-2000s, the scope of research on the output regulation problem has experienced a tremendous expansion along several directions summarized as follows. First, several new methods for constructing more general internal models have been developed. These new internal models have led to the solution of the output regulation problem for more general nonlinear plants and exosystems on one hand, and, on the other hand, have generated several new types of robust stabilization problems for more complex nonlinear systems, thus leading to some interesting results and techniques for global stabilization of uncertain nonlinear systems. Second, in order to handle uncertainty in exosystems, arbitrarily large exogenous signals and uncertain parameters, or unknown control direction, several adaptive control techniques have been incorporated into the original design framework for the output regulation problem, thus leading to the so-called adaptive output regulation problem. Third, extensive efforts have been made to solve real world practical control problems by output regulation theory. Some recent examples are asymptotic tracking of Chua's circuit, spacecraft attitude control, speed control of surface mounted motor, robust regulation of hyperchaotic Lozenz system, etc.

Given the rich new results of the nonlinear output regulation theory and applications obtained in recent years, and the need for studying the state-of-the-art techniques for handling the stabilization problem and the output regulation problem of nonlinear systems by graduate students and researchers in both academia and industries, we bring this book to readers. The book can be used as a textbook for graduate students in all engineering disciplines and applied mathematics. It can also be used as a reference book for both practitioners and theorists in broad areas of electrical engineering, aerospace engineering, mechanical engineering, and chemical engineering. Readers are assumed to have some knowledge of the fundamentals of linear algebra, advanced calculus, and linear systems.

The authors seek to strike a balance between the theoretical foundations of the output regulation problem and practical applications of the theory. The treatment is accompanied by many examples, including practical case studies with numerical simulations based on MATLAB. The book was typeset using LaTeX.

The development of this book would not have been possible without the support and help from many people. The authors are grateful to Alberto Isidori for his recommendation and encouragement for the publication of the book, and are indebted to the Series Editors Michael J. Grimble and Michael A. Johnson for their support and constructive comments. The Springer Editor Oliver Jackson, Managing Director Alexander Grossmann, and Senior Editorial Assistant Charlotte Cross are extremely helpful and enthusiastic in their advice and assistance. He Cai, Yi Dong, Wei Liu, and Maobin Lv, Ph.D. students of Jie Huang, Lijun Zhu and Haofei Meng, Ph.D. students of Zhiyong Chen, have proofread the manuscript. Some sections from Chaps. 6 to 9 are adapted from the joint publications with Lu Liu, Zhaowu Ping, Youfeng Su, Dabo Xu, and Xi Yang.

The bulk of this research was supported by the Australian Research Council under grants DP0878724 and DP130103039, the Hong Kong Research Grants Council under grants CUHK412810, CUHK412611, CUHK412612, and CUHK412813, and the National Science Foundation of China under grants 51120155001, 51328501, 61174049, and 61322304.

The authors would also like to thank Sharmila, Ishwarya Jayshree Krishna and Karin de Bie for their help in typesetting this book.

Newcastle, Australia, April 2014 — Zhiyong Chen
Hong Kong — Jie Huang

Contents

Symbols

$\|\cdot\|$	$\|x\|$	2-Norm of a vector x
$\|\cdot\|$	$\|A\|$	Induced 2-norm of a matrix A
$\mathbb{R}^n$	$x \in \mathbb{R}^n$	n-Dimensional Euclidean space $(\mathbb{R} = \mathbb{R}^1)$
$\mathbb{R}^{n \times m}$	$A \in \mathbb{R}^{n \times m}$	Set of $n \times m$ matrices with elements in $\mathbb{R}$
$\mathbb{R}^+$	$x \in \mathbb{R}^+$	Non-negative real number set
$\mathbb{C}^n$	$x \in \mathbb{C}^n$	n-Dimensional complex space $(\mathbb{C} = \mathbb{C}^1)$
$\mathbb{I}$	$x \in \mathbb{I}$	Set of integers
I_n(or I)	λI_n(or λI)	$n\times\ n$ identity matrix
$\in$	$\lambda \in \sigma(A)$	λ is a member of $\sigma(A)$
$\notin$	$\lambda \notin \sigma(A)$	λ is not a member of $\sigma(A)$
$\subset$	$A \subset B$	A set A is a subset of a set B
$\otimes$	$A \otimes B$	Kronecker product
$\oplus$	$A \oplus B$	Tracy-Singh product
$\circ$	$f_1 \circ f_2$	Composition of functions $(f_1 \circ f_2)(x) = f_1(f_2(x))$
$\mapsto$	$f : \mathbb{A} \mapsto \mathbb{B}$	A function f mapping a set $\mathbb{A}$ into a set $\mathbb{B}$
sup	$\sup_{t_1 \leq t \leq t_2} \|x(t)\|$	Supremum norm of $x(t)$ in $[t_1, t_2]$
col	$\mathrm{col}(x_1, \cdots, x_r)$	Vector stacked by $x_1 \in \mathbb{R}^{n_1}, \cdots, x_r \in \mathbb{R}^{n_r}$
diag	$\mathrm{diag}(A_1, \cdots, A_r)$	Matrix diagonalized by $A_1 \in \mathbb{R}^{n_1 \times n_1}, \cdots, A_r \in \mathbb{R}^{n_r \times n_r}$
T	A^{T}	Transpose of $A \in \mathbb{R}^{n \times m}$
lim sup	$\limsup_{k \to \infty} f(k)$	Limit superior
lim inf	$\liminf_{k \to \infty} f(k)$	Limit inferior

Acronyms

AG	Asymptotic gain
AS	Asymptotically stable
ATDRP	Attitude tracking and disturbance rejection problem
ES	Exponentially stable
GARP	Global adaptive regulation problem
GAS	Globally asymptotically stable
GASP	Global adaptive stabilization problem
GES	Globally exponentially stable
GRORP	Global robust output regulation problem
GRSP	Global robust stabilization problem
GS	Globally stable
iISS	Integral input-to-state stable
ISS	Input-to-state stable
RAG	Robustly asymptotic gain
RES	Robustly exponentially stable
RGES	Robustly globally exponentially stable
RGS	Robustly globally stable
RISS	Robustly input-to-state stable
RORP	Robust output regulation problem
RUAS	Robustly uniformly asymptotically stable
RUGAS	Robustly uniformly globally asymptotically stable
SISO	Single input and single output
UAS	Uniformly asymptotically stable
UGAS	Uniformly globally asymptotically stable
US	Uniformly stable

Chapter 1
Introduction

Nonlinear systems are ubiquitous in the real world. In this chapter, we first introduce the state space description of nonlinear systems, nonlinear control systems, and nonlinear control laws in Sect. 1.1. Then, in Sect. 1.2, we describe three specific nonlinear control systems and use them to illustrate the stabilization problem and the output regulation problem of nonlinear systems. Finally, in Sect. 1.3, we close this chapter with the organization of this book.

1.1 Nonlinear Systems

The state space representation of a nonlinear system is described by the following equation:

$$\dot{x}(t) = f(x(t), t), \ \ x(t_0) = x_0, \ \ t \geq t_0 \tag{1.1}$$

where $t_0 \geq 0$ is called the initial time, $x : [t_0, \infty) \mapsto \mathbb{R}^n$ is called the state trajectory with $x(t_0)$ called the initial state, and $f : \mathbb{R}^n \times [t_0, \infty) \mapsto \mathbb{R}^n$ is a nonlinear vector function. More specifically, the components of x and f are denoted by

$$x = [x_1 \ldots x_n]^{\mathrm{T}}, f = [f_1, \ldots, f_n]^{\mathrm{T}}.$$

The system (1.1) is called a *non-autonomous* or *time-varying* system. The linear system $\dot{x}(t) = A(t)x(t)$ where $A(t)$ is a square matrix of dimension n is a special case of (1.1) with $f(x(t), t) = A(t)x(t)$.

If the time t does not explicitly appear in the function f, then the system (1.1) reduces to

$$\dot{x}(t) = f(x(t)). \tag{1.2}$$

A dynamic system of the form (1.2) is called an *autonomous* or *time-invariant* system.

For convenience, the systems (1.1) and (1.2) can often be simplified as $\dot{x} = f(x, t)$ and $\dot{x} = f(x)$, respectively.

Z. Chen and J. Huang, *Stabilization and Regulation of Nonlinear Systems*,
Advanced Textbooks in Control and Signal Processing,
DOI 10.1007/978-3-319-08834-1_1

Example 1.1 Consider the model of a mass-spring system with a hardening spring subject to an external excitation, described by the differential equation

$$\ddot{y} + y + y^3 = \cos t, \quad t \geq t_0. \tag{1.3}$$

Denote $x = \text{col}(x_1, x_2)$ with $x_1 = y$ and $x_2 = \dot{y}$. Then the system (1.3) can be expressed in the form (1.1) with

$$f(x, t) = \begin{bmatrix} x_2 \\ -x_1 - x_1^3 + \cos t \end{bmatrix}$$

and $x(t_0) = \text{col}(y(t_0), \dot{y}(t_0))$.

A general non-autonomous nonlinear *control system* is described by the following equations:

$$\dot{x} = f(x, u, t), \quad x(t_0) = x_0, \quad t \geq t_0 \tag{1.4}$$

$$y = h(x, u, t) \tag{1.5}$$

$$y_{\text{m}} = h_{\text{m}}(x, u, t) \tag{1.6}$$

where x, x_0 and t_0 are defined as in (1.1), $u \in \mathbb{R}^m$ is the input, $y \in \mathbb{R}^p$ is the performance output, and $y_{\text{m}} \in \mathbb{R}^q$ is the measurement output. The functions $f : \mathbb{R}^n \times \mathbb{R}^m \times [t_0, \infty) \mapsto \mathbb{R}^n$, $h : \mathbb{R}^n \times \mathbb{R}^m \times [t_0, \infty) \mapsto \mathbb{R}^p$, and $h_{\text{m}} : \mathbb{R}^n \times \mathbb{R}^m \times [t_0, \infty) \mapsto \mathbb{R}^q$ satisfy $f(0, 0, t) = 0$, $h(0, 0, t) = 0$, and $h_{\text{m}}(0, 0, t) = 0$ for all $t \geq t_0$. A special case of (1.4)–(1.6) where t does not appear explicitly on the right hand side of these equations is as follows:

$$\dot{x} = f(x, u) \tag{1.7}$$

$$y = h(x, u) \tag{1.8}$$

$$y_{\text{m}} = h_{\text{m}}(x, u). \tag{1.9}$$

We call (1.7)–(1.9) an autonomous nonlinear control system.

A special case of the autonomous systems (1.7)–(1.9) is as follows

$$\begin{aligned} \dot{x} &= f(x) + g(x)u \\ y &= h(x) \\ y_{\text{m}} &= h_{\text{m}}(x) \end{aligned} \tag{1.10}$$

for some functions $f : \mathbb{R}^n \mapsto \mathbb{R}^n$, $g : \mathbb{R}^n \mapsto \mathbb{R}^{n \times m}$, $h : \mathbb{R}^n \mapsto \mathbb{R}^p$, and $h_{\text{m}} : \mathbb{R}^n \mapsto \mathbb{R}^q$. We call (1.10) an *affine* nonlinear control system. Note that $g(x)$ can be expanded as

$$g(x) = [g_1(x) \ \ldots \ g_m(x)]$$

where $g_i : \mathbb{R}^n \mapsto \mathbb{R}^n$ for $i = 1, \ldots, m$.

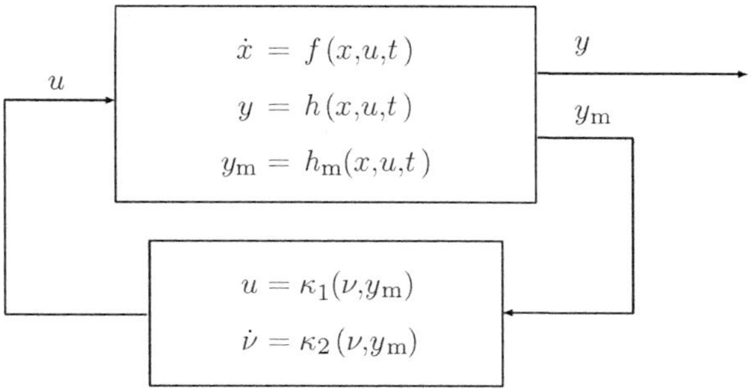

Fig. 1.1 Block diagram of a closed-loop system

A general nonlinear control law takes the following form

$$u = \kappa_1(\nu, y_m)$$
$$\dot{\nu} = \kappa_2(\nu, y_m) \tag{1.11}$$

where $\nu \in \mathbb{R}^s$ is the compensator state and $\kappa_1 : \mathbb{R}^s \times \mathbb{R}^q \mapsto \mathbb{R}^m$ and $\kappa_2 : \mathbb{R}^s \times \mathbb{R}^q \mapsto \mathbb{R}^s$ are two functions. The functions κ_1 and h_m need to satisfy the following algebraic equation:

$$u = \kappa_1(\nu, h_m(x, u, t)).$$

For simplicity, in the sequel, we assume the function h_m does not depend on u explicitly, that is,

$$y_m = h_m(x, t).$$

As a result,

$$u = \kappa_1(\nu, h_m(x, t)).$$

The controller (1.11) is called a *dynamic* measurement output feedback controller if $s > 0$ and a *static* measurement output feedback controller if $s = 0$, and it is called a *state feedback* controller if $y_m = x$ and an *output feedback* controller if $y_m = y$.

The concatenation of the control system (1.4)–(1.6) and the control law (1.11) is called the closed-loop system as shown in Fig. 1.1, which is a non-autonomous system of the form (1.1), i.e.,

$$\dot{x}_c = f_c(x_c, t) \tag{1.12}$$

where

$$x_c = \begin{bmatrix} x \\ \nu \end{bmatrix}, \quad f_c(x_c, t) = \begin{bmatrix} f(x, \kappa_1(\nu, h_m(x, t)), t) \\ \kappa_2(\nu, h_m(x, t)) \end{bmatrix}.$$

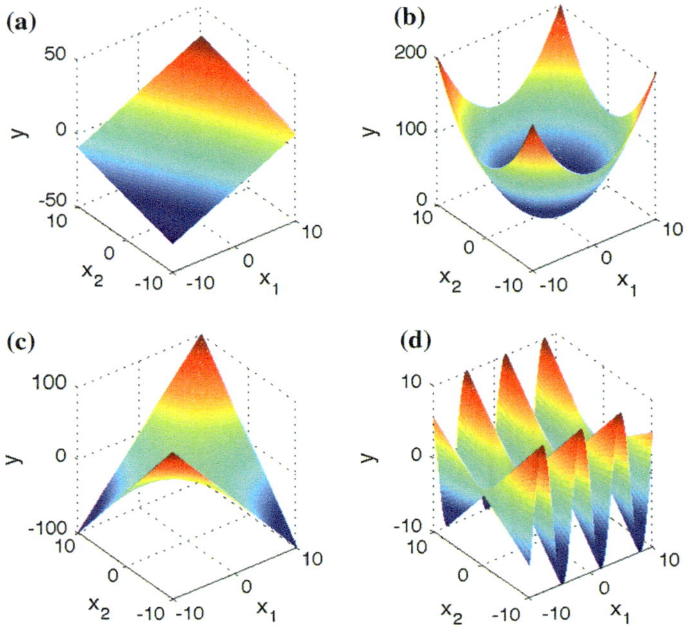

Fig. 1.2 Illustration of a linear function $y = 2x_1 + x_2$ **(a)** and various nonlinear functions $y = x_1^2 + x_2^2$, $y = x_1 x_2$, $y = \sin(x_1)x_2$ **(b–d)**

Figure 1.2 compares the graph of some nonlinear functions with the graph of a linear function. It can be seen that the variety of nonlinear functions is rather rich and, as a result, there are also many types of nonlinear systems. In the next section, we will introduce three examples of nonlinear control systems and use them to illustrate the stabilization problem and the output regulation problem of nonlinear control systems. More examples will be given in Chap. 3.

1.2 Examples of Nonlinear Control Systems

The Lorenz System: The Lorenz system, introduced in 1963 by Edward N. Lorenz, is a three-dimensional dynamic system shown below

$$\begin{aligned} \dot{\zeta}_1 &= \sigma(\zeta_2 - \zeta_1) \\ \dot{\zeta}_2 &= \zeta_1(\rho - \zeta_3) - \zeta_2 \\ \dot{\zeta}_3 &= \zeta_1\zeta_2 - \beta\zeta_3 \end{aligned} \tag{1.13}$$

where σ is called the Prandtl number, ρ is called the Rayleigh number, and β is a geometric factor. The state trajectory of the system shown in Fig. 1.3 displays a typical behavior of what is now known as chaotic motion. Chaotic motion is a special phenomenon of an unstable nonlinear system. This phenomenon may be harmful to the operation of a real system and it may often be desirable to apply a feedback control law to eliminate the chaotic motion in the system (1.13). For example, introducing a control input u into the right hand side of the second equation of (1.13) leads to the following nonlinear control system:

$$\begin{aligned}\dot{\zeta}_1 &= \sigma(\zeta_2 - \zeta_1)\\ \dot{\zeta}_2 &= \zeta_1(\rho - \zeta_3) - \zeta_2 + u\\ \dot{\zeta}_3 &= \zeta_1\zeta_2 - \beta\zeta_3\end{aligned} \tag{1.14}$$

which is called a controlled Lorenz system.

A circuit schematic that realizes the controlled Lorenz system is depicted in Fig. 1.4. Denote the voltages at the nodes labeled A, B, and C as $\zeta_1 = V_A$, $\zeta_2 = V_B$, and $\zeta_3 = V_C$, respectively. The control voltage u enters the circuit at the labeled node through a voltage adder. In the circuit,

$$\sigma = \frac{100\,\mathrm{k}\Omega}{R_a}, \quad \rho = \frac{V_a}{1\ \mathrm{volt}}, \quad \beta = \frac{100\,\mathrm{k}\Omega}{R_c}.$$

The problem of designing a control law of the form (1.11) such that the state trajectory starting from any initial state of the closed-loop system converges to the origin asymptotically is called a global stabilization problem. Clearly, if a control law solves the global stabilization problem of the controlled Lorenz system, then the control law eliminates the chaotic behavior in the controlled Lorenz system. If we further specify a performance output, say, $y = \zeta_2$, and we want to design a control law such that the state trajectory starting from any initial condition is bounded and the output can asymptotically track a given trajectory y_d, then the problem is called an asymptotic tracking problem. The output regulation problem as will be studied

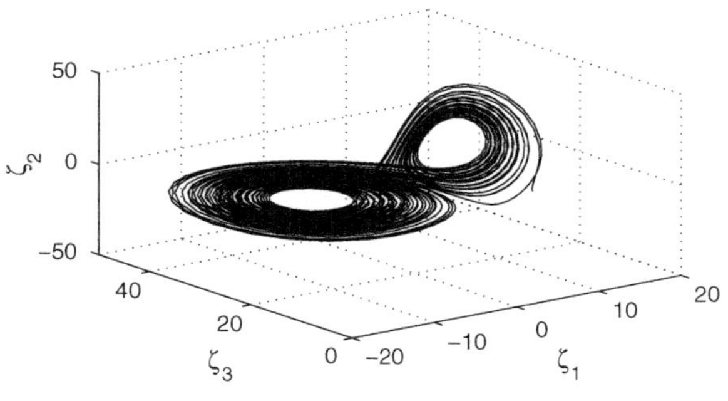

Fig. 1.3 Profile of the uncontrolled Lorenz system

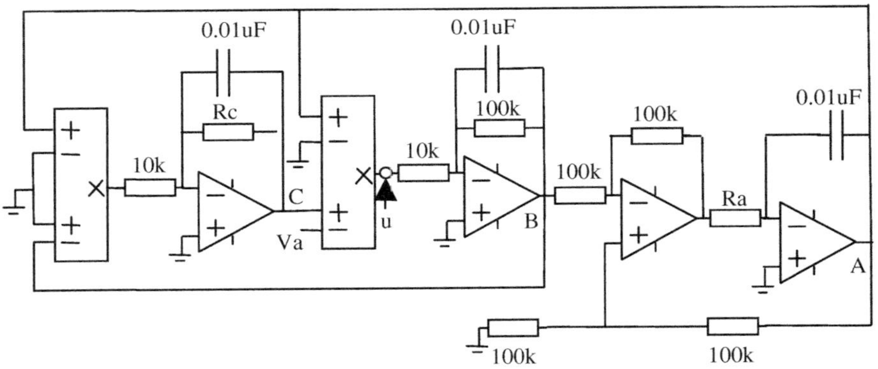

Fig. 1.4 Schematic diagram of a Lorenz circuit [2]

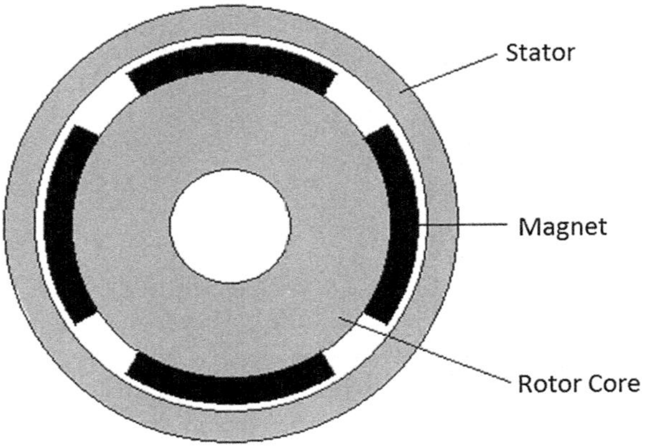

Fig. 1.5 The cross-sectional layout of a permanent magnet synchronous motor

in the second part of this book is one of the major tools for handling the asymptotic tracking problem.

Permanent Magnet Synchronous Motor: Permanent magnet (PM) synchronous motors have been used increasingly in various industrial applications in recent years because of their inherent advantages including high efficiency, good dynamics and small frame size compared to other kinds of motors with the same output. A PM synchronous motor does not have a field winding on the stator frame, and relies on permanent magnets to provide the magnetic field against which the rotor field interacts to produce torque (see Fig. 1.5). Conventionally, a two-phase equivalent circuit model (d-q model) has been used to analyze reluctance synchronous machines. This theory can also be applied in the analysis of PM synchronous motor. Following the convention of choosing the rotor reference frame, the direction of permanent magnet flux is chosen as the d-axis, while the q-axis is 90° ahead of the d-axis. Then, with the state $x = \text{col}(\omega_r, i_d, i_q)$, the motor has the following dynamics:

$$
\begin{aligned}
\dot{\omega}_r &= \frac{3p\Phi}{2J} i_q + \frac{3p}{2J}(L_d - L_q) i_d i_q - \frac{B}{J}\omega_r - \frac{1}{J}\tau_L \\
\dot{i}_d &= -\frac{R}{L_d} i_d + \frac{pL_q}{L_d} i_q \omega_r + \frac{1}{L_d} u_d \\
\dot{i}_q &= -\frac{R}{L_q} i_q - \frac{pL_d}{L_q} i_d \omega_r - \frac{p\Phi}{L_q}\omega_r + \frac{1}{L_q} u_q
\end{aligned} \tag{1.15}
$$

where ω_r is the rotor speed, τ_L is an unknown load torque, i_d and i_q are the d- and q-frame stator currents, u_d and u_q are the d- and q-frame stator voltages, respectively. The other parameters have the following physical meanings: p is the number of pole pairs, Φ the rotor flux, B the viscous friction coefficient, J the inertia, R the stator resistance, and L_d and L_q are the d- and q-axis inductances, respectively. The normal operation of the motor requires the motor rotor to rotate at a given reference speed $\omega_o(t)$ while keeping the current i_d at a given desirable current i_o. To achieve such an objective, one can design the controllers u_q and u_d such that, for the given reference speed $\omega_o(t)$ and reference current i_o, the rotor speed $\omega_r(t)$ converges to $\omega_o(t)$ and the current i_d converges to i_o, i.e., $\lim_{t\to\infty}(\omega_r(t)-\omega_o(t)) = 0$ and $\lim_{t\to\infty}(i_d(t) - i_o(t)) = 0$. Such a problem is another example of asymptotic tracking problem in control and can be handled by the output regulation problem with the performance output being specified as $y = \text{col}(w_r, i_d)$.

Rigid Spacecraft: The Hubble Space Telescope (HST) is an example for illustration of the stabilization and regulation requirements in control system design. The HST whirls around earth at a speed of 7.5 km/s at an altitude of 569 km, inclined 28.5° to the equator as shown in Fig. 1.6. In order to take images of distant and faint objects, the pointing control system was thus designed for extremely steady and accurate attitude tracking. It is able to lock onto a target without deviating more than

Fig. 1.6 Hubble space telescope (HST) in earth orbit. Photo from http://www.spacetelescope.org

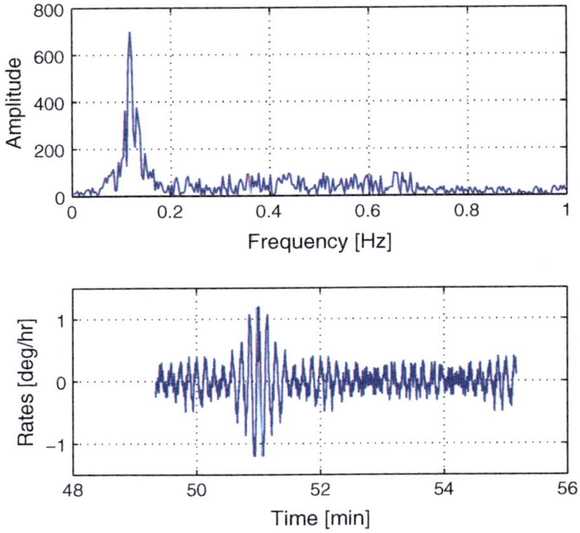

Fig. 1.7 Matlab generated curves imitating real rate gyroscope disturbance data in the HST. *Top* frequency spectrum; *bottom*: time history

7/1,000th of an arcsecond for the duration of an observation, which varies from a few seconds to a few hours. A more complicated situation is that the pointing control system is experiencing unexpectedly large disturbances. Several potential sources of the HST disturbances include the reaction wheel assemblies, data recorders, fine guidance sensors, high gain antennas, aperture door, magnetic torquers, thermal creak in the support systems module, the pointing control system, and the solar arrays. All of the potential sources were investigated systematically in view of the telemetry data which showed that the disturbances occurred mainly near 0.1 Hz as seen in Fig. 1.7. It is thus necessary to design a pointing control system to reject or attenuate the disturbance. From above, the pointing control system run in the 486 attitude-control computer of the HST was equipped with the attitude tracking and disturbance rejection algorithms. These algorithms were established on the ground of the fundamental control mechanisms for stabilization and regulation, including tracking and disturbance rejection. More specifically, the attitude kinematics and motion dynamics of a spacecraft can be modelled by the following equations, with the state $x = \mathrm{col}(q_v, q_4, \Omega)$,

$$\begin{aligned}\dot{q}_v &= \frac{1}{2}(q_4 I_3 + q_v^{\times})\Omega \\ \dot{q}_4 &= -\frac{1}{2}q_v^{\mathsf{T}}\Omega \\ J\dot{\Omega} &= -\Omega^{\times} J\Omega + u + d. \end{aligned} \tag{1.16}$$

In (1.16), $q = [q_1, q_2, q_3, q_4]^{\mathrm{T}}$ (with $q_v = [q_1, q_2, q_3]^{\mathrm{T}}$) is called the quaternion which is a typical mathematical representation of spacecraft attitude; $\Omega \in \mathbb{R}^3$ is the angular velocity. Also, $J \in \mathbb{R}^{3\times 3}$ is a constant, positive definite, symmetric, overall inertia matrix, $u \in \mathbb{R}^3$ is the control torque, and $d \in \mathbb{R}^3$ is the external disturbance. The notation $q_v^{\times}$ denotes a skew symmetric matrix whose precise definition is given in Chap. 10. The performance output is specified as $y = q$. For a fixed setting point $q_d = \mathrm{col}(q_{dv}, q_{d4})$ with $q_{dv} = [q_{d1}, q_{d2}, q_{d3}]^{\mathrm{T}}$ and trivial external disturbance $d = 0$, the stabilization problem is to design the controller u such that $\lim_{t\to\infty}(q(t) - q_d) = 0$ and $\lim_{t\to\infty} \Omega(t) = 0$. In a more practical situation, we suppose the desired attitude motion $q_d(t)$ is not a fixed setting point, but generated by the following reference system

$$\begin{aligned} \dot{q}_{dv} &= \frac{1}{2}(q_{d4}I_3 + q_{dv}^{\times})\Omega_d \\ \dot{q}_{d4} &= -\frac{1}{2}q_{dv}^{\mathrm{T}}\Omega_d \end{aligned} \tag{1.17}$$

where q_d is the unit quaternion representing the target attitude and $\Omega_d \in \mathbb{R}^3$ is the target angular velocity. The main control problem is to design the controller u to achieve $\lim_{t\to\infty}(q(t) - q_d(t)) = 0$ and $\lim_{t\to\infty}(\Omega(t) - \Omega_d(t)) = 0$ in the presence of the disturbance d. Such a problem is called attitude tracking and disturbance rejection and will be studied in detail in Chap. 10.

1.3 Organization of the Book

This book provides a self-contained exposition of recent research outcomes for two fundamental nonlinear control problems: stabilization and regulation, via both robust and adaptive techniques. The remaining chapters of the book are organized as follows (see Fig. 1.8).

Chapter 2: The chapter first introduces the fundamental properties of nonlinear systems including various concepts of stability, robust stability, and input-to-state stability. Then some fundamental technical tools for nonlinear control analysis and design are presented. These concepts and tools will be used in the subsequent chapters.

Chapter 3: The normal form and the zero dynamics of nonlinear systems are presented in this chapter. Three typical classes of nonlinear systems in lower triangular form, output feedback form, and filter extended form are described. Also, some examples of practical nonlinear control systems are given in this chapter.

Chapter 4: This chapter details a constructive design method for global robust stabilization of uncertain nonlinear systems via the changing supply function technique. The design method will be applied to solve the global robust stabilization problem of nonlinear systems containing both dynamic uncertainty and bounded static uncertainty. The stabilization results will be further used in handling the global robust

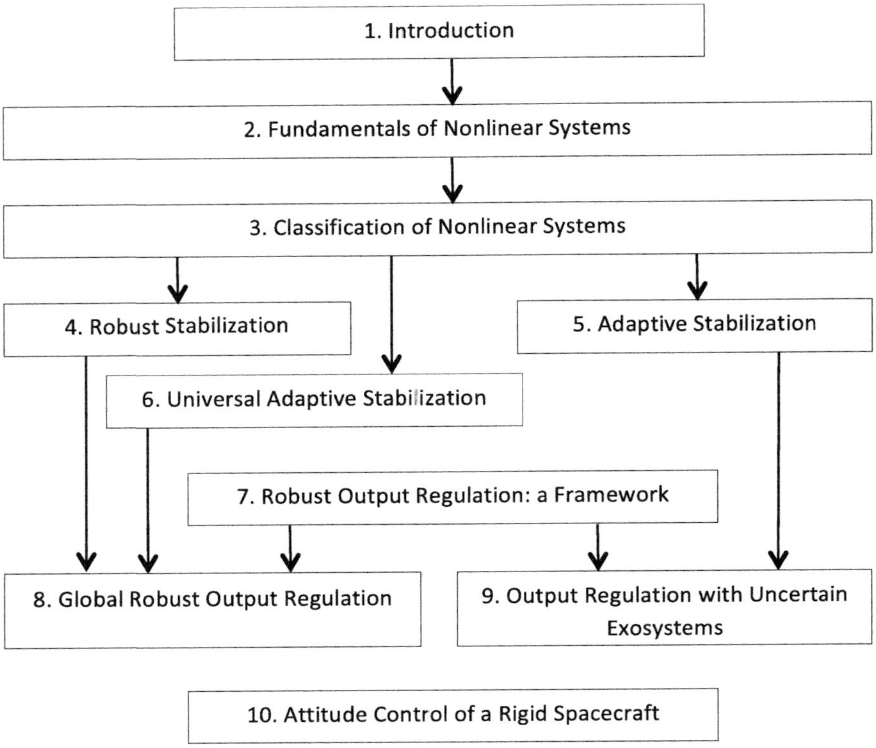

Fig. 1.8 Organization of the book chapters

output regulation problem of nonlinear systems in Chap. 8. The design method will be illustrated with the global stabilization problem of the well-known Chua's circuit.

Chapter 5: This chapter further introduces a few adaptive control techniques for handling linearly parameterized unknown parameters. Convergence of the estimated unknown parameters to actual values of the unknown parameters is discussed in detail. The techniques of this chapter will be further applied to the adaptive output regulation problem of nonlinear systems studied in Chap. 9. The design method will be illustrated with the adaptive stabilization problem of the Duffing equation.

Chapter 6: In this chapter, we first combine the robust control technique introduced in Chap. 4 with the universal control technique to handle nonlinear systems with unbounded uncertainties. Then, we further consider the case where the control direction is unknown by using the Nussbaum gain technique. The approach of this chapter will be illustrated with an asymptotic stabilization problem of the hyperchaotic Lorenz system. The stabilization results will be further used in handling the global robust output regulation problem for nonlinear systems with unbounded uncertainties in Chap. 8.

Chapter 7: Starting from this chapter, we turn to study the global robust output regulation problem for uncertain nonlinear systems. This chapter aims to establish a general framework for handling the nonlinear output regulation problem. This framework consists of two steps. In the first step, a general characterization of the concept of internal model is given. The internal model is defined as a dynamic compensator which together with the given plant constitutes a so-called augmented system. The internal model is such that the stabilization solution of the augmented system leads to the solution of the output regulation of the original plant. In short, this framework converts the robust output regulation problem of a given plant into the robust stabilization problem for the augmented system.

Chapter 8: In this chapter, we consider the global robust output regulation problem for various uncertain nonlinear systems with known exosystems. We first consider the case where both the uncertain parameters and external signals are within a compact set with a known boundary using the robust control approach studied in Chap. 4. Then we further consider the case where both the uncertain parameters and the external signals can be arbitrarily large using the approach studied in Chap. 6. The approach of this chapter will be illustrated with an asymptotic tracking problem of the Lorenz system.

Chapter 9: In this chapter, we further study the output regulation problem with uncertain exosystems. For this case, the augmented system contains a set of linearly parameterized uncertain parameters. We will apply the adaptive control technique developed in Chap. 5 to solve the adaptive stabilization problem of the augmented system, thus leading to the solution of the output regulation problem of the original plant. The approach of this chapter will be illustrated using the FitzHugh-Nagumo model.

Chapter 10: In this chapter, we first introduce an attitude control and disturbance rejection problem for a rigid body spacecraft system. Under the assumption that the disturbance is a multi-tone sinusoidal function with arbitrary unknown amplitudes, phases and frequencies, we first show that the problem can be reformulated as an adaptive stabilization problem of a class of multi-input, multi-output nonlinear system. Then we further apply the adaptive control techniques in Chap. 5 to solve the problem.

1.4 Notes and References

Three examples of nonlinear control systems are introduced in this chapter. The Lorenz system can find its original reference in [7]. The electrical circuit implementation is borrowed from [2]. The control of Lorenz system is studied in [6, 11–13]. The description of permanent magnet synchronous motor is from [1, 8, 14]. The description on Hubble Space Telescope is from [4, 10]. The mathematical model of a rigid spacecraft can be found in many books, e.g., [3, 5, 9].

1.5 Problems

Problem 1.1 Find the state space representation for the signals satisfying the following conditions.

(a) $v_1(t) = 2\sin(t + 30°)$, $v_2(t) = 2\cos(t + 30°)$;
(b) $v_1(t) = 2\sin(t + 30°)$, $v_2(t) = 2\cos(t + 60°)$;
(c) $x = H(s)u$, $y = \text{sat}(x)$ where $H(s) = 1/(s(s^2 + s + 1))$ is the transfer function from u to x.

Problem 1.2 Determine whether the following control systems are linear or nonlinear, autonomous or non-autonomous and determine the initial times and the initial states.

(a) $\dot{x}(t) = \sin(t)x(t) + \cos(t)u(t)$, $x(1) = 0$;
(b) $\dot{x}(t) = \sin(x(t)) + \cos(u(t))$, $x(1) = 0$;
(c) $\dot{x} = \begin{bmatrix} x_1^2 + x_2 \\ x_1x_2 + u \end{bmatrix}$, $x = \text{col}(x_1, x_2)$, $x(0) = 1$, $x_2(0) = 2$;
(d) $\ddot{x} + \dot{x} = \begin{bmatrix} x_1^2/(t + 1) + x_2 \\ x_1x_2 + u \end{bmatrix}$, $x = \text{col}(x_1, x_2)$, $x(0) = [1,\ 2]^{\mathrm{T}}$, $\dot{x}(0) = [0,\ 0]^{\mathrm{T}}$.

Problem 1.3 For the following control system

$$\begin{aligned} \dot{x}_1 &= -x_1 + x_2 \\ \dot{x}_2 &= -x_2 + x_3 \\ \dot{x}_3 &= x_1x_2 + u \end{aligned}$$

with the performance output $y = x_1$. Determine whether the following controllers are static or dynamic, state feedback or output feedback and write down the corresponding closed-loop systems.

(a) $u = -x_1x_2 - x_3$;
(b) $u = ax_1 + z$, $\dot{z} = k(z, x_1)$ for a constant a and a function k.

Problem 1.4 Derive the control system equation for the electric circuit given in Fig. 1.9 and determine the initial condition. The switch, S, is closed for a long time

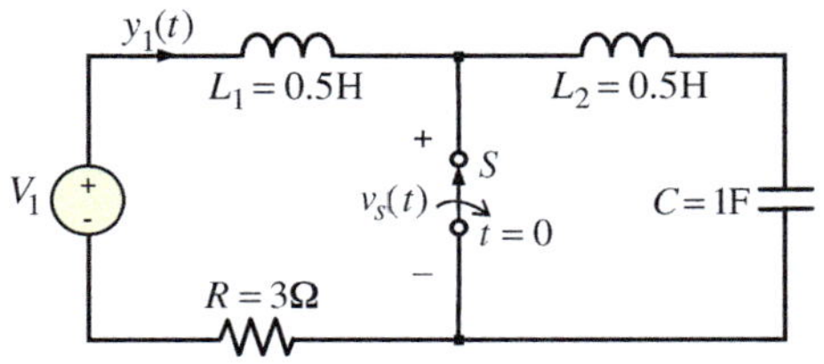

Fig. 1.9 Electric circuit of Problem 1.4

such that steady-state conditions are reached. It is opened instantaneously at time $t = 0$. The input is the voltage source V_1 and the output is the current across the inductor L_1. Repeat the problem if R is a nonlinear resistor whose electrical response from a voltage V to a current I is described by a nonlinear relation $I = s(V)V/R$ for a function s.

References

1. Baik IC, Kim KH, Youn MJ (2000) Robust nonlinear speed control of pm synchronous motor using boundary layer intergral sliding mode control technique. IEEE Trans Control Syst Technol 8:47–54
2. Corron NJ (2010) A simple circuit implementation of a chaotic Lorenz system. Creative consulting for research and education
3. Dixon WE, Behal A, Dawson DM, Nagarkatti SP (2003) Nonlinear control of engineering systems: a Lyapunov-based approach. Birkhäuser, Basel
4. Foster CL, Tinker ML, Nurre GS, Till WA (1995) The solar array-induced disturbance of the hubble space telescope pointing system. NASA Technical Paper 3556
5. Klumpp AR (1976) Singularity free extraction of a quaternion from a direction cosine matrix. J Spacecraft Rockets 13:754–755
6. Liang X, Zhang J, Xia X (2008) Adaptive synchronization for generalized Lorenz systems. IEEE Trans Autom Control 53:1740–1746
7. Lorenz EN (1963) Deterministic nonperiodic flow. J Atmos Sci 20:130–141
8. Ping Z, Huang J (2012) Speed tracking control of pm synchronous motor by internal model design. Int J Control 85:522–532
9. Sidi MJ (1997) Spacecraft dynamics and control. Cambridge University Press, Cambridge
10. Simpson DG (2011) UTC and the hubble space telescope flight software. AAS 11–673
11. Xu D, Huang J (2010) Global output regulation for output feedback systems with an uncertain exosystem and its application. Int J Robust Nonlinear Control 20:1678–1691
12. Xu D, Huang J (2010) Robust adaptive control of a class of nonlinear systems and its applications. IEEE Trans Circuits Syst II Express Briefs 57:691–702
13. Yu W (1999) Passive equivalence of chaos in Lorenz system. IEEE Trans Circuits Syst-I Fundam Theory Appl 46:876–878
14. Zhu G, Desaint LA, Akhrif O, Kaddouri A (2000) Speed tracking control of a permanent-magnet synchronous motor with state and load torque observer. IEEE Trans Industr Electron 47:346–355

Chapter 2
Fundamentals of Nonlinear Systems

In this chapter, we review some fundamental concepts and properties of nonlinear control systems that will be referred to in the subsequent chapters. In Sects. 2.1 and 2.2, we summarize the stability and robust stability concepts and the fundamental Lyapunov's stability theory. In Sect. 2.3, we establish some lemmas for the analysis of adaptive control systems. In Sect. 2.4, we introduce the concept of input-to-state stability of nonlinear control systems. In Sects. 2.5 and 2.6, we introduce the changing supply function technique and its applications to two classes of stabilzation approaches. In Sect. 2.7, we present the small gain method in the context of input-to-state stability. The notes and references are given in Sect. 2.8.

The materials in Sects. 2.1 and 2.2 are standard in nonlinear control literature and hence all proofs of results in these sections are omitted. Sections 2.3–2.7 do contain some new ingredients and detailed proofs are provided for those results which are considered non-standard.

2.1 Stability Concepts

In this section, we study the stability concepts for the general non-autonomous system described by (1.1) while viewing the autonomous systems as a special case of (1.1). To guarantee the existence of the unique solution $x(t)$ to the system (1.1) satisfying an initial condition (see Theorems 11.1 and 11.2 in the Appendix), it is assumed throughout the chapter that the function $f(x, t)$ in (1.1) is piecewise continuous in t and locally Lipschitz in x for all $t \geq t_0 \geq 0$ and all $x \in \mathbb{R}^n$. A constant vector $x_e \in \mathbb{R}^n$ is said to be an *equilibrium point* of the system (1.1) if

$$f(x_e, t) = 0, \ \forall t \geq t_0 \geq 0.$$

If a nonzero vector x_e is an equilibrium point of (1.1), then one can always introduce a new state variable $\hat{x} = x - x_e$ and define a new system $\dot{\hat{x}} = f(\hat{x} + x_e, t)$ which has $\hat{x} = 0$ as its equilibrium point. Thus, without loss of generality, one can assume

Z. Chen and J. Huang, *Stabilization and Regulation of Nonlinear Systems*,
Advanced Textbooks in Control and Signal Processing,
DOI 10.1007/978-3-319-08834-1_2

that the origin of $\mathbb{R}^n$, i.e., $x = 0$, is an equilibrium point of the system (1.1). With respect to the equilibrium point at the origin, various stability concepts are defined below.

Definition 2.1 The equilibrium point $x = 0$ of the system (1.1) is

(i) *Lyapunov stable (or stable)* at t_0 if for any $R > 0$, there exists $r(R, t_0) > 0$ such that $\|x(t)\| < R$ for all $t \geq t_0$, and all $\|x(t_0)\| < r(R, t_0)$.
(ii) *unstable* at t_0, if it is not stable at t_0.
(iii) *asymptotically stable (AS)* at t_0 if it is stable at t_0, and there exists $\delta(t_0) > 0$ such that $\|x(t)\| \to 0$ as $t \to \infty$ for all $\|x(t_0)\| < \delta(t_0)$.
(iv) *globally asymptotically stable (GAS)* at t_0 if it is stable at t_0 and $\|x(t)\| \to 0$ as $t \to \infty$ for all $x(t_0) \in \mathbb{R}^n$.

Definition 2.2 The equilibrium point $x = 0$ of the system (1.1) is

(i) *uniformly stable (US)* if for any $R > 0$, there exists $r(R) > 0$, independent of t_0, such that $\|x(t)\| < R$ for all $t \geq t_0$, and all $\|x(t_0)\| < r(R)$.
(ii) *uniformly asymptotically stable (UAS)* if it is uniformly stable, and there exists $\delta > 0$, independent of t_0, such that, for all $\|x(t_0)\| < \delta$, $\|x(t)\| \to 0$ as $t \to \infty$ uniformly in t_0, i.e., for any $\epsilon > 0$, there exists $T > 0$, independent of t_0, such that, for all $\|x(t_0)\| < \delta$, $\|x(t)\| < \epsilon$ whenever $t > t_0 + T$.
(iii) *uniformly globally asymptotically stable (UGAS)* if it is uniformly stable, and for any $\epsilon > 0$, and any $\delta > 0$, there exists $T > 0$, independent of t_0, such that, for all $\|x(t_0)\| < \delta$, $\|x(t)\| < \epsilon$ whenever $t > t_0 + T$.

In Definition 2.2, US, UAS, UGAS can be equivalently stated in terms of class $\mathcal{K}$, class $\mathcal{K}_\infty$ and class $\mathcal{KL}$ functions described as follows.

Definition 2.3 A continuous function $\gamma : [0, a) \mapsto [0, \infty)$ is said to belong to *class* $\mathcal{K}$ if it is strictly increasing and satisfies $\gamma(0) = 0$, and is said to belong to *class* $\mathcal{K}_\infty$ if, additionally, $a = \infty$ and $\lim_{r\to\infty} \gamma(r) = \infty$.

Definition 2.4 A continuous function $\beta : [0, a) \times [0, \infty) \mapsto [0, \infty)$ is said to belong to *class* $\mathcal{KL}$ if, for each fixed s, the function $\beta(\cdot, s)$ is a class $\mathcal{K}$ function defined on $[0, a)$, and, for each fixed r, the function $\beta(r, \cdot) : [0, \infty) \mapsto [0, \infty)$ is decreasing and $\lim_{s\to\infty} \beta(r, s) = 0$.

Definition 2.5 The equilibrium point $x = 0$ of the system (1.1) is

(i) *US* if there exist a class $\mathcal{K}$ function γ and $\delta > 0$, independent of t_0, such that $\|x(t)\| \leq \gamma(\|x(t_0)\|)$ for all $t \geq t_0$, and all $\|x(t_0)\| < \delta$.
(ii) *UAS* if there exist a class $\mathcal{KL}$ function β and $\delta > 0$, independent of t_0, such that $\|x(t)\| \leq \beta(\|x(t_0)\|, t - t_0)$ for all $t \geq t_0$, and all $\|x(t_0)\| < \delta$.
(iii) *UGAS* if there exists a class $\mathcal{KL}$ function β, independent of t_0, such that $\|x(t)\| \leq \beta(\|x(t_0)\|, t - t_0)$ for all $t \geq t_0$, and all $x(t_0) \in \mathbb{R}^n$.
(iv) *exponentially stable (ES)* or *globally exponentially stable (GES)* if it is UAS or UGAS with
$$\beta(r, s) = kre^{-\lambda s}, \; k \geq 1, \lambda > 0.$$

For an autonomous system of the form (1.2), if $x(t)$ is the solution of (1.2) satisfying the initial condition $x(t_0) = x_0$, then $\hat{x}(t) = x(t + t_0)$ is the solution of (1.2) satisfying the initial condition $\hat{x}(0) = x_0$. Therefore, one can always assume $t_0 = 0$ for an autonomous system. Moreover, for an autonomous system, if the equilibrium point is stable (or AS, GAS) at t_0, it is also US (or UAS, UGAS).

A typical non-autonomous system whose equilibrium point is GAS but not UGAS is given as follows.

Example 2.1 Consider a first order time-varying system

$$\dot{x} = -\frac{x}{1+t}, \quad x \in \mathbb{R}. \tag{2.1}$$

It can be verified that, for any initial state $x(t_0)$ with any initial time $t_0 \geq 0$, the solution of (2.1) is

$$x(t) = x(t_0)\frac{1+t_0}{1+t}, \quad \forall t \geq t_0.$$

Observe that the equilibrium point $x = 0$ is US and GAS. But, for given $\epsilon > 0$ and $\delta > 0$, in order to make $\|x(t)\| < \epsilon$ for all $\|x(t_0)\| < \delta$, t must be greater than $T = \delta(1 + t_0)/\epsilon - 1$. Since this T cannot be made independent of t_0, the equilibrium point is not UGAS.

There is another method to draw the above conclusion. Consider a class $\mathcal{KL}$ function

$$\beta(r, s) = r\frac{1+t_0}{1+t_0+s}$$

satisfying

$$\|x(t)\| = \beta\left(\|x(t_0)\|, t - t_0\right),$$

thus, the equilibrium point $x = 0$ is GAS. But the function β depends on t_0, and it is impossible to find another class $\mathcal{KL}$ function $\bar{\beta}$, independent of t_0, such that $\beta(r, s) \leq \bar{\beta}(r, s)$, which concludes that the equilibrium point is not UGAS. In fact, if this is not the case, we have

$$r\frac{1+t_0}{1+t_0+s} \leq \bar{\beta}(r, s),$$

and there exist a real number s^* satisfying $\bar{\beta}(r, s^*) \leq r/2$. Thus, for any $t_0 \geq 0$,

$$r\frac{1+t_0}{1+t_0+s^*} \leq \frac{r}{2},$$

i.e.,

$$t_0 \leq s^* - 1,$$

which contradicts that t_0 is an arbitrary nonnegative real number.

For a simple system such as (2.1), one may test the stability of its equilibrium point from the analytical expression of its solution. However, it is usually impossible to obtain the analytical solution to a complicated nonlinear system. Therefore, one may have to turn to other indirect methods to test a system's stability. Lyapunov's stability theory is one of the effective methods. Let us first introduce the Lyapunov's stability theory for the autonomous system (1.2). Suppose the function $f(x)$ is continuously differentiable with x in a neighborhood of the origin of $\mathbb{R}^n$. Define the Jacobian matrix of $f(x)$ at the origin as

$$F = \frac{\partial f}{\partial x}(0).$$

Then we have the following result.

Theorem 2.1 (Lyapunov's Linearization Theorem) *Consider the system (1.2). The equilibrium point $x = 0$ is AS if all the eigenvalues of the matrix F have negative real parts; and is unstable if at least one eigenvalue of the matrix F has positive real part.*

Remark 2.1 Theorem 2.1 cannot handle the case in which none of the eigenvalues of the matrix F has positive real part, but at least one of them has zero real part, and it does not tell whether the asymptotic stability is global or local.

Example 2.2 Consider a nonlinear system

$$\dot{x} = -\sin x + x^2. \tag{2.2}$$

The Jocobian matrix (scalar) is $F = -1$. So, the system is locally asymptotically stable. On the other hand, we have $\dot{x} = -\sin x + x^2 > 0$ if $x \geq 0.88$. Therefore, $x(t) > x(0) \geq 0.88,\ \forall t > 0$, if $x(0) \geq 0.88$. It implies that the system is not globally asymptotically stable. The state trajectories of the system (2.2) are illustrated in Fig. 2.1 with different initial state values. It shows that the state trajectory converges to the equilibrium point if the initial state is $x(0) = 0.80$ or $x(0) = 0.87$, but it diverges if the initial state is $x(0) = 0.88$.

On the other hand, the following Lyapunov's direct theorem can handle the two cases mentioned in Remark 2.1.

Theorem 2.2 (Lyapunov's Direct Theorem) *Consider the system (1.1). If there exists a continuously differentiable function $V : \mathbb{R}^n \times [t_0, \infty) \mapsto \mathbb{R}^+$ such that, for some class $\mathcal{K}$ functions $\bar{\alpha}$ and $\underline{\alpha}$, defined on $[0, \delta)$ for some $\delta > 0$,*

$$\underline{\alpha}(\|x\|) \leq V(x,t) \leq \bar{\alpha}(\|x\|) \tag{2.3}$$

$$\dot{V}(x,t) := \frac{\partial V(x,t)}{\partial t} + \frac{\partial V(x,t)}{\partial x} f(x,t) \leq 0,\ \forall \|x\| < \delta,\ \forall t \geq t_0, \tag{2.4}$$

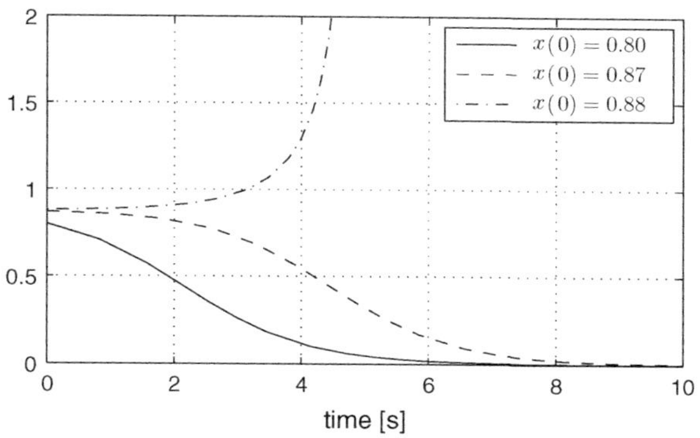

Fig. 2.1 Profile of state trajectories of the system in Example 2.2

then the equilibrium point $x = 0$ is US. If, (2.4) is replaced by

$$\dot{V}(x, t) \leq -\alpha(\|x\|), \ \forall \|x\| < \delta, \ \forall t \geq t_0, \tag{2.5}$$

where α is a class $\mathcal{K}$ function defined on $[0, \delta)$, then the equilibrium point $x = 0$ is UAS. Moreover, if $\delta = \infty$, and $\bar{\alpha}$ and $\underline{\alpha}$ are class $\mathcal{K}_\infty$ functions, then the equilibrium point $x = 0$ is UGAS.

A continuously differentiable function $V : \mathbb{R}^n \times [t_0, \infty) \mapsto \mathbb{R}^+$ satisfying (2.3) and (2.4) is called a Lyapunov function for (1.1).

Example 2.3 Consider the system (2.2) again. Let $V(x) = x^2$ whose derivative along the system trajectory satisfies

$$\dot{V}(x) = 2x(-\sin x + x^2) = -2x^2 + a(x)$$

for $a(x) = 2x(x - \sin x + x^2)$. There exists a constant δ such that $|a(x)/x^2| = 2|1 - \sin x/x + x| < 1$ for $|x| < \delta$. As a result, $|a(x)| < x^2$ for $|x| < \delta$, and hence $\dot{V}(x) \leq -x^2$. So, the system is locally asymptotically stable.

2.2 Robust Stability

In practice, a nonlinear system inevitably contains certain types of uncertainties such as external disturbances and parameter perturbations. To describe these uncertainties, we consider an uncertain nonlinear system of the following form:

$$\dot{x} = f(x, d(t)) \tag{2.6}$$

where d: $[t_0, \infty) \mapsto \mathbb{D} \subset \mathbb{R}^l$ with $\mathbb{D}$ a non-empty set represents external unpredictable disturbance and/or internal parameter variation. We assume that the function $f(x, d(t))$ is piecewise continuous in d and locally Lipschitz in x, and the function $d(t)$ is piecewise continuous in t. The system (2.6) is called a *non-autonomous/uncertain* system if d is time-varying. Taking into account the uncertainty $d(t)$, the control system (1.4)–(1.6) can be written as

$$\dot{x} = f(x, u, d(t)) \tag{2.7}$$

$$y = h(x, u, d(t)) \tag{2.8}$$

$$y_{\mathrm{m}} = h_{\mathrm{m}}(x, u, d(t)). \tag{2.9}$$

The closed-loop system composed of (2.7)–(2.9) with $y_{\mathrm{m}} = h_{\mathrm{m}}(x, d(t))$ and the controller (1.11) is

$$\dot{x}_{\mathrm{c}} = f_{\mathrm{c}}(x_{\mathrm{c}}, d(t)) \tag{2.10}$$

where

$$x_{\mathrm{c}} = \begin{bmatrix} x \\ v \end{bmatrix}, \quad f_{\mathrm{c}}(x_{\mathrm{c}}, d(t)) = \begin{bmatrix} f(x, \kappa_1(v, h_{\mathrm{m}}(x, d(t))), d(t)) \\ \kappa_2(v, h_{\mathrm{m}}(x, d(t))) \end{bmatrix}.$$

Example 2.4 In Example 1.1, if we replace the known external excitation $\cos t$ by an unknown external disturbance $d(t)$, then the system (1.3) can be expressed in the form (2.6) with

$$f(x, d(t)) = \begin{bmatrix} x_2 \\ -x_1 - x_1^3 + d(t) \end{bmatrix}.$$

For an uncertain system of the form (2.6), a constant vector $x_e \in \mathbb{R}^n$, independent of the signal $d(t)$, is said to be an *equilibrium point* of (2.6) if

$$f(x_{\mathrm{e}}, d(t)) = 0, \quad \forall t \geq t_0.$$

As explained for the equilibrium point of (1.1), without loss of generality, one only needs to consider the equilibrium point of the system (2.6) at the origin of $\mathbb{R}^n$, i.e., $x = 0$.

For each fixed $d(t)$, the uncertain system (2.6) reduces to the system (1.1). Therefore, various stability concepts described in Definition 2.5 for the system (1.1) also apply to the uncertain system (2.6). In this case, the functions β, γ, and the numbers δ, k, λ in Definition 2.5 may depend on specific $d(t)$. However, for some scenarios, one may find the functions β, γ, and the numbers δ, k, λ in Definition 2.5 which are independent of any $d(t) \in \mathbb{D}$. To distinguish these two scenarios, we further introduce the following *robust* stability concepts.

Definition 2.6 The equilibrium point $x = 0$ of the system (2.6) is *robustly uniformly asymptotically stable (RUAS)* or *robustly uniformly globally asymptotically stable*

(RUGAS) if it is UAS or UGAS with β and δ in Definition 2.5 independent of $d(t) \in \mathbb{D}$, or is *robustly exponentially stable (RES)* or *robustly globally exponentially stable (RGES)* if it is ES or GES with k and λ in Definition 2.5 independent of $d(t) \in \mathbb{D}$.

An uncertain system whose equilibrium point at the origin is UGAS but not RUGAS is given as follows.

Example 2.5 Consider the following system

$$\dot{x} = -\frac{1}{1+d^2}x, \quad x \in \mathbb{R}, \quad d \in \mathbb{R}. \tag{2.11}$$

It can be verified that, for any initial state $x(t_0)$ with any initial time $t_0 \geq 0$, the solution to (2.11) is

$$x(t) = x(t_0) \exp\left(\frac{-1}{1+d^2}(t-t_0)\right), \quad \forall t \geq t_0.$$

Observe that

$$\|x(t)\| = \beta(\|x(t_0)\|, t - t_0)$$

with

$$\beta(r, s) = r \exp\left(\frac{-1}{1+d^2}s\right).$$

Thus, the equilibrium point $x = 0$ is UGAS. But, the function β depends on d, and it is impossible to find another class $\mathcal{KL}$ function $\bar{\beta}$, independent of d, such that $\beta(r, s) \leq \bar{\beta}(r, s)$, which concludes that the equilibrium point is not RUGAS. In fact, if this is not the case, we have

$$r \exp\left(\frac{-1}{1+d^2}s\right) \leq \bar{\beta}(r, s),$$

and there exist a real number s^* satisfying $\bar{\beta}(r, s^*) \leq r/2$. Thus, for any $d \in \mathbb{R}$,

$$r \exp\left(\frac{-1}{1+d^2}s^*\right) \leq \frac{r}{2},$$

i.e.,

$$d^2 \leq \frac{s^*}{\ln 2} - 1,$$

which contradicts that d is an arbitrary real number.

Remark 2.2 It can be seen that if d is within a compact set, e.g., $d \in [-1, 1]$, then the origin of (2.11) is RUGAS, since $\|x(t)\| \leq \beta(\|x(t_0)\|, t - t_0)$ holds for $\beta(r, s) = re^{-0.5s}$, independent of d. In general, if the range of $d(t)$ is a compact set, then UAS, UGAS, and GES of the equilibrium point $x = 0$ of the system (2.6) imply RUAS, RUGAS, and RGES of the same equilibrium point, respectively.

The Lyapunov's stability theory, i.e., Theorems 2.1 and 2.2 can be generalized to the uncertain system (2.6). For example, the counterpart of Theorem 2.2 is stated as follows.

Theorem 2.3 (Lyapunov's Direct Theorem) *Consider the system (2.6). If there exists a continuously differentiable function $V : \mathbb{R}^n \times [t_0, \infty) \mapsto \mathbb{R}^+$ such that, for some class $\mathcal{K}$ functions $\bar{\alpha}$ and $\underline{\alpha}$, defined on $[0, \delta)$ for some $\delta > 0$,*

$$\underline{\alpha}(\|x\|) \leq V(x, t) \leq \bar{\alpha}(\|x\|) \tag{2.12}$$

$$\dot{V}(x, t) := \frac{\partial V(x, t)}{\partial t} + \frac{\partial V(x, t)}{\partial x} f(x, d(t)) \leq 0, \ \forall \|x\| < \delta, \ \forall t \geq t_0, \tag{2.13}$$

for all $d \in \mathbb{D}$, then the equilibrium point $x = 0$ is US. If, (2.13) is replaced by

$$\dot{V}(x, t) \leq -\alpha(\|x\|), \ \forall \|x\| < \delta, \ \forall t \geq t_0, \tag{2.14}$$

where α is a class $\mathcal{K}$ function defined on $[0, \delta)$, then the equilibrium point $x = 0$ is RUAS. Moreover, if $\delta = \infty$, and $\bar{\alpha}$ and $\underline{\alpha}$ are class $\mathcal{K}_\infty$ functions, then the equilibrium point $x = 0$ is RUGAS.

Theorem 2.4 *Suppose the conditions (2.12) and (2.14) in Theorem 2.3 are satisfied with*

$$\underline{\alpha}(\|x\|) = k_1\|x\|^c, \ \bar{\alpha}(\|x\|) = k_2\|x\|^c, \ \alpha(\|x\|) = k_3\|x\|^c,$$

for some positive constants k_1, k_2, k_3, and c. Then the equilibrium point $x = 0$ is RES. Moreover, if $\delta = \infty$, it is RGES.

2.3 Tools for Adaptive Control

In this section, we introduce some tools for adaptive control, including Barbalat's Lemma, LaSalle-Yoshizawa Theorem, persistent excitation criteria, and a parameter convergence lemma.

Lemma 2.1 (Barbalat's Lemma) *Let $\alpha : [t_0, \infty) \mapsto \mathbb{R}$ be a continuously differentiable scalar function. If $\alpha(t)$ has a finite limit as $t \to \infty$, and $\dot{\alpha}(t)$ is uniformly continuous over $[t_0, \infty)$, then*

$$\lim_{t \to \infty} \dot{\alpha}(t) = 0.$$

As an application of Barbalat's Lemma, we can obtain the following result.

Theorem 2.5 (LaSalle-Yoshizawa Theorem) *Consider the system (2.6) where $f(x, d(t))$ is locally Lipschitz in x uniformly in t. If there exists a continuously differentiable function $V(x, t) : \mathbb{R}^n \times [t_0, \infty) \mapsto \mathbb{R}^+$ such that*

$$\begin{aligned} W_1(x) &\leq V(x,t) \leq W_2(x) \\ \dot{V}(x,t) &\leq -\alpha(x) \leq 0, \ \forall x \in \mathbb{R}^n, \quad \forall t \geq t_0, \end{aligned} \tag{2.15}$$

where $W_1(x)$ and $W_2(x)$ are continuous positive definite and radially unbounded[1] *functions and $\alpha(x)$ is a continuous positive semidefinite function, then the state is bounded and satisfies*

$$\lim_{t\to\infty} \alpha(x(t)) = 0.$$

Moreover, if $\alpha(x)$ is positive definite, then the equilibrium point $x = 0$ is UGAS.

Remark 2.3 By stating that $f(x, d(t))$ is locally Lipschitz in x *uniformly in t*, it means that, for any $x^* \in \mathbb{R}^n$,

$$\|f(x, d(t)) - f(y, d(t)\| \leq L\|x - y\| \tag{2.16}$$

is satisfied for all $x, y \in \{x \in \mathbb{R}^n \mid \|x - x^*\| \leq r\}$ for some $r > 0$ and for all $t \geq t_0$. The Lipchitz constant L depends on x^*, but is independent of t. If $f(x, d(t))$ is locally Lipschitz in x, and $\mathbb{D}$ is a compact set, then, clearly, $f(x, d(t))$ is locally Lipschitz in x *uniformly in t*.

Remark 2.4 Theorem 2.5 holds with $W_1(x)$ and $W_2(x)$ replaced by two class $\mathcal{K}_\infty$ functions $\underline{\alpha}(\|x\|)$ and $\bar{\alpha}(\|x\|)$, respectively.

Example 2.6 Consider a second order nonlinear system

$$\begin{aligned} \dot{x}_1 &= -x_1 k(x_1) + x_2\omega(t) \\ \dot{x}_2 &= -x_1\omega(t) \end{aligned}$$

where $x = [x_1, x_2]^\top \in \mathbb{R}^2$ is the state and $\omega(t)$ is a bounded continuous function. The function k is assumed to be a continuously differentiable and strictly positive function, i.e., $k(x_1) > k_o > 0, \ \forall x_1 \in \mathbb{R}$. The asymptotic property of the system is analyzed as follows. Consider a lower bounded function $V(x) = \|x\|^2$. Along the trajectory of the system, the derivative of $V(x)$ satisfies

[1] A function $f(x)$ is called radially unbounded if $f(x) \to \infty$ as $\|x\| \to \infty$.

$$\dot{V}(x) = 2x_1\dot{x}_1 + 2x_2\dot{x}_2 = 2x_1(-x_1k(x_1) + x_2\omega) + 2x_2(-x_1\omega)$$
$$= -2x_1^2k(x_1) \leq 0. \tag{2.17}$$

Applying Theorem 2.5 and Remark 2.3 gives $\lim_{t\to\infty} x_1(t) = 0$.

The same conclusion can also be drawn by using Barbalat's Lemma. The inequality (2.17) implies that $V(x(t)) \leq V(x(0))$, $\forall t \geq 0$, and hence that $x(t)$ is bounded. Let $\alpha(t) = V(x(t))$. Its second derivative is $\ddot{\alpha}(t) = -[4x_1k(x_1) + 2x_1^2k'(x_1)][-x_1k(x_1) + x_2\omega]$ which is bounded since x and ω are bounded. Hence, $\dot{\alpha}(t)$ is uniformly continuous in t. By Lemma 2.1, $\lim_{t\to\infty} \dot{\alpha}(t) = 0$ and hence $\lim_{t\to\infty} x_1(t) = 0$.

The following material is concerned with the so-called persistent exciting property of a signal, which is widely used in the parameter convergence analysis in adaptive control.

Definition 2.7 A bounded piecewise continuous function $f : [0, \infty) \mapsto \mathbb{R}^n$ is said to be persistent exciting (PE) if there exist positive constants ϵ, t_0, T_0 such that, for any unit row vector c of dimension n,

$$\frac{1}{T_0}\int_t^{t+T_0} |cf(s)|ds \geq \epsilon, \quad \forall t \geq t_0. \tag{2.18}$$

Lemma 2.2 *A bounded piecewise continuous function $f : [0, \infty) \mapsto \mathbb{R}^n$ is PE if and only if there exist positive constants ϵ, t_0, T_0 such that*

$$\frac{1}{T_0}\int_t^{t+T_0} f(s)f^{\mathrm{T}}(s)ds \geq \epsilon^2 I, \quad \forall t \geq t_0. \tag{2.19}$$

Proof "Only if": By Jensen's inequality, i.e.,

$$(b-a)\int_a^b [g(s)]^2\, ds \geq \left(\int_a^b g(s)ds\right)^2,$$

for any integrable real-valued function g, we have

$$c\left[\frac{1}{T_0}\int_t^{t+T_0} f(s)f^{\mathrm{T}}(s)ds\right]c^{\mathrm{T}} = \frac{1}{T_0}\int_t^{t+T_0} [cf(s)]^2 ds \geq \left(\frac{1}{T_0}\int_t^{t+T_0} |cf(s)|ds\right)^2$$

for any unit row vector c of dimension n. As f is PE, one has (2.18), and hence

$$c\left[\frac{1}{T_0}\int_t^{t+T_0} f(s)f^{\mathsf{T}}(s)ds\right]c^{\mathsf{T}} \geq \epsilon^2 = c(\epsilon^2 I)c^{\mathsf{T}},$$

which implies (2.19).

"If": From (2.19), one has

$$\frac{1}{T_0}\int_t^{t+T_0} cf(s)f^{\mathsf{T}}(s)c^{\mathsf{T}}ds \geq \epsilon^2,$$

or

$$\int_t^{t+T_0} [cf(s)]^2\,ds \geq T_0\epsilon^2$$

for any unit row vector c of dimension n. Since the function f is bounded, so is $cf(s)$, i.e.,

$$|cf(s)| \leq R,\ \forall s \geq 0$$

for a constant R. Let $R_1 = \epsilon/\sqrt{2}$, $S_1 = \{s \mid |cf(s)| \geq R_1, t \leq s \leq t + T_0\}$ and $S_2 = \{s \mid |cf(s)| < R_1, t \leq s \leq t + T_0\}$. Then

$$S_1 \cup S_2 = [t, t + T_0],\ \ S_1 \cap S_2 = \emptyset.$$

Moreover, since $|cf(s)|$ is bounded and piecewise continuous, both S_1 and S_2 are Lebesgue measurable. Denote the length of a Lebesgue measurable set $S \subset [t, t+T_0]$ by $|S|$.[2] Then $0 \leq |S_i| \leq T_0$, $i = 1, 2$, and $|S_1 \cup S_2| = T_0$. Moreover,

$$\begin{aligned} T_0\epsilon^2 &\leq \int_t^{t+T_0} [cf(s)]^2\,ds = \int_{S_1} [cf(s)]^2\,ds + \int_{S_2} [cf(s)]^2\,ds \\ &\leq \int_{S_1} [cf(s)]^2\,ds + R_1^2|S_2|. \end{aligned}$$

The above inequality implies

[2] That is, $|S|$ is the Lebesgue measure on S [1].

$$|S_1|R^2 \geq \int\limits_{S_1} [cf(s)]^2\,ds \geq T_0\epsilon^2 - R_1^2|S_2| = (T_0 - |S_2|/2)\epsilon^2 \geq T_0\epsilon^2/2 > 0$$

and

$$|S_1| \geq T_0\epsilon^2/(2R^2) > 0.$$

Next, we have

$$\frac{1}{T_0}\int\limits_t^{t+T_0} |cf(s)|ds \geq \frac{1}{T_0}\int\limits_{S_1} |cf(s)|ds \geq \frac{|S_1|R_1}{T_0} \geq \frac{\epsilon^3}{2\sqrt{2}R^2}$$

which is (2.18) with ϵ replaced by another constant $\epsilon^3/(2\sqrt{2}R^2)$. From the definition, f is PE. The proof is thus completed. □

Example 2.7

1. Let $f(t)$ be a nonzero constant function for all $t \geq 0$. Then $f(t)$ is PE.
2. Let $f(t) = \sin\omega t$ with $\omega > 0$. Let $T_0 = 2\pi/\omega$. Then

$$\frac{1}{T_0}\int\limits_t^{t+T_0} |\sin\omega s|ds = \frac{2}{\pi}.$$

   Thus $f(t)$ is PE.
3. The function $f(t) = [\sin\omega t, \cos\omega t]^{\mathrm{T}}$ with $\omega > 0$ is PE while $f(t) = [\sin\omega t, \sin\omega t]^{\mathrm{T}}$ is not.

Next, we will show another criterion for the PE condition.

Lemma 2.3 *If a function $f : [0,\infty) \mapsto \mathbb{R}^n$ has spectral lines at frequencies $\omega_1, \cdots, \omega_n$, that is,*

$$\lim_{\delta\to\infty}\frac{1}{\delta}\int\limits_t^{t+\delta} f(s)e^{-j\omega_i s}ds = \hat{f}(\omega_i) \neq 0, \quad i = 1, \cdots, n$$

uniformly in t. Furthermore, $\hat{f}(\omega_i)$, $i = 1, \cdots, n$, are linearly independent in $\mathbb{C}^n$. Then, $f(t)$ is PE.

Proof Define the matrix

$$F(t,\delta) = \frac{1}{\delta}\int\limits_t^{t+\delta}\begin{bmatrix} e^{-j\omega_1 s} \\ \vdots \\ e^{-j\omega_n s}\end{bmatrix} f^{\mathrm{T}}(s)ds$$

and the matrix

$$F_0 = \begin{bmatrix} \hat{f}^{\mathsf{T}}(\omega_1) \\ \vdots \\ \hat{f}^{\mathsf{T}}(\omega_n) \end{bmatrix}.$$

The matrix F_0 is the limit of $F(t, \delta)$ as $\delta \to \infty$, uniformly in t. As F_0 is nonsingular by hypothesis, there exists a sufficiently large T_0, such that, for $\delta \geq T_0$, $F(t, \delta)$ is invertible and

$$\|F^{-1}(t, \delta)\| \leq 2\|F_0^{-1}\|, \ \forall t \geq 0.$$

Now, for any unit row vector $c \in \mathbb{R}^n$ and any $\omega \in \mathbb{R}$, we have

$$\begin{aligned} \frac{1}{\delta} \int_t^{t+\delta} [cf(s)]^2 ds &= \frac{1}{\delta} \int_t^{t+\delta} |cf(s)e^{-j\omega s}|^2 ds \\ &\geq \left| \frac{1}{\delta} \int_t^{t+\delta} cf(s)e^{-j\omega s} ds \right|^2 \end{aligned}$$

(by Jensen's inequality, see the proof of Lemma 2.2). For $\omega = \omega_1, \cdots, \omega_n$, one has

$$\begin{aligned} \frac{1}{\delta} \int_t^{t+\delta} [cf(s)]^2 ds &\geq \frac{1}{n} \sum_{i=1}^{n} \left| \frac{1}{\delta} \int_t^{t+\delta} cf(s)e^{-j\omega_i s} ds \right|^2 \\ &= \frac{1}{n} \|F(t, \delta)c^{\mathsf{T}}\|^2 \geq \frac{1}{n} \|F^{-1}(t, \delta)\|^{-2} \end{aligned}$$

and, for $\delta \geq T_0$,

$$\frac{1}{\delta} \int_t^{t+\delta} [cf(s)]^2 ds \geq \frac{1}{4n} \|F_0^{-1}\|^{-2}.$$

As a result,

$$c \left[\frac{1}{T_0} \int_t^{t+T_0} f(s) f^{\mathsf{T}}(s) ds - \epsilon^2 I \right] c^{\mathsf{T}} \geq 0, \ \epsilon = \|F_0^{-1}\|^{-1}/(2\sqrt{n})$$

for any unit row vector c of dimension n, which implies (2.19). The proof is thus completed. □

Example 2.8 Let

$$\tau(t) = \sum_{k=1}^{\ell} A_k \cos(\omega_k t + \phi_k)$$

for some $\ell > 0$ where A_k's are strictly positive real numbers and ω_k's are distinct strictly positive real numbers. Let

$$f(t) = [\tau(t), \dot{\tau}(t), \cdots, d^{(n-1)}\tau(t)/dt^{(n-1)}]^{\mathsf{T}}.$$

Then $f(t)$ is PE if $n \leq 2\ell$.

In fact, $\tau(t)$ and $f(t)$ can be rewritten as

$$\tau(t) = \sum_{k=1}^{\ell} A_k[e^{j(\omega_k t+\phi_k)} + e^{-j(\omega_k t+\phi_k)}]/2$$

and, respectively,

$$f(t) = \begin{bmatrix} \sum_{k=1}^{\ell} A_k[e^{j(\omega_k t+\phi_k)} + e^{-j(\omega_k t+\phi_k)}]/2 \\ \sum_{k=1}^{\ell} A_k[j\omega_k e^{j(\omega_k t+\phi_k)} + (-j\omega_k)e^{-j(\omega_k t+\phi_k)}]/2 \\ \vdots \\ \sum_{k=1}^{\ell} A_k[(j\omega_k)^{n-1} e^{j(\omega_k t+\phi_k)} + (-j\omega_k)^{n-1} e^{-j(\omega_k t+\phi_k)}]/2 \end{bmatrix}.$$

When $n \leq 2\ell$, we can pick n distinct frequencies $\hat{\omega}_i$ and the corresponding $\hat{\phi}_i$ and $\hat{A}_i$ as follows:

$$(\hat{\omega}_i, \hat{\phi}_i, \hat{A}_i) \in \{(\omega_1, \phi_1, A_1), (-\omega_1, -\phi_1, A_1), \cdots, (\omega_\ell, \phi_\ell, A_\ell), (-\omega_\ell, -\phi_\ell, A_\ell)\}, \quad i = 1, \cdots, n.$$

It is easy to check that

$$\hat{f}(\hat{\omega}_i) = \lim_{\delta\to\infty} \frac{1}{\delta} \int_t^{t+\delta} f(s) e^{-j\hat{\omega}_i s} ds = \begin{bmatrix} 1 \\ j\hat{\omega}_i \\ \vdots \\ (j\hat{\omega}_i)^{n-1} \end{bmatrix} \hat{A}_i e^{j\hat{\phi}_i}/2, \; i = 1, \cdots, n.$$

Since the frequencies $\hat{\omega}_i$, $i = 1, \cdots, n$, are distinct, the vectors $\hat{f}(\hat{\omega}_i), i = 1, \cdots, n$, are linearly independent in $\mathbb{C}^n$. Then, $f(t)$ is PE.

The PE property is useful in signal convergence analysis as illustrated in the following lemma. This lemma will be used in Chap. 5 for studying adaptive control.

Lemma 2.4 *Consider a continuously differentiable function* $g : [0, \infty) \mapsto \mathbb{R}^n$ *and a bounded piecewise continuous function* $f : [0, \infty) \mapsto \mathbb{R}^n$, *which satisfy*

$$\lim_{t\to\infty} g^{\mathrm{T}}(t) f(t) = 0. \tag{2.20}$$

Then,

$$\lim_{t\to\infty} g(t) = 0 \tag{2.21}$$

holds under the following two conditions:

(i) $\lim_{t\to\infty} \dot{g}(t) = 0$;
(ii) $f(t)$ *is PE.*

Proof Suppose (2.21) is not true. Then there exist a time sequence $s_1 < s_2 < \cdots$ satisfying $s_i \to \infty$ as $i \to \infty$ and a number $\delta_1 > 0$, such that $\|g(s_i)\| > \delta_1$. Under (2.20) and the condition (i), for any $\delta_2 > 0$ and $\delta_3 > 0$, there exists a time t_1, such that,

$$|\dot{g}_k(t)| \le \delta_2, \quad \forall t \ge t_1, \; k = 1, \cdots, n,$$

and

$$|g^{\mathrm{T}}(t) f(t)| \le \delta_3, \quad \forall t \ge t_1.$$

Also, under the condition (ii),

$$\int_t^{t+T_0} |cf(s)| ds \ge \epsilon_1 T_o, \quad \forall t \ge t_1 \tag{2.22}$$

for some constants T_0 and ϵ_1, independent of δ_3.

As a result, one has

$$|g_k(t+s) - g_k(t)| \le \int_t^{t+s} |\dot{g}_k(x)| dx \le \delta_2 T_0, \forall 0 \le s \le T_0, \; \forall t > t_1, \; k = 1, \cdots, n.$$

Let $\bar{f}$ be some real number such that $\|f(t)\| < \bar{f}, \forall t \ge 0$. Then, for any $s_i > t_1$,

$$\begin{aligned}\int_{s_i}^{s_i+T_0} |g^{\mathrm{T}}(s) f(s)| ds &\ge \int_{s_i}^{s_i+T_0} |g^{\mathrm{T}}(s_i) f(s)| ds - \int_{s_i}^{s_i+T_0} |[g(s) - g(s_i)]^{\mathrm{T}} f(s)| ds \\ &\ge T_0 \delta_1 \epsilon_1 - \delta_2 T_0^2 \bar{f}.\end{aligned}$$

Since δ_2 can be arbitrarily small,

$$\int_{s_i}^{s_i+T_0} |g^{\mathrm{T}}(s)f(s)|ds \geq \epsilon_2$$

for some positive ϵ_2 independent of δ_3. Thus, there exists a time $\bar{s}_i \in [s_i, s_i + T_0]$ such that

$$\delta_3 \geq |g^{\mathrm{T}}(\bar{s}_i)f(\bar{s}_i)| \geq \epsilon_2/T_0.$$

Noting δ_3 can be arbitrarily small leads to a contradiction. The proof is thus completed. □

This lemma gives the convergence condition of the function $g(t)$ to the origin based on the asymptotic condition (i) of $\dot{g}(t)$ and the PE condition (ii) of $f(t)$. Both conditions are indispensable as illustrated in the following examples.

Example 2.9 Let $f(t) = [\cos\omega t \quad \sin\omega t]^{\mathrm{T}}$ and $g(t) = [-\sin\omega t \quad \cos\omega t]^{\mathrm{T}}$. The condition $\lim_{t\to\infty} g^{\mathrm{T}}(t)f(t) = 0$ obviously holds and $f(t)$ is PE. However, $\lim_{t\to\infty} g(t) = 0$ is not true because $\lim_{t\to\infty} \dot{g}(t) = 0$ is not.

Example 2.10 Consider a continuously differentiable signal $g(t) = ca(t)$ for a constant vector $c \in \mathbb{R}^2$ and function $a(t)$ satisfying $\lim_{t\to\infty} \dot{a}(t) = 0$. Suppose $\lim_{t\to\infty} g^{\mathrm{T}}(t)f(t) = 0$ where $f(t)$ is a bounded piecewise continuous function.

If $f(t)$ is PE, we have $\lim_{t\to\infty} g(t) = 0$ for any c by Lemma 2.4.

If $f(t)$ is not PE, e.g., $f(t) = [\cos\omega t \quad \cos\omega t]^{\mathrm{T}}$, then $\lim_{t\to\infty} g(t) = 0$ is not necessarily true. For example, when $c = [1 \quad -1]^{\mathrm{T}}$ and $a(t) = 1$, the condition $\lim_{t\to\infty} g^{\mathrm{T}}(t)f(t) = 0a(t) = 0$ still holds. But $\lim_{t\to\infty} g(t) = [1 \quad -1]^{\mathrm{T}} \neq 0$.

2.4 Input-to-State Stability

In the previous sections, we have reviewed various stability concepts of the nonlinear systems described by (1.1) and (2.6), respectively. In this section, we will further consider the stability of the control systems described by (1.4) and (2.7). Since the response of the system (1.4) or (2.7) is excited not only by the initial state $x(t_0)$ but also by the input $u(t)$, we need to generalize the stability concepts about an equilibrium point to the so-called *input-to-state stability* of the system (2.7) while keeping in mind that the system (1.4) can be viewed as a special case of (2.7) by having $d(t) = t$.

Again, we assume the function $f(x, u, d(t))$ is piecewise continuous in d and locally Lipschitz in $\mathrm{col}(x, u)$, and the function $d(t)$ is piecewise continuous in t. And we use the notation L_∞^m to denote the set of all piecewise continuous bounded functions $u : [t_0, \infty) \mapsto \mathbb{R}^m$ with the supermum norm

$$\left\| u_{[t_0,\infty)} \right\| := \sup_{t \geq t_0} \|u(t)\|.$$

For convenience, we also denote the supermum norm of the truncation of $u(t)$ in $[t_1, t_2]$ with $t_0 \leq t_1 \leq t_2$ as follows,

$$\left\| u_{[t_1,t_2]} \right\| := \sup_{t_1 \leq t \leq t_2} \|u(t)\|.$$

Definition 2.8 The system (2.7) is said to be *input-to-state stable (ISS)* if there exist a class $\mathcal{KL}$ function β and a class $\mathcal{K}$ function γ, independent of t_0, such that for any initial state $x(t_0)$ and any input function $u \in L_\infty^m$, the solution $x(t)$ exists and satisfies

$$\|x(t)\| \leq \max\left\{\beta(\|x(t_0)\|, t - t_0), \gamma\left(\left\| u_{[t_0,t]} \right\|\right)\right\}, \ \forall t \geq t_0. \tag{2.23}$$

Since the control system (2.7) involves the uncertainty $d(t)$, the functions β and γ in Definition 2.8 may or may not depend on $d(t)$. If the functions β and γ can be made to be independent of the uncertainty $d(t)$, then we have the following robust input-to-state stability concept.

Definition 2.9 The system (2.7) is said to be *robustly input-to-state stable (RISS)* if it is ISS in the sense of Definition 2.8 with β and γ independent of $d(t) \in \mathbb{D}$.

Remark 2.5 We note that the functions β and γ are independent of t_0 in the definition of ISS or RISS. In other words, the concepts ISS and RISS implicitly include the fact that they are uniformly with respect to the initial time t_0. For an RISS system (2.7), when the input u is held at zero, the solution starting from any initial state $x(t_0)$ for any initial time t_0 satisfies

$$\|x(t)\| \leq \beta(\|x(t_0)\|, t - t_0), \ \forall t \geq t_0.$$

Thus, the equilibrium point at the origin of the unforced system $\dot{x} = f(x, 0, d(t))$ is RUGAS.

Remark 2.6 In (2.23), since, for any $x(t_0)$ and any t_0, $\beta(\|x(t_0)\|, t - t_0) \to 0$ as $t \to \infty$, one has

$$\lim_{t \to \infty} \left\| x_{[t,\infty)} \right\| \leq \gamma\left(\left\| u_{[t_0,\infty)} \right\|\right).$$

Due to this inequality, the class $\mathcal{K}$ function γ is called a *gain function* of (2.7).

Remark 2.7 There is an equivalent way to characterize the ISS property of (2.7) as follows. There exist a class $\mathcal{KL}$ function β and a class $\mathcal{K}$ function γ such that for any initial state $x(t_0)$ and any input function $u \in L_\infty^m$, the solution $x(t)$ exists and satisfies

$$\|x(t)\| \leq \beta(\|x(t_0)\|, t - t_0) + \gamma\left(\left\|u_{[t_0,t]}\right\|\right), \ \forall t \geq t_0.$$

This equivalence follows from the fact that $\max\{\beta, \gamma\} \leq \beta + \gamma \leq \max\{2\beta, 2\gamma\}$ for any pair $\beta \geq 0, \gamma \geq 0$.

The Lyapunov's direct theorem can also be generalized to analyze the ISS property of a system as described below.

Definition 2.10 Let $V : \mathbb{R}^n \times [t_0, \infty) \mapsto \mathbb{R}^+$ be a continuously differentiable function. It is called an *ISS-Lyapunov function* for the system (2.7) if there exist class $\mathcal{K}_\infty$ functions $\bar{\alpha}$, $\underline{\alpha}$, α, and a class $\mathcal{K}$ function ρ, such that

$$\begin{aligned} \underline{\alpha}(\|x\|) &\leq V(x,t) \leq \bar{\alpha}(\|x\|) \\ \dot{V}(x,t) &\leq -\alpha(\|x\|), \ \forall \|x\| \geq \rho(\|u\|) \end{aligned}$$

for all $x \in \mathbb{R}^n$, $u \in L_\infty^m$, $t \geq t_0$, and $d \in \mathbb{D}$.

Theorem 2.6 *If the system (2.7) has an ISS-Lyapunov function, then it is RISS with a gain function $\underline{\alpha}^{-1} \circ \bar{\alpha} \circ \rho$, i.e., there exist a class $\mathcal{KL}$ function β and a class $\mathcal{K}$ function $\gamma = \underline{\alpha}^{-1} \circ \bar{\alpha} \circ \rho$ such that for any initial state $x(t_0) \in \mathbb{R}^n$ and any input function $u \in L_\infty^m$, the solution $x(t)$ of (2.7) exists and satisfies (2.23).*

The proof of Theorem 2.6 can be found in [2] (see the proofs of Theorems 4.18 and 4.19). Suppose $V : \mathbb{R}^n \times [t_0, \infty) \mapsto \mathbb{R}^+$ is a continuously differentiable function, for all $x \in \mathbb{R}^n$, $u \in L_\infty^m$, $t \geq t_0$, and $d \in \mathbb{D}$, the derivative of V along the trajectory of $\dot{x} = f(x, u, d(t))$ satisfies

$$\dot{V}(x,t) \leq -\alpha(\|x\|) + \sigma(\|u\|) \tag{2.24}$$

where α is some class $\mathcal{K}_\infty$ function and σ some class $\mathcal{K}$ function. Let

$$\rho(s) = \alpha^{-1}(k\sigma(s))$$

with $k > 1$. Then

$$\|x\| \geq \rho(\|u\|) \ \Rightarrow \ \sigma(\|u\|) \leq \frac{1}{k}\alpha(\|x\|).$$

So, (2.24) gives

$$\dot{V}(x,t) \leq -\frac{k-1}{k}\alpha(\|x\|), \ \forall \|x\| \geq \rho(\|u\|)$$

for all $x \in \mathbb{R}^n$, $u \in L_\infty^m$, $t \geq t_0$, and $d \in \mathbb{D}$. Thus, $V(x,t)$ is an ISS Lyapunov function of (2.7). As a result, we obtain the following result.

Theorem 2.7 *Consider the system (2.7). If there exists a continuously differentiable function* $V : \mathbb{R}^n \times [t_0, \infty) \mapsto \mathbb{R}^+$ *such that, for some class* $\mathcal{K}_\infty$ *functions* $\bar{\alpha}$, $\underline{\alpha}$, α, *and some class* $\mathcal{K}$ *function* σ,

$$\begin{aligned} \underline{\alpha}(\|x\|) &\le V(x,t) \le \bar{\alpha}(\|x\|) \\ \dot{V}(x,t) &\le -\alpha(\|x\|) + \sigma(\|u\|) \end{aligned} \tag{2.25}$$

for all $x \in \mathbb{R}^n$, $u \in L_\infty^m$, $t \ge t_0$, *and* $d \in \mathbb{D}$, *then the system (2.7) is RISS with a gain function* $\underline{\alpha}^{-1} \circ \bar{\alpha} \circ \alpha^{-1} \circ k\sigma$ *for any* $k > 1$.

To simplify the presentation, we use the following notation

$$V(x,t) \sim \{\underline{\alpha}, \bar{\alpha}, \alpha, (\sigma_1, \cdots, \sigma_m) \mid \dot{x} = f(x,u,d)\} \tag{2.26}$$

to mean the following statement: there exist some class $\mathcal{K}_\infty$ functions $\bar{\alpha}$, $\underline{\alpha}$, α, and some class $\mathcal{K}$ functions σ_i, $i = 1, \cdots m$, such that,

$$\begin{aligned} \underline{\alpha}(\|x\|) &\le V(x,t) \le \bar{\alpha}(\|x\|) \\ \dot{V}(x,t) &\le -\alpha(\|x\|) + \sum_{i=1}^{m} \sigma_i(\|u_i\|) \end{aligned}$$

for all $x \in \mathbb{R}^n$, $u \in L_\infty^m$, $t \ge t_0$, and $d \in \mathbb{D}$. In particular, for a single input system $x = f(x,u,d)$, the notation (2.26) reduces to a simpler form

$$V(x,t) \sim \{\underline{\alpha}, \bar{\alpha}, \alpha, \sigma \mid \dot{x} = f(x,u,d)\}.$$

Example 2.11 Consider the system

$$\dot{x} = A(t)x + G(u,t), \quad t \ge t_0 \ge 0 \tag{2.27}$$

where $G(u,t)$ is a continuous function satisfying, for all $u \in \mathbb{R}$ and all $t \ge t_0$, $\|G(u,t)\| \le q(\|u\|)$ for some class $\mathcal{K}$ function q. Suppose the system $\dot{x} = A(t)x$ is UAS, i.e., there exist symmetric positive definite matrices $Q(t)$ and $P(t)$ satisfying $0 < \beta_1 I \le Q(t) \le \beta_2 I, \forall t \ge 0$ and $0 < \alpha_1 I \le P(t) \le \alpha_2 I, \forall t \ge 0$, such that

$$\dot{P}(t) + P(t)A(t) + A(t)^\top P(t) = -Q(t).$$

Let $V(x,t) = x^\top P(t)x$. Then, along the trajectory of (2.27),

$$\begin{aligned} \dot{V}(x,t) &\le -\|Q(t)\|\|x\|^2 + 2x^\top P(t)G(u,t) \\ &\le -(\|Q(t)\| - 1/\epsilon)\|x\|^2 + \epsilon\|P(t)G(u,t)\|^2 \\ &\le -(\beta_1 - 1/\epsilon)\|x\|^2 + \epsilon\alpha_2^2 q^2(\|u\|). \end{aligned}$$

Let ϵ be such that $l = \beta_1 - 1/\epsilon > 0$ and let $\sigma(\|u\|) = \epsilon\alpha_2^2 q^2(\|u\|)$. Then, we have

$$\dot{V}(x,t) < -l\|x\|^2 + \sigma(\|u\|). \tag{2.28}$$

Thus, the system (2.27) is ISS with a gain function

$$\gamma(s) = \sqrt{\frac{\alpha_2 k}{\alpha_1 l}\sigma(s)} = \sqrt{\frac{\alpha_2 k\epsilon}{\alpha_1 l}}\alpha_2 q(s)$$

for any $k > 1$.

Example 2.12 As a special case of the above example, a linear time-invariant system

$$\dot{x} = Ax + Bu, \quad t \geq 0$$

where A is a Hurwitz matrix is ISS. Also, since $\|Bu\| \leq b\|u\|$ for some $b > 0$, the gain function is $\gamma(s) = \sqrt{\alpha_2 k\epsilon/(\alpha_1 l)}\alpha_2 bs$, which is a linear function.

Example 2.13 The following scalar system $\dot{x} = -x + xu$ is not ISS. In fact, let $u(t) = 2$ for all $t \geq 0$. Then the response of the system with $x(0) = x_0$ is $x(t) = e^t x_0$, which shows that the inequality (2.23) cannot hold.

Next, we further introduce two other concepts for the system (2.7) as follows.

Definition 2.11 The system (2.7) is said to have the *robustly globally stable (RGS)* property, and the *robustly asymptotic gain (RAG)* property, respectively, if there exist class $\mathcal{K}$ functions γ_0 and γ, independent of $d(t)$, such that for any initial time t_0, any initial state $x(t_0) \in \mathbb{R}^n$, any $d(t) \in \mathbb{D}$, and any input function $u \in L_\infty^m$, the solution $x(t)$ exists and satisfies

$$\left\|x_{[t_0,\infty)}\right\| \leq \max\left\{\gamma_0(\|x(t_0)\|), \gamma\left(\left\|u_{[t_0,\infty)}\right\|\right)\right\},$$

and, respectively,

$$\lim_{t\to\infty}\left\|x_{[t,\infty)}\right\| \leq \gamma\left(\lim_{t\to\infty}\left\|u_{[t,\infty)}\right\|\right).$$

These two concepts are of particular interest to autonomous control systems, e.g., the system (2.7) with $d(t) = constant$, since it is possible to show, for autonomous control systems, the RISS property is equivalent to the RGS property plus the RAG property (see, e.g., [3]), i.e.,

$$\text{RISS} \iff \text{RGS} + \text{RAG}.$$

Example 2.14 Consider the linear time-invariant system

$$\dot{x} = Ax + Bu, \quad t \geq 0 \tag{2.29}$$

where A is Hurwitz and $u(t)$ is piecewise continuous in t and $\lim_{t\to\infty} u(t) = 0$. By Example 2.12, this system is ISS, and is thus of the asymptotic gain property. Therefore, for any initial condition $x(0)$, $\lim_{t\to\infty} x(t) = 0$.

However, for a non-autonomous control system (2.7), this equivalence does not hold any more. Specifically, the implication

$$\text{RISS} \Rightarrow \text{RGS} + \text{RAG}$$

is true, but the other direction is not, i.e.,

$$\text{RISS} \not\Leftarrow \text{RGS} + \text{RAG},$$

as shown in the following Example.

Example 2.15 Consider a non-autonomous system

$$\dot{x} = -\frac{x-u}{1+t}, \ x \in \mathbb{R}, \ u \in L_\infty^1. \tag{2.30}$$

It can be verified that, for any initial state $x(t_0)$ with any initial time $t_0 \geq 0$, the solution of (2.30) is

$$x(t) = \frac{1+t_0}{1+t}x(t_0) + \frac{1}{1+t}\int_{t_0}^{t} u(\tau)d\tau, \ \forall t \geq t_0.$$

On one hand,

$$|x(t)| \leq |x(t_0)| + \left\| u_{[t_0,\infty)} \right\|,$$

hence, $\left\| x_{[t_0,\infty)} \right\| \leq \max\left\{2|x(t_0)|, 2\left\| u_{[t_0,\infty)} \right\|\right\}$. That is, the system (2.30) has RGS property. On the other hand, for any $\epsilon > 0$, there exists $T_1 \geq t_0$ such that

$$\left\| u_{[T_1,\infty)} \right\| \leq \lim_{t\to\infty} \left\| u_{[t,\infty)} \right\| + \epsilon.$$

And there exists $T_2 \geq T_1$ such that

$$\frac{1+T_1}{1+T_2}x(T_1) \leq \epsilon.$$

Then, for any time $t \geq T_2$,

$$\begin{aligned}
|x(t)| &\leq \frac{1+T_1}{1+t}x(T_1) + \left\| u_{[T_1,\infty)} \right\| \\
&\leq \epsilon + \lim_{t\to\infty} \left\| u_{[t,\infty)} \right\| + \epsilon
\end{aligned}$$

hence, $\lim_{t\to\infty} \|x_{[t,\infty)}\| \le \lim_{t\to\infty} \|u_{[t,\infty)}\| + 2\epsilon$. Letting $\epsilon \to 0$ yields that the system (2.30) has the RAG property.

However, the system (2.30) is not RISS. If this were not the case, then

$$|x(t)| \le \max\left\{\beta(|x(t_0)|, t - t_0), \gamma\left(\left\|u_{[t_0,t]}\right\|\right)\right\}, \ \forall t \ge t_0.$$

for some class $\mathcal{KL}$ function β and class $\mathcal{K}$ function γ, independent of t_0. Let $u(t) = 1$ for all $t \ge t_0$, and $x(t_0) = 2\gamma(1)$. Then,

$$|x(t)| = \frac{1+t_0}{1+t}2\gamma(1) + \frac{t-t_0}{1+t} \le \max\{\beta(2\gamma(1), t-t_0), \gamma(1)\}.$$

Choose a finite real number s^* satisfying $\beta(2\gamma(1), s^*) \le \gamma(1)$. Then

$$|x(t_0+s^*)| = \frac{1+t_0}{1+t_0+s^*}2\gamma(1) + \frac{s^*}{1+t_0+s^*} \le \gamma(1),$$

hence, $(1+t_0)/(1+t_0+s^*) < 1/2$, i.e., $t_0 < s^* - 1$, which contradicts that t_0 is an arbitrary nonnegative real number.

2.5 Changing Supply Function

The ISS Lyapunov function $V(x,t)$ is also called a *supply function* or a *storage function*, and the pair (α, σ) is called a *supply pair*. The ISS Lyapunov function is not unique. It is possible to use the changing supply function technique to generate an alternative ISS Lyapunov function with exploitable property. In most scenarios encountered in this book, it is assumed that the range of uncertainties is represented by a *compact set* $\mathbb{D}$. In these scenarios, we usually consider an ISS Lyapunov function $V(x)$ not explicitly depending on t.

Lemma 2.5 (Changing Supply Function) *Suppose the system $\dot{x} = f(x,u,d)$ has an ISS Lyapunov function $V(x)$, i.e.,*

$$V(x) \sim \{\underline{\alpha}, \bar{\alpha}, \alpha, \sigma \mid \dot{x} = f(x,u,d)\}. \tag{2.31}$$

Let α' be a class $\mathcal{K}_\infty$ function such that $\alpha'(s) = \mathcal{O}[\alpha(s)]$ as $s \to 0^+$.[3] *Then the system $\dot{x} = f(x,u,d)$ has another ISS Lyapunov function $V'(x)$, i.e.,*

$$V'(x) \sim \{\underline{\alpha}', \bar{\alpha}', \alpha', \sigma' \mid \dot{x} = f(x,u,d)\}. \tag{2.32}$$

[3] The notation $\alpha'(s) = \mathcal{O}[\alpha(s)]$ as $s \to 0^+$ means $\limsup_{s\to 0^+}\left[\alpha'(s)/\alpha(s)\right] < \infty$.

In particular, the class $\mathcal{K}_\infty$ functions $\bar{\alpha}'$ and $\underline{\alpha}'$ and the class $\mathcal{K}$ function σ' are given by Algorithm 2.1.

Proof Let $\mathcal{SN}$ be the set of smooth non-decreasing functions defined over $[0, \infty)$ that satisfy $\rho(s) > 0,\ \forall s > 0$ for $\rho \in \mathcal{SN}$. Let

$$V'(x) = \int_0^{V(x)} \rho(s)ds \tag{2.33}$$

where $\rho \in \mathcal{SN}$. The statement (2.31) implies that

$$\underline{\alpha}'(\|x\|) \le \int_0^{\underline{\alpha}(\|x\|)} \rho(s)ds \le V'(x) \le \int_0^{\bar{\alpha}(\|x\|)} \rho(s)ds \le \bar{\alpha}'(\|x\|) \tag{2.34}$$

for some class $\mathcal{K}_\infty$ functions $\underline{\alpha}'$ and $\bar{\alpha}'$.

We now show that, along the trajectory of $\dot{x} = f(x, u, d)$,

$$\begin{aligned}\dot{V}'(x) &\le \rho(V(x))[-\alpha(\|x\|) + \sigma(\|u\|)]\\ &\le -\frac{1}{2}\rho(\underline{\alpha}(\|x\|))\alpha(\|x\|) + \rho(\bar{\alpha}(\alpha^{-1}(2\sigma(\|u\|))))\sigma(\|u\|).\end{aligned}$$

In fact, we consider the following two cases for the second inequality.

(i) $\alpha(\|x\|)/2 \ge \sigma(\|u\|)$: In this case, the claim follows from the fact that $\rho(V(x))[-\alpha(\|x\|) + \sigma(\|u\|)]$ is bounded from above by $-\rho(V(x))\alpha(\|x\|)/2$, and hence bounded from above by $-\rho\left(\underline{\alpha}(\|x\|)\right)\alpha(\|x\|)/2$.

(ii) $\alpha(\|x\|)/2 < \sigma(\|u\|)$: In this case, the following inequalities hold

$$\rho(V(x)) \le \rho(\bar{\alpha}(\|x\|)) \le \rho(\bar{\alpha}(\alpha^{-1}(2\sigma(\|u\|)))).$$

Since $\alpha'(s) = \mathcal{O}[\alpha(s)]$ as $s \to 0^+$, by Lemma 11.2 in the Appendix, it is always possible to find a function ρ such that

$$\frac{1}{2}\rho(\underline{\alpha}(s))\alpha(s) \ge \alpha'(s). \tag{2.35}$$

Also, there exists a class $\mathcal{K}$ function σ' such that

$$\sigma'(s) \ge \rho(\bar{\alpha}(\alpha^{-1}(2\sigma(s))))\sigma(s). \tag{2.36}$$

The proof is thus completed. □

Algorithm 2.1

INPUT: $\underline{\alpha}, \bar{\alpha}, \alpha, \sigma, \alpha'$
OUTPUT: $\underline{\alpha}', \bar{\alpha}', \sigma'$
STEP 1: Pick an $\mathcal{SN}$ function ρ satisfying (2.35).
STEP 2: Find the class $\mathcal{K}_\infty$ functions $\underline{\alpha}'$ and $\bar{\alpha}'$ from (2.34).
STEP 3: Find the class $\mathcal{K}$ function σ' from (2.36).
STEP 4: END

Corollary 2.1 *Suppose the system* $\dot{x} = f(x, u, d)$ *has an ISS Lyapunov function* $V(x)$, *i.e.,*

$$V(x) \sim \{\underline{\alpha}, \bar{\alpha}, \alpha, \sigma \mid \dot{x} = f(x, u, d)\}.$$

Then, for any smooth function Δ, *the system* $\dot{x} = f(x, u, d)$ *has another ISS Lyapunov function* $V'(x)$ *such that*

$$\begin{aligned} \underline{\alpha}'(\|x\|) \le V'(x) &\le \bar{\alpha}'(\|x\|) \\ \dot{V}'(x) &\le -\Delta(x)\alpha(\|x\|) + \varkappa(u)\sigma(\|u\|). \end{aligned} \tag{2.37}$$

for a smooth function $\varkappa$. *In particular, the class* $\mathcal{K}_\infty$ *functions* $\bar{\alpha}'$ *and* $\underline{\alpha}'$ *and the smooth function* $\varkappa$ *are given by Algorithm 2.2.*

Proof Following the proof of Lemma 2.5, it suffices to choose a function $\rho \in \mathcal{SN}$ such that

$$\frac{1}{2}\rho(\underline{\alpha}(\|x\|)) \ge \Delta(x) \tag{2.38}$$

and to choose a smooth function $\varkappa$ such that

$$\varkappa(u) \ge \rho(\bar{\alpha}(\alpha^{-1}(2\sigma(\|u\|)))). \tag{2.39}$$

The proof is thus completed. □

Algorithm 2.2

INPUT: $\underline{\alpha}, \bar{\alpha}, \alpha, \sigma, \Delta$
OUTPUT: $\underline{\alpha}', \bar{\alpha}', \varkappa$
STEP 1: Pick an $\mathcal{SN}$ function ρ satisfying (2.38).
STEP 2: Find the class $\mathcal{K}_\infty$ functions $\underline{\alpha}'$ and $\bar{\alpha}'$ from (2.34).
STEP 3: Find the smooth function $\varkappa$ from (2.39).
STEP 4: END

In many applications, we would like the supply pair α and σ to have the following properties, i.e.,

$$\limsup_{s\to 0^+} \frac{s^2}{\alpha(s)} < \infty, \ \limsup_{s\to 0^+} \frac{\sigma(s)}{s^2} < \infty. \tag{2.40}$$

For convenience, we make the following explicit assumption.

Assumption 2.1 The system $\dot{x} = f(x, u, d)$ has an ISS Lyapunov function $V(x)$, i.e.,

$$V(x) \sim \{\underline{\alpha}, \bar{\alpha}, \alpha, \sigma \mid \dot{x} = f(x, u, d)\}$$

and α and σ satisfy (2.40).

Remark 2.8 Let $\alpha(s) = \sum_{i=1}^{n} a_i s^{r_i}$ be a polynomial with $a_i \neq 0$ and $r_1 < \cdots < r_n$. Then, the condition $\limsup_{s\to 0^+} s^2/\alpha(s) < \infty$ is satisfied if and only if $r_1 \leq 2$ and the condition $\limsup_{s\to 0^+} \alpha(s)/s^2 < \infty$ is satisfied if and only if $r_1 \geq 2$.

Remark 2.9 Assumption 2.1 is slightly stronger than requiring the system $\dot{x} = f(x, u, d)$ be RISS viewing x as the state and u as the input. The RISS property only implies the asymptotic stability of the equilibrium point $x = 0$ of the undriven subsystem with $u = 0$. However, Assumption 2.1 may imply the exponential stability of the equilibrium point of the undriven subsystem if all the functions $\underline{\alpha}$, $\bar{\alpha}$, and α take the quadratic form.

Remark 2.10 Under Assumption 2.1, there exist smooth functions $\alpha_0(x)$ and $\sigma_0(u)$ such that

$$\alpha_0(x)\alpha(\|x\|) \geq \|x\|^2, \ \sigma_0(u)\|u\|^2 \geq \sigma(\|u\|). \tag{2.41}$$

On one hand, since α satisfies (2.40), there exits a constant $l_1 \geq 1$ such that $\alpha(\|x\|) \geq \|x\|^2/l_1^2$ for $\|x\| \leq 1$, and since α is of class $\mathcal{K}_\infty$, there exists a constant $l_2 > 0$ such that $\alpha(\|x\|) \geq l_2$ for $\|x\| \geq 1$. As a result, the first inequality of (2.41) holds for any α_0 satisfying

$$\alpha_0(x) \geq l_1^2 + \frac{1}{l_2}\|x\|^2.$$

On the other hand, since σ satisfies (2.40), we can define a function $l : [0, \infty) \mapsto [0, \infty)$ such that $l(s) = \sigma(s)/s^2$, $\forall s > 0$ and $l(0) = \lim_{s\to 0^+} l(s)$. Let $\sigma_0(u)$ be a smooth function such that $\sigma_0(u) \geq l(\|u\|)$. Then

$$\sigma_0(u)\|u\|^2 \geq l(\|u\|)\|u\|^2 = \sigma(\|u\|).$$

Under Assumption 2.1, Corollary 2.1 can be further specialized to the following corollary.

Corollary 2.2 *Under Assumption 2.1, for any smooth function $\Delta(x)$, there exists another ISS Lyapunov function $V'(x)$ satisfying*

$$\begin{aligned}\underline{\alpha}'(\|x\|) \leq V'(x) &\leq \bar{\alpha}'(\|x\|)\\ \dot{V}'(x) &\leq -\Delta(x)\|x\|^2 + \varkappa(u)\|u\|^2.\end{aligned} \tag{2.42}$$

for a smooth function $\varkappa$. In particular, the class $\mathcal{K}_\infty$ functions $\bar{\alpha}'$ and $\underline{\alpha}'$ and the smooth function $\varkappa$ are given by Algorithm 2.3.

Proof Under Assumption 2.1, we can find two smooth functions $\alpha_0(x)$ and $\sigma_0(u)$ such that (2.41) is satisfied. Letting

$$\bar{\Delta}(x) \geq \Delta(x)\alpha_0(x) \tag{2.43}$$

gives

$$\bar{\Delta}(x)\alpha(\|x\|) \geq \Delta(x)\alpha_0(x)\alpha(\|x\|) \geq \Delta(x)\|x\|^2. \tag{2.44}$$

By Corollary 2.1, for the smooth function $\bar{\Delta}(x)$, there exists some smooth function $\bar{\varkappa}$ such that the Lyapunov function $V'(x)$ defined in Corollary 2.1 satisfies

$$\dot{V}'(x) \leq -\bar{\Delta}(x)\alpha(\|x\|) + \bar{\varkappa}(u)\sigma(\|u\|)$$

which yields (2.42) upon letting

$$\varkappa(u) = \bar{\varkappa}(u)\sigma_0(u) \tag{2.45}$$

and using (2.41). □

Algorithm 2.3

INPUT: $\underline{\alpha}, \bar{\alpha}, \alpha, \sigma, \Delta$
OUTPUT: $\underline{\alpha}', \bar{\alpha}', \varkappa$
STEP 1: Pick two smooth functions $\alpha_0(x)$ and $\sigma_0(u)$ satisfying (2.41).
STEP 2: Pick the smooth function $\bar{\Delta}(x)$ satisfying (2.43).
STEP 3: Call $(\underline{\alpha}', \bar{\alpha}', \bar{\varkappa}) =$ ALGORITHM 2.2$(\underline{\alpha}, \bar{\alpha}, \alpha, \sigma, \bar{\Delta})$.
STEP 4: Find the smooth function $\varkappa$ from (2.45).
STEP 5: END

Remark 2.11 In Corollary 2.1, if $\alpha(\|x\|)$ is a quadratic function, say, $\alpha(\|x\|) = a\|x\|^2$ for some $a > 0$. Then, by letting Δ be a constant and $V'(x) = \Delta V(x)$, we have

$$\dot{V}'(x) \leq -a\Delta\|x\|^2 + \varkappa(u)\|u\|^2$$

where $\varkappa(u)$ is a smooth function such that $\varkappa(u)\|u\|^2 \geq \Delta\sigma(\|u\|)$. In particular, if $\sigma(\|u\|)$ is also a quadratic function, say, $\sigma(\|u\|) = b\|u\|^2$ for some $b > 0$, then $\varkappa(u) = b\Delta$ is a constant.

As an application of Corollary 2.2, we consider the global robust stabilization problem of the following class of nonlinear systems

$$\begin{aligned} \dot{z} &= q(z, x, d) \\ \dot{x} &= f(z, x, d) + b(d)u \end{aligned} \tag{2.46}$$

where $z \in \mathbb{R}^n$ and $x \in \mathbb{R}$ are the state variables, $u \in \mathbb{R}$ is the input, and $d : [t_0, \infty) \mapsto \mathbb{D}$ is a piecewise continuous function with $\mathbb{D}$ a compact subset of $\mathbb{R}^l$. The functions q and f are sufficiently smooth[4] with $q(0, 0, d) = 0$ and $f(0, 0, d) = 0$ for all $d \in \mathbb{D}$.

We need the following two assumptions.

Assumption 2.2 The function $b(d)$ is away from zero, e.g., $b(d) > 0, \forall d \in \mathbb{D}$.

Assumption 2.3 The subsystem $\dot{z} = q(z, x, d)$ has an ISS Lyapunov function $V(z)$, i.e.,

$$V(z) \sim \{\underline{\alpha}, \bar{\alpha}, \alpha, \sigma \mid \dot{z} = q(z, x, d)\}$$

and

$$\limsup_{s \to 0^+} \frac{s^2}{\alpha(s)} < \infty, \quad \limsup_{s \to 0^+} \frac{\sigma(s)}{s^2} < \infty.$$

In particular, the functions $\underline{\alpha}$, $\bar{\alpha}$, α and σ are known.

Theorem 2.8 *Consider the system (2.46) with a prescribed compact set $\mathbb{D}$. Under Assumptions 2.2 and 2.3, there exist a controller*

$$u = -\rho(x)x + \bar{u} \tag{2.47}$$

and an ISS Lyapunov function $W(z, x)$ satisfying

$$\underline{\beta}(\|col(z, x)\|) \leq W(z, x) \leq \bar{\beta}(\|col(z, x)\|)$$

for some class $\mathcal{K}_\infty$ functions $\underline{\beta}$ and $\bar{\beta}$, and, along the trajectory of the closed-loop system,

$$\dot{W}(z, x) \leq -\|z\|^2 - \|x\|^2 + \|\bar{u}\|^2. \tag{2.48}$$

As a result, the controller (2.47) with $\bar{u} = 0$ globally robustly stabilizes the system (2.46). In particular, the function ρ is given in Algorithm 2.4.

---

[4] A sufficiently smooth function means a function whose k-th derivatives exist for a sufficiently large integer k.

Proof Since $f(z, x, d)$ is a sufficiently smooth function, using (11.13) of the Appendix, one has

$$|f(z, x, d)| \leq m_1(z)\|z\| + m_2(x)|x|, \ \forall d \in \mathbb{D} \tag{2.49}$$

for some smooth positive functions m_1 and m_2. Let

$$\Delta(z) \geq 1 + m_1^2(z). \tag{2.50}$$

By Corollary 2.2, there exists a continuously differentiable function $V'(z, t)$ satisfying $\underline{\alpha}'(\|x_1\|) \leq V'(z, t) \leq \bar{\alpha}'(\|z\|)$ for some class $\mathcal{K}_\infty$ functions $\underline{\alpha}_1'$ and $\bar{\alpha}_1'$, and, along the trajectory of $\dot{z} = q(z, x, d)$,

$$\dot{V}'(z) \leq -\Delta(z)\|z\|^2 + \varkappa(x)x^2 \tag{2.51}$$

for a smooth function $\varkappa$. Since $b(d) > 0, \ \forall d \in \mathbb{D}$, there exist two constants $\bar{b}$ and $\underline{b}$ such that $\bar{b} \geq b(d) \geq \underline{b}, \ \forall d \in \mathbb{D}$. Then, define the function ρ such that

$$\rho(x) \geq [\varkappa(x) + m_2(x) + 5/4]/\underline{b} + \bar{b}/4 \tag{2.52}$$

and an ISS Lyapunov function candidate for the closed-loop system:

$$W(z, x) = V_1'(z) + x^2/2.$$

Direct calculation shows that the derivative of $W(z, x)$ along the trajectory of the closed-loop satisfies:

$$\begin{aligned} \dot{W}(z, x) &\leq -\Delta(z)\|z\|^2 + \varkappa(x)x^2 + x(f(z, x, d) + b(-\rho(x)x + \bar{u})) \\ &\leq -\Delta(z)\|z\|^2 + m_1^2(z)\|z\|^2 \\ &\quad + x^2[\varkappa(x) + 1/4 + m_2(x) - b\rho(x) + b^2/4] + \bar{u}^2 \\ &\leq -\|z\|^2 - \|x\|^2 + \bar{u}^2. \end{aligned}$$

The proof is thus completed by choosing the class $\mathcal{K}_\infty$ functions $\underline{\beta}$ and $\bar{\beta}$, using Lemma 11.3 of the Appendix, such that

$$\begin{aligned} \underline{\beta}(\|\mathrm{col}(z, x)\|) &\leq \underline{\alpha}_1'(\|z\|) + x^2/2 \\ \bar{\beta}(\|\mathrm{col}(z, x)\|) &\geq \bar{\alpha}_1'(\|z\|) + x^2/2. \end{aligned}$$ □

Algorithm 2.4

INPUT: $f, b, \underline{\alpha}, \bar{\alpha}, \alpha, \sigma, \mathbb{D}$

OUTPUT: ρ

STEP 1: Find the functions m_1 and m_2 from (2.49).

STEP 2: Pick the function Δ from (2.50) and call

$$(\underline{\alpha}', \bar{\alpha}', \varkappa) = \text{ALGORITHM 2.3 } (\underline{\alpha}, \bar{\alpha}, \alpha, \sigma, \Delta).$$

STEP 3: Calculate the function ρ from (2.52).
STEP 4: END

Remark 2.12 The function ρ selected in (2.52) is to satisfy (2.48) and thus solving the stabilization problem of the system (2.46). In fact, to solve the stabilization problem of the system (2.46), it suffices to pick

$$\rho(x) \geq [\varkappa(x) + m_2(x) + 5/4]/\underline{b} \tag{2.53}$$

so that

$$\begin{aligned}\dot{W}(z, x) &\leq -\Delta(z)\|z\|^2 + \varkappa(x)x^2 + x(f(z, x, d) + b(-\rho(x)x)) \\ &\leq -\Delta(z)\|z\|^2 + m_1^2(z)\|z\|^2 + x^2[\varkappa(x) + 1/4 + m_2(x) - b\rho(x)] \\ &\leq -\|z\|^2 - \|x\|^2.\end{aligned}$$

Remark 2.13 From (2.52), it can be seen that the validity of Theorem 2.8 requires that the functions $\varkappa$ and m_2 as well as the constants $\underline{b}$ and $\bar{b}$ be known precisely. This is possible since $\mathbb{D}$ is assumed to be a known compact set. The case where $\mathbb{D}$ is not a known compact set cannot be handled by Theorem 2.8, and will be studied in the next section.

Remark 2.14 If $b(d) < 0, \ \forall d \in \mathbb{D}$, Theorem 2.8 still works by rewriting the second equation of (2.46) as $\dot{x} = f(z, x, d) + \hat{b}(d)\hat{u}$ where $\hat{b}(d) = -b(d)$ and $\hat{u} = -u$.

Example 2.16 Consider a second order system

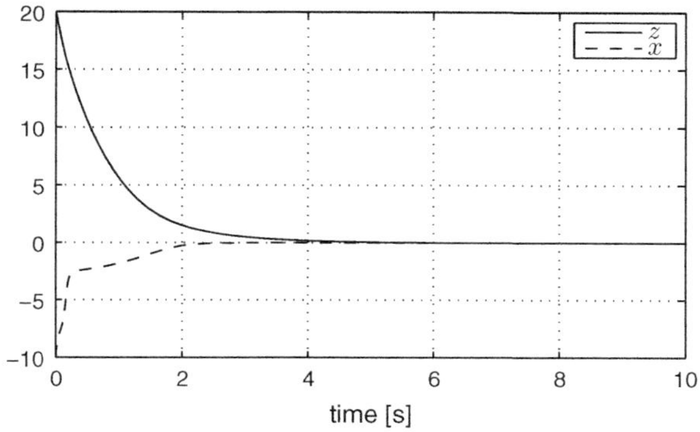

Fig. 2.2 Profile of aysmptoically stable state trajectories of the closed-loop system in Example 2.16

$$\begin{aligned}\dot{z} &= -z + x\\ \dot{x} &= w_1 z \sin x + w_2 x^3 + u\end{aligned}$$

where w_1 and w_2 are unknown parameters with $|w_1| \leq 2$ and $|w_2| \leq 1$. We will find a controller u for the global robust stabilization problem and a corresponding Lyapunov function of the closed-loop system.

First, it can be verified that the derivative of $V(z) = z^2$, along the trajectory of $\dot{z} = -z + x$, satisfies

$$\dot{V}(z) \leq -z^2 + x^2.$$

Assumption 2.3 is satisfied with $\underline{\alpha}(s) = \bar{\alpha}(s) = \alpha(s) = \sigma(s) = s^2$.

Observe that

$$|w_1 z \sin x + w_2 x^3| \leq 2|z| + x^2|x|,$$

so (2.49) is true for $m_1(z) = 2$ and $m_2(x) = x^2$. Using the inequality (2.50) gives $\Delta(z) = 1 + m_1^2 = 5$. Since Δ is constant, and both α and σ are quadratic, by Corollary 2.1 and Remark 2.11, letting $V'(z) = \Delta z^2$ shows (2.51) is satisfied with $\varkappa(x) = 5$. Now, we are ready to pick the following function according to (2.53)

$$\rho(x) = 5 + x^2 + 5/4 = x^2 + 6.25,$$

which gives the controller

$$u = -x^3 - 6.25x. \tag{2.54}$$

Moreover, the closed-loop system has a Lyapunov function

$$W(z, x) = V'(z) + x^2/2 = 5z^2 + x^2/2$$

whose derivative, along the state trajectory of the closed-loop system, satisfies

$$\dot{W}(z, x) \leq -5z^2 + 5x^2 + x(w_1 z \sin x + w_2 x^3 - x^3 - 6.25x) \leq -z^2 - x^2.$$

The global robust stabilization problem is thus solved. The controller (2.54) is designed for $|w_1| \leq 2$ and $|w_2| \leq 1$, and the simulation is conducted with $w_1 = 1.8$ and $w_2 = 1$. The initial state values are $z(0) = 20$ and $x(0) = -10$. The state of the closed-loop system converges to the origin as shown in Fig. 2.2. If the uncertainties are out of this range, the controller may fail as illustrated in Fig. 2.3 with $w_1 = 1.8$ but $w_2 = 2$.

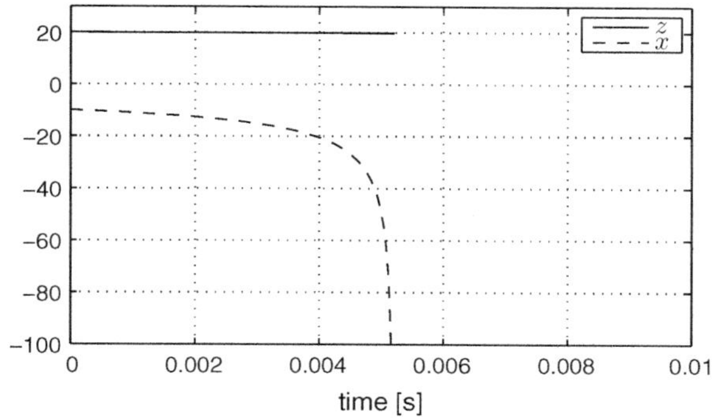

Fig. 2.3 Profile of unstable state trajectories of the closed-loop system in Example 2.16

2.6 Universal Adaptive Control

So far, we have assumed that the range of the uncertainty $d(t)$ belongs to a compact set $\mathbb{D}$ whose bound is known. In many cases, the range of $\mathbb{D}$ is unknown or $d(t)$ can be arbitrarily large, the robust control approach studied in the previous section cannot handle such uncertainty. In this section, we will further consider extending Theorem 2.8 to the case where the range of the uncertainty $d(t)$ is unknown or $d(t)$ can be arbitrarily large. In Theorem 2.8, it is known that for all $d \in \mathbb{D}$, there exist two known constants $\bar{b}$ and $\underline{b}$ such that $\bar{b} \geq b(d) \geq \underline{b}, \forall, d \in \mathbb{D}$. In this section, we assume b is an arbitrary unknown positive constant. More specifically, (2.46) is rewritten as follows

$$\begin{aligned} \dot{z} &= q(z, x, d) \\ \dot{x} &= f(z, x, d) + bu, \;\; b > 0. \end{aligned} \tag{2.55}$$

We first modify Assumption 2.1 to the following.

Assumption 2.4 The system $\dot{x} = f(x, u, d)$ has an ISS Lyapunov function $V(x)$, i.e.,

$$V(x) \sim \{\underline{\alpha}, \bar{\alpha}, \alpha, \hat{\sigma} \mid \dot{x} = f(x, u, d)\}$$

and

$$\limsup_{s \to 0^+} \frac{s^2}{\alpha(s)} < \infty, \;\; \limsup_{s \to 0^+} \frac{\hat{\sigma}(s)}{s^2} < \infty.$$

Moreover, the functions $\underline{\alpha}$, $\bar{\alpha}$, and α are known and the function $\hat{\sigma}$ is known up to a constant factor in the sense that there exist an unknown constant p and a known function σ such that $\hat{\sigma} = p\sigma$.

Remark 2.15 Assumption 2.4 is weaker than Assumption 2.1 since it allows the function $\hat{\sigma}$ to be known up to a constant factor. This assumption is more realistic when the range of $\mathbb{D}$ is unknown. For example, consider a scalar system $\dot{x} = -x + du$ where $u \in \mathbb{R}$ and d is an unknown constant. Let $V(x) = x^2$. Then, along the trajectory of the system $\dot{x} = -x + du$, we have

$$\dot{V}(x) \le -x^2 + pu^2 \tag{2.56}$$

for $p = d^2$. Letting $\alpha(s) = s^2$ and $\hat{\sigma}(s) = ps^2$ shows that Assumption 2.4 is satisfied, but Assumption 2.1 is not satisfied since p is unknown. In general, when the compact set $\mathbb{D}$ is unknown, all the functions $\underline{\alpha}, \bar{\alpha}, \alpha, \hat{\sigma}$ may only be known up to a constant factor. This more general case will be handled in Chap. 6.

Corresponding to Assumption 2.4, we can also modify Corollary 2.2 to the following.

Corollary 2.3 *Under Assumption 2.4, for any smooth function Δ, there exists another ISS Lyapunov function $V'(x)$ satisfying*

$$\begin{aligned} \underline{\alpha}'(\|x\|) \le V'(x) &\le \bar{\alpha}'(\|x\|) \\ \dot{V}'(x) &\le -\Delta(x)\|x\|^2 + p'\varkappa(u)\|u\|^2 \end{aligned} \tag{2.57}$$

for some unknown positive constant p' and some known smooth function $\varkappa$. In particular, the class $\mathcal{K}_\infty$ functions $\bar{\alpha}'$ and $\underline{\alpha}'$ and the smooth function $\varkappa$ are given by Algorithm 2.5.

Proof As shown in the proof of Lemma 2.5, for any smooth function $\Delta(x)$, there exists some ISS Lyapunov function $V'(x)$ for $\dot{x} = f(x, u, d)$ satisfying $\underline{\alpha}'(\|x\|) \le V'(x) \le \bar{\alpha}'(\|x\|)$ and the following inequality:

$$\dot{V}'(x) \le -\frac{1}{2}\rho(\underline{\alpha}(\|x\|))\alpha(\|x\|) + p\rho(\bar{\alpha}(\alpha^{-1}(2p\sigma(\|u\|))))\sigma(\|u\|).$$

for any $\rho \in \mathcal{SN}$. By Remark 2.10, there exist smooth functions $\alpha_0(x)$ and $\sigma_0(u)$ such that

$$\alpha_0(x)\alpha(\|x\|) \ge \|x\|^2, \ \sigma_0(u)\|u\|^2 \ge \sigma(\|u\|). \tag{2.58}$$

From the proof of Corollary 2.2, there exists a smooth function $\bar{\Delta}(x)$ satisfying (2.43).

Pick a function $\rho \in \mathcal{SN}$ such that

$$\frac{1}{2}\rho(\underline{\alpha}(\|x\|)) \ge \bar{\Delta}(x). \tag{2.59}$$

Then we have,

$$\dot{V}'(x) \leq -\bar{\Delta}(x)\alpha(\|x\|) + p\rho(\bar{\alpha}(\alpha^{-1}(2p\sigma(\|u\|))))\sigma(\|u\|). \quad (2.60)$$

By part (i) of Lemma 11.1 in the Appendix, one has

$$\rho(\bar{\alpha}(\alpha^{-1}(2\hat{\sigma}(\|u\|)))) = \rho(\bar{\alpha}(\alpha^{-1}(2p\sigma(\|u\|)))) \leq c(p)\bar{\varkappa}(u) \quad (2.61)$$

for some smooth functions $c(p) \geq 0$ and $\bar{\varkappa}(u) \geq 0$.

Letting p' be any unknown positive constant satisfying $p' \geq c(p)p$, and using (2.43) and (2.61) in (2.60) gives

$$\dot{V}'(x) \leq -\Delta(x)\alpha_0(x)\alpha(\|x\|) + p'\bar{\varkappa}(u)\sigma(\|u\|).$$

Letting

$$\varkappa(u) = \bar{\varkappa}(u)\sigma_0(u) \quad (2.62)$$

and using (2.58) completes the proof. □

Algorithm 2.5

INPUT: $\underline{\alpha}, \bar{\alpha}, \alpha, \sigma, \Delta$
OUTPUT: $\underline{\alpha}', \bar{\alpha}', \varkappa$
STEP 1: Pick the functions $\alpha_0(x)$ and $\sigma_0(u)$ satisfying (2.58).
STEP 2: Pick the smooth function $\bar{\Delta}(x)$ satisfying (2.43).
STEP 3: Pick an $\mathcal{SN}$ function ρ satisfying (2.59).
STEP 4: Find the class $\mathcal{K}_\infty$ functions $\underline{\alpha}'$ and $\bar{\alpha}'$ from (2.34).
STEP 5: Find the smooth function $\bar{\varkappa}$ from (2.61).
STEP 6: Find the smooth function $\varkappa$ from (2.62).
STEP 7: END

As pointed out in Remark 2.13, Theorem 2.8 cannot handle the system (2.46) when $\mathbb{D}$ is not a known compact set. We now modify Theorem 2.8 by using a so-called *universal adaptive control* technique. For this purpose, we modify Assumption 2.3 to the following.

Assumption 2.5 The subsystem $\dot{z} = q(z, x, d)$ has an ISS Lyapunov function $V(z)$, i.e.,

$$V(z) \sim \{\underline{\alpha}, \bar{\alpha}, \alpha, \hat{\sigma} \mid \dot{z} = q(z, x, d)\}$$

and

$$\limsup_{s\to 0^+} \frac{s^2}{\alpha(s)} < \infty, \quad \limsup_{s\to 0^+} \frac{\hat{\sigma}(s)}{s^2} < \infty.$$

Moreover, the functions $\underline{\alpha}$, $\bar{\alpha}$, and α are known and the function $\hat{\sigma}$ is known up to a constant factor in the sense that there exist an unknown constant p and a known function σ such that $\hat{\sigma} = p\sigma$.

Example 2.17 Consider a linear system

$$\dot{z} = Az + B(d)x \tag{2.63}$$

where $z \in \mathbb{R}^n$ and $x \in \mathbb{R}$ are the state variables, $d \in \mathbb{D}$ is the system uncertainty for an unknown compact set $\mathbb{D}$, and A is a Hurwitz matrix. Now, let P be a symmetric positive matrix such that $PA + A^{\top}P = -I$. Since $d \in \mathbb{D}$ for a compact set $\mathbb{D}$, we can pick a positive number $p \geq 2\|PB(d)\|^2$ which is not necessarily known because it depends on the size of $\mathbb{D}$. Let $V(z) = z^{\top}Pz$. Then its derivative along the system trajectory satisfies

$$\begin{aligned}\dot{V}(z) &= -\|z\|^2 + 2z^{\top}PB(d)x \leq -\|z\|^2/2 + 2\|PB(d)\|^2x^2 \\ &\leq -\alpha(\|z\|) + p\sigma(|x|)\end{aligned}$$

where $\alpha(s) = s^2/2$ and $\sigma(s) = s^2$ are known functions. So, Assumption 2.5 is satisfied for the system (2.63).

Theorem 2.9 *Consider the system (2.55) with any unknown compact set $\mathbb{D}$. Under Assumption 2.5, there exists a controller*

$$\begin{aligned}u &= -k\rho(x)x + \bar{u} \\ \dot{k} &= \lambda\rho(x)x^2, \ \lambda > 0\end{aligned} \tag{2.64}$$

such that the closed-loop system has an ISS Lyapunov function $W(z, x, \hat{k})$ where $\hat{k} = k - k^*$ for some constant $k^* > 0$, satisfying

$$\underline{\beta}(\|col(z, x, \hat{k})\|) \leq W(z, x, \hat{k}) \leq \bar{\beta}(\|col(z, x, \hat{k})\|)$$

for some class $\mathcal{K}_{\infty}$ functions $\underline{\beta}$ and $\bar{\beta}$, and, along the trajectory of the closed-loop system,

$$\dot{W}(z, x, \hat{k}) \leq -\|z\|^2 - \|x\|^2 + \|\bar{u}\|^2.$$

As a result, the controller (2.64) with $\bar{u} = 0$ globally stabilizes the system (2.55). In particular, the function ρ is given in Algorithm 2.6.

Proof By Corollary 2.3, for any given smooth function $\Delta(z) \geq 0$, there exists a continuously differentiable function $V'(z)$ satisfying $\underline{\alpha}'(\|z\|) \leq V'(z) \leq \bar{\alpha}'(\|z\|)$ for some class $\mathcal{K}_{\infty}$ functions $\underline{\alpha}'$ and $\bar{\alpha}'$, such that along the trajectory of the system $\dot{z} = q(z, x, d)$,

$$\dot{V}'(z) \leq -\Delta(z)\|z\|^2 + p'\varkappa(x)x^2 \tag{2.65}$$

for some unknown positive number p' and a known smooth function $\varkappa$. By Corollary 11.1 of the Appendix, there exist a positive number c, depending on the size of $\mathbb{D}$, and two positive and sufficiently smooth known functions m_1 and m_2, such that

$$|f(z, x, d)| \leq cm_1(z)\|z\| + cm_2(x)|x|, \ \forall d \in \mathbb{D}. \tag{2.66}$$

Let $\hat{k} = k - k^*$ with k^* a positive number to be specified later. Direct calculation shows that, along the trajectory of the closed-loop system, the derivative of

$$U(z, x) = V'(z) + x^2/2$$

satisfies,

$$\begin{aligned}\dot{U}(z, x) &\leq -\Delta(z)\|z\|^2 + p'\varkappa(x)x^2 + x[f(z, x, d) + bu] \\ &\leq -\Delta(z)\|z\|^2 + m_1^2(z)\|z\|^2 + x^2[p'\varkappa(x) + c^2/4 + cm_2(x) \\ &\quad + b^2/4 - bk\rho(x)] + \bar{u}^2 \\ &= -\Delta(z)\|z\|^2 + m_1^2(z)\|z\|^2 + x^2[p'\varkappa(x) + c^2/4 + cm_2(x) \\ &\quad + b^2/4 - bk^*\rho(x)] + \bar{u}^2 - b\hat{k}\rho(x)x^2.\end{aligned} \tag{2.67}$$

In (2.67), let

$$\Delta(z) \geq 1 + m_1^2(z) \tag{2.68}$$

$$\rho(x) \geq \max\{\varkappa(x), m_2(x), 1\} \tag{2.69}$$

and

$$k^* \geq (1 + p' + c^2/4 + c)/b + b/4. \tag{2.70}$$

One has

$$\dot{U}(z, x) \leq -\|z\|^2 - \|x\|^2 + \bar{u}^2 - b\hat{k}\rho(x)x^2. \tag{2.71}$$

Define a Lyapunov function candidate as follows:

$$W(z, x, \hat{k}) = U(z, x) + b\hat{k}^2/(2\lambda).$$

Direct calculation shows that the derivative of $W(z, x, \hat{k})$ along the trajectory of the closed-loop system satisfies,

$$\begin{aligned}\dot{W}(z, x, \hat{k}) &\leq -\|z\|^2 - \|x\|^2 + \bar{u}^2 - b\hat{k}\rho(x)x^2 + b(k - k^*)\dot{k}/\lambda \\ &\leq -\|z\|^2 - \|x\|^2 + \bar{u}^2.\end{aligned}$$

The proof is thus completed by choosing the class $\mathcal{K}_\infty$ functions $\underline{\beta}$ and $\bar{\beta}$, using Lemma 11.3 of the Appendix, such that

$$\begin{aligned}\underline{\beta}(\|\mathrm{col}(z,x,\hat{k})\|) &\le \underline{\alpha}'(\|z\|) + x^2/2 + b\hat{k}^2/(2\lambda)\\ \bar{\beta}(\|\mathrm{col}(z,x,\hat{k})\|) &\ge \bar{\alpha}'(\|z\|) + x^2/2 + b\hat{k}^2/(2\lambda).\end{aligned} \tag{2.72}$$

□

Algorithm 2.6

INPUT: $f, \underline{\alpha}, \bar{\alpha}, \alpha, \sigma$
OUTPUT: ρ
STEP 1: Find the functions m_1 and m_2 from (2.66).
STEP 2: Pick the function Δ from (2.68) and call

$$(\underline{\alpha}', \bar{\alpha}', \varkappa) = \text{ALGORITHM 2.5 } (\underline{\alpha}, \bar{\alpha}, \alpha, \sigma, \Delta).$$

STEP 3: Calculate the function ρ from (2.69).
STEP 4: END

Remark 2.16 If the size of the uncertainty $\mathbb{D}$ is known, then a real number k^* dominating the inequality (2.70) is known. One can pick a sufficiently large constant gain $k \ge k^*$ to be the controller gain. However, if the size of the uncertainty $\mathbb{D}$ is unknown, then k^* is unknown, either. One has to use a dynamic gain governed by (2.64). From (2.64), it can be seen that if the gain k is not large enough to achieve $\lim_{t\to\infty} x(t) = 0$, then it will increase until $\lim_{t\to\infty} x(t) = 0$. This type of adaptive approach for tuning the controller gain is called *universal adaptive control* or *self-tuning adaptive control.*

Example 2.18 Consider a second order nonlinear system

$$\begin{aligned}\dot{z} &= -z + w_3 x\\ \dot{x} &= w_1 z \sin x + w_2 x^3 + u\end{aligned}$$

which was studied in Example 2.16 assuming $w_3 = 1$ and the size of the unknown parameters w_1 and w_2 are known. Here, we consider the more general case where w_1, w_2, and w_3 can be any unknown real numbers. First, it can be verified that the derivative of $V(z) = z^2$, along the trajectory of $\dot{z} = -z + w_3 x$, satisfies

$$\dot{V}(z) \le -z^2 + px^2$$

where $p \ge w_3^2$ is any unknown constant. Thus, Assumption 2.5 is satisfied with $\underline{\alpha}(s) = \bar{\alpha}(s) = \alpha(s) = s^2$ and $\hat{\sigma}(s) = ps^2$.

Next, we note that

$$|w_1 z \sin x + w_2 x^3| \le c(|z| + x^2|x|)$$

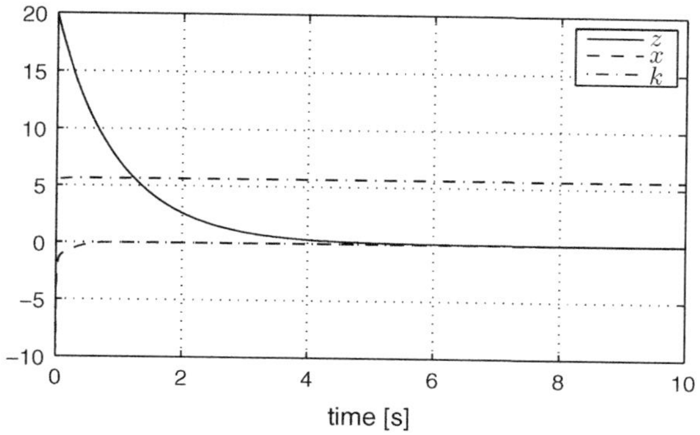

Fig. 2.4 Profile of state trajectories of the closed-loop system in Example 2.18

where $c \geq \max\{|w_1|, |w_2|\}$, that is, (2.66) is true for $m_1 = 2$ and $m_2(x) = x^2$. Let $\Delta(z) = 1 + m_1^2 = 5$. Since Δ is constant, and both α and $\hat{\sigma}$ are quadratic, by Corollary 2.3 and Remark 2.11, letting $V'(z) = \Delta z^2$ shows (2.65) is satisfied with $\varkappa(x) = 5$. Now, pick the following function according to (2.69)

$$\rho(x) = x^2 + 5 \geq \max\{5, x^2, 1\},$$

which leads to the controller (2.64). The performance of the controller is simulated and illustrated in Fig. 2.4 with $w_1 = 1.8$, $w_2 = 2$, and $w_3 = 1$. The initial state values are $z(0) = 20$, $x(0) = -10$, and $k(0) = 0$. It can be seen that both x and z approach 0 asymptotically while k approaches a finite constant asymptotically.

2.7 Small Gain Theorem

In this section, we introduce the small gain theorem to analyze the property of two inter-connected ISS systems. Let us consider the following two systems Σ_1 and Σ_2,

$$\begin{aligned} \Sigma_1 : \ \dot{x}_1 &= f_1(x_1, u_1, u_c, d), \\ \Sigma_2 : \ \dot{x}_2 &= f_2(x_2, u_2, u_c, d), \ t \geq t_0 \end{aligned} \tag{2.73}$$

where, for $i = 1, 2$, $x_i \in \mathbb{R}^{n_i}$ is the state, $u_i \in \mathbb{R}^{m_i}$, $u_c \in \mathbb{R}^{m_c}$ are the inputs of the subsystem Σ_i, and the function $f_i(x_i, u_i, u_c, d(t))$ is piecewise continuous in d and locally Lipschitz in $\text{col}(x_i, u_i, u_c)$ and $d(t) : [t_0, \infty) \mapsto \mathbb{D} \subset \mathbb{R}^l$ is piecewise continuous in t for a compact set $\mathbb{D}$.

Suppose $m_1 = n_2$ and $m_2 = n_1$, and consider the following connection (see Fig. 2.5),

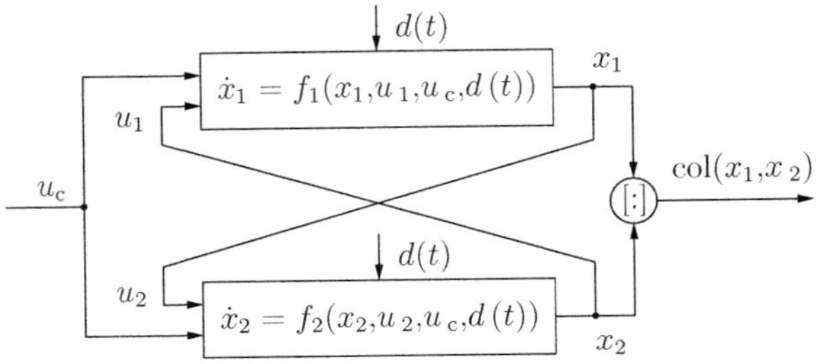

Fig. 2.5 Inter-connection of two ISS systems

$$\begin{aligned} u_1 &= x_2 \\ u_2 &= x_1. \end{aligned} \tag{2.74}$$

Under (2.74), the system (2.73) can be put in a compact form

$$\dot{x} = f(x, u_c, d)$$

with

$$x = \text{col}(x_1, x_2), \ \ f(x, u_c, d) = \text{col}(f_1(x_1, x_2, u_c, d), \ f_2(x_2, x_1, u_c, d)).$$

We will introduce two versions of the small gain theorem. The first version is formulated in terms of the ISS Lyapunov function of the individual subsystems, and this version will be frequently used in nonlinear controller design in the subsequent chapters.

Theorem 2.10 (Small Gain Theorem in terms of ISS Lyapunov Functions) *For $i = 1, 2$, assume the subsystem Σ_i of (2.73) is RISS with an ISS Lyapunov function $V_i(x_i)$, i.e.,*

$$V_i(x_i) \sim \{\underline{\alpha}_i, \bar{\alpha}_i, \alpha_i, (\sigma_i, \varsigma_i) \mid \dot{x}_i = f_i(x_i, u_i, u_c, d)\}.$$

Further assume

$$\alpha_1(s) - \sigma_2(s) \geq \delta_1(s), \ \ \alpha_2(s) - \sigma_1(s) \geq \delta_2(s), \ \ \forall s \geq 0 \tag{2.75}$$

for class $\mathcal{K}_\infty$ functions δ_i, $i = 1, 2$. Then the system (2.73) under the connection (2.74) is RISS with an ISS Lyapunov function $V(x)$, i.e.,

$$V(x) \sim \{\underline{\alpha}, \bar{\alpha}, \alpha, \varsigma \mid \dot{x} = f(x, u_c, d)\}.$$

Proof Let $V(x) = V_1(x_1) + V_2(x_2)$. Under (2.75), we have

$$\underline{\alpha}_1(\|x_1\|) + \underline{\alpha}_2(\|x_2\|) \le V(x) \le \bar{\alpha}_1(\|x_1\|) + \bar{\alpha}_2(\|x_2\|)$$
$$\dot{V}(x) \le -\delta_1(\|x_1\|) - \delta_2(\|x_2\|) + \varsigma_1(\|u_c\|) + \varsigma_2(\|u_c\|).$$

By Lemma 11.3 of the Appendix, there exist functions $\underline{\alpha}$, $\bar{\alpha}$, α, and ς such that

$$\underline{\alpha}(\|x\|) \le \underline{\alpha}_1(\|x_1\|) + \underline{\alpha}_2(\|x_2\|)$$
$$\bar{\alpha}(\|x\|) \ge \bar{\alpha}_1(\|x_1\|) + \bar{\alpha}_2(\|x_2\|)$$
$$\alpha(\|x\|) \le \delta_1(\|x_1\|) + \delta_2(\|x_2\|)$$
$$\varsigma(\|u_c\|) \ge \varsigma_1(\|u_c\|) + \varsigma_2(\|u_c\|).$$

The proof is thus completed. □

Example 2.19 Consider the following system

$$\dot{x}_1 = -x_1^3 + x_1 x_2$$
$$\dot{x}_2 = x_1^2 - a x_2 + u \tag{2.76}$$

where a is a real parameter. This system results from the connection (2.74) of the following two systems

$$\dot{x}_1 = -x_1^3 + x_1 u_1 \tag{2.77}$$
$$\dot{x}_2 = u_2^2 - a x_2 + u. \tag{2.78}$$

Let $V_1(x_1) = x_1^2/2$. We first show that $V_1(x_1)$ is an ISS-Lyapunov function for the system (2.77) with state x_1 and input u_1. Indeed, the derivative of $V(x_1)$ along the trajectory of (2.77) is

$$\dot{V}_1(x_1) \le -x_1^4 + x_1^2|u_1| \le -x_1^4/2 + u_1^2/2.$$

In other words, one has

$$V_1(x_1) \sim \{\underline{\alpha}_1, \bar{\alpha}_1, \alpha_1, \sigma_1 \mid \dot{x}_1 = -x_1^3 + x_1 u_1\}$$

for $\alpha_1(s) = s^4/2$ and $\sigma_1(s) = s^2/2$.

Next we consider the system (2.78) with state x_2 and input (u_2, u). Let $V_2(x_2) = x_2^2/2$. Then the derivative of $V_2(x_2)$ along (2.78) is

$$\dot{V}_2(x_2) \le |x_2|(|u_2|^2 - a|x_2| + |u|)$$
$$\le (1/(4\epsilon_1))u_2^4 + \epsilon_1 x_2^2 - a x_2^2 + \epsilon_2 x_2^2 + (1/(4\epsilon_2))u^2$$

for any $\epsilon_1, \epsilon_2 > 0$. In other words, one has

$$V_2(x_2) \sim \{\underline{\alpha}_2, \bar{\alpha}_2, \alpha_2, (\sigma_2, \varsigma_2) \mid \dot{x}_2 = u_2^2 - ax_2 + u\}$$

for $\alpha_2(s) = (a - \epsilon_1 - \epsilon_2)s^2$, $\sigma_2(s) = (1/(4\epsilon_1))s^4$, and $\varsigma_2(s) = (1/(4\epsilon_2))s^2$.

Obviously, if $a > 1$, there exist $\epsilon_1, \epsilon_2 > 0$ to satisfy

$$1/2 > 1/(4\epsilon_1), \;\; a - \epsilon_1 - \epsilon_2 > 1/2, \tag{2.79}$$

which implies the small gain condition (2.75). By Theorem 2.10, the inter-connected system, i.e., the original system (2.76) is ISS and admits an ISS Lyapunov function $V(x_1, x_2) = V_1(x_1) + V_1(x_2)$.

In Theorem 2.10, the inequalities in (2.75) are the small gain conditions. In real applications, the functions α_i and σ_i are usually modified using the changing supply function technique to make the conditions (2.75) satisfied. The following result will be used in robust controller design (see Chap. 4 and some other chapters).

Theorem 2.11 *Consider a nonlinear system*

$$\begin{aligned} \dot{x}_1 &= f_1(x_1, u, d) \\ \dot{x}_2 &= f_2(x_2, x_1, u, d). \end{aligned} \tag{2.80}$$

Assume both subsystems are RISS with ISS Lyapunov functions

$$\begin{aligned} V_1(x_1) &\sim \{\underline{\alpha}_1, \bar{\alpha}_1, \alpha_1, \sigma_1) \mid \dot{x}_1 = f_1(x_1, u, d)\}. \\ V_2(x_2) &\sim \{\underline{\alpha}_2, \bar{\alpha}_2, \alpha_2, (\varsigma, \sigma_2) \mid \dot{x}_2 = f_2(x_2, x_1, u, d)\}. \end{aligned}$$

Suppose the function α_i, σ_i, and ς satisfy the following properties:

$$\limsup_{s \to 0^+} \frac{s^2}{\alpha_i(s)} < \infty, \;\; \limsup_{s \to 0^+} \frac{\sigma_i(s)}{s^2} < \infty, \;\; i = 1, 2, \;\; \limsup_{s \to 0^+} \frac{\varsigma(s)}{s^2} < \infty.$$

Then there exists an ISS Lyapunov function $V(x)$, with $x = col(x_1, x_2)$, satisfying

$$\underline{\alpha}(\|x\|) \le V(x) \le \bar{\alpha}(\|x\|) \tag{2.81}$$

for some class $\mathcal{K}_\infty$ functions $\underline{\alpha}$ and $\bar{\alpha}$, and, along the trajectory of (2.80),

$$\dot{V}(x) \le -\|x\|^2 + \varkappa(u)\|u\|^2 \tag{2.82}$$

for some smooth function $\varkappa$.

Proof We first consider the x_1-subsystem. By Corollary 2.2, for any smooth function Δ, there exists an ISS Lyapunov function $V_1'(x_1)$ satisfying

$$\underline{\alpha}_1'(\|x_1\|) \leq V_1'(x_1) \leq \bar{\alpha}_1'(\|x_1\|)$$
$$\dot{V}_1'(x_1) \leq -\Delta(x_1)\|x_1\|^2 + \varkappa_1(u)\|u\|^2$$

for some class $\mathcal{K}_\infty$ functions $\underline{\alpha}_1'$ and $\bar{\alpha}_1'$ and some smooth function $\varkappa_1$. Then, we consider the x_2-subsystem. By Corollary 2.2 again, there exists an ISS Lyapunov function $V_2'(x_2)$ satisfying

$$\underline{\alpha}_2'(\|x_2\|) \leq V_2'(x_2) \leq \bar{\alpha}_2'(\|x_2\|)$$
$$\dot{V}_2'(x_2) \leq -\|x_2\|^2 + \varkappa(x_1, u)\|\mathrm{col}(x_1, u)\|^2$$

for some class $\mathcal{K}_\infty$ functions $\underline{\alpha}_2'$ and $\bar{\alpha}_2'$ and some smooth function $\varkappa$. Moreover, we have

$$\varkappa(x_1, u)\|\mathrm{col}(x_1, u)\|^2 \leq \varkappa_2(x_1)\|x_1\|^2 + \varkappa_3(u)\|u\|^2$$

for some smooth functions $\varkappa_2$ and $\varkappa_3$. Let $\Delta(x_1) = \varkappa_2(x_1) + 1$ and $V(x) = V_1'(x_1) + V_2'(x_2)$. One has (2.82) for any smooth function

$$\varkappa(u) \geq \varkappa_1(u) + \varkappa_3(u).$$

By Lemma 11.3 of the Appendix, there exist class $\mathcal{K}_\infty$ functions $\underline{\alpha}$ and $\bar{\alpha}$ such that

$$\underline{\alpha}(\|x\|) \leq \underline{\alpha}_1'(\|x_1\|) + \underline{\alpha}_2'(\|x_2\|)$$
$$\bar{\alpha}(\|x\|) \geq \bar{\alpha}_1'(\|x_1\|) + \bar{\alpha}_2'(\|x_2\|).$$

The inequalities in (2.81) are thus proved. □

Corollary 2.4 *Consider a nonlinear system*

$$\dot{x}_1 = f_1(x_1, u, d)$$
$$\dot{x}_2 = Ax_2 + \phi(x_1, u, d) \tag{2.83}$$

where the matrix A is Hurwitz and the function ϕ is sufficiently smooth with $\phi(0, 0, d) = 0$ for all $d \in \mathbb{D}$ with $\mathbb{D}$ a compact set. Assume the x_1-subsystem is RISS with an ISS Lyapunov function $V_1(x_1)$, i.e.,

$$V_1(x_1) \sim \{\underline{\alpha}_1, \bar{\alpha}_1, \alpha_1, \sigma_1) \mid \dot{x}_1 = f_1(x_1, u, d)\},$$

and the functions α_1 and σ_1 satisfy the following properties:

$$\limsup_{s\to 0^+} \frac{s^2}{\alpha_1(s)} < \infty, \ \limsup_{s\to 0^+} \frac{\sigma_1(s)}{s^2} < \infty.$$

Then (2.83) has an ISS Lyapunov function $V(x)$ with $x = col(x_1, x_2)$ satisfying

$$\underline{\alpha}(\|x\|) \leq V(x) \leq \bar{\alpha}(\|x\|)$$

for some class $\mathcal{K}_\infty$ functions $\underline{\alpha}$ and $\bar{\alpha}$, and, along the trajectory of (2.83),

$$\dot{V}(x) \leq -\|x\|^2 + \varkappa(u)\|u\|^2$$

for a smooth function $\varkappa$.

Proof By Theorem 2.11, it suffices to show the x_2-subsystem is RISS with an ISS Lyapunov function $V_2(x_2)$, i.e.,

$$V_2(x_2) \sim \{\underline{\alpha}_2, \bar{\alpha}_2, \alpha_2, (\varsigma, \sigma_2) \mid \dot{x}_2 = f_2(x_2, x_1, u, d)\}$$

and the function α_2, σ_2, and ς satisfy the following properties:

$$\limsup_{s\to 0^+} \frac{s^2}{\alpha_2(s)} < \infty, \ \limsup_{s\to 0^+} \frac{\sigma_2(s)}{s^2} < \infty, \ \limsup_{s\to 0^+} \frac{\varsigma(s)}{s^2} < \infty.$$

Let $V_2(x_2) = x_2^{\mathrm{T}} P x_2$ where P is a symmetric positive definite matrix satisfying the Lyapunov equation

$$PA + A^{\mathrm{T}} P = -I.$$

It can be seen that

$$\underline{\alpha}_2(\|x_2\|) = \lambda_{\min}\|x_2\|^2/2 \leq V_2(x_2) \leq \lambda_{\max}\|x_2\|^2 \leq \bar{\alpha}_2(\|x_2\|)$$

where $\lambda_{\min}$ and $\lambda_{\max}$ are the minimal and maximal eigenvalues of P, respectively. Since ϕ is sufficiently smooth with $\phi(0, 0, d) = 0$, by Corollary 11.1 of the Appendix,

$$|\phi(x_1, u, d)| \leq m_1(\|x_1\|)\|x_1\| + m_2(|u|)|u|, \ \forall d \in \mathbb{D},$$

for some smooth functions m_1 and m_2. Then, the derivative of $V_2(x_2)$ along the trajectory of the x_2-subsystem of (2.83) satisfies

$$\begin{aligned} \dot{V}_2(x_2) &= -\|x_2\|^2 + 2x_2^{\mathrm{T}} P\phi(x_1, u, d) \\ &\leq -\|x_2\|^2/2 + 4\|P\|^2 m_1^2(\|x_1\|)\|x_1\|^2 + 4\|P\|^2 m_2^2(|u|)u^2. \end{aligned}$$

Thus, the proof is completed with

$$\alpha_2(s) = s^2/2, \ \varsigma(s) \geq 4\|P\|^2 m_1^2(s)s^2, \ \sigma_2(s) \geq 4\|P\|^2 m_2^2(s)s^2.$$

□

The small gain theorem can also be given in terms of gain functions of the individual subsystems. The proof of the following theorem is given in the Appendix.

Theorem 2.12 (Small Gain Theorem) *For $i = 1, 2$, assume the subsystem Σ_i of (2.73) is RISS viewing x_i as state, $col(u_i, u_c)$ as input, i.e., there exist class $\mathcal{KL}$ functions β_i, class $\mathcal{K}$ functions γ_i^x, γ_i^u, independent of t_0 and $d(t)$, such that, for any initial state $x_i(t_0)$, and any input function $col(u_i, u_c) \in L_\infty^{m_i+m_c}$, the solution $x_i(t)$ of Σ_i exists and satisfies*

$$\|x_i(t)\| \le \max\left\{\beta_i(\|x_i(t_0)\|, t - t_0), \gamma_i^x\left(\left\|u_{i[t_0,t]}\right\|\right), \gamma_i^u\left(\left\|u_{c[t_0,t]}\right\|\right)\right\}, \quad \forall t \ge t_0. \tag{2.84}$$

Further assume

$$\gamma_1^x \circ \gamma_2^x(s) < s, \ \forall s > 0. \tag{2.85}$$

Then the system (2.73) under the connection (2.74) is RISS viewing $x = col(x_1, x_2)$ as state and u_c as input, i.e.,

$$\|x(t)\| \le \max\left\{\beta(\|x(t_0)\|, t - t_0), \gamma\left(\left\|u_{c[t_0,t]}\right\|\right)\right\}, \ \forall t \ge t_0, \tag{2.86}$$

for some class $\mathcal{KL}$ function β, and any class $\mathcal{K}$ function γ satisfying

$$\gamma(s) \ge \max\left\{2\gamma_1^x \circ \gamma_2^u(s), 2\gamma_1^u(s), 2\gamma_2^x \circ \gamma_1^u(s), 2\gamma_2^u(s)\right\}, \ \forall s > 0. \tag{2.87}$$

Corollary 2.5 *Consider the nonlinear system*

$$\begin{aligned} \dot{x}_1 &= f_1(x_1, u, d) \\ \dot{x}_2 &= f_2(x_2, x_1, u, d). \end{aligned}$$

Assume both the subsystems are RISS viewing x_1 as state, u as input, and viewing x_2 as state, $col(x_1, u)$ as input, respectively, Then, the overall system is RISS viewing $x = col(x_1, x_2)$ as state and u as input with its gain function given by any class $\mathcal{K}$ function γ satisfying

$$\gamma(s) \ge \max\left\{2\gamma_1^u(s), 2\gamma_2^u(s), 2\gamma_2^x \circ \gamma_1^u(s)\right\}, \ \forall s > 0 \tag{2.88}$$

Proof Note that the inequality (2.84) holds for $i = 1, 2$ with any class $\mathcal{K}$ function γ_1^x (noting $u_2 = x_1$ and $u_c = u$). In particular, let

$$\gamma_1^x(s) = \min\left\{(\gamma_2^x)^{-1}(s)/2, s\right\}.$$

Then, the inequality (2.84) holds with $i = 1$ and the inequality (2.85) holds with this γ_1^x. Thus, the inequality (2.87) reduces to (2.88). □

2.8 Notes and References

Most materials in this chapter are standard and can be found in many textbooks on nonlinear systems and control, e.g., [2, 4–8]. Theorems 11.1 and 11.2 (given in the Appendix) on the existence and uniqueness of the initial value solution to a nonlinear system can be found in Sect. 3.1 of [2]. A series of stability concepts are introduced in Sect. 2.1. The definitions are consistent with those commonly used in literature. When uncertainties are taken into consideration, the robust version of various stability concepts are introduced in Sect. 2.2. Some of the tools for adaptive control introduced in Sect. 2.3 can also been found in textbooks [6, 9–12]. Lemma 2.1 (Barbalat's Lemma) can be found in [2, 8]. Theorem 2.5 (LaSalle-Yoshizawa Theorem) due to LaSalle [13] and Yoshizawa [14] is from Theorem 2.1 of [6]. Lemma 2.3 is from Lemma 3.4 of [15]. Lemma 2.4 is taken from [16]. It can be viewed as an alternative form of Lemma 1 in [17] and can also be directly derived from Lemma 2 of [18]. For the linear adaptive control systems, the parameter convergence condition is established by showing, under the PE condition, the closed-loop system which is typically a linear time-varying system is uniform exponentially stable. However, for nonlinear adaptive control systems, the PE condition may not guarantee the uniform exponential stability of the closed-loop system which is typically a nonlinear time-varying system, see, e.g., [19]. Lemma 2.4 is of interest in that it only depends on the characteristics of the signals $g(t)$ and $f(t)$ without assuming that these signals are governed by some dynamical systems as in the literature of adaptive control of linear systems. Thus, it may also apply to the adaptive control of nonlinear systems. The notion of ISS discussed in Sect. 2.4 was first proposed by Sontag in [20–25], etc. It has become an effective tool in the analysis and design of nonlinear control systems. Theorems 2.6 and 2.7 are of particular interest for designing control laws for nonlinear systems. The time-invariant version of Theorems 2.6 and 2.7 can also be found in [5]. The technique of changing supply function was developed in [26]. Some variants of this technique are introduced in Sect. 2.5 and will be used in the subsequent chapters. The universal adaptive control technique in Sect. 2.6 has been used for handling static uncertainty with unknown boundary in, e.g., [9, 10, 27–29]. The small gain theorem was established in [30–32] in terms of a general inter-connection of two nonlinear subsystems. In Sect. 2.7, a more clear-cut version of the small gain theorem is introduced for a simpler inter-connection of two nonlinear subsystems. This simplified version is taken from [33].

2.9 Problems

Problem 2.1 Simulate the following systems starting from different initial conditions. Observe the stability, asymptotic stability, and globally asymptotic stability properties of the equilibrium point $x = 0$.

(a) $\dot{x} = ax + bx^2 + cx^3$, $a, b, c \in \mathbb{R}$;
(b) $\ddot{y} - \mu(1 - y^2)\dot{y} + y = 0$, $\mu \in \mathbb{R}$, $x = \mathrm{col}(y, \dot{y})$;

(c) $\dot{r} = r(1-r)$, $\dot{\theta} = \sin^2(\theta/2)$, $r = \sqrt{x_1^2 + x_2^2}$, $\theta = \arctan(x_2/x_1)$, $x = \text{col}(x_1, x_2)$;
(d) $\dot{x}_1 = -x_1 + 4x_2$, $\dot{x}_2 = -x_1 - x_2^3$, $x = \text{col}(x_1, x_2)$;
(e) $\dot{x}_1 = x_2$, $\dot{x}_2 = x_1 - \text{sat}(2x_1 + x_2)$, $x = \text{col}(x_1, x_2)$, where the saturation function is defined as

$$\text{sat}(s) = \begin{cases} s, & |s| \leq 1 \\ 1, & s > 1 \\ -1, & s < -1 \end{cases}.$$

Problem 2.2 Determine if the following functions belong to class $\mathcal{K}$ function, class $\mathcal{K}_\infty$ function, or class $\mathcal{KL}$ function.

(a) $\gamma(r) = 2r + r^2$, $r \in [0, \infty)$;
(b) $\gamma(r) = e^r$, $r \in [0, \infty)$;
(c) $\gamma(r) = \arctan(r)$, $r \in [0, 1)$;
(d) $\gamma(r, s) = re^{-2s}$, $r \in [0, \infty)$, $s \in [0, \infty)$;
(e) $\gamma(r, s) = (r^3 + r)/(s + 1)$, $r \in [0, \infty)$, $s \in [0, \infty)$.

Problem 2.3 Find the Jacobian matrices for the following systems at the origin $x = 0$ and determine the stability of their equilibrium points at the origin.

(a) $\dot{x} = -\sin x$;
(b) $\dot{x} = -x^3$;
(c) $\dot{x}_1 = \sin x_2$, $\dot{x}_2 = -x_1 + x_1 x_2$, $x = \text{col}(x_1, x_2)$;
(d) $\dot{x} = \begin{bmatrix} 5x_2 \\ 4x_1^2 - 2\sin(x_2 x_3) \\ x_2 x_3 \end{bmatrix}$, $x = \text{col}(x_1, x_2, x_3)$.

Problem 2.4 Use Lyapunov's linearization theorem to determine the stability of the equilibrium point of the mechanical system at the origin

$$m\ddot{y} + c\dot{y} + k_1 y + k_3 y^3 = 0, \ m, c > 0, \ k_1, k_2 \in \mathbb{R}.$$

Problem 2.5 Show the nonlinear system $\dot{x} = -c(x, t)$ is UGAS if

$$xc(x, t) \geq \alpha(|x|), \ \forall x \in \mathbb{R}$$

for a class $\mathcal{K}$ function α.

Problem 2.6 Consider the following system

$$\begin{aligned} \dot{x}_1 &= x_2 \\ \dot{x}_2 &= -x_1 - x_1^3 - x_2^3. \end{aligned}$$

(a) Find the Jacobian matrix at the origin; what does it say about the stability of the equilibrium point?

(b) Use Lyapunov's direct theorem to determine the stability of the equilibrium point at the origin.
Hint:

$$V(x_1, x_2) = \frac{x_2^2}{2} + \int_0^{x_1} (s + s^3)ds$$

Problem 2.7 Use Lyapunov's direct theorem to determine the stability of the equilibrium point at the origin for the systems in Problem 2.3.

Problem 2.8 Determine if $x = 0$ is an equilibrium point for the following systems, where $d(t)$ represents external disturbance.

(a) $\dot{x} = \sin x + d(t)$;
(b) $\dot{x} = -x + d(t)x^3$;
(c) $\dot{x}_1 = x_2,\ \dot{x}_2 = d(t)x_1^2 + x_2,\ x = \text{col}(x_1, x_2)$;
(d) $\dot{x} = (x + \cos x - 1)/(1 + d(t))$.

Problem 2.9 Find the Jacobian matrices for the following systems at $(x, d) = (0, 0)$ and determine the stability of their equilibrium points at $x = 0$, where d represents an unknown parameter.

(a) $\dot{x} = -(1 + d)\sin x$;
(b) $\dot{x} = -x^3/(2 + \sin(d))$;
(c) $\dot{x}_1 = \sin x_2,\ \dot{x}_2 = -x_1 + dx_1x_2, x = \text{col}(x_1, x_2)$;
(d) $\dot{x} = \begin{bmatrix} dx_2 \\ 4x_1^2 - 2\sin(x_2x_3) \\ x_2x_3 \end{bmatrix}, x = \text{col}(x_1, x_2, x_3)$.

Problem 2.10 Use Lyapunov's direct theorem to determine the stability of the equilibrium point at the origin for the systems in Problem 2.9.

Problem 2.11 Use LaSalle-Yoshizawa Theorem to show that the equilibrium point of the system in Problem 2.6 is GAS.

Problem 2.12 Consider the nonlinear system

$$\begin{aligned} \dot{x}_1 &= x_2 \\ \dot{x}_2 &= -\sin x_1 + (1 + d(t))x_2 \end{aligned}$$

where $d(t)$ represents external disturbance satisfying $|d(t)| < 0.1$. Use LaSalle-Yoshizawa Theorem (hint: $V(x_1, x_2) = 1 - \cos x_1 + x_2^2/2$) to show

$$\lim_{t\to\infty} x_2(t) = 0$$

for all initial state values $x_1(0), x_2(0) \in \mathbb{R}$.

Problem 2.13 Find ISS-Lyapunov functions for the following systems where $d(t)$ represents external disturbance.

(a) $\dot{x} = -|d(t)|x - x^3 + u$;
(b) $\dot{x} = -x^3 - x^2 u$;
(c) $\dot{x} = \begin{bmatrix} -x_1 - x_2 \\ x_1 - x_2^3 + d(t)u \end{bmatrix}, \ x = \mathrm{col}(x_1, x_2), \ 1 < d(t) < 2.$

Problem 2.14 Find the gain functions for the RISS systems in Problem 2.13.

Problem 2.15 The function V defined in Theorem 2.7 is an ISS Lyapunov function for the system (2.7). If the class $\mathcal{K}_\infty$ function α is relaxed by a continuous positive definite function α (not required to be unbounded), the function V is called an integral input-to-state stable (iISS) Lyapunov function for the system (2.7). The system (2.7) is said to be iISS if there exists an iISS Lyapunov function [34]. Show that the following systems are not ISS but iISS using the suggested iISS Lyapunov functions.

(a) $\dot{x} = -x + xu^2$, $V(x) = \ln(1 + x^2)$;
(b) $\dot{x} = -\arctan x + u$, $V(x) = x \arctan x$.

Problem 2.16 Write the ISS-Lyapunov functions in Problem 2.13 in the form

$$V(x) \sim \{\underline{\alpha}, \bar{\alpha}, \alpha, \sigma \mid \dot{x} = f(x, u, d)\}.$$

For each function Δ given below, find another ISS Lyapunov function $V'(x)$ satisfying (2.37) and the corresponding functions $\bar{\alpha}'$, $\underline{\alpha}'$ and $\varkappa$.

(a) $\Delta(x) = 2$;
(b) $\Delta(x) = \|x\|^2$;
(c) $\Delta(x) = 3\|x\|^2 + 2\|x\|^4$.

Problem 2.17 Determine if the functions $V'(x)$ given in Problem 2.16 satisfies (2.42). If yes, find the corresponding functions $\bar{\alpha}'$, $\underline{\alpha}'$ and $\varkappa$.

Problem 2.18 Suppose the system (2.7) is iISS with an iISS Lyapunov function $V(x)$ satisying (2.25) where α is a continuous positive definite function. Let α' be a continuous positive definite function such that $\alpha'(s) = \mathcal{O}[\alpha(s)]$ as $s \to 0^+$ and $\limsup_{s\to\infty} \left[\alpha'(s)/\alpha(s)\right] < \infty$ if α is not of class $\mathcal{K}_\infty$. Show that there exists another iISS Lyapunov function $V'(x)$ satisfying $\underline{\alpha}'(\|x\|) \le V'(x) \le \bar{\alpha}'(\|x\|)$ for some class $\mathcal{K}_\infty$ functions $\underline{\alpha}'$ and $\bar{\alpha}'$ such that, along the trajectory of (2.7),

$$\dot{V}'(x) \le -\alpha'(\|x\|) + \varkappa(u)$$

for some positive definite function $\varkappa$.

Problem 2.19 Consider the system (2.83) where the matrix A is Hurwitz and the function ϕ is sufficiently smooth with $\phi(0,0,d) = 0$ for all $d \in \mathbb{D}$ with $\mathbb{D}$ a compact set. Assume the x_1-subsystem is iISS with an iISS Lyapunov function $V_1(x_1)$ satisfying $\underline{\alpha}_1(\|x_1\|) \le V_1(x_1) \le \bar{\alpha}_1(\|x_1\|)$ for some class $\mathcal{K}_\infty$ functions $\underline{\alpha}_1$ and $\bar{\alpha}_1$ such that, along the trajectory of the x_1-subsystem,

$$\dot{V}(x_1) \le -\alpha_1(\|x_1\|) + \sigma_1(\|u\|)$$

for some positive definite function α_1 and some class $\mathcal{K}$ function σ_1 satisfying $\limsup_{s\to 0^+}\left[\sigma_1(s)/s^2\right] < \infty$.

Let ϕ be a continuous positive definite function such that $\phi(s) = \mathcal{O}[\alpha_1(s)]$ as $s \to 0^+$ and $\limsup_{s\to\infty}[\phi(s)/\alpha_1(s)] < \infty$ if α_1 is not of class $\mathcal{K}_\infty$. Show that there exists an iISS Lyapunov function $V(x)$ with $x = \mathrm{col}(x_1, x_2)$, satisfying $\underline{\alpha}(\|x\|) \le V(x) \le \bar{\alpha}(\|x\|)$ for some class $\mathcal{K}_\infty$ functions $\underline{\alpha}$ and $\bar{\alpha}$ such that, along the trajectory of (2.83),

$$\dot{V}(x) \le -\phi^2(\|x\|) + \varkappa(u)\|u\|^2$$

for some smooth function $\varkappa$.

Problem 2.20 Consider the system (2.46) with $\mathbb{D}$ a known compact set and $b(d)>0$. Assume the z-subsystem is iISS with an iISS Lyapunov function $V(z)$ satisfying $\underline{\alpha}(\|z\|) \le V(z) \le \bar{\alpha}(\|z\|)$ for some class $\mathcal{K}_\infty$ functions $\underline{\alpha}$ and $\bar{\alpha}$ such that, along the trajectory of the z-subsystem,

$$\dot{V}(z) \le -\alpha(\|z\|) + \sigma(|x|)$$

for some positive definite function α and some class $\mathcal{K}$ function σ satisfying $\limsup_{s\to 0^+}\left[\sigma(s)/s^2\right] < \infty$.

Let m_1 and m_2 be some smooth positive functions such that

$$|f(z,x,d)| \le m_1(\|z\|)\|z\| + m_2(|x|)|x|, \ \forall d \in \mathbb{D}.$$

Moreover, $m_1(s)s = \mathcal{O}[\alpha(s)]$ and $\limsup_{s\to\infty}[m_1(s)s/\alpha(s)] < \infty$ if α is not of class $\mathcal{K}_\infty$. Show that there exists a controller of the form $u = -\rho(x)x$ that globally robustly stabilizes the system (2.46).

Problem 2.21 For the inter-connected system

$$\begin{aligned}\dot{x}_1 &= f_1(x_1, x_2, d(t))\\ \dot{x}_2 &= f_2(x_1, x_2, d(t))\end{aligned}$$

where $d(t)$ represents external disturbance. Assume the system satisfies

$$\|x_1(t)\| \leq \max\left\{\beta_1(\|x_1(t_0)\|, t - t_0), \gamma_1\left(\left\|x_{2[t_0,t]}\right\|\right)\right\}$$
$$\|x_2(t)\| \leq \max\left\{\beta_2(\|x_2(t_0)\|, t - t_0), \gamma_2\left(\left\|x_{1[t_0,t]}\right\|\right)\right\}, \ \forall t \geq t_0$$

for some class $\mathcal{KL}$ functions β_1 and β_2 and some class $\mathcal{K}$ functions γ_1 and γ_2 given below. Use the small gain theorem to determine the asymptotic stability of the system for the following cases.

(a) $\gamma_1(s) = 2, \gamma_2(s) = 0.4$;
(b) $\gamma_1(s) = 0.8s^2, \gamma_2(s) = \sqrt{s}$;
(c) $\gamma_1(s) = \alpha^{-1}(s), \gamma_2(s) = s^2$ for $\alpha(s) = 2s^2 + s^4$.

Problem 2.22 For each of the systems given below with state $x = \text{col}(x_1, x_2)$, input u, and disturbance $1 < d(t) < 2$, find an ISS Lyapunov function $V(x)$ satisfying

$$\underline{\alpha}(\|x\|) \leq V(x) \leq \bar{\alpha}(\|x\|)$$
$$\dot{V}(x) \leq -\|x\|^2 + \varkappa(u)\|u\|^2.$$

Calculate the class $\mathcal{K}_\infty$ functions $\underline{\alpha}$ and $\bar{\alpha}$ and the smooth function $\varkappa$.

(a) $\dot{x}_1 = -x_1 + u, \ \dot{x}_2 = -x_2 + d(t)x_1^2 + u$;

(b) $\dot{x}_1 = \begin{bmatrix} 0 & 1 \\ -1 & -2 \end{bmatrix} x_1 + \begin{bmatrix} 1 \\ 2 \end{bmatrix} u, \ \dot{x}_2 = -x_2 + d(t)\|x_1\|^2 + u$;

(c) $\dot{x}_1 = -x_1 + u, \ \dot{x}_2 = -\begin{bmatrix} 0 & 1 \\ -1 & -2 \end{bmatrix} x_2 + \begin{bmatrix} 1 \\ 2 \end{bmatrix} (d(t)x_1^2 + u)$.

Problem 2.23 For each of the systems given below with state $x = \text{col}(x_1, x_2)$, input u, and disturbance $1 < d(t) < 2$, design a controller $u = -\rho(x_2)x_2 + \bar{u}$ and find an ISS Lyapunov function $V(x)$ satisfying

$$\underline{\alpha}(\|x\|) \leq V(x) \leq \bar{\alpha}(\|x\|)$$
$$\dot{V}(x) \leq -\|x\|^2 + \|\bar{u}\|^2.$$

Calculate the class $\mathcal{K}_\infty$ functions $\underline{\alpha}$ and $\bar{\alpha}$.

(a) $\dot{x}_1 = -x_1 + x_2, \ \dot{x}_2 = d(t)x_1x_2 + u$;

(b) $\dot{x}_1 = -x_1 + d(t)x_2, \ \dot{x}_2 = -x_2^3 + d(t)u$;

(c) $\dot{x}_1 = \begin{bmatrix} 0 & 1 \\ -1 & -2 \end{bmatrix} x_1 + \begin{bmatrix} 1 \\ 2 \end{bmatrix} x_2^2, \ \dot{x}_2 = d(t)\sin x_2 + u$.

Problem 2.24 For each of the systems in Problem 2.23, design a controller u to solve the global stabilization problem for $d(t) \in \mathbb{D}$ when $\mathbb{D}$ is an arbitrarily large compact set.

References

1. Rudin W (1976) Principles of mathematical analysis, 3rd edn. McGraw-Hill, New York
2. Khalil H (2002) Nonlinear systems. Prentice Hall, New Jersey
3. Chen Z, Huang J (2005) Robust input-to-state stability and small gain theorem for nonlinear systems containing time-varying uncertainty. Advanced robust and adaptive control-theory and applications. Springer, New York
4. Isidori A (1995) Nonlinear control systems, 3rd edn. Springer, New York
5. Isidori A (1999) Nonlinear control systems, vol II. Springer, New York
6. Krstic M, Kanellakopoulos I, Kokotović P (1995) Nonlinear and adaptive control design. Wiley, New York
7. Nijmeijer H, van der Schaft AJ (1990) Nonlinear dynamical control systems. Springer, New York
8. Slotine JJ, Li W (1991) Applied nonlinear control. Prentice Hall, Englewood Cliffs
9. Ilchmann A (1993) Non-identifier-based high-gain adaptive control. Springer, Berlin
10. Mareels I, Polderman JW (1996) Adaptive systems: an introduction. Birkhäuser, Boston
11. Marino R, Tomei P (1995) Nonlinear control design: geometric, adaptive and robust. Prentice Hall, Englewood Cliffs
12. Narendra KS, Annaswamy AM (1989) Stable adaptive systems. Printice-Hall, Englewood Cliffs
13. LaSalle JP (1968) Stability theory for ordinary differential equations. J Diff Equat 4:57–65
14. Yoshizawa T (1966) Stability theory by Lyapunov's second method. The Mathematical Society of Japan, Tokyo
15. Boyd S, Sastry S (1983) On parameter convergence in adaptive control. Syst Control Lett 3:311–319
16. Liu L, Chen Z, Huang J (2009) Parameter convergence and minimal internal model with an adaptive output regulation problem. Automatica 45:1306–1311
17. Ortega R, Fradkov A (1993) Asymptotic stability of a class of adaptive systems. Int J Adapt Control Signal Process 7:255–260
18. Yuan JS-C, Wonham WM (1977) Probing signals for model reference identification. IEEE Trans Autom Control 22:530–538
19. Loria A, Panteley E (2002) Uniform exponential stability of linear time-varying systems: revisited. Syst Control Lett 47:13–24
20. Sontag ED (1989) Smooth stabilization implies coprime factorization. IEEE Trans Autom Control 34:435–443
21. Sontag ED (1990) Further facts about input to state stabilization. IEEE Trans Autom Control 34:473–476
22. Sontag ED (1995) On the input-to-state stability property. Int J Control 1:24–36
23. Sontag ED, Wang Y (1996) New characterizations of input-to-state stability. IEEE Trans Autom Control 41:1283–1294
24. Sontag ED, Wang Y (1999) Notions of input to output stability. Syst Control Lett 38:351–359
25. Sontag ED, Wang Y (2001) Lyapunov characterizations of input to output stability. SIAM J Control Optim 39:226–249
26. Sontag ED, Teel A (1995) Changing supply function in input/state stable systems. IEEE Trans Autom Control 40:1476–1478
27. Ilchmann A, Ryan EP (1994) Universal λ-tracking for nonlinearly perturbed systems in the presence of noise. Automatica 30:337–346
28. Ryan EP (1994) A nonlinear universal servomechanism. IEEE Trans Autom Control 39:753–761
29. Ye XD, Huang J (2003) Decentralized adaptive output regulation for a class of large-scale nonlinear systems. IEEE Trans Autom Control 48:276–281
30. Jiang ZP, Mareels I (1997) A small-gain control method for nonlinear cascaded systems with dynamic uncertainties. IEEE Trans Autom Control 42:292–308

31. Jiang ZP, Mareels I, Wang Y (1996) A Lyapunov formulation of the nonlinear small gain theorem for interconnected ISS systems. Automatica 32:1211–1215
32. Jiang ZP, Teel AR, Praly L (1994) Small-gain theorem for ISS systems and applications. Math Control Signals Systems 7:95–120
33. Chen Z, Huang J (2005) A simplified small gain theorem for time-varying nonlinear systems. IEEE Trans Autom Control 50:1904–1908
34. Angeli D, Sontag ED, Wang Y (2000) A characterization of integral input-to-state stability. IEEE Trans Autom Control 45(6):1082–1097

Chapter 3
Classification of Nonlinear Control Systems

In this chapter, we study the classification of the affine nonlinear control systems of the form (1.10) repeated below:

$$\begin{aligned} \dot{x} &= f(x) + g(x)u \\ y &= h(x). \end{aligned} \tag{3.1}$$

We assume that all functions are sufficiently smooth and f and h vanish at the origin. In Sect. 3.1, we review the concepts of normal form and zero dynamics. In Sect. 3.2, we classify nonlinear systems into the so-called lower triangular systems, output feedback systems, and filter extended systems. In Sect. 3.3, we present some practical nonlinear control systems. The notes and references are given in Sect. 3.4.

3.1 Normal Form and Zero Dynamics

We begin by introducing some terminologies and notation.

Definition 3.1 Let $h : \mathbb{R}^n \mapsto \mathbb{R}$ be a sufficiently smooth scalar function and $f : \mathbb{R}^n \mapsto \mathbb{R}^n$ a vector field. Let $x = [x_1, \ldots, x_n]^{\mathsf{T}}$. Then,

$$\frac{\partial h}{\partial x} := \left[\frac{\partial h}{\partial x_1}, \ldots, \frac{\partial h}{\partial x_n}\right]$$

is called the *gradient* of h. For $k = 0, 1, \ldots,$ the operator $L_f^k : \mathbb{R}^n \mapsto \mathbb{R}$ is defined recursively as follows:

Z. Chen and J. Huang, *Stabilization and Regulation of Nonlinear Systems*,
Advanced Textbooks in Control and Signal Processing,
DOI 10.1007/978-3-319-08834-1_3

$$
\begin{aligned}
L_f^0 h &:= h \\
L_f^1 h &:= \sum_{i=1}^{n} \frac{\partial h}{\partial x_i} f_i = \frac{\partial h}{\partial x} f \\
&\vdots \\
L_f^k h &:= L_f(L_f^{k-1} h) = \frac{\partial L_f^{k-1} h}{\partial x} f.
\end{aligned}
$$

We call $L_f^k h$ the k-th *Lie derivative* of the function h along the vector field f.

Remark 3.1 The first Lie derivative of the function h along the vector field f, $L_f^1 h$, is simplified as $L_f h$. Let $g : \mathbb{R}^n \mapsto \mathbb{R}^n$ be a vector field. Then, for $k = 0, 1, \dots,$

$$
L_g L_f^k h := \frac{\partial L_f^k h}{\partial x} g.
$$

From the above definition, the gradient of the function h at the point x is

$$
\frac{\partial h(x)}{\partial x} := \left[\frac{\partial h(x)}{\partial x_1}, \dots, \frac{\partial h(x)}{\partial x_n} \right],
$$

and the k-th Lie derivative of the function h along the vector field f, at the point x, is

$$
L_f^k h(x) = \frac{\partial L_f^{k-1} h(x)}{\partial x} f(x).
$$

In what follows, we assume $p = m$, i.e., the number of the input equals the number of the output. We will use the following notation

$$
\begin{aligned}
g(x) &= [g_1(x) \ \dots \ g_m(x)] \\
h(x) &= \mathrm{col}(h_1(x) \ \dots \ h_m(x))
\end{aligned}
$$

where, for $i = 1, \dots, m$, $h_i : \mathbb{R}^n \mapsto \mathbb{R}$ is a sufficiently smooth scalar function and $g_i : \mathbb{R}^n \mapsto \mathbb{R}^n$ a vector field.

Definition 3.2 The system (3.1) with $m = p$ is said to have a (vector) relative degree $\{r_1, \dots, r_m\}$ at x_0 if

(i) $L_{g_j} L_f^k h_i(x) = 0$ for all $1 \le j \le m$, for all $k < r_i - 1$, for all $1 \le i \le m$, and for all x in a neighborhood of x_0,
(ii) the $m \times m$ matrix

$$A(x) = \begin{bmatrix} L_{g_1} L_f^{r_1 - 1} h_1(x) & \dots & L_{g_m} L_f^{r_1 - 1} h_1(x) \\ L_{g_1} L_f^{r_2 - 1} h_2(x) & \dots & L_{g_m} L_f^{r_2 - 1} h_2(x) \\ \vdots & \vdots & \vdots \\ L_{g_1} L_f^{r_m - 1} h_m(x) & \dots & L_{g_m} L_f^{r_m - 1} h_m(x) \end{bmatrix}$$

is nonsingular at $x = x_0$.

In particular, for a single input and single output (SISO) system of the form (3.1), i.e., $p = m = 1$, the definition of relative degree reduces to a simpler version given below.

Definition 3.3 The system (3.1) with $p = m = 1$ is said to have a relative degree r at x_0 if

$$L_g L_f^k h(x) = 0 \tag{3.2}$$

for all $k < r - 1$ and for all x in a neighborhood of x_0, and

$$L_g L_f^{r-1} h(x_0) \neq 0. \tag{3.3}$$

Example 3.1 Consider a three-dimensional system of the form (3.1) with $x = \text{col}(x_1, x_2, x_3)$, and

$$f(x) = \begin{bmatrix} 0 \\ x_1 + x_2 \\ a(x_1, x_2) + \theta x_3 \end{bmatrix}, \quad g(x) = \begin{bmatrix} k + x_2^2 \\ 0 \\ 0 \end{bmatrix}, \quad h(x) = x_2 \tag{3.4}$$

where k and θ are real numbers and a is a sufficiently smooth function vanishing at the origin. Simple calculation gives

$$L_g h(x) = 0, \quad L_g L_f h(x) = k + x_2^2.$$

When $k > 0$, by Definition 3.3, this system has a relative degree two at any point $x_0 \in \mathbb{R}^3$ as $k + x_2^2 > 0$. However, when $k \leq 0$, two different cases may occur. Define a set

$$\mathbb{X} = \{x = \text{col}(x_1, x_2, x_3) \in \mathbb{R}^3 \mid x_2^2 = -k\}.$$

(i) The system has a relative degree two at $x_0 \notin \mathbb{X}$ since $k + x_2^2 \neq 0$ for all x in a neighbourhood of x_0.
(ii) The system does not have a well defined relative degree at $x_0 \in \mathbb{X}$. In fact, one has

$$L_g L_f h(x_0) = 0,$$

but there exists no neighborhood of x_0 such that

$$L_g L_f h(x) = 0$$

in this neighborhood.

In what follows, we will focus on the SISO system and assume the system has a relative degree r at x_0. Then it can be verified that the trajectory of the system starting from any $x(0)$ sufficiently close to x_0 is such that, for sufficiently small t,

$$\begin{aligned} \dot{y}(t) &= L_f h(x(t)) \\ &\vdots \\ y^{(r-1)}(t) &= L_f^{r-1} h(x(t)) \\ y^{(r)}(t) &= L_f^r h(x(t)) + L_g L_f^{r-1} h(x(t)) u(t) \end{aligned} \tag{3.5}$$

with

$$L_g L_f^{r-1} h(x_0) \neq 0.$$

Solving the equation

$$\mathcal{U}(t) = L_f^r h(x(t)) + L_g L_f^{r-1} h(x(t)) u(t)$$

where $\mathcal{U} \in \mathbb{R}$ is viewed as a new input to the system (3.1) gives a state feedback controller of the form

$$u(t) = \frac{-L_f^r h(x(t)) + \mathcal{U}(t)}{L_g L_f^{r-1} h(x(t))}. \tag{3.6}$$

Applying (3.6) to the system (3.1) results in a new system whose input–output relation obeys, for all sufficiently small t,

$$y^{(r)}(t) = \mathcal{U}(t).$$

Example 3.2 In Example 3.1, direct calculation gives

$$\begin{aligned} \dot{y} &= x_1 + x_2 \\ \ddot{y} &= x_1 + x_2 + (k + x_2^2) u. \end{aligned}$$

So the controller

$$u = \frac{-x_1 - x_2 + \mathcal{U}}{k + x_2^2}$$

gives the relationship

$$\ddot{y} = \mathcal{U}.$$

Let x_0 be any initial condition of Example 3.1. Then, the controller is valid in a neighborhood of x_0 whenever $k + x_2^2(0) \neq 0$, or what is the same, Example 3.1 has a well defined relative degree at x_0.

The control law (3.6) is called an *input–output linearizing control law* as it results in a linear input–output relationship between the new input $\mathcal{U}$ and the output y. A further linear feedback control of the form

$$\mathcal{U}(t) = -\alpha_0 y - \alpha_1 \dot{y} - \cdots - \alpha_{r-1} y^{(r-1)} \tag{3.7}$$

where $\alpha_0, \alpha_1, \ldots, \alpha_{r-1}$ are such that

$$\lambda^r + \alpha_{r-1}\lambda^{r-1} + \cdots + \alpha_1 \lambda + \alpha_0$$

is a Hurwitz polynomial makes the output y satisfy a stable linear differential equation

$$y^{(r)} + \alpha_{r-1} y^{(r-1)} + \cdots + \alpha_1 \dot{y} + \alpha_0 y = 0. \tag{3.8}$$

Thus, the output $y(t)$ approaches 0 as $t \to \infty$. The composition of (3.6) and (3.7) yields a state feedback control law of form

$$u = \frac{-L_f^r h(x) - \sum_{i=0}^{r-1} \alpha_i L_f^i h(x)}{L_g L_f^{r-1} h(x)} \tag{3.9}$$

which is called an *output stabilizing control law*. However, this control law may not guarantee the asymptotic stability of the equilibrium point of the closed-loop system. To find the capability and limit of the control law, we need to further study the concepts of normal form and the zero dynamics for the system (3.1). For this purpose, we will introduce the concept of diffeomorphism and the inverse function theorem.

Definition 3.4 Let $\tau : \mathbb{X} \mapsto \mathbb{T}$ be a smooth mapping for two open subsets $\mathbb{X}, \mathbb{T} \subset \mathbb{R}^n$. It is said to be a local diffeomorphism on $\mathbb{X}$ if there exists a smooth mapping $\tau^{-1} : \mathbb{T} \mapsto \mathbb{X}$ such that $\tau^{-1}(\tau(x)) = x$ for all $x \in \mathbb{X}$. If $\mathbb{X} = \mathbb{T} = \mathbb{R}^n$, then τ is said to be a global diffeomorphism.

Theorem 3.1 (Inverse Function Theorem) *Let $\tau : \mathbb{X} \mapsto \mathbb{T}$ be a smooth mapping for two open subsets $\mathbb{X}, \mathbb{T} \subset \mathbb{R}^n$. Then τ is a local diffeomorphism on $\mathbb{X}$ if the matrix $T(x) = \partial\tau(x)/\partial x$ has a full rank, i.e.,*

$$rank(T(x)) = n, \quad \forall x \in \mathbb{X}.$$

If $\tau(x)$ is a diffeomorphism on $\mathbb{X} \subset \mathbb{R}^n$, then one can define a coordinate transformation $\mathcal{X} = \tau(x)$ on (3.1). Under the new state vector $\mathcal{X}$, the system (3.1) can be expressed as follows

$$\dot{\mathcal{X}} = \frac{\partial\tau}{\partial x}(f(x) + g(x)u)\Big|_{x=\tau^{-1}(\mathcal{X})}. \tag{3.10}$$

We will see that a suitable $\tau(x)$ may render (3.10) a desirable structure.

For this purpose, we first establish the following proposition.

Proposition 3.1 *If the system (3.1) has a relative degree r at $x = 0$, then the following row vectors*

$$\frac{\partial h}{\partial x}(0), \quad \frac{\partial L_f h}{\partial x}(0), \quad \ldots, \quad \frac{\partial L_f^{r-1} h}{\partial x}(0) \tag{3.11}$$

are linearly independent.

Proof Let

$$A = \frac{\partial f}{\partial x}(0), \ B = g(0), \ C = \frac{\partial h}{\partial x}(0).$$

Then it can be verified that

$$\begin{aligned} \frac{\partial L_f h(x)}{\partial x}\Big|_{x=0} &= \frac{\partial h}{\partial x}\frac{\partial f}{\partial x}\Big|_{x=0} = CA \\ \frac{\partial L_f^k h(x)}{\partial x}\Big|_{x=0} &= \frac{\partial L_f^{k-1} h(x)}{\partial x}\frac{\partial f}{\partial x}\Big|_{x=0} = CA^k, \quad k = 2, 3, \ldots, \end{aligned}$$

Therefore,

$$L_g L_f^k h(x)\Big|_{x=0} = \frac{\partial L_f^k h(x)}{\partial x} g(x)\Big|_{x=0} = CA^k B, \quad k = 0, 1, 2, \ldots.$$

Thus, the fact that $L_g L_f^k h(x) = 0$ for all $0 \le k < r - 1$ and for all x in a neighborhood of $x = 0$ implies that $CA^k B = 0$ for all $0 \le k < r - 1$, and the fact that $L_g L_f^{r-1} h(x)\Big|_{x=0} \neq 0$ implies $CA^{r-1} B \neq 0$.

Suppose there exist r real numbers $a_1, \dots, a_r$ such that

$$a_1 C + a_2 CA + \cdots + a_r CA^{r-1} = 0. \tag{3.12}$$

Post-multiplying (3.12) by B, AB, ..., $A^{r-1}B$ successively and using the fact that $CA^k B = 0$ for all $0 \le k < r-1$ and $CA^{r-1}B \neq 0$ yields that $a_r, a_{r-1}, \dots, a_1$ must be all equal to 0. Therefore, the r row vectors $C, CA, \dots, CA^{r-1}$ must be linearly independent. □

Now let

$$\tau_1(x) = h(x), \ \ \tau_2(x) = L_f h(x), \ \ \dots, \ \ \tau_r(x) = L_f^{r-1} h(x)$$

Since the r row vectors in (3.11) are linearly independent, there exist $n-r$ sufficiently smooth functions $\tau_{r+1}(x), \dots, \tau_n(x)$, vanishing at the origin, such that the n row vectors $(\partial\tau_i)/(\partial x)(0)$, $i = 1, \dots, n$, are linearly independent. As a result, $\tau(x)$ is a diffeomorphism on a neighborhood $\mathbb{X}$ of $x = 0$.

Under the new coordinates $\mathcal{X} = \tau(x)$, the system (3.10) admits the following specific form on $\mathbb{X}$:

$$\begin{aligned}
\dot{\mathcal{X}}_1 &= \mathcal{X}_2 \\
&\vdots \\
\dot{\mathcal{X}}_{r-1} &= \mathcal{X}_r \\
\dot{\mathcal{X}}_r &= \left(L_f^r h(x) + L_g L_f^{r-1} h(x) u\right)\Big|_{x=\tau^{-1}(\mathcal{X})} \\
\dot{\mathcal{X}}_{r+1} &= \left(L_f \tau_{r+1}(x) + L_g \tau_{r+1}(x) u\right)\big|_{x=\tau^{-1}(\mathcal{X})} \\
&\vdots \\
\dot{\mathcal{X}}_n &= \left(L_f \tau_n(x) + L_g \tau_n(x) u\right)\big|_{x=\tau^{-1}(\mathcal{X})} \\
y &= \mathcal{X}_1.
\end{aligned} \tag{3.13}$$

We call (3.13) the *normal form* of the system (3.1).

Example 3.3 If the system (3.1) has a relative degree r at $x = 0$, then the following row vectors

$$\frac{\partial h}{\partial x}(0), \ \ \frac{\partial L_f h}{\partial x}(0), \ \ \dots, \ \ \frac{\partial L_f^{r-1} h}{\partial x}(0)$$

are linearly independent. Therefore, there exist r components of x denoted by $x_{j_1}, \dots, x_{j_r}$ such that the matrix

$$T_1 = \begin{bmatrix} \frac{\partial h}{\partial x_{j_1}}(0) & \cdots & \frac{\partial h}{\partial x_{j_r}}(0) \\ \frac{\partial L_f h}{\partial x_{j_1}}(0) & \cdots & \frac{\partial L_f h}{\partial x_{j_r}}(0) \\ & \vdots & \\ \frac{\partial L_f^{r-1} h}{\partial x_{j_1}}(0) & \cdots & \frac{\partial L_f^{r-1} h}{\partial x_{j_r}}(0) \end{bmatrix}$$

is nonsingular. Denote the remaining $n - r$ components of x by $x_{j_{r+1}}, \ldots, x_{j_n}$. Let $\bar{x} = [x_{j_1}, \ldots, x_{j_n}]^{\mathrm{T}}$. Then $\bar{x} = Sx$ for some nonsingular matrix S. Let

$$\tau(x) = \left[h(x) \cdots L_f^{r-1} h(x) \; x_{j_{r+1}} \cdots x_{j_n} \right]^{\mathrm{T}}.$$

Then

$$\frac{\partial \tau}{\partial x}(0) = \frac{\partial \tau}{\partial \bar{x}}(0) S = \begin{bmatrix} T_1 & * \\ 0 & I_{n-r} \end{bmatrix} S$$

is nonsingular. Thus, τ is a diffeomorphism on a neighborhood of $x = 0$.

It should be noted that the normal form of the system (3.1) is not unique because of the freedom in selecting the functions $\tau_{r+1}(x), \ldots, \tau_n(x)$. In particular, if τ_i, $i = r+1, \ldots, n$, are such that

$$L_g \tau_i(x) = 0, \quad i = r+1, \ldots, n, \; \forall x \in \mathbb{X}, \tag{3.14}$$

then, the normal form (3.13) is simplified to

$$\begin{aligned} \dot{\mathcal{X}}_1 &= \mathcal{X}_2 \\ &\vdots \\ \dot{\mathcal{X}}_{r-1} &= \mathcal{X}_r \\ \dot{\mathcal{X}}_r &= \left(L_f^r h(x) + L_g L_f^{r-1} h(x) u \right) \Big|_{x=\tau^{-1}(\mathcal{X})} \\ \dot{\mathcal{X}}_{r+1} &= L_f \tau_{r+1}(x) \big|_{x=\tau^{-1}(\mathcal{X})} \\ &\vdots \\ \dot{\mathcal{X}}_n &= L_f \tau_n(x) \big|_{x=\tau^{-1}(\mathcal{X})} \\ y &= \mathcal{X}_1. \end{aligned} \tag{3.15}$$

Remark 3.2 Suppose the system (3.1) has a relative degree r at $x = 0$. By Frobenius Theorem (Theorem 11.3 of the Appendix), there exist smooth functions $\eta_i(x)$, $i = 1, \ldots, n-1$, vanishing at the origin, such that

$$L_g \eta_i(x) = 0, \quad i = 1, \ldots, n-1$$

on a neighborhood of $x = 0$. Moreover,

$$\frac{\partial \eta_1}{\partial x}(0), \ \ldots, \ \frac{\partial \eta_{n-1}}{\partial x}(0)$$

are linearly independent.

One can show that

$$T = \text{col}\left(\frac{\partial h}{\partial x}(0), \ \frac{\partial L_f h}{\partial x}(0), \ \ldots, \ \frac{\partial L_f^{r-1} h}{\partial x}(0), \ \frac{\partial \eta_1}{\partial x}(0), \ \ldots, \ \frac{\partial \eta_{n-1}}{\partial x}(0)\right)$$

has a rank n. In fact, if this is false, then $g(0)$ would be perpendicular to all row vectors in T which leads to the contradiction that $L_g L_f^{r-1} h(0) \neq 0$. By Proposition 3.1, the first r row vectors in T are linearly independent. Therefore, it is possible to select $n - r$ functions $\eta_{j_i}(x)$ for $i = r+1, \ldots, n$ and $1 \leq j_i \leq n-1$, such that

$$\text{col}\left(\frac{\partial h}{\partial x}(0), \ \frac{\partial L_f h}{\partial x}(0), \ \ldots, \ \frac{\partial L_f^{r-1} h}{\partial x}(0), \ \frac{\partial \eta_{j_{r+1}}}{\partial x}(0), \ \ldots, \ \frac{\partial \eta_{j_n}}{\partial x}(0)\right)$$

has a full rank. As a result,

$$\tau(x) = \begin{bmatrix} h(x) & \cdots & L_f^{r-1} h(x) & \eta_{j_{r+1}}(x) & \cdots & \eta_{j_n}(x) \end{bmatrix}^{\mathrm{T}}$$

is a diffeomorphism on a neighborhood of $x = 0$, denoted as $\mathbb{X}$.

Due to Remark 3.2, we can always assume that the normal form of an SISO system takes the form (3.15). Clearly, this normal form is more convenient for the development of nonlinear control methods.

To further investigate the characteristics of the system (3.15), we define a state feedback control law as follows

$$u = \left. \frac{-L_f^r h(x) + \mathcal{u}}{L_g L_f^{r-1} h(x)} \right|_{x=\tau^{-1}(\mathcal{X})}. \tag{3.16}$$

Then the closed-loop system composed of (3.15) and (3.16) is given by

$$\begin{aligned} \dot{\mathcal{X}}_i &= \mathcal{X}_{i+1}, \quad i = 1, \ldots, r-1 \\ \mathcal{X}_r &= \mathcal{u} \\ \dot{\mathcal{X}}_i &= L_f \tau_i(x)\big|_{x=\tau^{-1}(\mathcal{X})}, \quad i = r+1, \ldots, n \\ y &= \mathcal{X}_1. \end{aligned} \tag{3.17}$$

Consider the system (3.17) with a trivial input $\mathcal{u} = 0$, for all initial state $\mathcal{X}(0) = \text{col}(\mathcal{X}_1(0), \ldots, \mathcal{X}_n(0))$ satisfying $\mathcal{X}_1(0) = \mathcal{X}_2(0) = \cdots = \mathcal{X}_r(0) = 0$, the first r

components of the solution $\mathcal{X}(t)$ of (3.17) starting from $\mathcal{X}(0)$ are identically zero for sufficiently small t. Define a manifold $\mathbb{M}$ of dimension $n-r$ as follows:

$$\begin{aligned}\mathbb{M} &= \{x \in \mathbb{X} \mid H(x) = 0\}\\ H(x) &= \left[h(x)\ L_f h(x) \cdots L_f^{r-1} h(x) \right]^{\mathrm{T}}.\end{aligned} \tag{3.18}$$

Then we can see that, under the state feedback control

$$u = \frac{-L_f^r h(x)}{L_g L_f^{r-1} h(x)},$$

for all sufficiently small t, the solution $x(t)$ of (3.1) starting from any initial state $x(0) \in \mathbb{M}$ remains in $\mathbb{M}$. Thus, $\mathbb{M}$ is an invariant manifold of (3.1).

Let

$$\mathcal{X}_0 = \mathrm{col}(\mathcal{X}_{r+1}, \ldots, \mathcal{X}_n), \quad \vec{\mathcal{X}}_i = \mathrm{col}(\mathcal{X}_1, \ldots, \mathcal{X}_i), \quad i = 1, \ldots, r.$$

Then the system (3.17) becomes the following:

$$\begin{aligned}\dot{\mathcal{X}}_0 &= f_0(\mathcal{X}_0, \vec{\mathcal{X}}_r)\\ \dot{\mathcal{X}}_i &= \mathcal{X}_{i+1}, \quad i = 1, \ldots, r-1\\ \mathcal{X}_r &= u\\ y &= \mathcal{X}_1\end{aligned} \tag{3.19}$$

where

$$f_0(\mathcal{X}_0, \vec{\mathcal{X}}_r) = \begin{bmatrix} L_f \tau_{r+1}\left(\tau^{-1}(\mathcal{X})\right) \\ \vdots \\ L_f \tau_n \left(\tau^{-1}(\mathcal{X})\right) \end{bmatrix}.$$

In (3.19), the $n-r$ dimensional upper subsystem $\dot{\mathcal{X}}_0 = f_0(\mathcal{X}_0, \vec{\mathcal{X}}_r)$ is called the *inverse dynamics* of (3.1).[1] Based on the inverse dynamics, we can define a zero input system

$$\dot{\mathcal{X}}_0 = f_0(\mathcal{X}_0, 0) \tag{3.20}$$

which has an equilibrium point at $\mathcal{X}_0 = 0$. This system is precisely the system which governs the motion of the last $n-r$ components of $\mathcal{X}$ when the motion of the system

[1] Strictly speaking, we call $\dot{\mathcal{X}}_0 = f_0(\mathcal{X}_0, \vec{\mathcal{X}}_r)$ the inverse dynamics of (3.1) only if $\vec{\mathcal{X}}_r = \mathcal{X}_1$. However, for convenience, we will still call $\dot{\mathcal{X}}_0 = f_0(\mathcal{X}_0, \vec{\mathcal{X}}_r)$ the inverse dynamics of (3.1) even if $\vec{\mathcal{X}}_r \neq \mathcal{X}_1$.

(3.17) is restricted to the manifold $\mathbb{M}$. Since on the manifold $\mathbb{M}$, the system output and its derivatives are identically zero, the manifold $\mathbb{M}$ is called an *output zeroing manifold*. For this reason, the system (3.20) is called the *zero dynamics* of (3.1). Moreover, one calls (3.1) *minimum phase* if the equilibrium point $\mathcal{X}_0 = 0$ of the zero dynamics (3.20) is asymptotically stable, and *non-minimum phase* if unstable. It can be seen that the output stabilizing control law cannot even locally stabilize a non-minimum phase system.

Example 3.4 Let the Jacobian linearization of the system (3.1) at the origin be

$$\begin{aligned} \dot{x} &= Ax + Bu \\ y &= Cx. \end{aligned} \tag{3.21}$$

Then the transfer function of (3.21) is

$$P(s) = \frac{\det \begin{bmatrix} sI - A & -B \\ C & 0 \end{bmatrix}}{\det(sI - A)}.$$

On the other hand, it can be verified that the transfer function of (3.21) is also given by

$$P(s) = CA^{r-1}B \frac{\det(sI - Q)}{\det(sI - A)}$$

where Q is the Jacobian matrix of $f_0(\mathcal{X}_0, 0)$ at $\mathcal{X}_0 = 0$. Thus, if the triplet (A, B, C) is controllable and observable, then the eigenvalues of Q coincide with the zeros of (3.21). Therefore, the system (3.1) is minimum phase if all the eigenvalues of Q have negative real parts, or non-minimum phase if at least one eigenvalue of Q has positive real part. It can also be verified that the matrix Q is unaffected under the class of input–output linearization based control laws (3.9). Therefore, the input–output linearization based control laws can only stabilize a minimum phase nonlinear system.

Example 3.5 In Example 3.1, it can be verified that the zero dynamics of the system are $\dot{x}_3 = \theta x_3$. Therefore, the system is non-minimum phase when $\theta > 0$.

3.2 Typical Nonlinear Control Systems

In this section, we introduce several classes of nonlinear control systems that are frequently encountered in nonlinear control theory and applications.

Class I: Lower triangular systems

$$\begin{aligned}\dot{z}_i &= q_i(\vec{z}_i, \vec{x}_i, d)\\ \dot{x}_i &= f_i(\vec{z}_i, \vec{x}_i, d) + b_i(d)x_{i+1}, \quad i = 1, \ldots, r\\ y = x_1\end{aligned} \tag{3.22}$$

In the system (3.22), $z_i \in \mathbb{R}^{n_i}$ and $x_i \in \mathbb{R}$ for $i = 1, \ldots, r$, are the state variables, $u := x_{r+1} \in \mathbb{R}$ is the input, and $y \in \mathbb{R}$ is the output. Let $\vec{x}_i = \text{col}(x_1, \ldots, x_i)$ and $\vec{z}_i = \text{col}(z_1, \ldots, z_i)$. The variable d represents a function $d : [t_0, \infty) \mapsto \mathbb{D} \subset \mathbb{R}^l$, which is a piecewise continuous function in t with $\mathbb{D}$ a compact set of $\mathbb{R}^l$. The functions q_i and f_i are piecewise continuous in d and locally Lipschitz in the state variables and satisfy $q_i(0, 0, d) = 0$, and $f_i(0, 0, d) = 0$ for all $d \in \mathbb{D}$.

In (3.22), $d(t)$ represents any set of unknown parameters and/or disturbances, and is called the *static uncertainty*. On the other hand, the functions q_i's may not be known precisely, and/or the state z_i may not be available for feedback control. Thus the dynamics governing $\vec{z}_r$ are called *dynamic uncertainty* of the system (3.22) as opposed to the static uncertainty $d(t)$.

The system (3.22) is motivated from the system

$$\begin{aligned}\dot{x}_0 &= f_0(x_0, x_1)\\ \dot{x}_i &= x_{i+1}, \quad i = 1, \ldots, r\\ y &= x_1\end{aligned} \tag{3.23}$$

of the normal form (3.19) for an SISO affine nonlinear control system. The system (3.23) can be generalized by adding nonlinear perturbation terms $f_i(x_0, \vec{x}_i)$'s and non-unity control gains b_i's, that is,

$$\begin{aligned}\dot{x}_0 &= f_0(x_0, x_1)\\ \dot{x}_i &= f_i(x_0, \vec{x}_i) + b_i x_{i+1}, \quad i = 1, \ldots, r\\ y &= x_1.\end{aligned} \tag{3.24}$$

The system (3.24) can be further generalized to

$$\begin{aligned}\dot{x}_0 &= f_0(x_0, x_1, d)\\ \dot{z}_i &= q_i(x_0, \vec{z}_i, \vec{x}_i, d)\\ \dot{x}_i &= f_i(x_0, \vec{z}_i, \vec{x}_i, d) + b_i(d)x_{i+1}, \quad i = 1, \ldots, r\\ y &= x_1.\end{aligned} \tag{3.25}$$

where d is the static uncertainty and $\dot{z}_i = q_i(x_0, \vec{z}_i, \vec{x}_i, d)$ is the dynamic uncertainty. Since the x_0-dynamics and the z_1-dynamics have the same structure, we can combine the x_0-dynamics and the z_1-dynamics. As a result, the system (3.25) reduces to (3.22).

In the normal form (3.19), the performance output of the system is $y = x_1$. Typically, in the controller design for a lower triangular system containing uncertainties, the measurement output is specified as $y_{\mathrm{m}} = \vec{x}_r$. Such a controller is called a (partial) state feedback controller.

Class II: Output feedback systems

$$\begin{aligned} \dot{z} &= q(z, x_1, d) \\ \dot{x}_i &= f_i(z, x_1, d) + b_i x_{i+1}, \quad i = 1, \ldots, r-1 \\ \dot{x}_r &= f_r(z, x_1, d) + b_r u \\ y &= x_1 \end{aligned} \tag{3.26}$$

where $z \in \mathbb{R}^n$ and $x_i \in \mathbb{R}$ for $i = 1, \ldots, r$, are the state variables, $u \in \mathbb{R}$ is the input, and $y \in \mathbb{R}$ is the output. The functions q and f_i are piecewise continuous in d and locally Lipschitz in the state variables and satisfy $q(0, 0, d) = 0$ and $f_i(0, 0, d) = 0$ for all $d \in \mathbb{D}$. The variable d is defined as in the system (3.22) and $b_i \in \mathbb{R}$ is an unknown nonzero parameter.

As in the lower triangular system (3.22), the performance output of the system (3.26) is $y = x_1$. Therefore, the system (3.26) is a special case of the lower triangular system (3.22) in the sense that various nonlinear functions depend on z and y only. For this class of systems, we will consider the output feedback controller, i.e., with the measurement output $y_{\mathrm{m}} = y = x_1$ (the state variables $x_2, \ldots, x_r$ are not available for feedback).

Class III: Filter extended systems

$$\begin{aligned} \dot{z} &= q(z, x_1, d) \\ \dot{x}_1 &= f_1(z, x_1, d) + b x_2 \\ \dot{x}_i &= f_i(\vec{x}_i) + x_{i+1}, \quad i = 2, \ldots, r \\ y &= x_1 \end{aligned} \tag{3.27}$$

where $z \in \mathbb{R}^n$ and $x_i \in \mathbb{R}$ for $i = 1, \ldots, r$, are the state variables, $u := x_{r+1} \in \mathbb{R}$ is the input, and $y \in \mathbb{R}$ is the output. The functions q and f_i are piecewise continuous in d and locally Lipschitz in the state variables and satisfy $q(0, 0, d) = 0$, $f_1(0, 0, d) = 0$ for all $d \in \mathbb{D}$, and $f_i(0) = 0$. The variable d is defined as in the system (3.22) and $b \in \mathbb{R}$ is an unknown nonzero parameter.

The system (3.27) is a special case of the lower triangular system (3.22). An interesting feature of this class of lower triangular system is that the input dynamics (the mapping from u to x_2) contain no uncertainties. This feature may simplify the controller design.

As in the lower triangular system (3.22), the performance output of the system (3.27) is $y = x_1$, and the typical measurement output is specified as $y_{\rm m} = \overrightarrow{x}_r$ in a partial state feedback controller design scenario.

Next, we will show that, by an appropriate dynamic extension, the class of output feedback systems (3.26) can be converted to systems of the filter extended form (3.27). For this purpose, we first define a filter

$$\dot{\xi} = A\xi + Bu, \ \xi \in \mathbb{R}^r \tag{3.28}$$

where

$$A = \begin{bmatrix} -\lambda_1 & \\ \vdots & I_{r-1} \\ -\lambda_{r-1} & \\ -\lambda_r & 0 \end{bmatrix}, \quad B = \begin{bmatrix} 0 \\ \vdots \\ 0 \\ 1 \end{bmatrix}.$$

Also, let

$$C = [0 \ 1 \ 0 \ \cdots, \ 0], \ D = (b_1 \cdots b_r)^{-1} \begin{bmatrix} 1 & & & & \\ & b_1 & & & \\ & & b_1 b_2 & & \\ & & & \ddots & \\ & & & & b_1 \cdots b_{r-1} \end{bmatrix},$$

and

$$\epsilon = Dx - \xi$$

where $x = [x_1, \ldots, x_r]^{\mathrm{T}}$. Denote

$$\begin{aligned} \epsilon = \mathrm{col}(\epsilon_1, \ldots, \epsilon_r), \quad \lambda = \mathrm{col}(\lambda_1, \ldots, \lambda_r), \\ f(z, x_1, d) = \mathrm{col}(f_1(z, x_1, d), \ldots, f_r(z, x_1, d)). \end{aligned}$$

Proposition 3.2 *Under the following coordinate and input transformations:*

$$\begin{aligned} \hat{z} &= col(z, \epsilon) \\ \hat{x}_1 &= x_1 \\ \begin{bmatrix} \hat{x}_2 \\ \vdots \\ \hat{x}_r \end{bmatrix} &= \begin{bmatrix} C \\ CA \\ \vdots \\ CA^{r-2} \end{bmatrix} \xi \end{aligned} \tag{3.29}$$

and

$$\hat{u} = CA^{r-1}\xi + u, \tag{3.30}$$

the system composed of (3.26) and (3.28) can be put in the following form

$$\begin{aligned} \dot{\hat{z}} &= \hat{q}(\hat{z}, \hat{x}_1, d) \\ \dot{\hat{x}}_1 &= \hat{f}_1(\hat{z}, \hat{x}_1, d) + b\hat{x}_2 \\ \dot{\hat{x}}_i &= \hat{x}_{i+1}, \quad i = 2, \ldots, r, \end{aligned} \tag{3.31}$$

with $\hat{x}_{r+1} = \hat{u}$ *and*

$$\begin{aligned} b &= b_1 \cdots b_r \\ \hat{q}(\hat{z}, \hat{x}_1, d) &= \begin{bmatrix} q(z, x_1, d) \\ A\epsilon + (b_1 \cdots b_r)^{-1}\lambda x_1 + Df(z, x_1, d) \end{bmatrix} \\ \hat{f}_1(\hat{z}, \hat{x}_1, d) &= f_1(z, x_1, d) + (b_1 \cdots b_r)\epsilon_2. \end{aligned}$$

Proof It can be verified that ϵ is governed by the following dynamics

$$\dot{\epsilon} = A\epsilon + (b_1 \cdots b_r)^{-1}\lambda x_1 + Df(z, x_1, d).$$

Next, noting the fact that

$$CA^i B = 0, \quad i = 0, \ldots, r-3, \quad CA^{r-2}B = 1$$

completes the proof. □

It can be seen that the system (3.31) is of the filter extended form (3.27). Moreover, if a static feedback controller with the measurement output $y_{\mathrm{m}} = \mathrm{col}(\hat{x}_1, \ldots, \hat{x}_r)$ is designed for the system (3.31), then it will lead to a dynamic feedback controller with the measurement output $y_{\mathrm{m}} = x_1$ for the original system (3.26) (with ξ being the state of a dynamic compensator.) In other words,

> Partial state feedback controller for the filter extended system (3.27) + dynamic filter ⇒ Output feedback controller for the output feedback system (3.26).

In summary, in the aforementioned three classes of nonlinear systems, the class of lower triangular systems is of the most general form that includes the other two as special cases. The relationship of the three classes of systems is illustrated in Fig. 3.1. In the subsequent chapters, we will mainly study systems of the forms (3.22) and (3.27). It is noted that all nonlinear functions of (3.26) rely on the state z and the output x_1 only. It is this special structure that leads to the name of output feedback system and that facilitates the design of an output feedback controller.

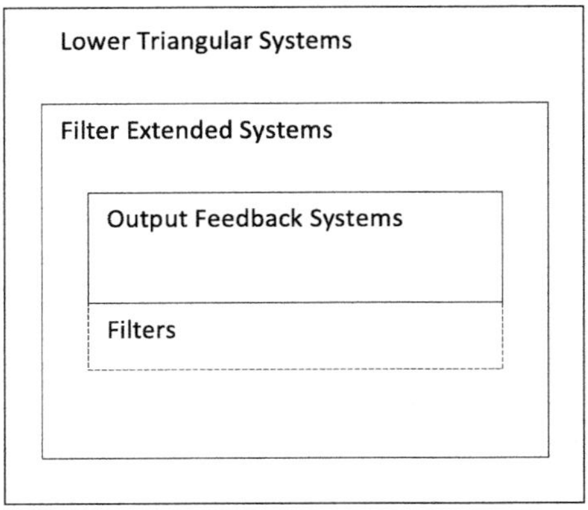

Fig. 3.1 Typical nonlinear control systems

As we will see in the subsequent chapters, whether or not the inverse dynamics of a system is ISS is the key for the solvability of the global stabilizability. Therefore, it is interesting to show that, if the output feedback system (3.26) has ISS inverse dynamics, so does the filter extended system (3.31) as summarized below.

Theorem 3.2 *Assume the z-subsystem of (3.26) is RISS with an ISS Lyapunov function $V(z)$, i.e.,*

$$V(z) \sim \{\underline{\alpha}, \bar{\alpha}, \alpha, \sigma \mid \dot{z} = q(z, x_1, d)\}$$

and

$$\limsup_{s \to 0^+} \frac{s^2}{\alpha(s)} < \infty, \quad \limsup_{s \to 0^+} \frac{\sigma(s)}{s^2} < \infty.$$

If all functions in (3.26) are sufficiently smooth and the matrix A in the filter (3.28) is Hurwitz, then, the $\hat{z}$-subsystem of (3.31) is RISS with an ISS Lyapunov function $W(\hat{z})$ satisfying

$$\underline{\beta}(\|\hat{z}\|) \leq W(\hat{z}) \leq \bar{\beta}(\|\hat{z}\|)$$

for some class $\mathcal{K}_\infty$ functions $\underline{\beta}$ and $\bar{\beta}$, and, along the trajectory of the $\hat{z}$-subsystem,

$$\dot{W}(\hat{z}) \leq -\|\hat{z}\|^2 + \varkappa(\hat{x}_1)\|\hat{x}_1\|^2$$

for some smooth function $\varkappa$.

Proof With $\hat{z} = \text{col}(z, \epsilon)$, the $\hat{z}$-subsystem of (3.31) has the form,

$$\begin{aligned}\dot{z} &= q(z, x_1, d)\\ \dot{\epsilon} &= A\epsilon + (b_1 \cdots b_r)^{-1}\lambda x_1 + Df(z, x_1, d).\end{aligned}$$

The proof is completed by directly applying Corollary 2.4. □

3.3 Examples of Nonlinear Control Systems

Many practical systems can be put in the lower triangular form (3.22), or the output feedback form (3.26). Some benchmark systems are given in this section.

3.3.1 The Duffing Equation

The controlled Duffing equation is a nonlinear second order differential equation:

$$\ddot{y} + \delta\dot{y} + \alpha y + \beta y^3 = \gamma\cos(\omega t + \phi) + u \tag{3.32}$$

where y is the displacement, u the control force, and δ, α, β, γ, ω and ϕ are constant parameters. The equation describes the motion of a damped oscillator with a more complicated potential than in simple harmonic motion. For example, it models a spring pendulum whose spring's stiffness does not exactly obey Hooke's law. When $u = 0$, the uncontrolled model was introduced by Duffing in 1918 and is thus named the Duffing equation. In recent years, the Duffing equation has become a benchmark system for testing various advanced nonlinear and/or adaptive control techniques.

Let $x = [x_1, x_2]^{\mathrm{T}} = [y, \dot{y}]^{\mathrm{T}}$ be the system state, $w = [\delta, \alpha, \beta]^{\mathrm{T}}$ the unknown parameter vector, and $v(t) = \gamma\cos(\omega t + \phi)$ the external disturbance. Then, the system (3.32) can be put in the following lower triangular form:

$$\begin{aligned}\dot{x}_1 &= x_2\\ \dot{x}_2 &= -\delta x_2 - \alpha x_1 - \beta x_1^3 + v(t) + u.\end{aligned} \tag{3.33}$$

which is in the form (3.22) with $r = 2$, $n_1 = n_2 = 0$, $b_1 = b_2 = 1$, $d(t) = \mathrm{col}(v(t), w)$, $f_1(x_1, d) = 0$ and $f_2(x_1, x_2, d) = -\delta x_2 - \alpha x_1 - \beta x_1^3 + v$.

3.3.2 The Lorenz System

The Lorenz system has been introduced in Chap. 1 whose dynamics (1.14) are repeated as follows:

$$\begin{aligned}\dot{\zeta}_1 &= \sigma(\zeta_2 - \zeta_1)\\ \dot{\zeta}_2 &= \zeta_1(\rho - \zeta_3) - \zeta_2 + u\\ \dot{\zeta}_3 &= \zeta_1\zeta_2 - \beta\zeta_3\end{aligned} \tag{3.34}$$

with all $\sigma, \rho, \beta > 0$. Let $z = [\zeta_1, \zeta_3]^{\mathsf{T}}$, $x_1 = \zeta_2$ and $y = x_1$. Then the system (3.34) can be rewritten as

$$\begin{aligned}
\dot{z} &= \begin{bmatrix} -\sigma & 0 \\ x_1 & -\beta \end{bmatrix} z + \begin{bmatrix} \sigma x_1 \\ 0 \end{bmatrix} \\
\dot{x}_1 &= [1,\ 0]z(\rho - [0,\ 1]z) - x_1 + u
\end{aligned} \tag{3.35}$$

which, together with $d = \mathrm{col}(\sigma, \beta, \rho)$, is in the output feedback form (3.26).

An extension of the model (3.34) is given as follows.

$$\begin{aligned}
\dot{\zeta}_1 &= \sigma(\zeta_2 - \zeta_1) \\
\dot{\zeta}_2 &= \zeta_1(\rho - \zeta_3) - \zeta_2 + \zeta_4 \\
\dot{\zeta}_3 &= \zeta_1\zeta_2 - \beta\zeta_3 \\
\dot{\zeta}_4 &= \beta\zeta_1 + \gamma\zeta_2\zeta_3 + u
\end{aligned} \tag{3.36}$$

where γ is a constant. This four-dimensional system is known as the controlled hyperchaotic Lorenz system.

Let $z = [\zeta_1, \zeta_3]^{\mathsf{T}}$, $x_1 = \zeta_2$ and $x_2 = \zeta_4$. Then, the system (3.36) can be rewritten as follows:

$$\begin{aligned}
\dot{z} &= \begin{bmatrix} -\sigma & 0 \\ x_1 & -\beta \end{bmatrix} z + \begin{bmatrix} \sigma x_1 \\ 0 \end{bmatrix} \\
\dot{x}_1 &= [1,\ 0]z(\rho - [0,\ 1]z) - x_1 + x_2 \\
\dot{x}_2 &= [\beta,\ \gamma x_1]z + u.
\end{aligned} \tag{3.37}$$

The system (3.37) is in the lower triangular form (3.22) with $r = 2$. Therefore it can be globally stabilized by a static feedback control with the measurement output $y_{\mathrm{m}} = \mathrm{col}(x_1, x_2)$ (i.e., both x_1 and x_2 are available for control) as we will see in Sect. 6.4.

On the other hand, the system (3.37) can also be viewed as the output feedback form (3.26) with $r = 2$. By Proposition 3.2, extending (3.37) by the following filter

$$\begin{aligned}
\dot{\xi}_1 &= -\lambda_1\xi_1 + \xi_2 \\
\dot{\xi}_2 &= -\lambda_2\xi_1 + u
\end{aligned} \tag{3.38}$$

where $\lambda_1, \lambda_2 > 0$ are two constants, will lead to a system of the form (3.27). In fact, let $z_1 = \zeta_1, z_2 = \zeta_3, z_3 = \zeta_2 - \xi_1, z_4 = \zeta_4 - \xi_2, x_1 = \zeta_2, x_2 = \xi_2$, and $\hat{u} = u - \lambda_2\xi_1$. The system composed of (3.36) and (3.38) can be put in the following form

$$\begin{aligned}
\dot{z} &= q(z, x_1, \sigma, \beta, \rho) \\
\dot{x}_1 &= z_1(\rho - z_2) - x_1 + z_4 + x_2 \\
\dot{x}_2 &= \hat{u}
\end{aligned} \tag{3.39}$$

where

$$q(z, x_1, \sigma, \beta, \rho) = \begin{bmatrix} \sigma(x_1 - z_1) \\ z_1 x_1 - \beta z_2 \\ z_1(\rho - z_2) - x_1 + z_4 + \lambda_1(x_1 - z_3) \\ \beta z_1 + \gamma x_1 z_2 + \lambda_2(x_1 - z_3) \end{bmatrix}.$$

Clearly, the system (3.39) is now of the filter extended form (3.27). Therefore, for the system (3.36), it is possible to obtain a dynamic feedback controller with the measurement output $y_{\mathrm{m}} = \zeta_2$.

3.3.3 Chua's Circuit

Chua's circuit was originally conceived in 1983 by Leon O. Chua. It is a simple electronic circuit that exhibits chaotic behavior. Later, researchers modified and added new features to the circuit to study such problems as chaos cancelation, synchronization, or stabilization, leading to what is called Chua's circuit family. In the circuit, the energy storage elements are two capacitors (labeled C_1 and C_2) and an inductor (labeled L). There is a linear resistor (labeled R) and a nonlinear resistor. Let the nonlinear resistor be defined by the nonlinear function $I = s(V)V/R$ where $s(V)$ is some function of V. Let V_1 and V_2 be the voltages across the capacitors C_1 and C_2, respectively, I the current flowing through the inductor L. Let V_c be the control voltage. As a result, the dynamic equations for the circuit can be derived as follows:

$$\begin{aligned} \dot{V}_1 &= \frac{V_2}{RC_1} - \left(\frac{1}{RC_1} + \frac{s(V_1)}{RC_1}\right) V_1 \\ \dot{V}_2 &= \frac{V_1 - V_2}{RC_2} + \frac{I}{C_2} \\ \dot{I} &= \frac{-V_2}{L} + \frac{V_c}{L}. \end{aligned} \tag{3.40}$$

Let $\zeta = [V_2, I]^{\mathrm{T}}$, $y = V_1$, and $u = V_c$. Then, the system (3.40) can be rewritten as

$$\begin{aligned} \dot{\zeta} &= \begin{bmatrix} -\beta & \gamma \\ -\sigma & 0 \end{bmatrix} \zeta + \begin{bmatrix} \beta \\ 0 \end{bmatrix} y + \begin{bmatrix} 0 \\ \sigma \end{bmatrix} u \\ \dot{y} &= [\alpha, \ 0]\zeta - \alpha(1 + s(y))y \end{aligned} \tag{3.41}$$

where $\alpha = 1/(RC_1)$, $\beta = 1/(RC_2)$, $\gamma = 1/C_2$, and $\sigma = 1/L$. At the first glance, the system (3.41) is not in any one of the three special forms studied in the previous section. However, it is easy to verify that the system has a global relative degree three. In fact, direct calculation shows

$$\begin{aligned}
\dot{y} &= s_1(y, \zeta_1) = \alpha\zeta_1 - \alpha(1 + s(y))y \\
\ddot{y} &= s_2(y, \zeta_1, \zeta_2) = \frac{\partial s_1(y, \zeta_1)}{\partial y} s_1(y, \zeta_1) + \alpha[-\beta\zeta_1 + \gamma\zeta_2 + \beta y] \\
y^{(3)} &= f(y, \zeta_1, \zeta_2) + \alpha\gamma\sigma u
\end{aligned}$$

for

$$f(y, \zeta_1, \zeta_2) = \frac{\partial s_2(y, \zeta_1, \zeta_2)}{\partial y} s_1(y, \zeta_1) + \frac{\partial s_2(y, \zeta_1, \zeta_2)}{\partial \zeta_1}[-\beta\zeta_1 + \gamma\zeta_2 + \beta y] - \alpha\gamma\sigma\zeta_1.$$

Thus, letting

$$x = \begin{bmatrix} x_1 \\ x_2 \\ x_3 \end{bmatrix} = \tau(y, \zeta_1, \zeta_2) = \begin{bmatrix} y \\ s_1(y, \zeta_1) \\ s_2(y, \zeta_1, \zeta_2) \end{bmatrix}$$

gives the normal form (3.13) of (3.41) as follows

$$\begin{aligned}
\dot{x}_1 &= x_2 \\
\dot{x}_2 &= x_3 \\
\dot{x}_3 &= f(y, \zeta_1, \zeta_2)|_{[y, \zeta_1, \zeta_2]^{\mathrm{T}} = \tau^{-1}(x)} + \alpha\gamma\sigma u.
\end{aligned} \tag{3.42}$$

Next, we extend (3.41) by the following input filter

$$\begin{aligned}
\dot{\xi}_1 &= -\lambda_1\xi_1 + \xi_2 \\
\dot{\xi}_2 &= -\lambda_2\xi_2 + u,
\end{aligned}$$

for any two constants $\lambda_1, \lambda_2 > 0$. Performing the coordinate transformation

$$z = \zeta - C(d)[\xi_1, \xi_2]^{\mathrm{T}} - h(d)y, \ \ x_1 = y, x_2 = \xi_1, x_3 = \xi_2$$

with $d = [\alpha, \beta, \gamma, \sigma]^{\mathrm{T}}$,

$$C(d) = \begin{bmatrix} \gamma\sigma & 0 \\ \lambda_2\sigma & \sigma \end{bmatrix}, \quad h(d) = \begin{bmatrix} -\beta + \lambda_1 + \lambda_2 \\ -\sigma + \lambda_1\lambda_2/\gamma \end{bmatrix} / \alpha,$$

on the extended system gives the following standard filter extended system:

$$\begin{aligned}
\dot{z} &= F(d)z + G(x_1, d) \\
\dot{x}_1 &= H(d)z + K(x_1, d) + bx_2 \\
\dot{x}_2 &= -\lambda_1 x_2 + x_3 \\
\dot{x}_3 &= -\lambda_2 x_3 + u
\end{aligned} \tag{3.43}$$

where

$$\begin{aligned}
F(d) &= \begin{bmatrix} -\lambda_1 - \lambda_2 & \gamma \\ -\lambda_1\lambda_2/\gamma & 0 \end{bmatrix}, \quad H(d) = [\alpha \ 0] \\
K(x_1, d) &= [-\beta + \lambda_1 + \lambda_2 - \alpha(1 + s(x_1))]x_1 \\
G(x_1, d) &= \begin{bmatrix} -(\lambda_1 + \lambda_2)^2 + (\alpha + \beta)(\lambda_1 + \lambda_2) - \sigma\gamma + \lambda_1\lambda_2 \\ -\lambda_1\lambda_2(\lambda_1 + \lambda_2)/\gamma - \alpha\sigma + (\alpha + \beta)\lambda_1\lambda_2/\gamma \end{bmatrix} x_1/\alpha \\
&\quad + \begin{bmatrix} -\beta + \lambda_1 + \lambda_2 \\ -\sigma + \lambda_1\lambda_2/\gamma \end{bmatrix} s(x_1)x_1 \\
b &= \alpha\gamma\sigma.
\end{aligned}$$

3.3.4 The FitzHugh-Nagumo Model

The following system with $u = 0$ is called the FitzHugh-Nagumo model:

$$\begin{aligned}
\dot{y} &= y - y^3/3 - \phi + \varphi + I(t) + u \\
\dot{\phi} &= \epsilon(y + a - b\phi) \\
\dot{\varphi} &= \varepsilon(-y + c - d\varphi)
\end{aligned} \tag{3.44}$$

where the parameters a, b, c, d, ϵ and ε are positive. The model was suggested by Richard FitzHugh in 1961 and the equivalent circuit was created by J. Nagumo et al. in the following year. It is an idealized nerve membrane model which exhibits the bursting phenomenon. This model is capable of generating spike of membrane voltage, represented by y, in a neuron after stimulation by an external input current $I(t)$. The first FitzHugh-Nagumo model was a second order system with $u = 0$ and $\varphi = 0$. The additional φ dynamics are introduced later to describe the slow modulation of the current, and the control u is introduced to control the behavior of the system.

Performing the following coordinate transformation on (3.44)

$$z = \begin{bmatrix} \phi - a/b \\ \varphi - c/d \end{bmatrix}$$

gives

$$\begin{aligned}
\dot{z} &= \begin{bmatrix} -\epsilon b & 0 \\ 0 & -\varepsilon d \end{bmatrix} z + \begin{bmatrix} \epsilon \\ -\varepsilon \end{bmatrix} y \\
\dot{y} &= y - y^3/3 + [-1 \ 1]z - a/b + c/d + I(t) + u.
\end{aligned} \tag{3.45}$$

This system is in the form (3.27) with $x_1 = y$, $w = [a, b, c, d, \epsilon, \varepsilon]$ and $d(t) = \text{col}(I(t), w)$.

3.4 Notes and References

The notions of zero dynamics and normal form summarized in Sect. 3.1 were introduced by Byrnes and Isidori [1]. The main references include Chap. 4 of [2], Chap. 2 of [3], Chap. 13 of [4], etc. The three typical nonlinear systems discussed in Sect. 3.2 are extensively studied in literature including [5–13], etc. The references for the examples of nonlinear control systems in Sect. 3.3 include [14–16] for the Duffing equation, [17–20] for the Lorenz system, [21–23] for Chua's circuit, and [19, 24] for the FitzHugh-Nagumo model.

3.5 Problems

Problem 3.1 Determine the relative degree for each of the systems given below at each x_0 of the space.

(a) $\dot{x}_1 = x_2, \dot{x}_2 = -x_1^2 + \theta\sin(x_1x_2) + u,\ y = x_1, \theta > 0$;
(b) $\dot{x}_1 = x_2, \dot{x}_2 = -x_1^2 + \theta\sin(x_1x_2) + u,\ y = x_2, \theta > 0$;
(c) $\dot{x}_1 = x_2, \dot{x}_2 = -x_1^2 + \theta\sin(x_1x_2) + u,\ y = x_1^2 + x_2^2, \theta > 0$;
(d) $\dot{x}_1 = -x_1 + u, \dot{x}_2 = -x_2 + 2 - x_1x_3, \dot{x}_3 = x_1x_2,\ y = x_3$.

Problem 3.2 For each of the systems given in Problem 3.1, find the input–ouput linearizing control law at the x_0 where a relative degree is well defined.

Problem 3.3 Determine the relative degree of the following system at $\mathrm{col}(x_1, x_2, x_3) = 0$,

$$\begin{aligned}\dot{x}_1 &= x_3 - x_2^3\\ \dot{x}_2 &= -x_2 - u\\ \dot{x}_3 &= x_1^2 - x_3 + u\\ y &= x_1;\end{aligned}$$

determine its normal form and zero dynamics.

Problem 3.4 Determine the normal form of the system

$$\begin{aligned}\dot{x}_1 &= -x_1 + (0.5\cos x_3 + 1)u\\ \dot{x}_2 &= x_3\\ \dot{x}_3 &= x_1x_3 + u\end{aligned}$$

with the output specified below on a neighbourhood of $\mathrm{col}(x_1, x_2, x_3) = 0$.

(a) $y = x_1$;
(b) $y = x_2$;
(c) $y = x_3$.

Problem 3.5 Find new coordinates such that the system in Problem 3.3 is rewritten in the form (3.15); find the zero dynamics of the system and determine if the system is minimum phase.

Problem 3.6 Consider the following output feedback system with state $x = \text{col}(x_1, x_2, x_3)$, input u, and performance output $y = x_1$,

$$\begin{aligned} \dot{z} &= -a(z, d) + x_1 \\ \dot{x}_1 &= x_2 \\ \dot{x}_2 &= dzx_1 + 2x_3 \\ \dot{x}_3 &= dzx_1^2 + u, \end{aligned}$$

which is affected by dynamic uncertainty governed by the z-dynamics and static uncertainty d. Find a filter and a set of new coordinate which convert the system into a filter extended system in the form (3.27).

References

1. Byrnes CI, Isidori A (1984) A frequency domain philosophy for nonlinear systems. In: Proceedings of the 23rd IEEE conference on decision and control, vol 23, pp 1569–1573
2. Isidori A (1995) Nonlinear control systems, 3rd edn. Springer, New York
3. Huang J (2004) Nonlinear output regulation problem: theory and Applications. SIAM, Philadelphia
4. Khalil H (2002) Nonlinear systems. Prentice Hall, Upper Saddle River
5. Chen Z, Huang J (2002) Global robust stabilization of cascaded polynomial systems. Syst Control Lett 47:445–453
6. Chen Z, Huang J (2004) Dissipativity, stabilization, and regulation of cascade-connected systems. IEEE Trans Autom Control 49:635–650
7. Ding Z (2003) Universal disturbance rejection for nonlinear systems in output feedback form. IEEE Trans Autom Control 48:1222–1226
8. Huang J, Chen Z (2004) A general framework for tackling the output regulation problem. IEEE Trans Autom Control 49:2203–2218
9. Jiang ZP, Mareels I (1997) A small-gain control method for nonlinear cascaded systems with dynamic uncertainties. IEEE Trans Autom Control 42:292–308
10. Jiang ZP, Praly L (1993) Stabilization by output feedback for systems with ISS inverse dynamics. Syst Control Lett 21:19–33
11. Khalil H (1994) Robust servomechanism output feedback controllers for feedback linearizable systems. Automatica 30:1587–1589
12. Lin W, Gong Q (2003) A remark on partial-state feedback stabilization of cascade systems using small gain theorem. IEEE Trans Autom Control 48:497–500
13. Serrani A, Isidori A, Marconi L (2001) Semiglobal nonlinear output regulation with adaptive internal model. IEEE Trans Autom Control 46:1178–1194

14. Liu L, Huang J (2008) Asymptotic disturbance rejection of the Duffing's system by adaptive output feedback control. IEEE Trans Circuits Syst II Express Briefs 55:1030–1066
15. Jiang ZP (2002) Advanced feedback control of the chaotic duffing equation. IEEE Trans Circuits Syst I Fundam Theory Appl 49:241–249
16. Nijmeijer H, Berghuis H (1995) On Lyapunov control of the Duffing equation. IEEE Trans Circuits Syst I Fundam Theory Appl 42:473–477
17. Liang X, Zhang J, Xia X (2008) Adaptive synchronization for generalized Lorenz systems. IEEE Trans Autom Control 53:1740–1746
18. Xu D, Huang J (2010) Global output regulation for output feedback systems with an uncertain exosystem and its application. Int J Robust Nonlinear Control 20:1678–1691
19. Xu D, Huang J (2010) Robust adaptive control of a class of nonlinear systems and its applications. IEEE Trans Circuits Syst I Regul Pap 57:691–702
20. Yu W (1999) Passive equivalence of chaos in Lorenz system. IEEE Trans Circuits Syst I Fundam Theory Appl 46:876–878
21. Liao TL, Chen FW (1998) Control of Chua's circuit with a cubic nonlinearity via nonlinear linearization technique. Circuits Syst Signal Process 17:719–731
22. Liu L, Huang J (2006) Adaptive robust stabilization of output feedback systems with application to Chua's circuit. IEEE Trans Circuits Syst II Express Briefs 53:926–930
23. Zhong GQ (1994) Implementation of Chua's circuit with a cubic nonlinearity. IEEE Trans Circuits Syst I Fundam Theory Appl 41:934–941
24. Rinzel J (1987) A formal classification of bursting mechanisms in excitable systems. In: Teramoto E, Yamaguti M (eds) Mathematical topics in population biology, morphogenesis and neurosciences. Springer, Berlin, pp 267–281

Chapter 4
Robust Stabilization

In this chapter, we study the global robust stabilization problem for uncertain nonlinear control systems with the emphasis on systems in the lower triangular form (3.22). In Sect. 4.1, we introduce a recursive design procedure for constructing a static controller. Then this recursive design procedure is applied to output feedback systems and general lower triangular systems in Sects. 4.2 and 4.3, respectively. In Sect. 4.4, we further discuss how to reduce the control gain. A case study on the robust stabilization of Chua's circuit is studied in Sect. 4.5. The notes and references are given in Sect. 4.6.

4.1 An Overview of the Approach

The global robust stabilization problem to be studied in this chapter is formulated as follows.

Global Robust Stabilization Problem (GRSP): *Given a nonlinear control system of the form* (2.7–2.9) *satisfying* $f(0, 0, d) = 0$ *for all* $d \in \mathbb{D}$, *design a controller of the form* (1.11), *such that the equilibrium point* $x_c = 0$ *of the closed-loop system* (2.10) *is RUGAS.*

In this section, we outline the technique for achieving GRSP for lower triangular systems of the form (3.22), i.e.,

$$\begin{aligned} \dot{z}_i &= q_i(\vec{z}_i, \vec{x}_i, d(t)) \\ \dot{x}_i &= f_i(\vec{z}_i, \vec{x}_i, d(t)) + b_i(d(t))x_{i+1}, \quad i = 1, \ldots, r. \end{aligned} \tag{4.1}$$

In the system, $\vec{z}_i = \text{col}(z_1, \ldots, z_i)$ with $z_i \in \mathbb{R}^{n_i}$ and $\vec{x}_i = \text{col}(x_1, \ldots, x_i)$ with $x_i \in \mathbb{R}$ are the state variables, $u := x_{r+1} \in \mathbb{R}$ is the input, $d(t) \in \mathbb{D} \subset \mathbb{R}^l$ represents system uncertainties in a known compact set $\mathbb{D}$. It is also assumed that the functions q_i, f_i, and b_i in the system (4.1) are sufficiently smooth satisfying $q_i(0, 0, d) = 0$

Z. Chen and J. Huang, *Stabilization and Regulation of Nonlinear Systems*,
Advanced Textbooks in Control and Signal Processing,
DOI 10.1007/978-3-319-08834-1_4

and $f_i(0, 0, d) = 0$, and $d(t)$ is a piecewise continuous function in t. We consider the scenario that the measurement output is $y_{\mathrm{m}} = \vec{x}_r$.

The general approach is to construct a control law for the system (4.1) and the corresponding Lyapunov function for the closed-loop system recursively.

To start with, we introduce a coordinate transformation:

$$\begin{aligned} \mathcal{X}_1 &= x_1 \\ \mathcal{X}_{i+1} &= x_{i+1} - s_i\left(\vec{\mathcal{X}}_i\right), \quad i = 1, \ldots, r \end{aligned} \tag{4.2}$$

where $\vec{\mathcal{X}}_i = \mathrm{col}(\mathcal{X}_1, \ldots, \mathcal{X}_i)$ and s_i is a sufficiently smooth function to be determined recursively. Then, we define a coordinate transformation as follows:

$$\vec{\mathcal{X}}_i = \tau_i\left(\vec{x}_i\right) \tag{4.3}$$

where

$$\begin{aligned} \tau_1(x_1) &= x_1 \\ \tau_i\left(\vec{x}_i\right) &= \left[x_1 \; x_2 - s_1(\tau_1(x_1)) \; \cdots \; x_i - s_{i-1}\left(\tau_{i-1}\left(\vec{x}_{i-1}\right)\right)\right]^{\mathsf{T}}, \\ & \quad i = 2, \ldots, r. \end{aligned} \tag{4.4}$$

Moreover, let

$$\mathcal{S}_i\left(\vec{x}_i\right) = s_i\left(\tau_i\left(\vec{x}_i\right)\right) = s_i\left(\vec{\mathcal{X}}_i\right), \quad i = 1, \ldots, r. \tag{4.5}$$

Then, the coordinate transformation (4.2) is equivalent to

$$\begin{aligned} \mathcal{X}_1 &= x_1 \\ \mathcal{X}_{i+1} &= x_{i+1} - \mathcal{S}_i\left(\vec{x}_i\right), \quad i = 1, \ldots, r. \end{aligned} \tag{4.6}$$

Under the transformation (4.2), the system (4.1) in the new coordinate $\mathrm{col}(\vec{z}_r, \vec{\mathcal{X}}_r)$ can be put into the following form

$$\begin{aligned} \dot{z}_i &= \mathcal{Q}_i\left(\vec{z}_i, \vec{\mathcal{X}}_i, d\right) \\ \dot{\mathcal{X}}_i &= \mathcal{F}_i\left(\vec{z}_i, \vec{\mathcal{X}}_i, d\right) + b_i(d) s_i\left(\vec{\mathcal{X}}_i\right) + b_i(d)\mathcal{X}_{i+1}, \quad i = 1, \ldots, r, \end{aligned} \tag{4.7}$$

where various functions are defined as follows,

$$\begin{aligned}
\mathcal{Q}_i\left(\vec{z}_i, \vec{\mathcal{X}}_i, d\right) &= q_i\left(\vec{z}_i, \vec{x}_i, d\right), \quad i = 1, \ldots, r \\
\mathcal{F}_1\left(\vec{z}_1, \vec{\mathcal{X}}_1, d\right) &= f_1\left(\vec{z}_1, \vec{x}_1, d\right) \\
\mathcal{F}_i\left(\vec{z}_i, \vec{\mathcal{X}}_i, d\right) &= f_i\left(\vec{z}_i, \vec{x}_i, d\right) - \sum_{j=1}^{i-1} \frac{\partial\left(s_{i-1}\left(\vec{\mathcal{X}}_{i-1}\right)\right)}{\partial \mathcal{X}_j} \times \left[\mathcal{F}_j\left(\vec{z}_j, \vec{\mathcal{X}}_j, d\right)\right. \\
&\quad \left. + b_j(d) s_j\left(\vec{\mathcal{X}}_j\right) + b_j(d)\mathcal{X}_{j+1}\right], \quad i = 2, \ldots, r. \qquad (4.8)
\end{aligned}$$

For the convenience of presentation, we denote the subsystem governing $\zeta_i = \mathrm{col}(\vec{z}_i, \vec{\mathcal{X}}_i)$ in a compact form

$$\dot{\zeta}_i = \varphi_i(\zeta_i, \mathcal{X}_{i+1}, d), \quad i = 1, \ldots, r. \qquad (4.9)$$

The above coordinate transformation actually suggests a recursive procedure to construct a static state feedback control law to stabilize the system (4.1) as indicated by the following result.

Proposition 4.1 *If there exist sufficiently smooth functions s_i or $\mathcal{S}_i$, $i = 1, \ldots, r$, vanishing at the origin, such that, the equilibrium point $\zeta_r = 0$ of the system $\dot{\zeta}_r = \varphi_r(\zeta_r, 0, d)$ is RUGAS, then the GRSP for the system* (4.1) *is solved by a static controller*

$$\begin{aligned}
u &= s_r\left(\vec{\mathcal{X}}_r\right) \\
\mathcal{X}_i &= x_i - s_{i-1}\left(\vec{\mathcal{X}}_{i-1}\right), \quad i = r, \ldots, 2 \\
\mathcal{X}_1 &= x_1. \qquad (4.10)
\end{aligned}$$

Remark 4.1 In terms of the notation (4.5), the control law (4.10) can be put in the following compact form

$$u = \mathcal{S}_r\left(\vec{x}_r\right). \qquad (4.11)$$

By Proposition 4.1, what we need to do is to construct the function s_i or the function $\mathcal{S}_i$, for $i = 1, \ldots, r$, to make the equilibrium point $\mathrm{col}(\vec{z}_r, \vec{\mathcal{X}}_r) = 0$ of the system (4.7) with $\mathcal{X}_{r+1} = 0$ RUGAS. Taking advantage of the lower triangular structure of the system (4.1), these functions can be recursively constructed as illustrated in the following simple example.

Example 4.1 Consider a second order system

$$\begin{aligned}
\dot{x}_1 &= x_2 \\
\dot{x}_2 &= \sin(x_1 x_2) + u. \qquad (4.12)
\end{aligned}$$

First, we try a function

$$W(x_1, x_2) = \frac{1}{2}x_1^2 + \frac{1}{2}x_2^2$$

whose derivative along the state trajectory of (4.12) is

$$\dot{W}(x_1, x_2) = x_2(x_1 + \sin(x_1 x_2) + u).$$

There does not exist any $u = \kappa(x_1, x_2)$ such that $\dot{W}(x_1, x_2)$ is negative definite since, for all x_1,

$$\dot{W}(x_1, x_2) = x_2(x_1 + \sin(x_1 x_2) + \kappa(x_1, x_2)) = 0$$

whenever $x_2 = 0$.

Next, we turn to a different approach to designing the Lyapunov function in two steps.

Step 1: Let $\mathcal{X}_1 = x_1$ and $\mathcal{X}_2 = x_2 - s_1(x_1)$ with $s_1(x_1) = -x_1$. Define a new function

$$W(\mathcal{X}_1, \mathcal{X}_2) = \frac{1}{2}\mathcal{X}_1^2 + \frac{1}{2}\mathcal{X}_2^2$$

whose derivative along the state trajectory of the closed-loop system is

$$\begin{aligned}\dot{W}(\mathcal{X}_1, \mathcal{X}_2) &= x_1 x_2 + (x_1 + x_2)(x_2 + \sin(x_1 x_2) + \kappa(x_1, x_2)) \\ &= -x_1^2 + (x_1 + x_2)(x_1 + x_2 + \sin(x_1 x_2) + \kappa(x_1, x_2)).\end{aligned}$$

Step 2: Pick the function $\kappa(x_1, x_2) = -2(x_1 + x_2) - \sin(x_1 x_2)$. As a result,

$$\dot{W}(\mathcal{X}_1, \mathcal{X}_2) = -\mathcal{X}_1^2 - \mathcal{X}_2^2.$$

That is, the control $u = \kappa(x_1, x_2)$ guarantees the RUGAS property of the equilibrium point $x = [x_1, x_2]^{\mathrm{T}} = 0$ of the closed-loop system. Figure 4.1 shows a state response of the closed-loop system with the initial state values $x_1(0) = -4$ and $x_2(0) = 8$.

In the above example, we see that $W(\mathcal{X}_1, \mathcal{X}_2) = (\mathcal{X}_1^2 + \mathcal{X}_2^2)/2$ is a Lyapunov function for the closed-loop system (4.12) under controller $u = \kappa(x_1, x_2)$. This idea can be extended, through a recursive design, to more general systems such as systems in output feedback form and lower triangular form.

4.2 Output Feedback Systems

We first consider a special case of (4.1), i.e.,

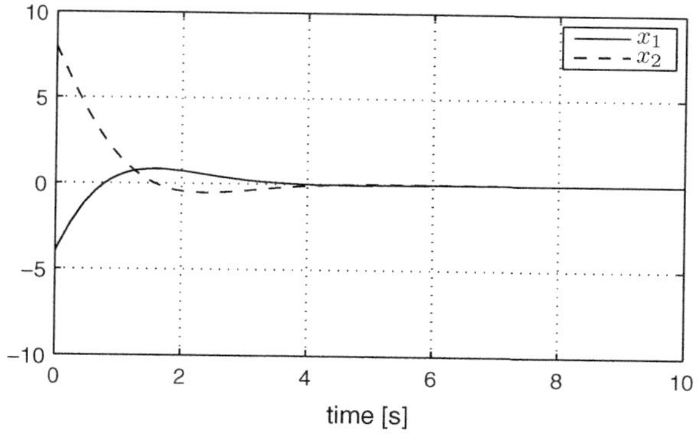

Fig. 4.1 Profile of state trajectories of the closed-loop system in Example 4.1

$$\begin{aligned} \dot{z} &= q(z, x_1, d) \\ \dot{x}_i &= f_i(z, x_1, d) + b_i x_{i+1}, \quad i = 1, \ldots, r-1 \\ \dot{x}_r &= f_r(z, x_1, d) + b_r u \\ y &= x_1 \end{aligned} \tag{4.13}$$

where $z \in \mathbb{R}^n$ and $x_i \in \mathbb{R}$, $i = 1, \ldots, r$, are the state variables and $u \in \mathbb{R}$ is the input.

It is known that (4.13) is called the output feedback system with relative degree r. The system (4.13) includes the case with $r = 1$ as a special case by ignoring the second equation of (4.13). The global robust stabilization problem of this special case has been studied in Theorem 2.8. Thus, in what follows, we assume $r \geq 2$. We first make the following assumption.

Assumption 4.1 The subsystem $\dot{z} = q(z, x_1, d)$ has an ISS Lyapunov function $V(z)$, i.e.,

$$V(z) \sim \left\{\underline{\alpha}, \bar{\alpha}, \alpha, \sigma \mid \dot{z} = q(z, x_1, d)\right\}$$

and

$$\limsup_{s \to 0^+} \frac{s^2}{\alpha(s)} < \infty, \quad \limsup_{s \to 0^+} \frac{\sigma(s)}{s^2} < \infty.$$

It has been discussed in Sect. 3.2 that, with the filter (3.28), i.e.,

$$\dot{\xi} = A\xi + Bu, \quad \xi \in \mathbb{R}^r \tag{4.14}$$

the system (4.13) with $b_i \neq 0$, $i = 1, \ldots, r$, can be converted into the so-called filter extended system (3.31), i.e.,

$$\begin{aligned}\dot{\hat{z}} &= \hat{q}\left(\hat{z}, \hat{x}_1, d\right)\\ \dot{\hat{x}}_1 &= \hat{f}_1\left(\hat{z}, \hat{x}_1, d\right) + b\hat{x}_2\\ \dot{\hat{x}}_i &= \hat{x}_{i+1}, \quad i = 2, \ldots, r\end{aligned} \tag{4.15}$$

with $\hat{x}_{r+1} = \hat{u}$. Any partial state feedback control $\hat{u} = \kappa(\hat{x}_1, \ldots, \hat{x}_r)$ for the system (4.15) is equivalent to an output feedback controller

$$\begin{aligned}u &= \kappa\left(y, C\xi, \ldots, CA^{r-2}\xi\right) - CA^{r-1}\xi\\ \dot{\xi} &= A\xi + Bu\end{aligned}$$

for the original system (4.13), according to the explicit coordinate transformation proposed in Sect. 3.2.

From the above manipulation, the output feedback stabilization problem for the system (4.13) is equivalent to the partial state feedback stabilization problem for the system (4.15). The system (4.15) can be viewed as a special case of the following system:

$$\begin{aligned}\dot{z} &= q(z, x_1, d)\\ \dot{x}_1 &= f_1(z, x_1, d) + bx_2\\ \dot{x}_i &= f_i\left(\overrightarrow{x}_i\right) + x_{i+1}, \quad i = 2, \ldots, r\end{aligned} \tag{4.16}$$

where $x_{r+1} = u$ and $f_i(\overrightarrow{x}_i), i = 2, \cdots, r$, are sufficiently smooth vanishing at the origin. We will now focus on the stabilization problem of (4.16). It can be seen that the system (4.16) is in the lower triangular form (4.1) and has a relative degree r.

By Theorem 3.2, under Assumption 4.1, the z-dynamics of (4.16) also satisfy an assumption similar to Assumption 4.1. To save notation, in what follows, we assume Assumption 4.1 also applies to the z-dynamics of (4.16). In this section, we assume b is known to simplify the controller design. Without loss of generality, we assume $b > 0$. The more general case for uncertain b can be regarded as a special lower triangular system which will be discussed in the next section.

As we did in Theorem 2.8 for the case with $r = 1$, we have the following inequality

$$|f_1(z, x_1, d)| \leq m_1(z)\|z\| + m_2(x_1)|x_1|, \quad \forall d \in \mathbb{D}, \tag{4.17}$$

for some sufficiently smooth and non-negative functions m_1 and m_2. Let

$$\Delta(z) \geq 1 + rm_1^2(z). \tag{4.18}$$

By using the changing supply function technique (Corollary 2.2), there exists a continuously differentiable function $V'(z)$ satisfying $\underline{\alpha}'(\|z\|) \leq V'(z) \leq \bar{\alpha}'(\|z\|)$ for some class $\mathcal{K}_\infty$ functions $\underline{\alpha}'$ and $\bar{\alpha}'$, and its derivative along the trajectory of $\dot{z} = q(z, x_1, d)$ satisfies

$$\dot{V}'(z) \leq -\Delta(z)\|z\|^2 + \varkappa(x_1)x_1^2 \tag{4.19}$$

for a smooth function $\varkappa$. Let $\varrho(x_1)$ be a sufficiently smooth function satisfying

$$\varrho(x_1) \geq \varkappa(x_1) + m_2(x_1) + (r-1)m_2^2(x_1). \tag{4.20}$$

Then we can specify the functions $\mathcal{S}_i(\vec{x}_i)$, $i = 1, \ldots, r$, in the coordinate transformation (4.6) as follows:

$$\begin{aligned} \mathcal{S}_1(x_1) &= -\rho_1(x_1)\mathcal{X}_1 \\ \mathcal{S}_i\left(\vec{x}_i\right) &= -\rho_i\left(\vec{x}_i\right)\mathcal{X}_i + \nu_i\left(\vec{x}_i\right), \quad i = 2, \ldots, r \end{aligned} \tag{4.21}$$

where

$$\begin{aligned} \rho_1(x_1) &= [\varrho(x_1) + 5/4]/b + b/4 \\ \rho_i(\vec{x}_i) &\geq \left(\frac{\partial \mathcal{S}_{i-1}\left(\vec{x}_{i-1}\right)}{\partial x_1}\right)^2 \Big/ 2 + 9/4 \\ \nu_i(\vec{x}_i) &= -f_i\left(\vec{x}_i\right) + \frac{\partial \mathcal{S}_{i-1}(\vec{x}_{i-1})}{\partial x_1} b x_2 \\ &\quad + \sum_{j=2}^{i-1} \frac{\partial \mathcal{S}_{i-1}\left(\vec{x}_{i-1}\right)}{\partial x_j}\left[f_j\left(\vec{x}_j\right) + x_{j+1}\right], \quad i = 2, \ldots, r. \end{aligned} \tag{4.22}$$

We have the following result.

Theorem 4.1 *Consider the system* (4.16) *with* $\mathbb{D}$ *a prescribed compact set. Under Assumption* 4.1, *let*

$$W_r\left(z, \vec{\mathcal{X}}_r\right) = V'(z) + \sum_{j=1}^{r} \mathcal{X}_j^2/2.$$

Then, with $\mathcal{S}_r$ *being given by* (4.21) *and* (4.22),

$$\dot{W}_r\left(z, \vec{\mathcal{X}}_r\right) \leq -\|z\|^2 - \left\|\vec{\mathcal{X}}_r\right\|^2 + \|\mathcal{X}_{r+1}\|^2. \tag{4.23}$$

As a result, the GRSP of (4.16) *is solved by the following controller*

$$u = \mathcal{S}_r\left(\vec{x}_r\right). \tag{4.24}$$

Proof Consider the coordinate transformation (4.6) with $\mathcal{S}_r$ being given by (4.21) and (4.22). For $i = 1, \ldots, r$, let

$$W_i\left(z, \vec{\mathcal{X}}_i\right) = V'(z) + \sum_{j=1}^{i} \mathcal{X}_j^2/2.$$

We will prove that, along the state trajectory of the $\mathrm{col}(z, \vec{\mathcal{X}}_i)$-subsystem,

$$\begin{aligned} \dot{W}_i\left(z, \vec{\mathcal{X}}_i\right) \leq & -(r-i)m_1^2(z)\|z\|^2 - (r-i)m_2^2(\mathcal{X}_1)\mathcal{X}_1^2 \\ & - \|z\|^2 - \sum_{j=1}^{i} \mathcal{X}_j^2 + \mathcal{X}_{i+1}^2. \end{aligned} \tag{4.25}$$

For $i = 1$, direct calculation shows that the derivative of $W_1(z, \mathcal{X}_1)$ along the state trajectory satisfies:

$$\begin{aligned} \dot{W}_1(z, \mathcal{X}_1) \leq & -\Delta(z)\|z\|^2 + \varkappa(\mathcal{X}_1)\mathcal{X}_1^2 + \mathcal{X}_1(f_1(z, \mathcal{X}_1, d) + b(\mathcal{X}_2 - \mathcal{X}_1\rho_1(\mathcal{X}_1))) \\ \leq & -\Delta(z)\|z\|^2 + m_1^2(z)\|z\|^2 \\ & + \mathcal{X}_1^2\left[\varkappa(\mathcal{X}_1) + 1/4 + m_2(\mathcal{X}_1) + b^2/4 - b\rho_1(\mathcal{X}_1)\right] + \mathcal{X}_2^2. \end{aligned}$$

Using $\Delta(z)$ and $\rho_1(\mathcal{X}_1)$ defined in (4.18) and (4.22) gives

$$\dot{W}_1(z, \mathcal{X}_1) \leq -(r-1)m_1^2(z)\|z\|^2 - (r-1)m_2^2(\mathcal{X}_1)\mathcal{X}_1^2 - \|z\|^2 - \mathcal{X}_1^2 + \mathcal{X}_2^2$$

which is (4.25) with $i = 1$.

Next, we will show that the inequality (4.25) is true for $2 \leq i \leq r$ if it is true for $(i-1)$ by using the relationship

$$\dot{W}_i\left(z, \vec{\mathcal{X}}_i\right) = \dot{W}_{i-1}\left(z, \vec{\mathcal{X}}_{i-1}\right) + \mathcal{X}_i\dot{\mathcal{X}}_i.$$

In fact, we note

$$\begin{aligned} \mathcal{X}_i\dot{\mathcal{X}}_i = & \mathcal{X}_i\left[f_i\left(\vec{x}_i\right) + \mathcal{X}_{i+1} + \mathcal{S}_i\left(\vec{x}_i\right) - \dot{\mathcal{S}}_{i-1}\left(\vec{x}_{i-1}\right)\right] \\ \leq & \mathcal{X}_i\left[f_i\left(\vec{x}_i\right) + \mathcal{X}_{i+1} + \mathcal{S}_i\left(\vec{x}_i\right)\right] + \left[\mathcal{X}_i \frac{\partial \mathcal{S}_{i-1}\left(\vec{x}_{i-1}\right)}{\partial x_1}\right]^2 \Big/ 2 \\ & + m_1^2(z)\|z\|^2 + m_2^2(x_1)x_1^2 - \mathcal{X}_i \frac{\partial \mathcal{S}_{i-1}\left(\vec{x}_{i-1}\right)}{\partial x_1} b x_2 \end{aligned}$$

$$
\begin{aligned}
& - \mathcal{X}_i \sum_{j=2}^{i-1} \frac{\partial \mathcal{S}_{i-1}\left(\vec{x}_{i-1}\right)}{\partial x_j} \left[f_j\left(\vec{x}_j\right) + x_{j+1} \right] \\
\leq & \mathcal{X}_i \left[f_i\left(\vec{x}_i\right) + \mathcal{S}_i\left(\vec{x}_i\right) \right] + \mathcal{X}_i^2/4 + \mathcal{X}_i^2 \left[\frac{\partial \mathcal{S}_{i-1}\left(\vec{x}_{i-1}\right)}{\partial x_1} \right]^2 \Big/ 2 \\
& - \mathcal{X}_i \frac{\partial \mathcal{S}_{i-1}\left(\vec{x}_{i-1}\right)}{\partial x_1} b x_2 - \mathcal{X}_i \sum_{j=2}^{i-1} \frac{\partial \mathcal{S}_{i-1}\left(\vec{x}_{i-1}\right)}{\partial x_j} \left[f_j\left(\vec{x}_j\right) + x_{j+1} \right] \\
& + m_1^2(z)\|z\|^2 + m_2^2(\mathcal{X}_1)\mathcal{X}_1^2 + \mathcal{X}_{i+1}^2 \\
\leq & - 2\mathcal{X}_i^2 + m_1^2(z)\|z\|^2 + m_2^2(\mathcal{X}_1)\mathcal{X}_1^2 + \mathcal{X}_{i+1}^2
\end{aligned}
$$

where the function $\mathcal{S}_i(\vec{x}_i)$ is defined in (4.21) and $\dot{\mathcal{S}}_{i-1}(\vec{x}_{i-1})$ is calculated as follows

$$
\begin{aligned}
\dot{\mathcal{S}}_{i-1}\left(\vec{x}_{i-1}\right) = & \frac{\partial \mathcal{S}_{i-1}\left(\vec{x}_{i-1}\right)}{\partial x_1} [f_1(z, x_1, d(t)) + b x_2] \\
& + \sum_{j=2}^{i-1} \frac{\partial \mathcal{S}_{i-1}\left(\vec{x}_{i-1}\right)}{\partial x_j} \left[f_j\left(\vec{x}_j\right) + x_{j+1} \right].
\end{aligned}
$$

As a result,

$$
\begin{aligned}
\dot{W}_i\left(z, \vec{\mathcal{X}}_i\right) \leq & - (r - i + 1) m_1^2(z)\|z\|^2 - (r - i + 1) m_2^2(\mathcal{X}_1)\mathcal{X}_1^2 - \|z\|^2 \\
& - \sum_{j=1}^{i-1} \mathcal{X}_j^2 + \mathcal{X}_i^2 - 2\mathcal{X}_i^2 + m_1^2(z)\|z\|^2 + m_2^2(\mathcal{X}_1)\mathcal{X}_1^2 + \mathcal{X}_{i+1}^2 \\
\leq & - (r - i) m_1^2(z)\|z\|^2 - (r - i) m_2^2(\mathcal{X}_1)\mathcal{X}_1^2 - \|z\|^2 - \sum_{j=1}^{i} \mathcal{X}_j^2 + \mathcal{X}_{i+1}^2
\end{aligned}
$$

which is (4.25). Letting $i = r$ in the inequality (4.25) gives (4.23).

Finally, letting $\mathcal{X}_{r+1} = 0$ in (4.23) gives the controller (4.24) and

$$
\dot{W}_r(z, \vec{\mathcal{X}}_r) \leq -\|z\|^2 - \|\vec{\mathcal{X}}_r\|^2. \tag{4.26}
$$

By Theorem 2.3, the equilibrium point $\mathrm{col}(z, \vec{\mathcal{X}}_r) = 0$ is RUGAS. By Proposition 4.1, the GRSP of (4.16) is solved by (4.24). □

The following algorithm summarizes the steps of constructing the controller.

Algorithm 4.1

INPUT: $f_i,\ i = 1, \ldots, r, b, \underline{\alpha}, \bar{\alpha}, \alpha, \sigma, \mathbb{D}$
OUTPUT: $\mathcal{S}_i, i = 1, \ldots, r$
STEP 1: For the given $f_1(z, x_1, d)$, find m_1 and m_2 from (4.17).
STEP 2: Pick the function Δ from (4.18) and call

$$(\underline{\alpha}', \bar{\alpha}', \varkappa) = \text{ALGORITHM } 2.3(\underline{\alpha}, \bar{\alpha}, \alpha, \sigma, \Delta).$$

STEP 3: Let $i = 1$; find $\mathcal{S}_i$ from (4.21) and (4.22).
STEP 4: IF $i = r$, GO TO STEP 6, ELSE GO TO STEP 5.
STEP 5: Let $i = i + 1$; find $\mathcal{S}_i$ from (4.21) and (4.22); GO TO STEP 4
STEP 6: END

Example 4.2 Consider the following third order system

$$\begin{aligned}\dot{z} &= -z + x_1\\ \dot{x}_1 &= w_1 z \sin x_1 + w_2 x_1^3 + x_2\\ \dot{x}_2 &= c(x_1, x_2) + u\end{aligned} \tag{4.27}$$

where w_1 and w_2 are unknown parameters with $|w_1| \leq 2$ and $|w_2| \leq 1$ and c is some smooth function vanishing at the origin. We will construct a controller u for (4.27).

First, we note that Assumption 4.1 is satisfied with $V(z) = z^2$. In fact, along the trajectory of $\dot{z} = -z + x_1$,

$$\dot{V}(z) \leq -z^2 + x_1^2.$$

Observe that

$$\left| w_1 z \sin x_1 + w_2 x_1^3 \right| \leq 2|z| + x_1^2 |x_1|,$$

so (4.17) is true for $m_1(z) = 2$ and $m_2(x_1) = x_1^2$. Let $\Delta = 1 + r m_1^2 = 9$. Since Δ is constant, by Corollary 2.1 and Remark 2.11, letting $V'(z) = \Delta z^2$ shows (4.19) is satisfied with $\varkappa(x) = 9$. Then, we have $W_1(z, x_1) = 9z^2 + x_1^2/2$. Let $\rho_1(x_1) = 10.5 + x_1^2 + x_1^4$ and

$$\begin{aligned}\mathcal{X}_1 &= x_1\\ \mathcal{X}_2 &= x_2 + \left(10.5 + x_1^2 + x_1^4\right) x_1.\end{aligned}$$

Along the state trajectory of the $\text{col}(z, \mathcal{X}_1)$-subsystem, the derivative of $W_1(z, \mathcal{X}_1) = 9z^2 + \mathcal{X}_1^2/2$ satisfies

$$\begin{aligned}\dot{W}_1(z,\mathcal{X}_1) \le & -9z^2+9\mathcal{X}_1^2+\mathcal{X}_1\left(w_1 z\sin\mathcal{X}_1+w_2\mathcal{X}_1^3\right)\\ & -\left(10.5\mathcal{X}_1^2+\mathcal{X}_1^4+\mathcal{X}_1^6\right)+\mathcal{X}_1^2/4+\mathcal{X}_2^2\\ \le & -5z^2-\mathcal{X}_1^2-\mathcal{X}_1^6+\mathcal{X}_2^2.\end{aligned}$$

Then, the $\mathcal{X}_2$-dynamics become

$$\dot{\mathcal{X}}_2=c(x_1,x_2)+u+\left(10.5+3\mathcal{X}_1^2+5\mathcal{X}_1^4\right)\left(w_1 z\sin x_1+w_2x_1^3+x_2\right).$$

So, the derivative of $W_2(z,\mathcal{X}_1,\mathcal{X}_2)=9z^2+\mathcal{X}_1^2/2+\mathcal{X}_2^2/2$ along the state trajectory of the system (4.27) satisfies

$$\begin{aligned}\dot{W}_2(z,\mathcal{X}_1,\mathcal{X}_2) \le & -5z^2-\mathcal{X}_1^2-\mathcal{X}_1^6+\mathcal{X}_2^2+\mathcal{X}_2\dot{\mathcal{X}}_2\\ \le & -5z^2-\mathcal{X}_1^2-\mathcal{X}_1^6+\mathcal{X}_2^2+\mathcal{X}_2c(x_1,x_2)+\mathcal{X}_2u\\ & +\mathcal{X}_2^2\left(10.5+3\mathcal{X}_1^2+5\mathcal{X}_1^4\right)^2/2+4z^2+\mathcal{X}_1^6\\ & +\mathcal{X}_2\left(10.5+3\mathcal{X}_1^2+5\mathcal{X}_1^4\right)x_2\\ \le & -z^2-\mathcal{X}_1^2-\mathcal{X}_2^2\end{aligned}$$

with the controller

$$\begin{aligned}u= & -2x_2-2\left(10.5+x_1^2+x_1^4\right)x_1-c(x_1,x_2)\\ & -\left[x_2+\left(10.5+x_1^2+x_1^4\right)x_1\right]\left(10.5+3x_1^2+5x_1^4\right)^2\Big/2\\ & -\left(10.5+3x_1^2+5x_1^4\right)x_2.\end{aligned}$$

In other words, this controller solves the GRSP of (4.27). Figure 4.2 shows the response of the closed-loop system with the initial condition $z(0)=9$, $x_1(0)=-5$, and $x_2(0)=20$ and the parameters $w_1=1.8$ and $w_2=1$. It is observed that the state of the closed-loop system converges to the equilibrium point asymptotically.

4.3 Lower Triangular Systems

In this section, we will consider the system in the lower triangular form (4.1) which is more general than the system in the output feedback form (4.13). Under the coordinate transformation (4.2), the system (4.1) can be transformed into (4.9), i.e., $\dot{\zeta}_i=\varphi_i(\zeta_i,\mathcal{X}_{i+1},d),\ i=1,\ldots,r$. Our objective is to make $\dot{\zeta}_r=\varphi_r(\zeta_r,\mathcal{X}_{r+1},d)$ input-to-state stable with ζ_r as the state and $\mathcal{X}_{r+1}$ as the input. As a result, the equi-

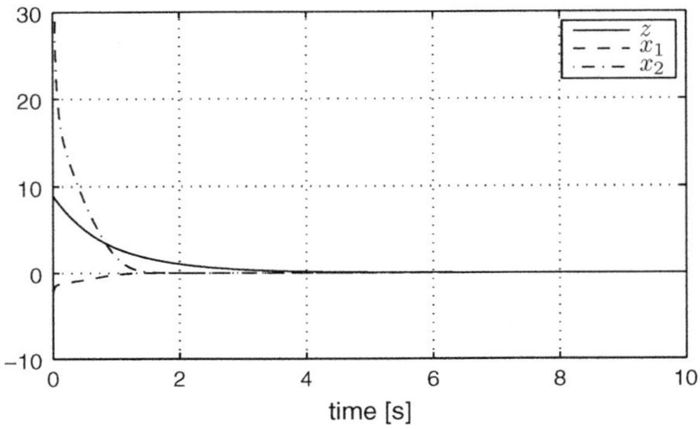

Fig. 4.2 Profile of state trajectories of the closed-loop system in Example 4.2

librium point $\zeta_r = 0$ of $\dot{\zeta}_r = \varphi_r(\zeta_r, 0, d)$ is RUGAS. Like in the previous section, we will adopt a recursive approach to designing the control law. For this purpose, decompose the system $\dot{\zeta}_i = \varphi_i(\zeta_i, \mathcal{X}_{i+1}, d)$ into the following form:

$$\begin{aligned} \dot{\zeta}_{i-1} &= \varphi_{i-1}(\zeta_{i-1}, \mathcal{X}_i, d) \\ \dot{z}_i &= \mathcal{Q}_i\left(\overrightarrow{z}_i, \overrightarrow{\mathcal{X}}_i, d\right) \\ \dot{\mathcal{X}}_i &= \mathcal{F}_i\left(\overrightarrow{z}_i, \overrightarrow{\mathcal{X}}_i, d\right) - b_i(d)s_i\left(\overrightarrow{\mathcal{X}}_i\right) + b_i(d)\mathcal{X}_{i+1}. \end{aligned} \tag{4.28}$$

Based on the structure of (4.28), we will find functions s_i, $i = 1, \ldots, r$, such that certain ISS property of the system $\dot{\zeta}_i = \varphi_i(\zeta_i, \mathcal{X}_{i+1}, d)$ can be inferred based on Theorem 2.10 (small gain theorem in Lyapunov function form). To this end, we need some assumptions on the control gain b_i and the dynamic uncertainty $\dot{z}_i = q_i(\overrightarrow{z}_i, \overrightarrow{x}_i, d)$ as follows.

Assumption 4.2 For $i = 1, \ldots, r$, the function $b_i(d)$ is away from zero, e.g., $b_i(d) > 0$, $\forall d \in \mathbb{D}$.

Assumption 4.3 For $i = 1, \ldots, r$, the subsystem $\dot{z}_i = q_i(\overrightarrow{z}_i, \overrightarrow{x}_i, d)$ has an ISS Lyapunov function $V_i(z_i)$, i.e.,

$$V_i(z_i) \sim \left\{\underline{\alpha}_i, \bar{\alpha}_i, \alpha_i, \sigma_i \mid \dot{z}_i = q_i\left(\overrightarrow{z}_i, \overrightarrow{x}_i, d\right)\right\}$$

and

$$\lim_{s\to 0^+} \sup \frac{s^2}{\alpha_i(s)} < \infty, \quad \lim_{s\to 0^+} \sup \frac{\sigma_i(s)}{s^2} < \infty.$$

Remark 4.2 Under Assumption 4.2, one has $\bar{b}_i \geq b_i(d) \geq \underline{b}_i$ for some positive numbers $\bar{b}_i$ and $\underline{b}_i$. Assumption 4.3 plays the same role as Assumption 4.1 by requiring the subsystem $\dot{z}_i = q_i(\overrightarrow{z}_i, \overrightarrow{x}_i, d)$ be RISS viewing z_i as the state and $\text{col}(\overrightarrow{z}_{i-1}, \overrightarrow{x}_i)$ as the input.

Also, we can see that if the system $\dot{z}_i = q_i(\overrightarrow{z}_i, \overrightarrow{x}_i, d)$ satisfies Assumption 4.3, then the system $\dot{z}_i = \mathcal{Q}_i(\overrightarrow{z}_i, \overrightarrow{\mathcal{X}}_i, d)$ also satisfies Assumption 4.3 as summarized in the following lemma.

Lemma 4.1 *Under Assumption 4.3, if the function s_i in the coordinate transformation (4.2) takes the form $s_i(\overrightarrow{\mathcal{X}}_i) = -\rho_i(\mathcal{X}_i)\mathcal{X}_i$ for some sufficiently smooth function ρ_i, then the subsystem $\dot{z}_i = \mathcal{Q}_i(\overrightarrow{z}_i, \overrightarrow{\mathcal{X}}_i, d)$ has an ISS Lyapunov functions V_i, i.e.,*

$$V_i(z_i) \sim \left\{\underline{\alpha}_i, \bar{\alpha}_i, \alpha_i, \tilde{\sigma}_i \mid \dot{z}_i = \mathcal{Q}_i\left(\overrightarrow{z}_i, \overrightarrow{\mathcal{X}}_i, d\right)\right\}$$

and $\tilde{\sigma}_i$ satisfies

$$\lim_{s\to 0^+} \sup \frac{\tilde{\sigma}_i(s)}{s^2} < \infty.$$

Proof Let $\theta = \text{col}(\overrightarrow{z}_{i-1}, \overrightarrow{x}_i)$ and $\vartheta = \text{col}(\overrightarrow{z}_{i-1}, \overrightarrow{\mathcal{X}}_i)$. It suffices to prove the existence of a class $\mathcal{K}$ function $\tilde{\sigma}_i$ satisfying

$$\tilde{\sigma}_i(\|\vartheta\|) \geq \sigma_i(\|\theta\|).$$

In fact, we have

$$\theta = G(\vartheta)\vartheta$$

where

$$G(\vartheta) = \begin{bmatrix} 1 & & & \\ & \ddots & & \\ & -\rho_1(\mathcal{X}_1) & 1 & \\ & & \ddots & \ddots \\ & & & -\rho_{i-1}(\mathcal{X}_{i-1}) \; 1 \end{bmatrix}.$$

Clearly, there exists a non-decreasing continuous function g such that

$$g(\|\vartheta\|) \geq \|G(\vartheta)\|. \tag{4.29}$$

Let $\tilde{\sigma}_i(s) = \sigma_i(g(s)s)$ be a class $\mathcal{K}$ function. As a result,

$$\tilde{\sigma}_i(\|\vartheta\|) = \sigma_i(g(\|\vartheta\|)\|\vartheta\|) \geq \sigma_i(\|G(\vartheta)\vartheta\|) = \sigma_i(\|\theta\|)$$

and

$$\limsup_{s\to 0^+} \frac{\tilde{\sigma}_i(s)}{s^2} = \limsup_{s\to 0^+} \frac{\sigma_i(g(s)s)}{[g(s)s]^2}\frac{[g(s)s]^2}{s^2} = g^2(0) \limsup_{s\to 0^+} \frac{\sigma_i(g(s)s)}{[g(s)s]^2} < \infty. \qquad \square$$

Now, we are ready to solve the GRSP for the system (4.1) using a sufficiently smooth partial state feedback controller. We will use a recursive approach to synthesize the controller. The recursive approach is illustrated in the constructive proof of Theorem 4.2 given below. From Theorem 4.1, we recall that the Lyapunov function for the closed-loop system is recursively constructed as

$$W_i\left(z, \vec{\mathcal{X}}_i\right) = W_{i-1}(z, \vec{\mathcal{X}}_{i-1}) + \mathcal{X}_i^2/2.$$

But, the construction of the Lyapunov function in Theorem 4.2 is more complicated and takes the following form:

$$W_i\left(\vec{z}_i, \vec{\mathcal{X}}_i\right) = W'_{i-1}\left(\vec{z}_{i-1}, \vec{\mathcal{X}}_{i-1}\right) + V'_i(z_i) + \mathcal{X}_i^2/2$$

where $W'_{i-1}(\vec{z}_{i-1}, \vec{\mathcal{X}}_{i-1})$ and $V'_i(z_i)$ are obtained from $W_{i-1}(\vec{z}_{i-1}, \vec{\mathcal{X}}_{i-1})$ and $V_i(z_i)$, respectively, using the changing supply function approach. The result is summarized as follows.

Theorem 4.2 *Consider the system* (4.1) *with* $\mathbb{D}$ *a prescribed compact set. Under Assumptions* 4.2 *and* 4.3, *there exist sufficiently smooth functions* ρ_i, $i = 1, \ldots, r$, *such that the GRSP for the system* (4.1) *is solved by the following static controller*

$$\begin{aligned} u &= \mathcal{S}_r\left(\vec{x}_r\right) = s_r\left(\vec{\mathcal{X}}_r\right) \\ \mathcal{X}_i &= x_i - s_{i-1}\left(\vec{\mathcal{X}}_{i-1}\right), \quad i = r, \ldots, 2 \\ \mathcal{X}_1 &= x_1, \end{aligned}$$

with

$$s_i\left(\vec{\mathcal{X}}_i\right) = -\rho_i\left(\mathcal{X}_i\right)\mathcal{X}_i.$$

In particular, the functions ρ_i, $i = 1, \ldots, r$, *are given in Algorithm* 4.2.

Proof We will first show that, for any $1 \leq i \leq r$, there exist sufficiently smooth and non-negative functions $\rho_1, \ldots, \rho_i$ such that the system $\dot{\zeta}_i = \varphi_i(\zeta_i, \mathcal{X}_{i+1}, d)$ has an ISS Lyapunov function $W_i(\vec{z}_i, \vec{\mathcal{X}}_i)$, i.e.,

$$W_i\left(\vec{z}_i, \vec{x}_i\right) \sim \left\{\underline{\beta}_i, \bar{\beta}_i, \beta_i, \varsigma_i, \mid \dot{\zeta}_i = \varphi_i(\zeta_i, x_{i+1}, d)\right\},$$

and

$$\lim_{s\to 0^+} \sup \frac{s^2}{\beta_i(s)} < \infty, \quad \lim_{s\to 0^+} \sup . \frac{\varsigma_i(s)}{s^2} < \infty.$$

When $i = 1$, the statement is proved in Theorem 2.8 by identifying (z, x, u) of (2.46) with (z_1, x_1, x_2) of (4.1), and (W, ρ) of Theorem 2.8 with (W_1, ρ_1) here. In particular,

$$|f_1(z_1, x_1, d)| \le m_1(z_1)\|z_1\| + m_2(x_1)|x_1|, \quad \forall d \in \mathbb{D} \tag{4.30}$$

for some sufficiently smooth and non-negative functions m_1 and m_2. For

$$\Delta(z_1) \ge 1 + m_1^2(z_1), \tag{4.31}$$

there exists a continuously differentiable function $V_1'(z_1)$ satisfying

$$\dot{V}_1'(z_1) \le -\Delta(z_1)\|z_1\|^2 + \varkappa(x_1)x_1^2 \tag{4.32}$$

for a smooth function $\varkappa$. So, the function ρ_1 is as follows:

$$\rho_1(x_1) = [\varrho(x_1) + 5/4]/\underline{b}_1 + \bar{b}_1/4 \tag{4.33}$$

for a sufficiently smooth function

$$\varrho_1(x_1) \ge \varkappa(x_1) + m_2(x_1). \tag{4.34}$$

For any $2 \le i \le r$, we will show that if the statement is true for $(i - 1)$, then it is also true for i. We note that the system $\dot{\zeta}_i = \varphi_i(\zeta_i, x_{i+1}, d)$ consists of three subsystems as indicated in (4.28), and we will find the ISS Lyapunov function for each of them.

First, under the induction assumption, by Corollary 2.2, for any smooth function Δ_0, the subsystem governing ζ_{i-1} has an ISS Lyapunov function $W_{i-1}'(\vec{z}_{i-1}, \vec{x}_{i-1})$ satisfying

$$\underline{\beta}_{i-1}'(\|\zeta_{i-1}\|) \le W_{i-1}'\left(\vec{z}_{i-1}, \vec{x}_{i-1}\right) \le \bar{\beta}_{i-1}'(\|\zeta_{i-1}\|)$$
$$\dot{W}_{i-1}'\left(\vec{z}_{i-1}, \vec{x}_{i-1}\right) \le -\Delta_0(\zeta_{i-1})\|\zeta_{i-1}\|^2 + \varkappa_0(x_i)x_i^2$$

for a positive and sufficiently smooth function $\varkappa_0$.

Secondly, under Assumption 4.3, by Lemma 4.1 and Corollary 2.2, for any smooth function Δ, there exists an ISS Lyapunov function $V_i'(z_i)$ satisfying

$$\underline{\alpha}_i'(\|z_i\|) \leq V_i'(z_i) \leq \bar{\alpha}_i'(\|z_i\|)$$
$$\dot{V}_i'(z_i) \leq -\Delta(z_i)\|z_i\|^2 + \varkappa(\zeta_{i-1}, \mathcal{X}_i)\|\mathrm{col}(\zeta_{i-1}, \mathcal{X}_i)\|^2$$

for a positive and sufficiently smooth function $\varkappa$. Moreover, by (11.13) of the Appendix, one has

$$\varkappa(\zeta_{i-1}, \mathcal{X}_i)\|\mathrm{col}(\zeta_{i-1}, \mathcal{X}_i)\|^2 \leq \varkappa_1(\zeta_{i-1})\|\zeta_{i-1}\|^2 + \varkappa_2(\mathcal{X}_i)\mathcal{X}_i^2 \tag{4.35}$$

for positive and sufficiently smooth functions $\varkappa_1$ and $\varkappa_2$.

Thirdly, by (11.13) of the Appendix again, there exist some smooth positive functions m_0, m_1, and m_2 such that

$$\left|\mathcal{F}_i\left(\vec{z}_i, \vec{\mathcal{X}}_i, d\right)\right| \leq m_0(\zeta_{i-1})\|\zeta_{i-1}\| + m_1(z_i)\|z_i\| + m_2(\mathcal{X}_i)|\mathcal{X}_i|. \tag{4.36}$$

Then, the derivative of $\mathcal{X}_i^2/2$ is

$$\begin{aligned}
\mathcal{X}_i\dot{\mathcal{X}}_i \leq& \mathcal{X}_i\left[\mathcal{F}_i\left(\vec{z}_i, \vec{\mathcal{X}}_i, d\right) - b_i\mathcal{X}_i\rho_i(\mathcal{X}_i) + b_i\mathcal{X}_{i+1}\right]\\
\leq& m_0^2(\zeta_{i-1})\|\zeta_{i-1}\|^2 + m_1^2(z_i)\|z_i\|^2\\
&+ \mathcal{X}_i^2\left[1/2 + m_2(\mathcal{X}_i) + b_i^2/4 - b_i\rho_i(\mathcal{X}_i)\right] + \mathcal{X}_{i+1}^2.
\end{aligned}$$

Next, we define a function

$$W_i\left(\vec{z}_i, \vec{\mathcal{X}}_i\right) = W_{i-1}'\left(\vec{z}_{i-1}, \vec{\mathcal{X}}_{i-1}\right) + V_i'(z_i) + \mathcal{X}_i^2/2.$$

Let

$$\Delta_0(\zeta_{i-1}) \geq 1 + \varkappa_1(\zeta_{i-1}) + m_0^2(\zeta_{i-1}) \tag{4.37}$$
$$\Delta(z_i) \geq 1 + m_1^2(z_i). \tag{4.38}$$

Also, let $\varrho_i(\mathcal{X}_i)$ and $\rho_i(\mathcal{X}_i)$ be sufficiently smooth functions such that

$$\varrho_i(\mathcal{X}_i) \geq \varkappa_0(\mathcal{X}_i) + \varkappa_2(\mathcal{X}_i) + m_2(\mathcal{X}_i) \tag{4.39}$$

and

$$\rho_i(\mathcal{X}_i) = [\varrho_i(\mathcal{X}_i) + 3/2]/\underline{b}_i + \bar{b}_i/4. \tag{4.40}$$

Then, the derivative of $W_i(\vec{z}_i, \vec{\mathcal{X}}_i)$ along the state trajectory of the $\mathrm{col}(\vec{z}_i, \vec{\mathcal{X}}_i)$-subsystem satisfies

$$\begin{aligned}\dot{W}_i(\vec{z}_i, \vec{\mathcal{X}}_i) \leq & -\Delta_0(\zeta_{i-1})\|\zeta_{i-1}\|^2 + \varkappa_0(\mathcal{X}_i)\mathcal{X}_i^2 \\ & -\Delta(z_i)\|z_i\|^2 + \varkappa_1(\zeta_{i-1})\|\zeta_{i-1}\|^2 + \varkappa_2(\mathcal{X}_i)\mathcal{X}_i^2 \\ & + m_0^2(\zeta_{i-1})\|\zeta_{i-1}\|^2 + m_1^2(z_i)\|z_i\|^2 \\ & + \mathcal{X}_i^2\left[1/2 + m_2(\mathcal{X}_i) + b_i^2/4 - b_i\rho_i(\mathcal{X}_i)\right] + \mathcal{X}_{i+1}^2 \\ \leq & -\|\zeta_{i-1}\|^2 - \|z_i\|^2 - \mathcal{X}_i^2 + \mathcal{X}_{i+1}^2 \\ = & -\|\zeta_i\|^2 + \mathcal{X}_{i+1}^2.\end{aligned}$$

Moreover, using Lemma 11.3 of the Appendix, one has

$$\begin{aligned}\underline{\beta}_i(\|\zeta_i\|) \leq \underline{\beta}'_{i-1}(\|\zeta_{i-1}\|) + \underline{\alpha}'_i(\|z_i\|) + \mathcal{X}_i^2 \leq W_i\left(\vec{z}_i, \vec{\mathcal{X}}_i\right) \\ \leq \bar{\beta}'_{i-1}(\|\zeta_{i-1}\|) + \bar{\alpha}'_i(\|z_i\|) + \mathcal{X}_i^2 \leq \bar{\beta}_i(\|\zeta_i\|)\end{aligned} \tag{4.41}$$

for some class $\mathcal{K}_\infty$ functions $\underline{\beta}_i$ and $\bar{\beta}_i$. That is, the statement is true for i with

$$\beta_i(s) = \varsigma_i(s) = s^2. \tag{4.42}$$

Finally, the ISS Lyapunov function $W_r(\vec{z}_r, \vec{\mathcal{X}}_r)$ implies that the equilibrium point $\zeta_r = 0$ of the system $\dot{\zeta}_r = \varphi_r(\zeta_r, 0, d)$ (with $\mathcal{X}_{r+1} = 0$) is RUGAS. The GRSP for the system (4.1) is thus solved by using Theorem 2.3 and Proposition 4.1. □

Algorithm 4.2

INPUT: $f_i, b_i, \underline{\alpha}_i, \bar{\alpha}_i, \alpha_i, \sigma_i, i = 1, \ldots, r, \mathbb{D}$
OUTPUT: $\rho_i, i = 1, \ldots, r$
STEP 1: Let $i = 1$; find the functions m_1 and m_2 from (4.30).
STEP 2: Pick the function Δ from (4.31) and call

$$(\underline{\alpha}'_1, \bar{\alpha}'_1, \varkappa) = \text{ALGORITHM } 2.3(\underline{\alpha}_1, \bar{\alpha}_1, \alpha_1, \sigma_1, \Delta).$$

STEP 3: Find the function ρ_1 from (4.33).
STEP 4: IF $i = r$, GO TO STEP 10, ELSE let $i = i + 1$.
STEP 5: Calculate $\mathcal{F}_i(\vec{z}_i, \vec{\mathcal{X}}_i, d)$ from (4.8).
STEP 6: Find the functions m_0, m_1, and m_2 from (4.36).
STEP 7: Pick the function Δ from (4.38); define a function g as in (4.29) and let $\tilde{\sigma}_i(s) = \sigma_i(g(s)s)$; call

$$(\underline{\alpha}'_i, \bar{\alpha}'_i, \varkappa) = \text{ALGORITHM } 2.3(\underline{\alpha}_i, \bar{\alpha}_i, \alpha_i, \tilde{\sigma}_i, \Delta);$$

and pick the functions $\varkappa_1$ and $\varkappa_2$ from (4.35).
STEP 8: Pick the function Δ_0 from (4.37) and call

$$(\underline{\beta}'_{i-1}, \bar{\beta}'_{i-1}, \varkappa_0) = \text{ALGORITHM } 2.3(\underline{\beta}_{i-1}, \bar{\beta}_{i-1}, \beta_{i-1}, \varsigma_{i-1}, \Delta_0).$$

STEP 9: Find the function ρ_i from (4.40); GO TO STEP 4.
STEP 10: END

Example 4.3 Consider the following second order system

$$\begin{aligned}\dot{x}_1 &= w_1 x_1 + x_2\\ \dot{x}_2 &= w_2 x_1^2 + u.\end{aligned} \tag{4.43}$$

This system is a special case of the system (4.1) with $r = 2$, $z_i \in \mathbb{R}^0$, $i = 1, 2$. We will find a controller u to solve the GRSP for $-1 \le w_1, w_2 \le 1$.

We note $|w_1 x_1| \le |x_1|$ and define a function $\rho_1 = 2.5$ from (4.33). With $\mathcal{X}_1 = x_1$ and $\mathcal{X}_2 = x_2 + 2.5x_1$, the derivative of $W_1(\mathcal{X}_1) = \mathcal{X}_1^2/2$ along the state trajectory of the $\mathcal{X}_1$-subsystem satisfies

$$\dot{W}_1(\mathcal{X}_1) \le \mathcal{X}_1(w_1\mathcal{X}_1 - 2.5\mathcal{X}_1 + \mathcal{X}_2) \le -\mathcal{X}_1^2 + \mathcal{X}_2^2.$$

The $\mathcal{X}_2$-dynamics are

$$\dot{\mathcal{X}}_2 = w_2\mathcal{X}_1^2 + 2.5(w_1\mathcal{X}_1 - 2.5\mathcal{X}_1 + \mathcal{X}_2) + u.$$

Observe that

$$\left| w_2\mathcal{X}_1^2 + 2.5(w_1\mathcal{X}_1 - 2.5\mathcal{X}_1 + \mathcal{X}_2)\right| \le (|\mathcal{X}_1| + 8.75)|\mathcal{X}_1| + 2.5|\mathcal{X}_2|.$$

Then, with $u = -\mathcal{X}_2\rho_2(\mathcal{X}_2) + \mathcal{X}_3$, the derivative of $\mathcal{X}_2^2/2$ along the state trajectory of the $\mathcal{X}_2$-subsystem satisfies

$$\begin{aligned}\mathcal{X}_2\dot{\mathcal{X}}_2 &= \mathcal{X}_2[w_2\mathcal{X}_1^2 + 2.5(w_1\mathcal{X}_1 - 2.5\mathcal{X}_1 + \mathcal{X}_2) - \mathcal{X}_2\rho_2(\mathcal{X}_2) + \mathcal{X}_3]\\ &\le (|\mathcal{X}_1| + 8.75)^2\mathcal{X}_1^2 + \mathcal{X}_2^2[1/2 + 2.5 - \rho_2(\mathcal{X}_2)] + \mathcal{X}_3^2.\end{aligned}$$

Let $\Delta_0(\mathcal{X}_1) = 2\mathcal{X}_1^2 + 155 > (|\mathcal{X}_1| + 8.75)^2 + 1$. The derivative of

$$W_1'(\mathcal{X}_1) = \int_0^{W_1(\mathcal{X}_1)} (8s + 310)ds$$

along the state trajectory of the $\mathcal{X}_1$-subsystem satisfies

$$\begin{aligned}\dot{W}_1'(\mathcal{X}_1) &\le \left(4\mathcal{X}_1^2 + 310\right)\left(-\mathcal{X}_1^2 + \mathcal{X}_2^2\right)\\ &\le -\left(2\mathcal{X}_1^2 + 155\right)\mathcal{X}_1^2 + \left(8\mathcal{X}_2^2 + 310\right)\mathcal{X}_2^2.\end{aligned}$$

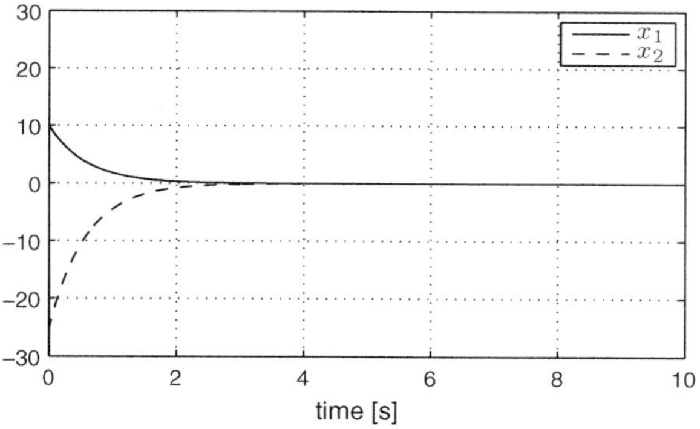

Fig. 4.3 Profile of state trajectories of the closed-loop system in Example 4.3

Let $W_2'(\mathcal{X}_1, \mathcal{X}_2) = W_1'(\mathcal{X}_1) + \mathcal{X}_2^2/2$ whose derivative along the system trajectory satisfies

$$\begin{aligned}\dot{W}_2'(\mathcal{X}_1, \mathcal{X}_2) \le & -\left(2\mathcal{X}_1^2 + 155\right)\mathcal{X}_1^2 + \left(8\mathcal{X}_2^2 + 310\right)\mathcal{X}_2^2 \\ & + (|\mathcal{X}_1| + 8.75)^2\mathcal{X}_1^2 + \mathcal{X}_2^2[1/2 + 2.5 - \rho_2(\mathcal{X}_2)] + \mathcal{X}_3^2 \\ \le & -\mathcal{X}_1^2 + \mathcal{X}_2^2\left[8\mathcal{X}_2^2 + 310 + 1/2 + 2.5 - \rho_2(\mathcal{X}_2)\right] + \mathcal{X}_3^2 \\ \le & -\mathcal{X}_1^2 - \mathcal{X}_2^2 + \mathcal{X}_3^2\end{aligned}$$

for $\rho_2(\mathcal{X}_2) = 8\mathcal{X}_2^2 + 315$.

Finally, letting $\mathcal{X}_3 = 0$ gives $u = -8(x_2 + 2.5x_1)^3 - 315(x_2 + 2.5x_1)$ which solves the GRSP for (4.43). The performance of the controller is shown in Fig. 4.3 with $w_1 = 0.8$ and $w_2 = -1$ and the initial condition $x_1(0) = 10$ and $x_2(0) = -20$. It is observed that the state of the closed-loop system converges to the equilibrium point asymptotically.

More complex examples will be given in Chap. 8 when we study the output regulation problem.

4.4 Reduction of Control Gain

Typically, the robust control method uses a sufficiently large function to dominate an unknown function in a compact set as can be observed, e.g., from (2.52). The function $\rho(x)$ determines the control power. In practice, an excessively large control power may be harmful and costly, and it is interesting to reduce the control gain if possible. From (2.52), we can see that two functions, i.e., $\varkappa(x)$ and $m_2(x)$, contribute to the gain function $\rho(x)$. In many cases, it is possible to use some technique to reduce the control power. We first use one simple system to illustrate the idea. Consider the following system with relative degree one:

$$\begin{aligned}\dot{z} &= Mz + \phi(x_1, d) \\ \dot{x}_1 &= Hz + \psi(x_1, d) + bu, \ \ b > 0\end{aligned} \tag{4.44}$$

where the matrix M is Hurwitz.

Decompose the two unknown functions ϕ and ψ to the following

$$\begin{aligned}\phi(x_1, d) &= \phi^o(x_1) + \tilde{\phi}(x_1, d) \\ \psi(x_1, d) &= \psi^o(x_1) + \tilde{\psi}(x_1, d)\end{aligned}$$

where $\phi^o(x_1)$ and $\psi^o(x_1)$ represent the exactly known components (called nominal terms) of $\phi(x_1, d)$ and $\psi(x_1, d)$, respectively, and $\tilde{\phi}(x_1, d)$ and $\tilde{\psi}(x_1, d)$ represent the uncertain components of $\phi(x_1, d)$ and $\psi(x_1, d)$, respectively. Without loss of generality, we assume $\phi^o(0) = 0$, $\tilde{\phi}(0, d) = 0$, $\psi^o(0) = 0$, and $\tilde{\psi}(0, d) = 0$.

We will apply a dynamic transformation and a static transformation, respectively, to reduce the control gain. In order to reduce the dynamic uncertainty, introduce a dynamic coordinate transformation as follows:

$$\begin{aligned}\tilde{z} &= z - z^o \\ \dot{z}^o &= Mz^o + \phi^o(x_1).\end{aligned} \tag{4.45}$$

As a result, (4.44) becomes

$$\begin{aligned}\dot{\tilde{z}} &= M\tilde{z} + \tilde{\phi}(x_1, d) \\ \dot{x}_1 &= f_1^o\left(z^o, x_1\right) + \tilde{f}_1\left(\tilde{z}, x_1, d\right) + bu\end{aligned} \tag{4.46}$$

where $f_1^o(z^o, x_1) = Hz^o + \psi^o(x_1)$ and $\tilde{f}_1(\tilde{z}, x_1, d) = H\tilde{z} + \tilde{\psi}(x_1, d)$. In order to reduce the static uncertainty, we further introduce a static coordinate transformation as follows:

$$u = \bar{u} - f_1^o\left(z^o, x_1\right)/b \tag{4.47}$$

which converts (4.46) to

$$\begin{aligned}\dot{\tilde{z}} &= M\tilde{z} + \tilde{\phi}(x_1, d) \\ \dot{x}_1 &= \tilde{f}_1\left(\tilde{z}, x_1, d\right) + b\bar{u}.\end{aligned} \tag{4.48}$$

The system (4.48) is in the same form as the system (4.44) and both of them are in the form of the unity relative degree systems studied in Theorem 2.8. In particular, the $\tilde{z}$-subsystem is ISS with an ISS-Lyapunov function $V(\tilde{z})$ satisfying

$$\dot{V}(\tilde{z}) \le -\alpha(\|\tilde{z}\|) + \sigma(|x_1|)$$

where $\alpha(s) = s^2$ and σ is some class $\mathcal{K}$ function. Thus, the GRSP of (4.48) is solved by a static control law of the following form

$$\bar{u} = -\rho(x_1)x_1,$$

where the function $\rho(x_1)$ can be constructed by the approach detailed in Theorem 2.8. As a result, the following controller

$$\begin{aligned} u &= -\rho(x_1)x_1 - f_1^o\left(z^o, x_1\right)/b \\ \dot{z}^o &= Mz^o + \phi^o(x_1) \end{aligned}$$

solves the GRSP of (4.44).

If the two uncertain functions $\tilde{\phi}(x_1, d)$ and $\tilde{\psi}(x_1, d)$ are significantly smaller than the two uncertain functions $\phi(x_1, d)$ and $\psi(x_1, d)$, the control gain for the transformed system (4.48) will also be significantly smaller than the control gain for the original system (4.44).

We now apply this idea to a more general system as follows:

$$\begin{aligned} \dot{z} &= Mz + \phi(x_1, d) \\ \dot{x}_1 &= Hz + \psi(x_1, d) + bx_2, \quad b > 0 \\ \dot{x}_i &= f_i(\vec{x}_i) + x_{i+1}, \quad i = 2, \ldots, r \end{aligned} \tag{4.49}$$

with $u = x_{r+1}$. Under the coordinate transformation (4.45), one has

$$\begin{aligned} \dot{\tilde{z}} &= M\tilde{z} + \tilde{\phi}(x_1, d) \\ \dot{x}_1 &= f_1^o(z^o, x_1) + \tilde{f}_1\left(\tilde{z}, x_1, d\right) + bx_2 \\ \dot{x}_i &= f_i(\vec{x}_i) + x_{i+1}, \quad i = 2, \ldots, r. \end{aligned} \tag{4.50}$$

The system (4.50) is in a special form of the output feedback systems (4.16) with relative degree r. The GRSP of (4.50) can be solved by Theorem 4.1 viewing z^o as a known function. However, to further reduce the static uncertainty, it is worthwhile to further perform the static transformation (4.47) on the system (4.50). Since the relative degree of (4.50) is greater than one, the static transformation (4.47) will incur more coordinate transformations in the subsequent steps for the controller design. The whole process will be a slight modification of the proof of Theorem 4.1. We detail this process as follows.

First, we note that, like in (4.17), we have

$$\left|\tilde{f}_1\left(\tilde{z}, x_1, d\right)\right| \leq m_1\|\tilde{z}\| + m_2(x_1)|x_1|, \quad \forall d \in \mathbb{D} \tag{4.51}$$

for some constant m_1 and some smooth positive function m_2. Then, like in (4.18), we choose Δ such that

$$\Delta \geq 1 + m_1^2 + (r-1)m_1^2/\lambda.$$

It is noted that the above inequality is the same as (4.18) if $\lambda = 1$. However, as we will see later, having $\lambda > 0$ as a design parameter will give us additional freedom for tuning the control gain more precisely.

Since Δ is constant, and $\alpha(s) = s^2$, by Corollary 2.2 and Remark 2.11, letting $V'(\tilde{z}) = \Delta V(\tilde{z})$ shows the derivative satisfies

$$\dot{V}'(\tilde{z}) \leq -\Delta \|\tilde{z}\|^2 + \varkappa(x_1)x_1^2 \tag{4.52}$$

where $\varkappa$ is some smooth positive function. Like in (4.20), define a gain function

$$\varrho(x_1) \geq \varkappa(x_1) + m_2(x_1) + (r-1)m_2^2(x_1)/\lambda. \tag{4.53}$$

Again, if $\lambda = 1$, then (4.53) is the same as (4.20).

Then we can specify the following coordinate transformation

$$\begin{aligned} \mathcal{X}_1 &= x_1 \\ \mathcal{X}_i &= x_i - \mathcal{S}_{i-1}\left(z^o, \vec{x}_{i-1}\right), \quad i = 2, \ldots, r+1. \end{aligned}$$

The functions $\mathcal{S}_i(z^o, \vec{x}_i)$, $i = 1, \ldots, r$, are given as follows

$$\begin{aligned} \mathcal{S}_1\left(z^o, x_1\right) &= -\rho_1(x_1)x_1 + \nu_1\left(z^o, x_1\right) \\ \mathcal{S}_i\left(z^o, \vec{x}_i\right) &= -\rho_i\left(z^o, \vec{x}_i\right)\mathcal{X}_i + \nu_i\left(z^o, \vec{x}_i\right), \quad i = 2, \ldots, r \end{aligned} \tag{4.54}$$

where

$$\begin{aligned} \rho_1(x_1) &= [\varrho(x_1) + 5/4]/b + b/4 \\ \nu_1\left(z^o, x_1\right) &= -f_1^o\left(z^o, x_1\right)/b \end{aligned}$$

and

$$\begin{aligned} \rho_i\left(z^o, \vec{x}_i\right) &\geq \lambda \left(\frac{\partial \mathcal{S}_{i-1}\left(z^o, \vec{x}_{i-1}\right)}{\partial x_1}\right)^2 \Big/ 2 + 9/4 \\ \nu_i\left(z^o, \vec{x}_i\right) &= -f_i\left(\vec{x}_i\right) + \frac{\partial \mathcal{S}_{i-1}\left(z^o, \vec{x}_{i-1}\right)}{\partial x_1}\left[f_1^o\left(z^o, x_1\right) + bx_2\right] \\ &\quad + \sum_{j=2}^{i-1} \frac{\partial \mathcal{S}_{i-1}\left(z^o, \vec{x}_{i-1}\right)}{\partial x_j}\left[f_j\left(\vec{x}_j\right) + x_{j+1}\right] \end{aligned}$$

$$+\frac{\partial \mathcal{S}_{i-1}\left(z^o, \overrightarrow{x}_{i-1}\right)}{\partial z^o}\left[Mz^o + \phi^o(x_1)\right], \quad i = 2, \ldots, r.$$

We have the following result.

Theorem 4.3 *Consider the system* (4.49) *with* $\mathbb{D}$ *a prescribed compact set. Suppose* M *is Hurwitz. Then the GRSP is solved by the controller*

$$\begin{aligned} u &= \mathcal{S}_r\left(z^o, \overrightarrow{x}_r\right) \\ \dot{z}^o &= Mz^o + \phi^o(x_1) \end{aligned} \tag{4.55}$$

with $\mathcal{S}_r$ *given in* (4.54).

Proof The proof is similar to that of Theorem 4.1. For $i = 1, \ldots, r$, let

$$W_i\left(\tilde{z}, \overrightarrow{\mathcal{X}}_i\right) = V'(\tilde{z}) + \sum_{j=1}^{i} \mathcal{X}_j^2/2.$$

We will show that, along the state trajectory of the $\text{col}(\tilde{z}, \overrightarrow{\mathcal{X}}_i)$-subsystem of (4.50),

$$\begin{aligned} \dot{W}_i\left(\tilde{z}, \overrightarrow{\mathcal{X}}_i\right) \leq &-(r-i)m_1^2\|\tilde{z}\|^2/\lambda - (r-i)m_2^2(\mathcal{X}_1)\mathcal{X}_1^2/\lambda \\ &- \|\tilde{z}\|^2 - \sum_{j=1}^{i}\mathcal{X}_j^2 + \mathcal{X}_{i+1}^2. \end{aligned} \tag{4.56}$$

For $i = 1$, direct calculation shows that the derivative of $W_1(\tilde{z}, \mathcal{X}_1)$ along the state trajectory satisfies:

$$\begin{aligned} \dot{W}_1(\tilde{z}, \mathcal{X}_1) \leq &-\Delta\|\tilde{z}\|^2 + \varkappa(\mathcal{X}_1)\mathcal{X}_1^2 + \mathcal{X}_1\left(\tilde{f}_1\left(\tilde{z}, \mathcal{X}_1, \tilde{d}\right) + b\left(\mathcal{X}_2 - \mathcal{X}_1\rho_1(\mathcal{X}_1)\right)\right) \\ \leq &-\Delta\|\tilde{z}\|^2 + m_1^2\|\tilde{z}\|^2 \\ &+ \mathcal{X}_1^2\left[\varkappa(\mathcal{X}_1) + 1/4 + m_2(\mathcal{X}_1) + b^2/4 - b\rho_1(\mathcal{X}_1)\right] + \mathcal{X}_2^2 \\ \leq &-(r-1)m_1^2\|\tilde{z}\|^2/\lambda - (r-1)m_2^2(\mathcal{X}_1)\mathcal{X}_1^2/\lambda - \|\tilde{z}\|^2 - \mathcal{X}_1^2 + \mathcal{X}_2^2 \end{aligned}$$

which is (4.56) with $i = 1$.

Next, we will show the inequality (4.56) is true for $2 \leq i \leq r$ if it is true for $(i-1)$ by using the relationship

$$\dot{W}_i\left(\tilde{z}, \overrightarrow{\mathcal{X}}_i\right) = \dot{W}_{i-1}\left(\tilde{z}, \overrightarrow{\mathcal{X}}_{i-1}\right) + \mathcal{X}_i\dot{\mathcal{X}}_i.$$

In fact, we have

$$
\begin{aligned}
\mathcal{X}_i \dot{\mathcal{X}}_i &= \mathcal{X}_i \left[f_i\left(\overrightarrow{x}_i\right) + \mathcal{X}_{i+1} + \mathcal{S}_i\left(z^o, \overrightarrow{x}_i\right) - \dot{\mathcal{S}}_{i-1}\left(z^o, \overrightarrow{x}_{i-1}\right) \right] \\
&\le \mathcal{X}_i \left[f_i\left(\overrightarrow{x}_i\right) + \mathcal{X}_{i+1} + \mathcal{S}_i\left(z^o, \overrightarrow{x}_i\right) \right] + \lambda \left(\mathcal{X}_i \frac{\partial \mathcal{S}_{i-1}(\overrightarrow{x}_{i-1})}{\partial x_1} \right)^2 \Big/ 2 \\
&\quad + m_1^2 \|\tilde{z}\|^2/\lambda + m_2^2(x_1) x_1^2/\lambda - \mathcal{X}_i \frac{\partial \mathcal{S}_{i-1}\left(z^o, \overrightarrow{x}_{i-1}\right)}{\partial x_1} \left[f_1^o\left(z^o, x_1\right) + b x_2 \right] \\
&\quad - \mathcal{X}_i \sum_{j=2}^{i-1} \frac{\partial \mathcal{S}_{i-1}\left(z^o, \overrightarrow{x}_{i-1}\right)}{\partial x_j} \left[f_j\left(\overrightarrow{x}_j\right) + x_{j+1} \right] \\
&\quad - \mathcal{X}_i \frac{\partial \mathcal{S}_{i-1}\left(z^o, \overrightarrow{x}_{i-1}\right)}{\partial z^o} \left[M z^o + \phi^o(x_1) \right] \\
&\le -\mathcal{X}_i^2 \rho_i\left(z^o, \overrightarrow{x}_i\right) + \mathcal{X}_i^2/4 + \lambda \mathcal{X}_i^2 \left(\frac{\partial \mathcal{S}_{i-1}\left(\overrightarrow{x}_{i-1}\right)}{\partial x_1} \right)^2 \Big/ 2 \\
&\quad + m_1^2 \|\tilde{z}\|^2/\lambda + m_2^2(\mathcal{X}_1) \mathcal{X}_1^2/\lambda + \mathcal{X}_{i+1}^2 \\
&\le -2\mathcal{X}_i^2 + m_1^2 \|\tilde{z}\|^2/\lambda + m_2^2(\mathcal{X}_1) \mathcal{X}_1^2/\lambda + \mathcal{X}_{i+1}^2
\end{aligned}
$$

where the function $\mathcal{S}_i(z^o, \overrightarrow{x}_i)$ is defined in (4.54) and $\dot{\mathcal{S}}_{i-1}(z^o, \overrightarrow{x}_{i-1})$ is calculated as follows

$$
\begin{aligned}
\dot{\mathcal{S}}_{i-1}\left(z^o, \overrightarrow{x}_{i-1}\right) &= \frac{\partial \mathcal{S}_{i-1}\left(z^o, \overrightarrow{x}_{i-1}\right)}{\partial x_1} \left[\tilde{f}_1\left(\tilde{z}, x_1, \tilde{d}\right) \right] \\
&\quad + \frac{\partial \mathcal{S}_{i-1}\left(z^o, \overrightarrow{x}_{i-1}\right)}{\partial x_1} \left[f_1^o\left(z^o, x_1\right) + b x_2 \right] \\
&\quad + \sum_{j=2}^{i-1} \frac{\partial \mathcal{S}_{i-1}\left(z^o, \overrightarrow{x}_{i-1}\right)}{\partial x_j} \left[f_j\left(\overrightarrow{x}_j\right) + x_{j+1} \right] \\
&\quad + \frac{\partial \mathcal{S}_{i-1}\left(z^o, \overrightarrow{x}_{i-1}\right)}{\partial z^o} \left[M z^o + \phi^o(x_1) \right]
\end{aligned}
$$

As a result,

$$
\begin{aligned}
\dot{W}_i\left(\tilde{z}, \overrightarrow{\mathcal{X}}_i\right) \le &- (r - i + 1) m_1^2 \|\tilde{z}\|^2/\lambda - (r - i + 1) m_2^2(\mathcal{X}_1) \mathcal{X}_1^2/\lambda - \|\tilde{z}\|^2 \\
&- \sum_{j=1}^{i-1} \mathcal{X}_j^2 + \mathcal{X}_i^2 - 2\mathcal{X}_i^2 + m_1^2 \|\tilde{z}\|^2/\lambda + m_2^2(\mathcal{X}_1) \mathcal{X}_1^2/\lambda + \mathcal{X}_{i+1}^2
\end{aligned}
$$

$$\leq -(r-i)m_1^2\|\tilde{z}\|^2/\lambda - (r-i)m_2^2(\mathcal{X}_1)\mathcal{X}_1^2/\lambda - \|\tilde{z}\|^2$$
$$-\sum_{j=1}^{i}\mathcal{X}_j^2 + \mathcal{X}_{i+1}^2$$

which is (4.56). Letting $i = r$ and $x_{r+1} = 0$ gives

$$\dot{W}_r\left(\tilde{z}, \vec{\mathcal{X}}_r\right) \leq -\|\tilde{z}\|^2 - \|\vec{\mathcal{X}}_r\|^2.$$

So, the GRSP of (4.50) is solved by the controller $u = \mathcal{S}_r(z^o, \vec{x}_r)$. In other words, the GRSP of (4.49) is solved by (4.55). □

As we mentioned before, the smaller the two uncertain functions $\tilde{\phi}(x_1, d)$ and $\tilde{\psi}(x_1, d)$ are, the smaller the two bounding functions $\varkappa(x_1)$ and $m_2(x_1)$ and therefore the gain function $\rho(x_1)$ is. In the special case where $\tilde{\phi}(x_1, d) = 0$ and $\tilde{\psi}(x_1, d) = 0$, we have $m_1 = 0, m_2(x_1) = 0$, and $\varkappa(x_1) = 0$. In this case, it can be verified that the GRSP of (4.49) is solved by (4.55) with

$$\rho_1(x_1) = \rho_1 = 5/(4b) + b/4$$
$$\rho_i(z^o, \vec{x}_i) = \rho_i \geq 9/4, \quad i = 2, \ldots, r.$$

The closed-loop system is of the form

$$\begin{aligned}
\dot{z} &= Mz + \phi^o(\mathcal{X}_1)\\
\dot{\mathcal{X}}_1 &= -b\rho_1\mathcal{X}_1 + b\mathcal{X}_2\\
\dot{\mathcal{X}}_i &= -\rho_i\mathcal{X}_i + \mathcal{X}_{i+1}, \quad i = 2, \ldots, r.
\end{aligned} \tag{4.57}$$

Thus, when the system is known exactly, the controller reduces to a simple input to output linearization controller.

4.5 Robust Stabilization of Chua's Circuit

In this section, we consider the Chua's circuit (3.43) with the parameters $R = 1\,\Omega$, $C_1 = C_2 = 1$ F, and $L = 1$ H. With $\lambda_1 = 1$ and $\lambda_2 = 1$, the system can be put in the following form:

$$\begin{aligned}
\dot{z} &= \begin{bmatrix} -2 & 1 \\ -1 & 0 \end{bmatrix} z + \begin{bmatrix} s(x_1) \\ -1 \end{bmatrix} x_1\\
\dot{x}_1 &= [1\ 0]z - s(x_1)x_1 + x_2\\
\dot{x}_2 &= -x_2 + x_3\\
\dot{x}_3 &= -x_3 + u.
\end{aligned} \tag{4.58}$$

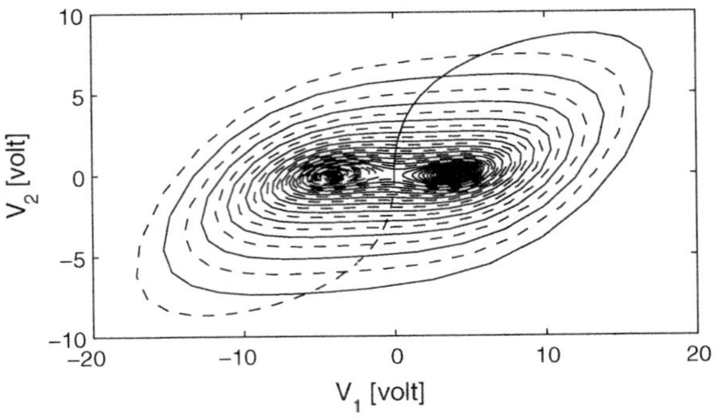

Fig. 4.4 Profile of chaotic waveforms of the uncontrolled Chua's circuit

And the original state variables in (3.40) are $V_1 = x_1$ volts, $V_2 = z_1 + x_2 + x_1$ volts, $I = z_2 + x_2 + x_3$ amps, and $V_c = u$ volts. Consider a nonlinear function $s(x_1) = -w_1 + w_2|x_1|$ containing unknown parameter $w_1 = 1.2 \pm 0.1$ and $w_2 = 0.02 \pm 0.005$.

The responses of the open-loop system with $u = 0$ are shown in Fig. 4.4 with $w_1 = 1.1$ and $w_2 = 0.025$. The solid curve starts from the initial condition $\mathrm{col}(z, x_1, x_2, x_3) = [0, 10, 0, 0, 0]^{\mathrm{T}}$ with the dashed curve from $\mathrm{col}(z, x_1, x_2, x_3) = [0, -10, 0, 0, 0]^{\mathrm{T}}$. The chaotic behavior can be observed from the figure. To eliminate the chaotic behavior, we will design a controller u such that the equilibrium point at the origin of the closed-loop system is RUGAS. For this purpose, let

$$V(z) = z^{\mathrm{T}} \begin{bmatrix} 1 & -1 \\ -1 & 3 \end{bmatrix} z$$

whose derivative along the state trajectory of the z-subsystem satisfies

$$\begin{aligned} \dot{V}(z) &= -2\|z\|^2 + 2z^{\mathrm{T}} \begin{bmatrix} s(x_1) + 1 \\ -s(x_1) - 3 \end{bmatrix} x_1 \\ &\le -\|z\|^2 + 7.4x_1^2 + 0.0025x_1^4. \end{aligned}$$

So, Assumption 4.1 is satisfied and hence the solvability of the GRSP is guaranteed by Theorem 4.1. The controller construction is given in Algorithm 4.1. To be specific, we define some quantities from the following inequality:

$$|[1\ 0]z - s(x_1)x_1| \le \|z\| + (1.3 + 0.025|x_1|)|x_1|.$$

Let $\Delta = 4$ and $V'(z) = 4V(z)$. By Corollary 2.1 and Remark 2.11, we have

$$\dot{V}'(z) \le -4\|z\|^2 + \left(29.6 + 0.01x_1^2\right) x_1^2.$$

Then, the gain function can be selected as

$$\begin{aligned} \varrho(x_1) &= 38 + 0.025x_1^2 \\ &\geq 29.6 + 0.01x_1^2 + 1.3 + 0.025|x_1| + 2(1.3 + 0.025|x_1|)^2. \end{aligned} \tag{4.59}$$

Next, we calculate the functions $\mathcal{S}_1(x_1)$, $\mathcal{S}_2(\vec{x}_2)$, and $\mathcal{S}_3(\vec{x}_3)$ in order. First, we have

$$\mathcal{S}_1(x_1) = -(\varrho(x_1) + 1.5)x_1 = -\left(39.5 + 0.025x_1^2\right)x_1.$$

Thus,

$$\begin{aligned} \mathcal{S}_{1x}(x_1) &= \frac{\partial \mathcal{S}_1(x_1)}{\partial x_1} = -39.5 - 0.075x_1^2 \\ \mathcal{S}_{1xx}(x_1) &= \frac{\partial \mathcal{S}_{1x}(x_1)}{\partial x_1} = -0.15x_1. \end{aligned}$$

For $i = 2$, we have

$$\begin{aligned} \rho_2(x_1) &= 0.5\mathcal{S}_{1x}^2(x_1) + 2.25 \\ \nu_2(x_1, x_2) &= x_2 + s_{1x}(x_1)x_2 \end{aligned}$$

and

$$\mathcal{S}_2(x_1, x_2) = -\rho_2(x_1)(x_2 - \mathcal{S}_1(x_1)) + \nu_2(x_1, x_2).$$

Let

$$\begin{aligned} \mathcal{S}_{2x1}(x_1, x_2) &= \frac{\partial \mathcal{S}_2(x_1, x_2)}{\partial x_1} \\ &= -(\mathcal{S}_{1x}(x_1)\mathcal{S}_{1xx}(x_1))(x_2 - \mathcal{S}_1(x_1)) + \rho_2(x_1)\mathcal{S}_{1x}(x_1) \\ &\quad + s_{1xx}(x_1)x_2 \\ \mathcal{S}_{2x2}(x_1, x_2) &= \frac{\partial \mathcal{S}_2(x_1, x_2)}{\partial x_2} = -\rho_2(x_1) + 1 + s_{1x}(x_1). \end{aligned}$$

Finally, for $i = 3$, we have

$$\begin{aligned} \rho_3(x_1, x_2) &= 0.5\mathcal{S}_{2x1}^2(x_1, x_2) + 2.25 \\ \nu_3(x_1, x_2, x_3) &= x_3 + \mathcal{S}_{2x1}(x_1, x_2)x_2 + \mathcal{S}_{2x2}(x_1, x_2)(-x_2 + x_3). \end{aligned}$$

Thus

$$\begin{aligned} \mathcal{S}_3(x_1, x_2, x_3) = &- \rho_3(x_1, x_2)(x_3 - \mathcal{S}_2(x_1, x_2)) \\ &+ \nu_3(x_1, x_2, x_3). \end{aligned}$$

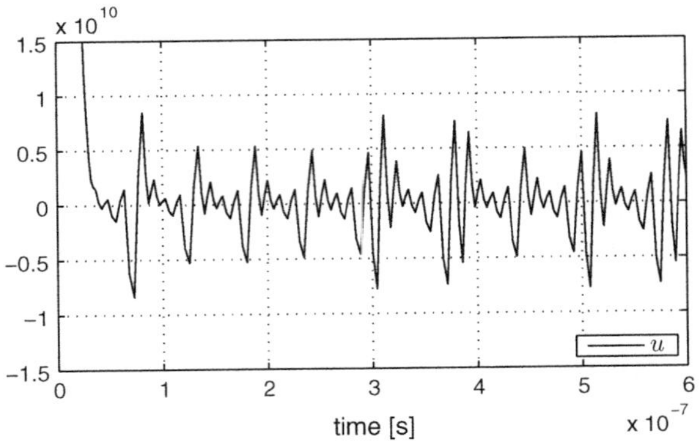

Fig. 4.5 Profile of control input of the controlled Chua's circuit (high gain)

By Theorem 4.1, the overall controller is given as follows:

$$u = \mathcal{S}_3(x_1, x_2, x_3).$$

The simulation is conducted with $w_1 = 1.1$, $w_2 = 0.025$, and the initial state values $z(0) = [20,\ 2]^{\mathrm{T}}$, $x_1(0) = -1$, $x_2(0) = 0$, and $x_3(0) = 0$. However, it is observed that the control signal u depicted in Fig. 4.5 is unacceptably large.

It is possible to reduce the gain function using the approach introduced in Sect. 4.4 The details are given as follows.

Design a dynamic compensator

$$\dot{z}^o = q^o(z^o, x_1) = \begin{bmatrix} -2 & 1 \\ -1 & 0 \end{bmatrix} z^o + \begin{bmatrix} -1.2 + 0.02|x_1| \\ -1 \end{bmatrix} x_1.$$

Then, the dynamics governing the error $\tilde{z} = z - z^o$ are

$$\dot{\tilde{z}} = \begin{bmatrix} -2 & 1 \\ -1 & 0 \end{bmatrix} \tilde{z} + \begin{bmatrix} -\bar{w}_1 + \bar{w}_2|x_1| \\ 0 \end{bmatrix} x_1$$

where $|\bar{w}_1| \leq 0.1$ and $|\bar{w}_2| \leq 0.005$. Define a function

$$V(\tilde{z}) = (\tilde{z})^{\mathrm{T}} \begin{bmatrix} 1 & -1 \\ -1 & 3 \end{bmatrix} \tilde{z}$$

whose derivative along the state trajectory of the $\tilde{z}$-subsystem satisfies

$$\dot{V}(\tilde{z}) = -2\|\tilde{z}\|^2 + 2(\tilde{z})^{\mathrm{T}} \begin{bmatrix} -\bar{w}_1 + \bar{w}_2|x_1| \\ \bar{w}_1 - \bar{w}_2|x_1| \end{bmatrix} x_1$$

$$\leq -\|\tilde{z}\|^2 + 0.04x_1^2 + 10^{-4}x_1^4.$$

Equation (4.51) becomes

$$|[1\ 0]\tilde{z} + \bar{w}_1 x_1 - \bar{w}_2|x_1|x_1| \leq \|\tilde{z}\| + (0.1 + 0.005|x_1|)|x_1|$$

for $m_1 = 1$ and $m_2(x_1) = 0.1 + 0.005|x_1|$. Let $\lambda = 0.2$ and $\Delta = 1 + m_1^2 + (r-1)m_1^2/\lambda = 12$. Then, the gain function can be selected as

$$\begin{aligned}\varrho(x_1) =& 0.79 + 0.005x_1^2 \geq 12\left(0.04 + 10^{-4}x_1^2\right)\\ &+ (0.1 + 0.005|x_1|) + 2(0.1 + 0.005|x_1|)^2/0.2.\end{aligned} \tag{4.60}$$

Compared with the gain function $\varrho(x_1)$ in (4.59), the one in (4.60) is significantly reduced.

Next, we calculate the functions $\mathcal{S}_1(z^o, x_1)$, $\mathcal{S}_2(z^o, \vec{x}_2)$, and $\mathcal{S}_3(z^o, \vec{x}_3)$ in order. First, we have

$$\begin{aligned}\mathcal{S}_1\left(z^o, x_1\right) &= -(\varrho(x_1) + 1.5)x_1 - [1\ 0]z^o + (-1.2 + 0.02|x_1|)x_1\\ &= -\left(2.29 + 0.005x_1^2\right)x_1 - [1\ 0]z^o + (-1.2 + 0.02|x_1|)x_1.\end{aligned}$$

Thus,

$$\begin{aligned}\mathcal{S}_{1x}(x_1) &= \frac{\partial \mathcal{S}_1\left(z^o, x_1\right)}{\partial x_1} = -3.49 - 0.015x_1^2 + 0.04|x_1|\\ \mathcal{S}_{1xx}(x_1) &= \frac{\partial \mathcal{S}_{1x}(x_1)}{\partial x_1} = -0.03x_1 + 0.04\operatorname{sgn}(x_1).\end{aligned}$$

Also, we define

$$\begin{aligned}\psi^o(x_1) &= -(-1.2 + 0.02|x_1|)x_1\\ \psi_x^o(x_1) &= \frac{\partial \psi_x^o(x_1)}{\partial x_1} = 1.2 - 0.04|x_1|.\end{aligned}$$

For $i = 2$, we have

$$\begin{aligned}\rho_2(x_1) &= 0.1\mathcal{S}_{1x}^2(x_1) + 2.25\\ \nu_2\left(z^o, x_1, x_2\right) &= x_2 + s_{1x}(x_1)\left([1\ 0]z^o + \psi^o(x_1) + x_2\right) - [-2\ 1]z^o + \psi^o(x_1)\end{aligned}$$

and

$$\mathcal{S}_2\left(z^o, x_1, x_2\right) = -\rho_2(x_1)\left(x_2 - \mathcal{S}_1\left(z^o, x_1\right)\right) + \nu_2\left(z^o, x_1, x_2\right).$$

Let

$$
\begin{aligned}
s_{2x1}\left(z^o, x_1, x_2\right) &= \frac{\partial s_2\left(z^o, x_1, x_2\right)}{\partial x_1} \\
&= -\left(0.2 s_{1x}(x_1) s_{1xx}(x_1)\right)\left(x_2 - s_1\left(z^o, x_1\right)\right) + \rho_2(x_1) s_{1x}(x_1) \\
&\quad + s_{1xx}(x_1)\left([1\ 0]z^o + \psi^o(x_1) + x_2\right) \\
&\quad + s_{1x}(x_1)\psi_x^o(x_1) + \psi_x^o(x_1) \\
s_{2x2}(x_1, x_2) &= \frac{\partial s_2\left(z^o, x_1, x_2\right)}{\partial x_2} = -\rho_2(x_1) + 1 + s_{1x}(x_1) \\
s_{2z}(x_1) &= \frac{\partial s_2\left(z^o, x_1, x_2\right)}{\partial z^o} = -\rho_2(x_1)[1\ 0] + s_{1x}(x_1)[1\ 0] - [-2\ 1].
\end{aligned}
$$

Finally, for $i = 3$, we have

$$
\begin{aligned}
\rho_3\left(z^o, x_1, x_2\right) &= 0.1 s_{2x1}^2\left(z^o, x_1, x_2\right) + 2.25 \\
\nu_3\left(z^o, x_1, x_2, x_3\right) &= x_3 + s_{2x1}\left(z^o, x_1, x_2\right)\left([1\ 0]z^o + \psi^o(x_1) + x_2\right) \\
&\quad + s_{2x2}(x_1, x_2)(-x_2 + x_3) + s_{2z}(x_1) q^o\left(z^o, x_1\right).
\end{aligned}
$$

Thus,

$$
\begin{aligned}
s_3\left(z^o, x_1, x_2, x_3\right) = &- \rho_3\left(z^o, x_1, x_2\right)\left(x_3 - s_2\left(z^o, x_1, x_2\right)\right) \\
&+ \nu_3(z^o, x_1, x_2, x_3).
\end{aligned}
$$

By Theorem 4.3, the overall controller is given as follows:

$$
\begin{aligned}
u &= s_3\left(z^o, x_1, x_2, x_3\right) \\
\dot{z}^o &= q^o(z^o, x_1) = \begin{bmatrix} -2 & 1 \\ -1 & 0 \end{bmatrix} z^o + \begin{bmatrix} -1.2 + 0.02|x_1| \\ -1 \end{bmatrix} x_1.
\end{aligned}
$$

The simulation result is shown in Figs. 4.6 and 4.7 with $w_1 = 1.1$, $w_2 = 0.025$, and the initial condition $z(0) = [20,\ 2]^{\mathrm{T}}$, $x_1(0) = -1$, $x_2(0) = 0$, $x_3(0) = 0$. It is observed that all state variables of (4.58) asymptotically converge to the origin. The corresponding control signal u is illustrated in Fig. 4.8 which is much smaller than the one shown in Fig. 4.5.

4.6 Notes and References

Over the past two decades, the GRSP for nonlinear cascaded systems has been extensively studied under various characterizations of system complexity. In the most simple case where the system involves neither dynamic uncertainty nor static uncertainty, the problem was first solved using the now well known backstepping technique in a series of papers [1–3]. When the system contains only the static

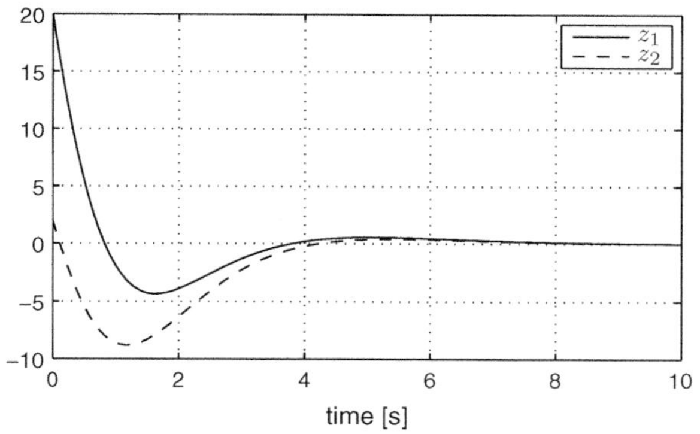

Fig. 4.6 Profile of state trajectories of the controlled Chua's circuit (part 1)

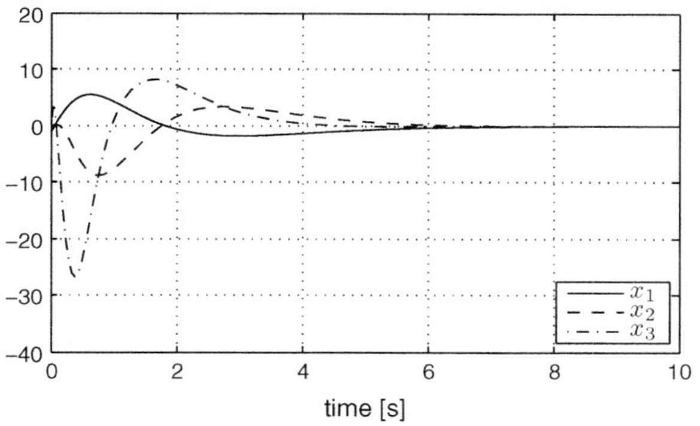

Fig. 4.7 Profile of state trajectories of the controlled Chua's circuit (part 2)

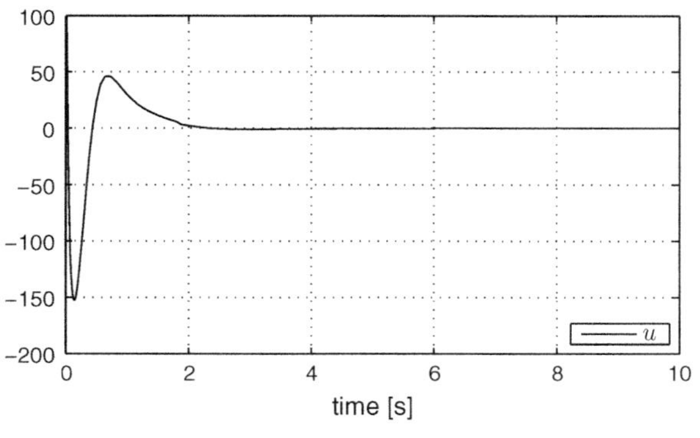

Fig. 4.8 Profile of control input of the controlled Chua's circuit (with gain reduction)

uncertainty, the problem was dealt with using the adaptive backstepping approach under various assumptions on the uncertain function and the way how the uncertainty enters the system in, e.g., [4–6]. When the system contains the dynamic uncertainty, the technique for stabilization becomes more complicated. There are basically two ways to handle the GRSP for nonlinear systems with dynamic uncertainty. One is based on the small gain theorem developed by Jiang et al. [7, 8]. The method was also used in [9, 10] in dealing with the output regulation problem. The other one is based on the Lyapunov's direct approach. Some main references are [11–13]. Other relevant references can be found in [4–6, 14], etc. This chapter is mainly adapted from [12] and the example on Chua's circuit is from [15].

4.7 Problems

Problem 4.1 Consider the following system with state $x = \text{col}(x_1, x_2, x_3)$, input u, and uncertainty $d(t) \in [1, 2]$,

$$\begin{aligned}\dot{x}_1 &= x_2\\ \dot{x}_2 &= x_1x_2 + d(t)x_3\\ \dot{x}_3 &= \sin(x_3) + d(t)u,\end{aligned}$$

and the controller

$$\begin{aligned}\mathcal{X}_1 &= x_1\\ \mathcal{X}_2 &= x_2 + \mathcal{X}_1\rho_1(\mathcal{X}_1)\\ \mathcal{X}_3 &= x_3 + \mathcal{X}_2\rho_2(\mathcal{X}_2)\\ u &= -\mathcal{X}_3\rho_3(\mathcal{X}_3).\end{aligned}$$

Put the system in the new coordinate $\mathcal{X} = \text{col}(\mathcal{X}_1, \mathcal{X}_2, \mathcal{X}_3)$ for

(a) $\rho_1(\mathcal{X}_1) = 1$, $\rho_2(\mathcal{X}_2) = 2$, $\rho_3(\mathcal{X}_3) = \mathcal{X}_3^2 + 1$;
(b) $\rho_1(\mathcal{X}_1) = 1$, $\rho_2(\mathcal{X}_2) = 3\mathcal{X}_2^2 + 5$, $\rho_3(\mathcal{X}_3) = 10\mathcal{X}_3^4 + 15$.

Simulate the closed-loop system for $d(t) = 1$ starting from different initial conditions.

Problem 4.2 For each function $f(x, y, d)$ given below, find functions $a(x)$ and $b(y)$ such that $|f(x, y, d)| \leq a(x)\|x\| + b(y)\|y\|$ (the inequality is often used in robust stablization).

(a) $f(x, y, d) = (x + dy)^2, \quad x, y \in \mathbb{R}, \quad |d| \leq 1$;
(b) $f(x, y, d) = (x + 2x^3 + dy)^2, \quad x, y \in \mathbb{R}, \quad 1 < d < 2$;
(c) $f(x, y, d) = x\cos(dy), \quad x, y \in \mathbb{R}, \quad |d| < 10$;
(d) $f(x, y, d) = dxe^y, \quad x, y \in \mathbb{R}, \quad |d| < 2$;

(e) $f(x, y, d) = \sin(dx^\top y), \quad x, y \in \mathbb{R}^2, \quad |d| \leq 2.$

Problem 4.3 Find a controller u for the GRSP of each of the following systems (the state z represents dynamic uncertainty) and construct a corresponding Lyapunov function of the closed-loop system. The unknown parameters w_1 and w_2 satisfy $|w_1| \leq 2$ and $|w_2| \leq 1$.

(a) $\dot{z} = -z + 0.5x, \ \dot{x} = w_1|zx| + 2u;$
(b) $\dot{z} = -(2 + w_1)z + x, \ \dot{x} = z^2 + 2w_2x^2 + 2u;$
(c) $\dot{z} = \begin{bmatrix} 0 & 1 \\ -1 & -2 \end{bmatrix} z + \begin{bmatrix} 0 \\ 0.5w_1 \end{bmatrix} x, \ \dot{x} = w_2\|z\|x + u.$

Simulate the closed-loop system for $w_1 = 1.5$ and $w_2 = 0.5$ starting from different initial conditions.

Problem 4.4 Repeat Problem 4.3 for the systems with an input filter added, i.e.,

(a) $\dot{z} = -z + 0.5x_1, \ \dot{x}_1 = w_1|zx_1| + 2x_2, \dot{x}_2 = u;$
(b) $\dot{z} = -(2 + w_1)z + x_1, \ \dot{x}_1 = z^2 + 2w_2x_1^2 + 2x_2, \dot{x}_2 = u;$
(c) $\dot{z} = \begin{bmatrix} 0 & 1 \\ -1 & -2 \end{bmatrix} z + \begin{bmatrix} 0 \\ 0.5w_1 \end{bmatrix} x_1, \ \dot{x} = w_2\|z\|x_1 + x_2, \dot{x}_2 = u.$

Problem 4.5 Show each of the following systems $\dot{z} = q(z, \zeta, x, d)$ is RISS viewing z as the state, $\text{col}(\zeta, x)$ as the input, and $-1 < d < 1$ as the external disturbance, and find the ISS Lyapunov function.

(a) $\dot{z} = -z - z^3 + \zeta^2 + dx^2;$
(b) $\dot{z} = -z + \zeta^2 + d\|x\|^2, \ x = \text{col}(x_1, x_2);$
(c) $\dot{z} = \begin{bmatrix} 0 & 1 \\ -1 & -2 \end{bmatrix} z + \begin{bmatrix} |\zeta| + x_1 \\ \sin(x_2) \end{bmatrix}, \ x = \text{col}(x_1, x_2).$

Problem 4.6 Put the ISS Lyapunov functions found in Problem 4.5 in the form

$$V(z) \sim \left\{\underline{\alpha}, \bar{\alpha}, \alpha, \sigma \mid \dot{z} = q(z, \zeta, x, d)\right\},$$

and verify

$$\lim_{s \to 0^+} \sup \frac{s^2}{\alpha(s)} < \infty, \quad \lim_{s \to 0^+} \sup \frac{\sigma(s)}{s^2} < \infty.$$

Problem 4.7 Show each of the systems (a)–(c) in Problem 4.5 is RISS viewing z as the state, $\text{col}(\zeta, \mathcal{X})$ as the input, and $-1 < d < 1$ as the external disturbance, under the following coordinate transformation, respectively; verify it admits the same ISS Lyapunov function found in Problem 4.5.

(a) $\mathcal{X} = 2x$;
(b) $\mathcal{X} = \text{col}(\mathcal{X}_1, \mathcal{X}_2), \mathcal{X}_1 = x_1, \ \mathcal{X}_2 = x_2 + 3x_1$;
(c) $\mathcal{X} = \text{col}(\mathcal{X}_1, \mathcal{X}_2), \mathcal{X}_1 = x_1, \ \mathcal{X}_2 = x_2 + x_1(2 + x_1^2)$.

Problem 4.8 Repeat Problem 4.3 for the systems with u replaced by $d(t)u$ for $1 \le d(t) \le 2$.

Problem 4.9 Repeat Problem 4.3 for the systems with one dimension added ($1 \le d(t) \le 2$), i.e.,

(a) $\dot{z} = -z + 0.5x_1, \ \dot{x}_1 = w_1|zx_1| + 2x_2, \dot{x}_2 = x_2^2 + d(t)u$;
(b) $\dot{z} = -(2 + w_1)z + x_1, \ \dot{x}_1 = z^2 + 2w_2x_1^2 + 2x_2, \dot{x}_2 = x_2^2 + d(t)u$;
(c) $\dot{z} = \begin{bmatrix} 0 & 1 \\ -1 & -2 \end{bmatrix} z + \begin{bmatrix} 0 \\ 0.5w_1 \end{bmatrix} x_1, \ \dot{x} = w_2\|z\|x_1 + x_2, \dot{x}_2 = x_2^2 + d(t)u$.

Problem 4.10 Find a static controller u for the GRSP of each of the following systems (the state z represents dynamic uncertainty) and construct a corresponding Lyapunov function of the closed-loop system. The unknown parameters are $w_1 = 3 \pm 0.1$ and $w_2 = 0.2 \pm 0.1$.

(a) $\dot{z} = -z + x_1, \ \dot{x}_1 = w_1x_1^2 + w_2z + x_2, \ \dot{x}_2 = u$;
(b) $\dot{z} = -z + x_2, \ \dot{x}_1 = w_1x_1^2 + x_2, \ \dot{x}_2 = w_2z + u$.

Simulate the closed-loop system for $w_1 = 5$ and $w_2 = 0.2$ starting from different initial conditions.

Problem 4.11 Find a dynamic controller with smaller gains for Problem 4.10.

Problem 4.12 The GRSP for the system (4.16) has been solved in Sect. 4.2 for a known parameter $b > 0$. Extend the result to the case that the parameter $b > 0$ is unknown (Note: the extension is a special case of (4.1)).

References

1. Byrnes CI, Isidori A (1989) New results and examples in nonlinear feedback stabilization. Syst Control Lett 12:437–442
2. Kanellakopoulos I, Kokotović P, Morse AS (1992) A toolkit for nonlinear feedback design. Syst Control Lett 18:83–92
3. Tsinias J (1989) Sufficient Lyapunov-like conditions for stabilization. Math Control Signals Syst 2:343–357
4. Chen Z, Huang J (2002) Global robust stabilization of cascaded polynomial systems. Syst Control Lett 47:445–453
5. Jiang ZP, Praly L (1998) Design of robust adaptive controllers for nonlinear systems with dynamic uncertainties. Automatica 34:825–840
6. Lin W, Gong Q (2003) A remark on partial-state feedback stabilization of cascade systems using small gain theorem. IEEE Trans Autom Control 48:497–500

7. Jiang ZP, Mareels I (1997) A small-gain control method for nonlinear cascaded systems with dynamic uncertainties. IEEE Trans Autom Control 42:292–308
8. Jiang ZP, Teel AR, Praly L (1994) Small-gain theorem for ISS systems and applications. Math Control Signals Syst 7:95–120
9. Chen Z, Huang J (2005) Global robust output regulation for output feedback systems. IEEE Trans Autom Control 50:117–121
10. Huang J, Chen Z (2004) A general framework for tackling the output regulation problem. IEEE Trans Autom Control 49:2203–2218
11. Chen Z (2009) Global stabilization of nonlinear cascaded systems with a Lyapunov function in superposition form. Automatica 45:2041–2045
12. Chen Z, Huang J (2005) A general formulation and solvability of the global robust output regulation problem. IEEE Trans Autom Control 50:448–462
13. Chen Z, Huang J (2008) A Lyapunov's direct method for the global robust stabilization of nonlinear cascaded systems. Automatica 44:745–752
14. Isidori A (1999) Nonlinear control systems, vol II. Springer, New York
15. Liu L, Huang J (2006) Adaptive robust stabilization of output feedback systems with application to Chua's circuit. IEEE Trans Circuits Syst II Express Briefs 53:926–930

Chapter 5
Adaptive Stabilization

In Chap. 4, the plant uncertainty, represented by $d(t)$, must be ranged in a compact set $\mathbb{D}$. As a result, a high gain feedback control strategy can be used to dominate uncertainty so that the stability of the closed-loop system is maintained regardless of the change of $d(t)$ as long as $d(t) \in \mathbb{D}$ for all $t \geq 0$. This type of control is also called robust control. In many cases, $d(t)$ can be unbounded which cannot be handled by the approach in Chap. 4. In this chapter, we will consider another type of uncertainty satisfying the so-called "linear parameterization" property which roughly means that an uncertain function can be represented as a linear function of some unknown constant parameters. We will introduce the so-called adaptive control scheme to handle this type of uncertainty. An adaptive control law is also a type of nonlinear control law that usually includes a mechanism to estimate the unknown parameters. This chapter consists of five sections. In Sect. 5.1, we introduce the adaptive control problem via a simple nonlinear system. In Sect. 5.2, we formalize the method so that it applies to a class of lower triangular uncertain nonlinear systems. In Sect. 5.3, we combine the robust control method studied in Chap. 4 and the adaptive control method studied in Sect. 5.2 leading to the so-called robust adaptive control method. This method can handle nonlinear systems containing both bounded time-varying uncertainty and unbounded constant uncertainty satisfying the linear parameterization property. The case study on the Duffing equation is illustrated in Sect. 5.4. The notes and references are given in Sect. 5.5.

5.1 A Motivating Example

To introduce the adaptive control problem, in this section, we first consider a simple nonlinear system of the following form:

$$\dot{x} = f^{\mathsf{T}}(x, t)\mu + u \tag{5.1}$$

where $x \in \mathbb{R}$ is the state, $u \in \mathbb{R}$ is the input, $f : \mathbb{R} \times [t_0, \infty] \mapsto \mathbb{R}^l$ is a known function satisfying locally Lipchitz condition with respect to x uniformly in t, and

Z. Chen and J. Huang, *Stabilization and Regulation of Nonlinear Systems*,
Advanced Textbooks in Control and Signal Processing,
DOI 10.1007/978-3-319-08834-1_5

$\mu \in \mathbb{R}^l$ is an unknown constant parameter vector. We allow each component of μ to take any real value. Therefore, if we define

$$g(x, d(t)) = f^{\mathrm{T}}(x, t)\mu \tag{5.2}$$

with $d(t) = (t, \mu)$, then d is unbounded and the approach in Chap. 4 does not work.

An uncertain function of the form (5.2) is said to satisfy the linear parameterization property since, for any real number γ, $f^{\mathrm{T}}(x, t)(\gamma\mu) = \gamma(f^{\mathrm{T}}(x, t)\mu)$. On the other hand, the uncertain function $\sin(\mu x)$ where $x \in \mathbb{R}$ and μ is a constant real number does not satisfy the linear parameterization property.

In the system (5.1), if μ were known, then the control law

$$u = -f^{\mathrm{T}}(x, t)\mu - \rho x, \quad \rho > 0 \tag{5.3}$$

would lead to a closed-loop system $\dot{x} = -\rho x$ which is an asymptotically stable linear system. Since μ is unknown, the control law (5.3) cannot be implemented. We will instead consider the following control law

$$u = -f^{\mathrm{T}}(x, t)\hat{\mu} - \rho x, \quad \rho > 0 \tag{5.4}$$

where $\hat{\mu} \in \mathbb{R}^l$ is a constant vector viewed as an estimation of μ. Under this control law, the closed-loop system is

$$\dot{x} = -f^{\mathrm{T}}(x, t)\tilde{\mu} - \rho x \tag{5.5}$$

where $\tilde{\mu} = \hat{\mu} - \mu$ is called the parameter estimation error. Now consider the simplest case of (5.5) with $l = 1$ and $f(x, t) = x$. In this case, (5.5) reduces to the following linear system

$$\dot{x} = -(\tilde{\mu} + \rho)x. \tag{5.6}$$

Now since μ can take arbitrarily large value, no matter how large ρ is chosen, one cannot guarantee the stability of (5.6) because $(\tilde{\mu} + \rho) > 0$ cannot be guaranteed. Thus, a static state feedback control law of the form (5.4) does not work for this case.

Next we turn to the following dynamic state feedback control law

$$\begin{aligned} u &= -f^{\mathrm{T}}(x, t)\hat{\mu} - \rho x, \ \rho > 0 \\ \dot{\hat{\mu}} &= \Lambda x f(x, t) \end{aligned} \tag{5.7}$$

where $\hat{\mu} \in \mathbb{R}^l$ is not a constant vector but a function of time governed by the second equation of (5.7) and $\Lambda \in \mathbb{R}^{l \times l}$ is a symmetric and positive definite constant matrix. We call $\hat{\mu}$ as the dynamic estimation of μ or simply an estimation of μ and call the second equation of (5.7) as the parameter update law or simply update law. The composition of the plant (5.1) and the control law (5.7) is called the closed-loop system and takes the following form

$$\begin{aligned} \dot{x} &= -f^{\mathrm{T}}(x,t)\tilde{\mu} - \rho x \\ \dot{\hat{\mu}} &= \Lambda x f(x,t). \end{aligned} \tag{5.8}$$

We now show that the closed-loop system has the property that, for any initial condition $x(0)$ and $\hat{\mu}(0)$, the solution of the closed-loop system (5.8) is bounded and $\lim_{t\to\infty} x(t) = 0$. For this purpose, we consider a Lyapunov function candidate

$$W(x,\tilde{\mu}) = x^2/2 + \tilde{\mu}^{\mathrm{T}}\Lambda^{-1}\tilde{\mu}/2 \tag{5.9}$$

for the closed-loop system (5.8). The derivative of $W(x,\tilde{\mu})$ along the state trajectory of the closed-loop system is

$$\begin{aligned} \dot{W}(x,\tilde{\mu}) &= x\dot{x} + \tilde{\mu}^{\mathrm{T}}\Lambda^{-1}\dot{\tilde{\mu}} \\ &= -\rho x^2 - \tilde{\mu}^{\mathrm{T}} x f(x,t) + \tilde{\mu}^{\mathrm{T}}\Lambda^{-1}\dot{\hat{\mu}} \\ &= -\rho x^2 \le 0. \end{aligned} \tag{5.10}$$

Thus, the state variables x and $\hat{\mu}$ are bounded. Moreover, by Theorem 2.5, $\lim_{t\to\infty} x(t) = 0$.

A few remarks are in order.

Remark 5.1 (i) The control law (5.7) is a special type of nonlinear dynamic state feedback control law. Since the state $\hat{\mu}$ is interpreted as a dynamic estimation of the unknown parameter vector μ and its value is adapted according to the parameter update law, we call the control law (5.7) a state feedback adaptive control law. The matrix Λ is called the adaptation gain matrix.

(ii) If $f(0,t) = 0$ for all $t \ge 0$, then all such points as $(x,\hat{\mu}) = (0,\hat{\mu})$ are equilibrium points of the closed-loop system. Thus, the control law (5.7) does not make the origin of the closed-loop system asymptotically stable. Therefore, we call the above problem global adaptive stabilization problem instead of global stabilization problem introduced in Chap. 4. A formal definition of global adaptive stabilization problem for the system of the form (2.7)–(2.9) will be introduced in the next section.

(iii) The control law design philosophy revealed in the above example is called certainty equivalence principle. Indeed, the first equation of the control law (5.7) is obtained as if μ were known by mimicking the control law (5.3). The second equation of the control law (5.7) is designed such that the closed-loop system admits a Lyapunov function of the form (5.9) whose derivative along the trajectory of the closed-loop system is negative semi-definite as shown in (5.10). More specifically, suppose the parameter update law is given by $\dot{\hat{\mu}} = \tau(x,t)$ for some function τ. Then the derivative of the Lyapunov function (5.9) along the trajectory of the closed-loop system is

$$\begin{aligned}\dot{W}(x,\tilde{\mu}) &= x\dot{x} + \tilde{\mu}^{\mathsf{T}}\Lambda^{-1}\dot{\tilde{\mu}} \\ &= -\rho x^2 - \tilde{\mu}^{\mathsf{T}} x f(x,t) + \tilde{\mu}^{\mathsf{T}}\Lambda^{-1}\dot{\hat{\mu}} \\ &= -\rho x^2 + \tilde{\mu}^{\mathsf{T}}\left(-x f(x,t) + \Lambda^{-1}\tau(x,t)\right). \end{aligned} \tag{5.11}$$

Letting $\dot{W}(x,\tilde{\mu}) = -\rho x^2$ yields $\tau(x,t) = \Lambda x f(x,t)$.

(iv) The success of the certainty equivalence principle relies on the linear parameterization property of the uncertain function which leads to two properties, i.e., $f(x,t)^{\mathsf{T}}\hat{\mu} - f(x,t)^{\mathsf{T}}\mu = f(x,t)^{\mathsf{T}}\tilde{\mu}$ and $\dot{\hat{\mu}} = \dot{\tilde{\mu}}$.

(v) Even though $\hat{\mu}$ is interpreted as the dynamic estimation of the unknown parameter vector μ, the solvability of the adaptive stabilization problem does not have to imply the convergence of $\hat{\mu}(t)$ to μ as t tends to infinity. The quantity $\hat{\mu}$ may not even converge at all. The limiting behavior of $\hat{\mu}(t)$ is known as the parameter convergence issue and will be further studied later in Sect. 5.3 and Chap. 9

To close this section, we summarize the result of this section as follows.

Theorem 5.1 *The global adaptive stabilization problem for the system* (5.1) *is solved by the controller*

$$\begin{aligned} u &= -f^{\mathsf{T}}(x,t)\hat{\mu} - \rho x, \ \rho > 0 \\ \dot{\hat{\mu}} &= \Lambda x f(x,t), \ \Lambda = \Lambda^{\mathsf{T}} > 0. \end{aligned} \tag{5.12}$$

5.2 Adaptive Stabilization: Tuning Functions Design

In the previous section, we illustrated the adaptive stabilization problem for a simple scalar system. In this section, we will further study the adaptive stabilization problem for nonlinear systems of the form (2.7)–(2.9). Let us first give a formal description of the adaptive stabilization problem as follows.

Global Adaptive Regulation Problem (GARP): *Given a nonlinear control system of the form* (2.7)–(2.9), *design a controller of the form* (1.11), *such that, for any initial condition* $x_c(0)$ of the closed-loop system (2.10), *the state trajectory* $x_c(t)$ *of the closed-loop system is bounded and* $\lim_{t\to\infty} y(t) = 0$.

In the description of GARP, we only require the performance output y to approach the origin asymptotically. If $y = x$, then we further call GARP as a global adaptive stabilization problem (GASP). Clearly, the system (5.1) is a special case of (2.7)–(2.9) with $y_m = y = x$.

In this section, we will first study the adaptive stabilization problem for a class of lower triangular nonlinear systems containing linearly parameterized uncertainties described as follows:

$$\dot{x}_i = f_i^{\mathsf{T}}\left(\vec{x}_i, t\right)\mu + x_{i+1}, \quad i = 1,\ldots,r \tag{5.13}$$

where $\vec{x}_i = \text{col}(x_1, \ldots, x_i)$ with $x_i \in \mathbb{R}$ is the state, $u := x_{r+1} \in \mathbb{R}$ the input, and $\mu \in \mathbb{R}^l$ the unknown constant parameter vector. It is also assumed that all functions in the system (5.13) are sufficiently smooth. The system (5.13) can be viewed as a special case of (2.7)–(2.9) with $y_m = y = x$. On the other hand, the system (5.1) is a special case of (5.13) with $r = 1$. Like the special case with $r = 1$ explained in the previous section, the basic idea for solving the GASP for (5.13) is to find a controller and an update law such that the derivative of a suitable Lyapunov function along the state trajectory of the closed-loop system is negative semi-definite. However, when $r > 1$, the construction of the control law is recursive and we will detail this procedure as follows.

The design procedure consists of r steps. At each step i, $i = 1, \ldots, r$, two functions $\mathcal{S}_i$ and τ_i will be designed based on previous steps. If the relative degree of the system were equal to i, then these two functions $\mathcal{S}_i$ and τ_i would constitute the control law and the parameter update law, respectively. Otherwise, the process will continue recursively until $i = r$ so that we obtain the actual control law $x_{r+1} = \mathcal{S}_r(\cdot)$ and the actual parameter update law $\dot{\hat{\mu}} = \tau_r(\cdot)$. Since the functions τ_i are called tuning functions in the literature, this design approach is called the tuning functions approach which can be viewed as a generalization of Proposition 4.1.

More specifically, define the following dynamic coordinate transformation:

$$\begin{aligned}
\mathcal{X}_1 &= x_1 \\
\mathcal{X}_{i+1} &= x_{i+1} - \mathcal{S}_i\left(\vec{x}_i, \hat{\mu}, t\right), \quad i = 1, \ldots, r \\
\dot{\hat{\mu}} &= \tau_r\left(\vec{x}_r, \hat{\mu}, t\right)
\end{aligned} \tag{5.14}$$

for some sufficiently smooth functions $\mathcal{S}_i, i = 1, \cdots, r$, and τ_r. Let $\zeta_i = \text{col}(\vec{\mathcal{X}}_i, \hat{\mu})$ and $\vec{\mathcal{X}}_i = \text{col}(\mathcal{X}_1, \ldots, \mathcal{X}_i)$. Denote the system composed of the plant (5.13) and the transformation (5.14) by $\dot{\zeta}_r = \varphi_r(\zeta_r, \mathcal{X}_{r+1}, t)$. Then, Proposition 4.1 can be generalized to the following version.

Proposition 5.1 *If there exist sufficiently smooth functions $\mathcal{S}_i$, $i = 1, \ldots, r$, and τ_r, such that, for any initial condition $\zeta_r(0)$, the state ζ_r of the system $\dot{\zeta}_r = \varphi_r(\zeta_r, 0, t)$ is bounded and $\lim_{t\to\infty} \vec{\mathcal{X}}_r = 0$, then the GARP with $y = x_1$ for the system* (5.13) *is solved by the following dynamic state feedback controller*

$$\begin{aligned}
u &= \mathcal{S}_r(\vec{x}_r, \hat{\mu}, t) \\
\dot{\hat{\mu}} &= \tau_r(\vec{x}_r, \hat{\mu}, t).
\end{aligned} \tag{5.15}$$

Moreover, if $\mathcal{S}_i(0, \hat{\mu}, t) = 0$, $i = 1, \ldots, r-1$, then the GASP with $y = \vec{x}_r$ for the system (5.13) *is solved by the same controller.*

We now introduce a recursive procedure to find the functions $\mathcal{S}_i$ and τ_i for $i = 1, \ldots, r$. The construction of $\mathcal{S}_1$ and τ_1 is motivated from the previous section, i.e.,

$$\begin{aligned} \mathcal{S}_1\left(x_1, \hat{\mu}, t\right) &= -f_1^{\mathrm{T}}(x_1, t)\hat{\mu} - \rho_1 x_1, \ \rho_1 \geq 5/4 \\ \tau_1\left(x_1, \hat{\mu}, t\right) &= \Lambda x_1 f_1(x_1, t), \ \Lambda = \Lambda^{\mathrm{T}} > 0. \end{aligned} \tag{5.16}$$

The remaining functions are recursively constructed as follows, for $i = 2, \ldots, r$,

$$\mathcal{S}_i\left(\vec{x}_i, \hat{\mu}, t\right) = -\varrho_i^{\mathrm{T}}\left(\vec{x}_i, \hat{\mu}, t\right)\hat{\mu} - \rho_i \mathcal{X}_i + \nu_i\left(\vec{x}_i, \hat{\mu}, t\right), \rho_i \geq 9/4 \tag{5.17}$$

$$\tau_i\left(\vec{x}_i, \hat{\mu}, t\right) = \tau_{i-1}\left(\vec{x}_{i-1}, \hat{\mu}, t\right) + \Lambda \mathcal{X}_i \varrho_i\left(\vec{x}_i, \hat{\mu}, t\right) \tag{5.18}$$

where the functions ν_i and ϱ_i are defined as follows

$$\begin{aligned} \nu_i\left(\vec{x}_i, \hat{\mu}, t\right) = &\sum_{j=1}^{i-1} \frac{\partial \mathcal{S}_{i-1}\left(\vec{x}_{i-1}, \hat{\mu}, t\right)}{\partial x_j} x_{j+1} + \frac{\partial \mathcal{S}_{i-1}\left(\vec{x}_{i-1}, \hat{\mu}, t\right)}{\partial t} \\ &+ \frac{\partial \mathcal{S}_{i-1}\left(\vec{x}_{i-1}, \hat{\mu}, t\right)}{\partial \hat{\mu}} \tau_i\left(\vec{x}_i, \hat{\mu}, t\right) \\ &+ \sum_{j=1}^{i-2} \mathcal{X}_{j+1} \frac{\partial \mathcal{S}_j\left(\vec{x}_j, \hat{\mu}, t\right)}{\partial \hat{\mu}} \Lambda \varrho_i\left(\vec{x}_i, \hat{\mu}, t\right) \end{aligned} \tag{5.19}$$

$$\varrho_i\left(\vec{x}_i, \hat{\mu}, t\right) = f_i\left(\vec{x}_i, t\right) - \sum_{j=1}^{i-1} \frac{\partial \mathcal{S}_{i-1}\left(\vec{x}_{i-1}, \hat{\mu}, t\right)}{\partial x_j} f_j\left(\vec{x}_j, t\right). \tag{5.20}$$

We now state the main result of this section as follows.

Theorem 5.2 *The GARP for the system* (5.13) *with the performance output* $y = x_1$ *is solved by the controller* (5.15) *where the functions* $\mathcal{S}_r$ *and* τ_r *are defined in* (5.16)–(5.18) *and also summarized in Algorithm 5.1. Moreover, if* $f_i(0, t) = 0$, $i = 1, \ldots, r-1$, *then the GASP for the system* (5.13) *with the performance output* $y = \vec{x}_r$ *is solved by the same controller.*

Proof Consider the system (5.13) under the coordinate transformation (5.14). We will show that, for $1 \leq i \leq r$, along the state trajectory of the $\vec{\mathcal{X}}_i$-subsystem, the derivative of the following function

$$W_i\left(\vec{\mathcal{X}}_i, \hat{\mu}\right) = \sum_{j=1}^{i} \mathcal{X}_j^2/2 + \tilde{\mu}^{\mathrm{T}}\Lambda^{-1}\tilde{\mu}/2 \tag{5.21}$$

satisfies

$$
\dot{W}_i\left(\vec{\chi}_i,\hat{\mu}\right) \le -\sum_{j=1}^{i}\chi_j^2 + \chi_{i+1}^2
$$

$$
+\left(\tilde{\mu}^{\mathsf{T}}\Lambda^{-1} - \sum_{j=1}^{i-1}\chi_{j+1}\frac{\partial\varsigma_j\left(\vec{x}_j,\hat{\mu},t\right)}{\partial\hat{\mu}}\right)\left[\dot{\hat{\mu}} - \tau_i\left(\vec{x}_i,\hat{\mu},t\right)\right] \tag{5.22}
$$

where $\tilde{\mu} = \hat{\mu} - \mu$ is the parameter estimation error.

For $i = 1$, direct calculation shows that the derivative of $W_1\left(\chi_1,\hat{\mu}\right) = \chi_1^2/2 + \tilde{\mu}^{\mathsf{T}}\Lambda^{-1}\tilde{\mu}/2$ along the state trajectory of the χ_1-subsystem satisfies

$$
\begin{aligned}
\dot{W}_1\left(\chi_1,\hat{\mu}\right) &= \chi_1\left(f_1(\chi_1,t)^{\mathsf{T}}\mu + \varsigma_1\left(x_1,\hat{\mu},t\right) + \chi_2\right) + \tilde{\mu}^{\mathsf{T}}\Lambda^{-1}\dot{\hat{\mu}} \\
&= \chi_1\left(-f_1(\chi_1,t)^{\mathsf{T}}\tilde{\mu} - \rho_1\chi_1 + \chi_2\right) + \tilde{\mu}^{\mathsf{T}}\Lambda^{-1}\dot{\hat{\mu}} \\
&\le -\chi_1^2 + \chi_2^2 + \tilde{\mu}^{\mathsf{T}}\Lambda^{-1}\left[\dot{\hat{\mu}} - \tau_1\left(x_1,\hat{\mu},t\right)\right].
\end{aligned} \tag{5.23}
$$

If $r = 1$ as in the case of Theorem 5.1, then letting $\chi_2 = 0$ gives the actual control $x_2 = \varsigma_1\left(x_1,\hat{\mu},t\right)$. And the update function is chosen as $\dot{\hat{\mu}} = \tau_1(x_1,\hat{\mu},t)$ to make $\dot{W}_1\left(\chi_1,\hat{\mu}\right) \le -\chi_1^2$. But, when $r > 1$, x_2 is not the actual control in this theorem, so we should retain τ_1 as the first tuning function and allow the presence of $\dot{\hat{\mu}}$ in $\dot{W}_1\left(\chi_1,\hat{\mu}\right)$ at this stage.

Next, we will prove that the statement is true for any $1 < i \le r$ if it is true for $i - 1$. To this end, we note the relationship

$$
W_i\left(\vec{\chi}_i,\hat{\mu}\right) = W_{i-1}\left(\vec{\chi}_{i-1},\hat{\mu}\right) + \chi_i^2/2.
$$

The derivative of $\chi_i^2/2$ along the state trajectory of the χ_i-subsystem satisfies

$$
\begin{aligned}
\chi_i\dot{\chi}_i &= \chi_i\dot{x}_i - \chi_i\dot{\varsigma}_{i-1}\left(\vec{x}_{i-1},\hat{\mu},t\right) \\
&= \chi_i f_i\left(\vec{x}_i,t\right)^{\mathsf{T}}\mu + \chi_i\varsigma_i\left(\vec{x}_i,\hat{\mu},t\right) + \chi_i\chi_{i+1} \\
&\quad - \chi_i\sum_{j=1}^{i-1}\frac{\partial\varsigma_{i-1}\left(\vec{x}_{i-1},\hat{\mu},t\right)}{\partial x_j}\left(f_j\left(\vec{x}_j,t\right)^{\mathsf{T}}\mu + x_{j+1}\right) \\
&\quad - \chi_i\frac{\partial\varsigma_{i-1}\left(\vec{x}_{i-1},\hat{\mu},t\right)}{\partial t} - \chi_i\frac{\partial\varsigma_{i-1}\left(\vec{x}_{i-1},\hat{\mu},t\right)}{\partial\hat{\mu}}\dot{\hat{\mu}}.
\end{aligned}
$$

By using the definition of $\varrho_i^{\mathrm{T}}\left(\overrightarrow{x}_i, \hat{\mu}, t\right)$ and $\mathcal{S}_i\left(\overrightarrow{x}_i, \hat{\mu}, t\right)$, we have

$$
\begin{aligned}
\mathcal{X}_i\dot{\mathcal{X}}_i \le{}& \mathcal{X}_i\varrho_i^{\mathrm{T}}\left(\overrightarrow{x}_i, \hat{\mu}, t\right)\mu + \mathcal{X}_i\left[-\varrho_i^{\mathrm{T}}\left(\overrightarrow{x}_i, \hat{\mu}, t\right)\hat{\mu} - \rho_i\mathcal{X}_i + \nu_i\left(\overrightarrow{x}_i, \hat{\mu}, t\right)\right] \\
&+ \mathcal{X}_i^2/4 + \mathcal{X}_{i+1}^2 - \mathcal{X}_i\sum_{j=1}^{i-1}\frac{\partial\mathcal{S}_{i-1}\left(\overrightarrow{x}_{i-1}, \hat{\mu}, t\right)}{\partial x_j}x_{j+1} \\
&- \mathcal{X}_i\frac{\partial\mathcal{S}_{i-1}\left(\overrightarrow{x}_{i-1}, \hat{\mu}, t\right)}{\partial t} - \mathcal{X}_i\frac{\partial\mathcal{S}_{i-1}\left(\overrightarrow{x}_{i-1}, \hat{\mu}, t\right)}{\partial\hat{\mu}}\dot{\hat{\mu}} \\
\le{}& -\mathcal{X}_i\varrho_i^{\mathrm{T}}\left(\overrightarrow{x}_i, \hat{\mu}, t\right)\tilde{\mu} - 2\mathcal{X}_i^2 + \mathcal{X}_{i+1}^2 + \mathcal{X}_i\nu_i\left(\overrightarrow{x}_i, \hat{\mu}, t\right) \\
&- \mathcal{X}_i\sum_{j=1}^{i-1}\frac{\partial\mathcal{S}_{i-1}\left(\overrightarrow{x}_{i-1}, \hat{\mu}, t\right)}{\partial x_j}x_{j+1} - \mathcal{X}_i\frac{\partial\mathcal{S}_{i-1}\left(\overrightarrow{x}_{i-1}, \hat{\mu}, t\right)}{\partial t} \\
&- \mathcal{X}_i\frac{\partial\mathcal{S}_{i-1}\left(\overrightarrow{x}_{i-1}, \hat{\mu}, t\right)}{\partial\hat{\mu}}\dot{\hat{\mu}}.
\end{aligned}
$$

Then, using the definition of $\nu_i\left(\overrightarrow{x}_i, \hat{\mu}, t\right)$ and $\tau_i\left(\overrightarrow{x}_i, \hat{\mu}, t\right)$ gives

$$
\begin{aligned}
\mathcal{X}_i\dot{\mathcal{X}}_i \le{}& -\mathcal{X}_i\varrho_i^{\mathrm{T}}\left(\overrightarrow{x}_i, \hat{\mu}, t\right)\tilde{\mu} - 2\mathcal{X}_i^2 + \mathcal{X}_{i+1}^2 \\
&+ \mathcal{X}_i\sum_{j=1}^{i-2}\mathcal{X}_{j+1}\frac{\partial\mathcal{S}_j\left(\overrightarrow{x}_j, \hat{\mu}, t\right)}{\partial\hat{\mu}}\Lambda\varrho_i\left(\overrightarrow{x}_i, \hat{\mu}, t\right) \\
&- \mathcal{X}_i\frac{\partial\mathcal{S}_{i-1}\left(\overrightarrow{x}_{i-1}, \hat{\mu}, t\right)}{\partial\hat{\mu}}\left[\dot{\hat{\mu}} - \tau_i\left(\overrightarrow{x}_i, \hat{\mu}, t\right)\right] \\
={}& -2\mathcal{X}_i^2 + \mathcal{X}_{i+1}^2 - \mathcal{X}_i\frac{\partial\mathcal{S}_{i-1}\left(\overrightarrow{x}_{i-1}, \hat{\mu}, t\right)}{\partial\hat{\mu}}\left[\dot{\hat{\mu}} - \tau_i\left(\overrightarrow{x}_i, \hat{\mu}, t\right)\right] \\
&+ \left(\tilde{\mu}^{\mathrm{T}}\Lambda^{-1} - \sum_{j=1}^{i-2}\mathcal{X}_{j+1}\frac{\partial\mathcal{S}_j\left(\overrightarrow{x}_j, \hat{\mu}, t\right)}{\partial\hat{\mu}}\right) \\
&\times\left[\tau_{i-1}\left(\overrightarrow{x}_{i-1}, \hat{\mu}, t\right) - \tau_i\left(\overrightarrow{x}_i, \hat{\mu}, t\right)\right].
\end{aligned}
$$

Now, we are ready to show

$$
\dot{W}_i\left(\overrightarrow{\mathcal{X}}_i, \hat{\mu}\right) = \dot{W}_{i-1}\left(\overrightarrow{\mathcal{X}}_{i-1}, \hat{\mu}\right) + \mathcal{X}_i\dot{\mathcal{X}}_i
$$

$$
\begin{aligned}
&\leq -\sum_{j=1}^{i-1} \mathcal{X}_j^2 + \mathcal{X}_i^2 - 2\mathcal{X}_i^2 + \mathcal{X}_{i+1}^2 \\
&\quad - \mathcal{X}_i \frac{\partial \mathcal{S}_{i-1}\left(\vec{x}_{i-1}, \hat{\mu}, t\right)}{\partial \hat{\mu}} \left[\dot{\hat{\mu}} - \tau_i\left(\vec{x}_i, \hat{\mu}, t\right)\right] \\
&\quad + \left(\tilde{\mu}^{\mathrm{T}}\Lambda^{-1} - \sum_{j=1}^{i-2} \mathcal{X}_{j+1} \frac{\partial \mathcal{S}_j\left(\vec{x}_j, \hat{\mu}, t\right)}{\partial \hat{\mu}}\right) \left[\dot{\hat{\mu}} - \tau_{i-1}\left(\vec{x}_i, \hat{\mu}, t\right)\right] \\
&\quad + \left(\tilde{\mu}^{\mathrm{T}}\Lambda^{-1} - \sum_{j=1}^{i-2} \mathcal{X}_{j+1} \frac{\partial \mathcal{S}_j\left(\vec{x}_j, \hat{\mu}, t\right)}{\partial \hat{\mu}}\right) \\
&\quad \times \left[\tau_{i-1}\left(\vec{x}_{i-1}, \hat{\mu}, t\right) - \tau_i\left(\vec{x}_i, \hat{\mu}, t\right)\right] \\
&= -\sum_{j=1}^{i} \mathcal{X}_j^2 + \mathcal{X}_{i+1}^2 \\
&\quad + \left(\tilde{\mu}^{\mathrm{T}}\Lambda^{-1} - \sum_{j=1}^{i-1} \mathcal{X}_{j+1} \frac{\partial \mathcal{S}_j\left(\vec{x}_j, \hat{\mu}, t\right)}{\partial \hat{\mu}}\right) \left[\dot{\hat{\mu}} - \tau_i\left(\vec{x}_i, \hat{\mu}, t\right)\right].
\end{aligned}
$$

The proof of the statement (5.23) is thus completed by mathematical induction.

Finally, consider the closed-loop system composed of the system (5.13) and the controller (5.15), which results from letting $\mathcal{X}_{r+1} = 0$ and $\dot{\hat{\mu}} = \tau_r\left(\vec{x}_r, \hat{\mu}, t\right)$. Then, the statement (5.22) with $i = r$ implies

$$\dot{W}_r\left(\vec{\mathcal{X}}_r, \hat{\mu}\right) \leq -\sum_{j=1}^{r} \mathcal{X}_j^2.$$

With $W_r\left(\vec{\mathcal{X}}_r, \hat{\mu}\right)$ as the Lyapunov function for the closed-loop system, the stability of the closed-loop system can be established by invoking Theorem 2.5. In particular, one has $\lim_{t\to\infty} \vec{\mathcal{X}}_r(t) = 0$. That is, the GARP for the system (5.13) with the performance output $y = x_1$ is solved by Proposition 5.1. To show the moreover part, it suffices to note that $\mathcal{S}_i\left(0, t, \hat{\mu}\right) = 0$ if $f_i(0, t) = 0$. □

Algorithm 5.1

INPUT: $f_i, i = 1, \ldots, r$
OUTPUT: $\mathcal{S}_r$ and τ_r
STEP 1: Let $i = 1$; find the functions $\mathcal{S}_1$ and τ_1 from (5.16).
STEP 2: IF $i = r$, GO TO STEP 5, ELSE GO TO STEP 3.

STEP 3: Let $i = i + 1$; calculate the functions ϱ_i, τ_i, ν_i, and $\mathcal{S}_i$ in order, from (5.17)–(5.20).

STEP 4: GO TO STEP 2.

STEP 5: END

Example 5.1 Consider a second order nonlinear system

$$\begin{aligned} \dot{x}_1 &= x_1^2 \mu_1 + x_2 \\ \dot{x}_2 &= \sin(x_1 x_2)\mu_2 + u \end{aligned} \tag{5.24}$$

where $\mu = [\mu_1 \ \mu_2]^{\mathrm{T}}$ is an unknown constant parameter vector. The objective is to find the controller u to solve the GASP with the performance output $y = [x_1, x_2]^{\mathrm{T}}$.

We follow Algorithm 5.1 to explicitly construct the controller. The first step is to pick the following functions based on (5.16),

$$\begin{aligned} \mathcal{S}_1 \left(x_1, \hat{\mu}\right) &= -x_1^2 \hat{\mu}_1 - 2x_1 \\ \tau_1 \left(x_1, \hat{\mu}\right) &= [x_1^3 \ 0]^{\mathrm{T}}. \end{aligned}$$

Next we will calculate ϱ_2, τ_2, ν_2, and $\mathcal{S}_2$ in order using the formulae (5.17)–(5.20). First, we have

$$\varrho_2 \left(\vec{x}_2, \hat{\mu}\right) = f_2\left(\vec{x}_2\right) - \frac{\partial \mathcal{S}_1 \left(\vec{x}_1, \hat{\mu}\right)}{\partial x_1} f_1\left(\vec{x}_1\right) = \left[\left(2x_1\hat{\mu}_1 + 2\right) x_1^2 \ \ \sin(x_1 x_2)\right]^{\mathrm{T}}$$

and then,

$$\begin{aligned} \tau_2 \left(\vec{x}_2, \hat{\mu}\right) &= \tau_1 \left(x_1, \hat{\mu}\right) + \left[x_2 - \mathcal{S}_1\left(x_1, \hat{\mu}\right)\right] \varrho_2 \left(\vec{x}_2, \hat{\mu}\right) \\ &= \begin{bmatrix} x_1^3 + \left(2x_1\hat{\mu}_1 + 2\right) x_1^2 \left(x_2 + x_1^2\hat{\mu}_1 + 2x_1\right) \\ \sin(x_1 x_2)\left(x_2 + x_1^2 \hat{\mu}_1 + 2x_1\right) \end{bmatrix}. \end{aligned}$$

With the available functions, we are ready to obtain

$$\begin{aligned} \nu_2 \left(\vec{x}_2, \hat{\mu}\right) &= \frac{\partial \mathcal{S}_1 \left(\vec{x}_1, \hat{\mu}\right)}{\partial x_1} x_2 + \frac{\partial \mathcal{S}_1 \left(\vec{x}_1, \hat{\mu}\right)}{\partial \hat{\mu}} \tau_2 \left(\vec{x}_2, \hat{\mu}\right) \\ &= -\left(2x_1\hat{\mu}_1 + 2\right) x_2 - \left[x_1^2 \ 0\right] \\ &\quad \times \begin{bmatrix} x_1^3 + \left(2x_1\hat{\mu}_1 + 2\right) x_1^2 \left(x_2 + x_1^2\hat{\mu}_1 + 2x_1\right) \\ \sin(x_1 x_2)\left(x_2 + x_1^2 \hat{\mu}_1 + 2x_1\right) \end{bmatrix} \\ &= -\left(2x_1\hat{\mu}_1 + 2\right) x_2 - x_1^4 \left(x_1 + \left(2x_1\hat{\mu}_1 + 2\right)\left(x_2 + x_1^2\hat{\mu}_1 + 2x_1\right)\right) \end{aligned}$$

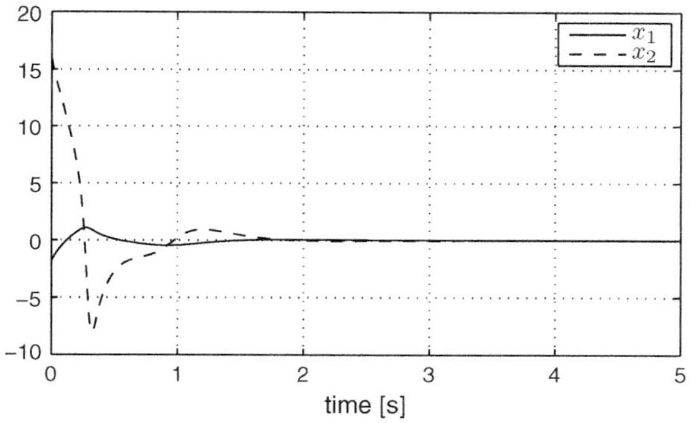

Fig. 5.1 Profile of state trajectories of the closed-loop system in Example 5.1

and

$$\begin{aligned}\mathcal{S}_2\left(\vec{x}_2, \hat{\mu}\right) &= -\varrho_2^{\mathrm{T}}\left(\vec{x}_2, \hat{\mu}\right)\hat{\mu} - \rho_2\left[x_2 - \mathcal{S}_1\left(\vec{x}_1, \hat{\mu}\right)\right] + \nu_2\left(\vec{x}_2, \hat{\mu}\right)\\ &= -\left[\left(2x_1\hat{\mu}_1 + 2\right)x_1^2 \ \sin(x_1x_2)\right]\hat{\mu} - 9/4\left(x_2 + x_1^2\hat{\mu}_1 + 2x_1\right)\\ &\quad - \left(2x_1\hat{\mu}_1 + 2\right)x_2 - x_1^4\left(x_1 + \left(2x_1\hat{\mu}_1 + 2\right)\left(x_2 + x_1^2\hat{\mu}_1 + 2x_1\right)\right).\end{aligned}$$

Now, the controller is of the form

$$\begin{aligned}u &= \mathcal{S}_2\left(\vec{x}_2, \hat{\mu}\right)\\ \dot{\hat{\mu}} &= \tau_2\left(\vec{x}_2, \hat{\mu}\right).\end{aligned}$$

The performance of the controller is shown in Figs. 5.1 and 5.2 with $\mu = [2, 4]^{\mathrm{T}}$. The initial state values are $x(0) = [-2, 18]^{\mathrm{T}}$ and $\hat{\mu}(0) = [0, 10]^{\mathrm{T}}$. It is seen that the state x of the closed-loop system asymptotically converges to the equilibrium point at the origin and the estimated parameter $\hat{\mu}$ is bounded, but not convergent to the real value of the parameter μ. The condition under which the estimated parameter will converge to its real value will be studied in the next section.

5.3 Robust Adaptive Stabilization

In this section, we combine the robust and adaptive techniques to deal with a class of uncertain nonlinear systems containing both disturbances and unknown parameters. The systems are described as follows:

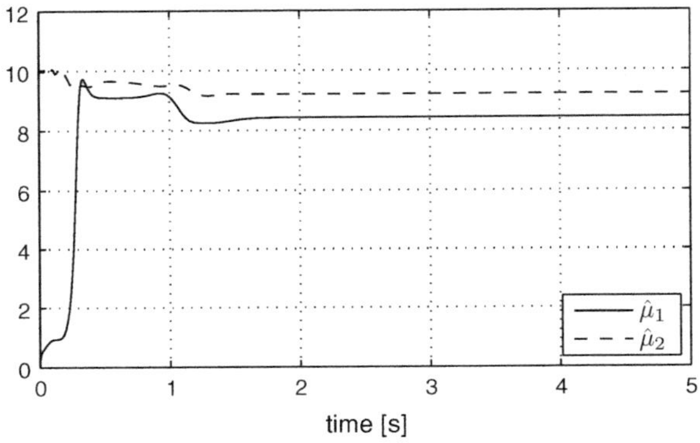

Fig. 5.2 Profile of estimated parameters of the closed-loop system in Example 5.1

$$
\begin{aligned}
\dot{z} &= q(z, x_1, d(t)) \\
\dot{x}_1 &= f_1(z, x_1, d(t)) + bf_a^{\mathrm{T}}(x_1, t)\mu + bx_2 \\
\dot{x}_i &= f_i\left(\vec{x}_i\right) + x_{i+1}, i = 2, \ldots, r
\end{aligned} \tag{5.25}
$$

where $z \in \mathbb{R}^n$ and $\vec{x}_i = \mathrm{col}(x_1, \ldots, x_i)$ with $x_i \in \mathbb{R}$, $i = 1, \ldots, r$, are the state variables, $u := x_{r+1}$ is the input, $\mu \in \mathbb{R}^{l_1}$ is the unknown constant parameter, and $d : [0, \infty) \mapsto \mathbb{D} \subset \mathbb{R}^{l_2}$ with $\mathbb{D}$ a compact set represents unknown parameters and/or disturbances. The functions q, f_a and f_i, $i = 1, \ldots, r$, are sufficiently smooth with $q(0, 0, d) = 0$, $f_1(0, 0, d) = 0$, for all $d \in \mathbb{D}$ and $f_i(0) = 0$, $i = 2, \ldots, r$. The subscript a means the term $f_a^{\mathrm{T}}(x_1, t)\mu$ will be dealt with by adaptive control. In (5.25), the function q may not be known precisely, and/or the state z may not be used for feedback control. Thus, the dynamics governing z are regarded as the dynamic uncertainty. When $\mu = 0$, the system (5.25) reduces to the filter extended form studied in the previous chapter. The parameter b in the system is an unknown positive number satisfying the following assumption.

Assumption 5.1 There exist two known positive constants $\bar{b}$ and $\underline{b}$ such that $\bar{b} \geq b \geq \underline{b}$.

5.3.1 Systems with Relative Degree One

We first consider the system (5.25) with the relative degree $r = 1$, i.e.,

$$
\begin{aligned}
\dot{z} &= q(z, x, d(t)) \\
\dot{x} &= f(z, x, d(t)) + bf_a^{\mathrm{T}}(x, t)\mu + bu.
\end{aligned} \tag{5.26}
$$

In the system (5.26), the stabilization problem with $\mu = 0$ has been studied in Chap. 4 by the robust control technique. The special case of (5.26) with $z \in \mathbb{R}^0$ and $b = 1$ has been studied in Sect. 5.2. To handle the nontrivial inverse dynamics with $y = x$, we need the following assumption which plays the same role as Assumption 4.1.

Assumption 5.2 The subsystem $\dot{z} = q(z, x, d)$ has an ISS Lyapunov function $V(z)$, i.e.,

$$V(z) \sim \left\{\underline{\alpha}, \bar{\alpha}, \alpha, \sigma \mid \dot{z} = q(z, x, d)\right\}$$

and

$$\lim_{s \to 0^+} \sup \frac{s^2}{\alpha(s)} < \infty, \quad \lim_{s \to 0^+} \sup \frac{\sigma(s)}{s^2} < \infty.$$

The first step is to repeat the procedure in Sect. 4.2. Specifically, by (11.13) of the Appendix, we can find two sufficiently smooth and non-negative functions m_1 and m_2 such that

$$|f(z, x, d)| \leq m_1(z)\|z\| + m_2(x)|x|, \ \forall d \in \mathbb{D}. \tag{5.27}$$

Pick a sufficiently smooth function

$$\Delta(z) \geq 1 + m_1^2(z). \tag{5.28}$$

By Corollary 2.2, there exists a continuously differentiable function $V'(z)$ satisfying $\underline{\alpha}'(\|z\|) \leq V'(z) \leq \bar{\alpha}'(\|z\|)$ for some class $\mathcal{K}_\infty$ functions $\underline{\alpha}'$ and $\bar{\alpha}'$, such that, along the state trajectory of $\dot{z} = q(z, x, d)$,

$$\dot{V}'(z) \leq -\Delta(z)\|z\|^2 + \varkappa(x)x^2 \tag{5.29}$$

for a sufficiently smooth function $\varkappa$. Then, we define a sufficiently smooth function $\rho(x)$ such that

$$\rho(x) \geq [\varkappa(x) + m_2(x) + 5/4]/\underline{b} \tag{5.30}$$

with which the controller can be constructed in the following theorem.

Theorem 5.3 *Consider the system* (5.26) *with* $\mathbb{D}$ *a prescribed compact set. Under Assumptions* 5.1 *and* 5.2, *the GASP of* (5.26) *with the performance output* $y = \text{col}(z, x)$ *is solved by the following controller*:

$$\begin{aligned} u &= -f_a^{\mathsf{T}}(x, t)\hat{\mu} - \rho(x)x \\ \dot{\hat{\mu}} &= \Lambda x f_a(x, t), \ \Lambda = \Lambda^{\mathsf{T}} > 0 \end{aligned} \tag{5.31}$$

where the function ρ *is given in* (5.30).

Proof Define a Lyapunov function candidate for the closed-loop system as follows:

$$W(z, x, \hat{\mu}) = V'(z) + x^2/2 + b\tilde{\mu}^{\mathsf{T}}\Lambda^{-1}\tilde{\mu}/2$$

where $\tilde{\mu} = \hat{\mu} - \mu$ is the parameter estimation error. Direct calculation shows that the derivative of $W(z, x, \hat{\mu})$ along the state trajectory of the closed-loop system satisfies

$$\begin{aligned}
\dot{W}(z, x, \hat{\mu}) &\leq -\Delta(z)\|z\|^2 + \varkappa(x)x^2 + x[f(z, x, d) + bf_a^{\mathsf{T}}(x, t)\mu + bu] + b\tilde{\mu}^{\mathsf{T}}\Lambda^{-1}\dot{\hat{\mu}} \\
&\leq -\Delta(z)\|z\|^2 + \varkappa(x)x^2 + x[f(z, x, d) - b\rho(x)x] \\
&\quad + bx[f_a^{\mathsf{T}}(x, t)\mu - f_a^{\mathsf{T}}(x, t)\hat{\mu}] + b\tilde{\mu}^{\mathsf{T}}\Lambda^{-1}\dot{\hat{\mu}} \\
&\leq -\Delta(z)\|z\|^2 + m_1^2(z)\|z\|^2 + x^2[\varkappa(x) + 1/4 + m_2(x) - b\rho(x)] \\
&\quad + b\tilde{\mu}^{\mathsf{T}}(-xf_a(x, t) + \Lambda^{-1}\dot{\hat{\mu}}) \\
&\leq -\|z\|^2 - x^2. \qquad (5.32)
\end{aligned}$$

Now the stability of the closed-loop system can be established by invoking Theorem 2.5. In particular, one has $\lim_{t\to\infty} \mathrm{col}(z(t), x(t)) = 0$. The GASP is thus solved. □

Finally, we will consider the convergence issue of the parameter $\hat{\mu}$ as illustrated by the following corollary.

Corollary 5.1 *In Theorem 5.3, the closed-loop system composed of the given system* (5.26) *and the controller* (5.31) *has the properties that, for any initial state,*

$$\lim_{t\to\infty} \dot{\hat{\mu}}(t) = 0, \qquad (5.33)$$

and, if $d(t)$, $\dot{d}(t)$, $f_a(x, t)$, $\frac{\partial f_a(x, t)}{\partial t}$, *and* $\frac{\partial f_a(x, t)}{\partial x}$ *are bounded for all* $t \geq 0$, *then*

$$\lim_{t\to\infty} f_a^{\mathsf{T}}(0, t)\left(\hat{\mu}(t) - \mu\right) = 0. \qquad (5.34)$$

Moreover, if $f_a(0, t)$ *is PE, then*

$$\lim_{t\to\infty} \hat{\mu}(t) = \mu. \qquad (5.35)$$

Proof In Theorem 5.3, it has been proved that $\lim_{t\to\infty} \mathrm{col}(z(t), x(t)) = 0$ which implies (5.33). Since $d(t), \dot{d}(t)$, $f_a(x, t)$, $\frac{\partial f_a(x, t)}{\partial t}$, and $\frac{\partial f_a(x, t)}{\partial x}$ are bounded, $\ddot{x}$ is bounded and hence $\dot{x}$ is uniformly continuous. By Barbalat's Lemma, $\dot{x}(t)$ approaches zero as $t \to \infty$, which implies (5.34). Finally, by Lemma 2.4, (5.35) holds if $f_a(0, t)$ is PE. □

5.3.2 Systems with High Relative Degree

In this section, we consider the general system (5.25) with $r > 1$. When $r > 1$, x_2 in (5.25) is not a real control, but the controller (5.31) in Theorem 5.3 motivates a coordinate transformation as follows:

$$\mathcal{X}_2 = x_2 - \mathcal{S}_1\left(x_1, \hat{\mu}, t\right), \ \mathcal{S}_1\left(x_1, \hat{\mu}, t\right) = -f_a^{\mathrm{T}}(x_1, t)\hat{\mu} - \rho_1(x_1)x_1$$

where $\mathcal{X}_2$ is governed by the following equation:

$$\begin{aligned}\dot{\mathcal{X}}_2 &= f_2\left(\vec{x}_2\right) - \dot{\mathcal{S}}_1\left(x_1, \hat{\mu}, t\right) + x_3\\ &= f_2\left(\vec{x}_2\right) - \frac{\partial \mathcal{S}_1\left(x_1, \hat{\mu}, t\right)}{\partial x_1}\left[f_1(z, x_1, d) + bf_a^{\mathrm{T}}(x_1, t)\mu + bx_2\right]\\ &\quad - \frac{\partial \mathcal{S}_1\left(x_1, \hat{\mu}, t\right)}{\partial \hat{\mu}}\dot{\hat{\mu}} - \frac{\partial \mathcal{S}_1\left(x_1, \hat{\mu}, t\right)}{\partial t} + x_3.\end{aligned} \tag{5.36}$$

In (5.36), besides μ, the parameter b is also unknown and should be estimated. Let $\hat{b}$ be the estimation of b, and $\tilde{b} = \hat{b} - b$ the estimation error.

With the introduction of the estimation $\hat{b}$, various functions that define the control law may also rely on $\hat{b}$. Thus, we define the dynamic coordinate transformation as follows:

$$\begin{aligned}\mathcal{X}_1 &= x_1\\ \mathcal{X}_{i+1} &= x_{i+1} - \mathcal{S}_i\left(\vec{x}_i, \hat{\mu}, \hat{b}, t\right), \quad i = 1, \ldots, r\\ \dot{\hat{\mu}} &= \tau_r\left(\vec{x}_r, \hat{\mu}, \hat{b}, t\right)\\ \dot{\hat{b}} &= \varpi_r\left(\vec{x}_r, \hat{\mu}, \hat{b}, t\right)\end{aligned} \tag{5.37}$$

for some sufficiently smooth functions $\mathcal{S}_i, i = 1, \ldots, r$, τ_r and ϖ_r, to be specified. Let $\zeta_i = \mathrm{col}(\vec{\mathcal{X}}_i, \hat{\mu}, \hat{b})$. Denote the system governing ζ_r by $\dot{\zeta}_r = \varphi_r(\zeta_r, \mathcal{X}_{r+1}, t)$. Then, Proposition 4.1 can be generalized to the following form.

Proposition 5.2 *If there exist sufficiently smooth functions $\mathcal{S}_i$, $i = 1, \ldots, r$, and τ_r, such that, the state ζ_r of the system $\dot{\zeta}_r = \varphi_r(\zeta_r, 0, t)$ is bounded and $\lim_{t\to\infty} \vec{\mathcal{X}}_r = 0$, then the GARP with $y = x_1$ for the system (5.25) is solved by the following dynamic controller*

$$\begin{aligned}u &= \mathcal{S}_r\left(\vec{x}_r, \hat{\mu}, \hat{b}, t\right)\\ \dot{\hat{\mu}} &= \tau_r\left(\vec{x}_r, \hat{\mu}, \hat{b}, t\right)\\ \dot{\hat{b}} &= \varpi_r\left(\vec{x}_r, \hat{\mu}, \hat{b}, t\right).\end{aligned} \tag{5.38}$$

Moreover, if $\mathcal{S}_i\left(0, \hat{\mu}, \hat{b}, t\right) = 0,\ i = 1, \ldots, r-1$, *then the GASP with* $y = \vec{x}_r$ *for the system (5.25) is solved by the same controller.*

Like in Sect. 5.2, we need to find the functions $\mathcal{S}_r$, τ_r, and ϖ_r recursively. The first step is motivated from the previous case with $r = 1$. Again we note that, by (11.13) of the Appendix, one has

$$|f_1(z, x_1, d)| \le m_1(z)\|z\| + m_2(x_1)|x_1|,\ \forall d \in \mathbb{D} \tag{5.39}$$

for some sufficiently smooth and non-negative functions m_1 and m_2. Let

$$\Delta(z) \ge 1 + rm_1^2(z). \tag{5.40}$$

Under Assumption 5.2, by using the changing supply function technique (Corollary 2.2), there exists a continuously differentiable function $V'(z)$ satisfying $\underline{\alpha}'(\|z\|) \le V'(z) \le \bar{\alpha}'(\|z\|)$ for some class $\mathcal{K}_\infty$ functions $\underline{\alpha}'$ and $\bar{\alpha}'$, such that, along the state trajectory of $\dot{z} = q(z, x_1, d)$,

$$\dot{V}'(z) \le -\Delta(z)\|z\|^2 + \varkappa(x_1)x_1^2 \tag{5.41}$$

for a sufficiently smooth and non-negative function $\varkappa$. Let

$$\rho_1(x_1) \ge [\varkappa(x_1) + m_2(x_1) + 5/4 + (r-1)m_2^2(x_1)]/\underline{b} + \bar{b}/4. \tag{5.42}$$

Then we define

$$\begin{aligned}
\mathcal{S}_1\left(x_1, \hat{\mu}, \hat{b}, t\right) &= -f_a^{\mathrm{T}}(x_1, t)\hat{\mu} - \rho_1(x_1)x_1 \\
\tau_1\left(x_1, \hat{\mu}, \hat{b}, t\right) &= \Lambda_\mu x_1 f_a(x_1, t) \\
\varpi_1\left(x_1, \hat{\mu}, \hat{b}, t\right) &= 0.
\end{aligned} \tag{5.43}$$

The remaining functions are constructed as follows, for $i = 2, \ldots, r$,

$$\begin{aligned}
\mathcal{S}_i\left(\vec{x}_i, \hat{\mu}, \hat{b}, t\right) &= \varrho_i\left(\vec{x}_{i-1}, \hat{\mu}, \hat{b}, t\right) - \rho_i\left(\vec{x}_{i-1}, \hat{\mu}, \hat{b}, t\right)\mathcal{X}_i + \nu_i\left(\vec{x}_i, \hat{\mu}, \hat{b}, t\right) \\
\tau_i\left(\vec{x}_i, \hat{\mu}, \hat{b}, t\right) &= \tau_{i-1}\left(\vec{x}_{i-1}, \hat{\mu}, \hat{b}, t\right) \\
&\quad -\Lambda_\mu \mathcal{X}_i \frac{\partial \mathcal{S}_{i-1}\left(\vec{x}_{i-1}, \hat{\mu}, \hat{b}, t\right)}{\partial x_1} f_a(x_1, t) \\
\varpi_i\left(\vec{x}_i, \hat{\mu}, \hat{b}, t\right) &= \varpi_{i-1}\left(\vec{x}_{i-1}, \hat{\mu}, \hat{b}, t\right) \\
&\quad -\Lambda_b \mathcal{X}_i \frac{\partial \mathcal{S}_{i-1}\left(\vec{x}_{i-1}, \hat{\mu}, \hat{b}, t\right)}{\partial x_1}\left[f_a^{\mathrm{T}}(x_1, t)\hat{\mu} + x_2\right]
\end{aligned} \tag{5.44}$$

where

$$\begin{aligned}
\varrho_i\left(\vec{x}_{i-1}, \hat{\mu}, \hat{b}, t\right) =& \left(-\sum_{j=1}^{i-2} x_{j+1} \frac{\partial s_j\left(\vec{x}_j, \hat{\mu}, \hat{b}, t\right)}{\partial \hat{\mu}} \Lambda_\mu\right) \\
&\times \frac{\partial s_{i-1}\left(\vec{x}_{i-1}, \hat{\mu}, \hat{b}, t\right)}{\partial x_1} f_a(x_1, t) \\
&+ \left(\hat{b} - \sum_{j=1}^{i-2} x_{j+1} \frac{\partial s_j\left(\vec{x}_j, \hat{\mu}, \hat{b}, t\right)}{\partial \hat{b}} \Lambda_b\right) \\
&\times \frac{\partial s_{i-1}\left(\vec{x}_{i-1}, \hat{\mu}, \hat{b}, t\right)}{\partial x_1} \left[f_a^{\mathsf{T}}(x_1, t)\hat{\mu} + x_2\right] \\
\rho_i\left(\vec{x}_{i-1}, \hat{\mu}, \hat{b}, t\right) \geq& \left[\frac{\partial s_{i-1}\left(\vec{x}_{i-1}, \hat{\mu}, \hat{b}, t\right)}{\partial x_1}\right]^2 /2 + 9/4,
\end{aligned}$$

and

$$\begin{aligned}
\nu_i\left(\vec{x}_i, \hat{\mu}, \hat{b}, t\right) =& -f_i\left(\vec{x}_i\right) + \sum_{j=2}^{i-1} \frac{\partial s_{i-1}\left(\vec{x}_{i-1}, \hat{\mu}, \hat{b}, t\right)}{\partial x_j} \left[f_j\left(\vec{x}_j\right) + x_{j+1}\right] \\
&+ \frac{\partial s_{i-1}\left(\vec{x}_{i-1}, \hat{\mu}, \hat{b}, t\right)}{\partial \hat{\mu}} \tau_i\left(\vec{x}_i, \hat{\mu}, \hat{b}, t\right) \\
&+ \frac{\partial s_{i-1}\left(\vec{x}_{i-1}, \hat{\mu}, \hat{b}, t\right)}{\partial \hat{b}} \varpi_i\left(\vec{x}_i, \hat{\mu}, \hat{b}, t\right) \\
&+ \frac{\partial s_{i-1}\left(\vec{x}_{i-1}, \hat{\mu}, \hat{b}, t\right)}{\partial t}.
\end{aligned}$$

In the above equations, the matrix $\Lambda_\mu = \Lambda_\mu^{\mathsf{T}} > 0$ and the scalar $\Lambda_b > 0$ can be arbitrarily selected. It is noted that, for $i \geq 2$, the function $s_i\left(\vec{x}_i, \hat{\mu}, \hat{b}, t\right)$ is split into three terms for the convenience of proving the following theorem.

Theorem 5.4 *Consider the system* (5.25) *with* $\mathbb{D}$ *a prescribed compact set. Under Assumptions* 5.1 *and* 5.2, *the GARP of* (5.25) *with the performance output* $y = \text{col}(z, x_1)$ *is solved by the controller* (5.38) *where the functions* s_r *and* τ_r *are defined in* (5.43)–(5.44) *and also summarized in Algorithm* 5.2. *Moreover, if* $f_a(0, t) = 0$, *then the GASP of* (5.25) *with the performance output* $y = \text{col}(z, \vec{x}_r)$ *is solved by the same controller.*

Proof Consider the system (5.25) under the coordinate transformation (5.37). Let

$$W_1\left(z, \mathcal{X}_1, \hat{\mu}, \hat{b}\right) = V'(z) + \mathcal{X}_1^2/2 + b\tilde{\mu}^{\mathsf{T}}\Lambda_\mu^{-1}\tilde{\mu}/2 + \Lambda_b^{-1}\tilde{b}^2/2$$

and, for $2 \le i \le r$,

$$W_i\left(z, \vec{\mathcal{X}}_i, \hat{\mu}, \hat{b}\right) = W_{i-1}\left(z, \vec{\mathcal{X}}_{i-1}, \hat{\mu}, \hat{b}\right) + \mathcal{X}_i^2/2. \tag{5.45}$$

Then we will show that the derivative of the function $W_i\left(z, \vec{\mathcal{X}}_i, \hat{\mu}, \hat{b}\right)$ along the state trajectory of the closed-loop system satisfies

$$\begin{aligned}\dot{W}_i\left(z, \vec{\mathcal{X}}_i, \hat{\mu}, \hat{b}\right) \le & -(r-i)m_1^2(z)\|z\|^2 - (r-i)m_2^2(\mathcal{X}_1)\|\mathcal{X}_1\|^2 \\ & -\|z\|^2 - \sum_{j=1}^{i}\mathcal{X}_j^2 + \mathcal{X}_{i+1}^2 + \nabla_i\left(z, \vec{x}_i, \hat{\mu}, \hat{b}, t\right),\end{aligned} \tag{5.46}$$

where

$$\begin{aligned}\nabla_i\left(z, \vec{x}_i, \hat{\mu}, \hat{b}, t\right) = & \left(b\tilde{\mu}^{\mathsf{T}}\Lambda_\mu^{-1} - \sum_{j=1}^{i-1}\mathcal{X}_{j+1}\frac{\partial \mathcal{S}_j\left(\vec{x}_j, \hat{\mu}, \hat{b}, t\right)}{\partial\hat{\mu}}\right) \\ & \times\left[\dot{\hat{\mu}} - \tau_i\left(\vec{x}_i, \hat{\mu}, \hat{b}, t\right)\right] \\ & + \left(\Lambda_b^{-1}\tilde{b} - \sum_{j=1}^{i-1}\mathcal{X}_{j+1}\frac{\partial \mathcal{S}_j\left(\vec{x}_j, \hat{\mu}, \hat{b}, t\right)}{\partial\hat{b}}\right) \\ & \times\left[\dot{\hat{b}} - \varpi_i\left(\vec{x}_i, \hat{\mu}, \hat{b}, t\right)\right].\end{aligned}$$

The first step is to verify (5.46) for $i = 1$. Direct calculation shows that the derivative of $W_1\left(z, \mathcal{X}_1, \hat{\mu}, \hat{b}\right)$ along the state trajectory satisfies

$$\begin{aligned}\dot{W}_1\left(z, \mathcal{X}_1, \hat{\mu}, \hat{b}\right) \le & -\Delta(z)\|z\|^2 + \varkappa(x_1)x_1^2 + x_1\left[f_1(z, x_1, d(t)) + bf_a^{\mathsf{T}}(x_1, t)\mu\right. \\ & \left. + b\mathcal{S}_1\left(x_1, t, \hat{\mu}\right) + b\mathcal{X}_2\right] + b\tilde{\mu}^{\mathsf{T}}\Lambda_\mu^{-1}\dot{\hat{\mu}} + \Lambda_b^{-1}\tilde{b}\dot{\hat{b}} \\ \le & -(r-1)m_1^2(z)\|z\|^2 - (r-1)m_2^2(\mathcal{X}_1)\|\mathcal{X}_1\|^2 - \|z\|^2 \\ & - \mathcal{X}_1^2 + \mathcal{X}_2^2 + b\tilde{\mu}^{\mathsf{T}}\Lambda_\mu^{-1}\left[\dot{\hat{\mu}} - \tau_1\left(x_1, \hat{\mu}, \hat{b}, t\right)\right] + \Lambda_b^{-1}\tilde{b}\dot{\hat{b}}.\end{aligned}$$

Thus (5.46) is verified with $i = 1$.

Next, we will prove (5.46) for any $1 < i \le r$ assuming it is true for $i - 1$. To this end, note that the derivative of $\mathcal{X}_i^2/2$ along the state trajectory of the $\mathcal{X}_i$-dynamics satisfies

$$
\begin{aligned}
\mathcal{X}_i\dot{\mathcal{X}}_i &= \mathcal{X}_i\dot{x}_i - \mathcal{X}_i\dot{\mathcal{S}}_{i-1}\left(\overrightarrow{x}_{i-1}, \hat{\mu}, \hat{b}, t\right) \\
&\le \mathcal{X}_i f_i\left(\overrightarrow{x}_i\right) + \mathcal{X}_i\mathcal{S}_i\left(\overrightarrow{x}_i, \hat{\mu}, \hat{b}, t\right) + \mathcal{X}_i^2/4 + \mathcal{X}_{i+1}^2 \\
&\quad + \left[\mathcal{X}_i\frac{\partial\mathcal{S}_{i-1}\left(\overrightarrow{x}_{i-1}, \hat{\mu}, \hat{b}, t\right)}{\partial x_1}\right]^2/2 + m_1^2(z)\|z\|^2 + m_2^2(x_1)x_1^2 \\
&\quad - \mathcal{X}_i\frac{\partial\mathcal{S}_{i-1}\left(\overrightarrow{x}_{i-1}, \hat{\mu}, \hat{b}, t\right)}{\partial x_1}\left[bf_a^{\mathrm{T}}(x_1, t)\mu + bx_2\right] \\
&\quad - \mathcal{X}_i\sum_{j=2}^{i-1}\frac{\partial\mathcal{S}_{i-1}\left(\overrightarrow{x}_{i-1}, \hat{\mu}, \hat{b}, t\right)}{\partial x_j}\left[f_j\left(\overrightarrow{x}_j\right) + x_{j+1}\right] \\
&\quad - \mathcal{X}_i\frac{\partial\mathcal{S}_{i-1}\left(\overrightarrow{x}_{i-1}, \hat{\mu}, \hat{b}, t\right)}{\partial\hat{\mu}}\dot{\hat{\mu}} - \mathcal{X}_i\frac{\partial\mathcal{S}_{i-1}\left(\overrightarrow{x}_{i-1}, \hat{\mu}, \hat{b}, t\right)}{\partial\hat{b}}\dot{\hat{b}} \\
&\quad - \mathcal{X}_i\frac{\partial\mathcal{S}_{i-1}\left(\overrightarrow{x}_{i-1}, \hat{\mu}, \hat{b}, t\right)}{\partial t} \\
&\le \mathcal{X}_i f_i\left(z, \overrightarrow{x}_i\right) + \mathcal{X}_i\mathcal{S}_i\left(\overrightarrow{x}_i, \hat{\mu}, \hat{b}, t\right) + \mathcal{X}_i^2/4 + \mathcal{X}_{i+1}^2 \\
&\quad + \left[\mathcal{X}_i\frac{\partial\mathcal{S}_{i-1}\left(\overrightarrow{x}_{i-1}, \hat{\mu}, \hat{b}, t\right)}{\partial x_1}\right]^2/2 + m_1^2(z)\|z\|^2 + m_2^2(x_1)x_1^2 \\
&\quad - \mathcal{X}_i\sum_{j=2}^{i-1}\frac{\partial\mathcal{S}_{i-1}\left(\overrightarrow{x}_{i-1}, \hat{\mu}, \hat{b}, t\right)}{\partial x_j}\left[f_j\left(\overrightarrow{x}_j\right) + x_{j+1}\right] \\
&\quad - \mathcal{X}_i\frac{\partial\mathcal{S}_{i-1}\left(\overrightarrow{x}_{i-1}, \hat{\mu}, \hat{b}, t\right)}{\partial\hat{\mu}}\tau_i\left(\overrightarrow{x}_i, \hat{\mu}, \hat{b}, t\right) \\
&\quad - \mathcal{X}_i\frac{\partial\mathcal{S}_{i-1}\left(\overrightarrow{x}_{i-1}, \hat{\mu}, \hat{b}, t\right)}{\partial\hat{b}}\varpi_i\left(\overrightarrow{x}_i, \hat{\mu}, \hat{b}, t\right) \\
&\quad - \mathcal{X}_i\frac{\partial\mathcal{S}_{i-1}\left(\overrightarrow{x}_{i-1}, \hat{\mu}, \hat{b}, t\right)}{\partial t} - \mathcal{X}_i\Gamma_i(\overrightarrow{x}_i, \hat{\mu}, \hat{b}, t)
\end{aligned}
$$

where

$$
\Gamma_i\left(\overrightarrow{x}_i, \hat{\mu}, \hat{b}, t\right) = \frac{\partial\mathcal{S}_{i-1}\left(\overrightarrow{x}_{i-1}, \hat{\mu}, \hat{b}, t\right)}{\partial x_1}b\left[f_a^{\mathrm{T}}(x_1, t)\hat{\mu} + x_2\right]
$$

$$+\frac{\partial \mathcal{S}_{i-1}\left(\overrightarrow{x}_{i-1},\hat{\mu},\hat{b},t\right)}{\partial x_1}\left[-bf_a^{\mathsf{T}}(x_1,t)\tilde{\mu}\right]$$
$$+\frac{\partial \mathcal{S}_{i-1}\left(\overrightarrow{x}_{i-1},\hat{\mu},\hat{b},t\right)}{\partial \hat{\mu}}\left[\dot{\hat{\mu}}-\tau_i\left(\overrightarrow{x}_{i},\hat{\mu},\hat{b},t\right)\right]$$
$$+\frac{\partial \mathcal{S}_{i-1}\left(\overrightarrow{x}_{i-1},\hat{\mu},\hat{b},t\right)}{\partial \hat{b}}\left[\dot{\hat{b}}-\varpi_i\left(\overrightarrow{x}_{i},\hat{\mu},\hat{b},t\right)\right].$$

By using the definition of $\mathcal{S}_i\left(\overrightarrow{x}_{i},\hat{\mu},\hat{b},t\right)$, we have

$$\begin{aligned}\mathcal{X}_i\dot{\mathcal{X}}_i &\le \mathcal{X}_i\varrho_i\left(\overrightarrow{x}_{i-1},\hat{\mu},\hat{b},t\right)-2\mathcal{X}_i^2+\mathcal{X}_{i+1}^2+m_1^2(z)\|z\|^2\\ &\quad+m_2^2(x_1)x_1^2-\mathcal{X}_i\Gamma_i\left(\overrightarrow{x}_{i},\hat{\mu},\hat{b},t\right).\end{aligned}$$

Now we are ready to verify that

$$\begin{aligned}\dot{W}_i\left(z,\overrightarrow{\mathcal{X}}_i,\hat{\mu},\hat{b}\right) &= \dot{W}_{i-1}\left(z,\overrightarrow{\mathcal{X}}_{i-1},\hat{\mu},\hat{b}\right)+\mathcal{X}_i\dot{\mathcal{X}}_i\\
&\le -(r-i+1)m_1^2(z)\|z\|^2-(r-i+1)m_2^2(\mathcal{X}_1)\|\mathcal{X}_1\|^2\\
&\quad -\|z\|^2-\sum_{j=1}^{i-1}\mathcal{X}_j^2+\mathcal{X}_i^2+\nabla_{i-1}\left(z,\overrightarrow{\mathcal{X}}_{i-1},\hat{\mu},\hat{b},t\right)\\
&\quad +\mathcal{X}_i\varrho_i\left(\overrightarrow{x}_{i-1},\hat{\mu},\hat{b},t\right)-2\mathcal{X}_i^2+\mathcal{X}_{i+1}^2+m_1^2(z)\|z\|^2\\
&\quad +m_2^2(x_1)x_1^2-\mathcal{X}_i\Gamma_i\left(\overrightarrow{x}_{i},\hat{\mu},\hat{b},t\right)\\
&\le -(r-i)m_1^2(z)\|z\|^2-(r-i)m_2^2(\mathcal{X}_1)\|\mathcal{X}_1\|^2\\
&\quad -\|z\|^2-\sum_{j=1}^{i}\mathcal{X}_j^2+\mathcal{X}_{i+1}^2+\nabla_{i-1}\left(z,\overrightarrow{\mathcal{X}}_{i-1},\hat{\mu},\hat{b},t\right)\\
&\quad +\mathcal{X}_i\varrho_i\left(\overrightarrow{x}_{i-1},\hat{\mu},\hat{b},t\right)-\mathcal{X}_i\Gamma_i\left(\overrightarrow{x}_{i},\hat{\mu},\hat{b},t\right).\end{aligned}$$

To complete the proof of (5.46), it suffices to verify that

$$\begin{aligned}\mathcal{X}_i\varrho_i\left(\overrightarrow{x}_{i-1},\hat{\mu},\hat{b},t\right) =&\ \nabla_i\left(z,\overrightarrow{\mathcal{X}}_i,\hat{\mu},\hat{b},t\right)-\nabla_{i-1}\left(z,\overrightarrow{\mathcal{X}}_{i-1},\hat{\mu},\hat{b},t\right)\\ &+\mathcal{X}_i\Gamma_i\left(\overrightarrow{x}_{i},\hat{\mu},\hat{b},t\right).\end{aligned}$$

In fact, direct calculation shows

$$\begin{aligned}
&\nabla_i\left(z,\vec{\mathcal{X}}_i,\hat{\mu},\hat{b},t\right)-\nabla_{i-1}\left(z,\vec{\mathcal{X}}_{i-1},\hat{\mu},\hat{b},t\right)\\
&=-\mathcal{X}_i\frac{\partial \mathcal{S}_{i-1}\left(\vec{x}_{i-1},\hat{\mu},\hat{b},t\right)}{\partial\hat{\mu}}\left[\dot{\hat{\mu}}-\tau_i\left(\vec{x}_i,\hat{\mu},\hat{b},t\right)\right]\\
&+\left(b\tilde{\mu}^{\mathsf{T}}\Lambda_\mu^{-1}-\sum_{j=1}^{i-2}\mathcal{X}_{j+1}\frac{\partial \mathcal{S}_j\left(\vec{x}_j,\hat{\mu},\hat{b},t\right)}{\partial\hat{\mu}}\right)\\
&\times\left[-\tau_i\left(\vec{x}_i,\hat{\mu},\hat{b},t\right)+\tau_{i-1}\left(\vec{x}_{i-1},\hat{\mu},\hat{b},t\right)\right]\\
&-\mathcal{X}_i\frac{\partial \mathcal{S}_{i-1}\left(\vec{x}_{i-1},\hat{\mu},\hat{b},t\right)}{\partial\hat{b}}\left[\dot{\hat{b}}-\varpi_i\left(\vec{x}_i,\hat{\mu},\hat{b},t\right)\right]\\
&+\left(\Lambda_b^{-1}\tilde{b}-\sum_{j=1}^{i-2}\mathcal{X}_{j+1}\frac{\partial \mathcal{S}_j\left(\vec{x}_j,\hat{\mu},\hat{b},t\right)}{\partial\hat{b}}\right)\\
&\times\left[-\varpi_i\left(\vec{x}_i,\hat{\mu},\hat{b},t\right)+\varpi_{i-1}\left(\vec{x}_{i-1},\hat{\mu},\hat{b},t\right)\right].
\end{aligned}$$

Using the equations in (5.44), one has

$$\begin{aligned}
&\nabla_i\left(z,\vec{\mathcal{X}}_i,\hat{\mu},\hat{b},t\right)-\nabla_{i-1}\left(z,\vec{\mathcal{X}}_{i-1},\hat{\mu},\hat{b},t\right)\\
&=-\mathcal{X}_i\frac{\partial \mathcal{S}_{i-1}\left(\vec{x}_{i-1},\hat{\mu},\hat{b},t\right)}{\partial\hat{\mu}}\left[\dot{\hat{\mu}}-\tau_i\left(\vec{x}_i,\hat{\mu},\hat{b},t\right)\right]\\
&+\left(b\tilde{\mu}^{\mathsf{T}}-\sum_{j=1}^{i-2}\mathcal{X}_{j+1}\frac{\partial \mathcal{S}_j\left(\vec{x}_j,\hat{\mu},\hat{b},t\right)}{\partial\hat{\mu}}\Lambda_\mu\right)\mathcal{X}_i\frac{\partial \mathcal{S}_{i-1}\left(\vec{x}_{i-1},\hat{\mu},\hat{b},t\right)}{\partial x_1}f_a(x_1,t)\\
&-\mathcal{X}_i\frac{\partial \mathcal{S}_{i-1}\left(\vec{x}_{i-1},\hat{\mu},\hat{b},t\right)}{\partial\hat{b}}\left[\dot{\hat{b}}-\varpi_i\left(\vec{x}_i,\hat{\mu},\hat{b},t\right)\right]\\
&+\left(\tilde{b}-\sum_{j=1}^{i-2}\mathcal{X}_{j+1}\frac{\partial \mathcal{S}_j\left(\vec{x}_j,\hat{\mu},\hat{b},t\right)}{\partial\hat{b}}\Lambda_b\right)\\
&\times\mathcal{X}_i\frac{\partial \mathcal{S}_{i-1}\left(\vec{x}_{i-1},\hat{\mu},\hat{b},t\right)}{\partial x_1}\left[f_a^{\mathsf{T}}(x_1,t)\hat{\mu}+x_2\right]\\
&=\mathcal{X}_i\varrho_i\left(\vec{x}_{i-1},\hat{\mu},\hat{b},t\right)-\mathcal{X}_i\Gamma_i\left(\vec{x}_i,\hat{\mu},\hat{b},t\right).
\end{aligned}$$

By mathematical induction, (5.46) holds for $i=1,\ldots,r$.

Finally, we consider the closed-loop system composed of (5.25) and the controller (5.38), i.e. $\mathcal{X}_{r+1}=0$, $\dot{\hat{\mu}}=\tau_r\left(\vec{x}_r,\hat{\mu},\hat{b},t\right)$, and $\dot{\hat{b}}=\varpi_r\left(\vec{x}_r,\hat{\mu},\hat{b},t\right)$. Then, the inequality (5.46) with $i = r$ becomes

$$\dot{W}_r\left(z, \vec{\mathcal{X}}_r, \hat{\mu}, \hat{b}\right) \leq -\|z\|^2 - \sum_{j=1}^{r} \mathcal{X}_j^2.$$

With $W_r\left(z, \vec{\mathcal{X}}_r, \hat{\mu}, \hat{b}\right)$ as the Lyapunov function for the closed-loop system, the stability of the closed-loop system is established by invoking Theorem 2.5 again. In particular, $\lim_{t\to\infty} \text{col}(z(t), \vec{\mathcal{X}}_r(t)) = 0$. That is, the GARP for the system (5.25) with the performance output $y = \text{col}(z, x_1)$ is solved by Proposition 5.2. To show the moreover part, it suffices to note $\mathcal{S}_i\left(0, \hat{\mu}, \hat{b}, t\right) = 0$ if $f_1(0, 0, d) = 0$, $f_a(0, t) = 0$, and $f_i(0) = 0$. □

Algorithm 5.2

INPUT: $f_i, i = 1, \cdots, r, f_a, \bar{b}, \underline{b}, \underline{\alpha}, \bar{\alpha}, \alpha, \sigma, \mathbb{D}$
OUTPUT: $\mathcal{S}_r$, τ_r, and ϖ_r
STEP 1: For the given $f_1(z, x_1, d)$, find m_1 and m_2 from (5.39).
STEP 2: Pick the function Δ from (5.40) and call

$$(\underline{\alpha}', \bar{\alpha}', \varkappa) = \text{ALGORITHM 2.3 } (\underline{\alpha}, \bar{\alpha}, \alpha, \sigma, \Delta).$$

STEP 3: Let $i = 1$; find the functions $\mathcal{S}_1$ and τ_1 from (5.43).
STEP 4: IF $i = r$, GO TO STEP 7, ELSE GO TO STEP 5.
STEP 5: Let $i = i + 1$; calculate the functions $\mathcal{S}_i$, τ_i, and ϖ_i from (5.44).
STEP 6: GO TO STEP 4.
STEP 7: END

Remark 5.2 When the parameter b is known, e.g., $b = 1$, the controller (5.38) can be simplified. In particular, the functions in (5.44) reduce to

$$\mathcal{S}_i\left(\vec{x}_i, \hat{\mu}, t\right) = \varrho_i\left(\vec{x}_{i-1}, \hat{\mu}, t\right) - \rho_i\left(\vec{x}_{i-1}, \hat{\mu}, t\right)\mathcal{X}_i + \nu_i\left(\vec{x}_i, \hat{\mu}, t\right)$$

$$\tau_i\left(\vec{x}_i, \hat{\mu}, t\right) = \tau_{i-1}\left(\vec{x}_{i-1}, \hat{\mu}, t\right) - \Lambda\mathcal{X}_i \frac{\partial \mathcal{S}_{i-1}\left(\vec{x}_{i-1}, \hat{\mu}, t\right)}{\partial x_1} f_a(x_1, t)$$

where

$$\varrho_i\left(\vec{x}_{i-1}, \hat{\mu}, t\right) = \left(-\sum_{j=1}^{i-2} \mathcal{X}_{j+1} \frac{\partial \mathcal{S}_j\left(\vec{x}_j, \hat{\mu}, t\right)}{\partial \hat{\mu}} \Lambda\right) \frac{\partial \mathcal{S}_{i-1}\left(\vec{x}_{i-1}, \hat{\mu}, t\right)}{\partial x_1} f_a(x_1, t)$$
$$+ \frac{\partial \mathcal{S}_{i-1}\left(\vec{x}_{i-1}, \hat{\mu}, t\right)}{\partial x_1} \left[f_a^{\mathsf{T}}(x_1, t)\hat{\mu} + x_2\right]$$

$$\rho_i\left(\overrightarrow{x}_{i-1}, \hat{\mu}, t\right) \geq \left[\frac{\partial \mathcal{S}_{i-1}\left(\overrightarrow{x}_{i-1}, \hat{\mu}, t\right)}{\partial x_1}\right]^2 /2 + 9/4$$

$$\begin{aligned}\nu_i\left(\overrightarrow{x}_{i}, \hat{\mu}, t\right) &= -f_i\left(\overrightarrow{x}_i\right) + \sum_{j=2}^{i-1} \frac{\partial \mathcal{S}_{i-1}\left(\overrightarrow{x}_{i-1}, \hat{\mu}, t\right)}{\partial x_j}\left[f_j\left(\overrightarrow{x}_j\right) + x_{j+1}\right] \\ &\quad + \frac{\partial \mathcal{S}_{i-1}\left(\overrightarrow{x}_{i-1}, \hat{\mu}, t\right)}{\partial \hat{\mu}} \tau_i\left(\overrightarrow{x}_i, \hat{\mu}, t\right) \\ &\quad + \frac{\partial \mathcal{S}_{i-1}\left(\overrightarrow{x}_{i-1}, \hat{\mu}, t\right)}{\partial t}.\end{aligned}$$

In the above equations, the matrix $\Lambda = \Lambda^{\mathsf{T}} > 0$ can be arbitrarily selected.

Now we will consider the convergence issue of the parameter $\hat{\mu}$ by the following corollary.

Corollary 5.2 *In Theorem 5.4, the closed-loop system composed of the given system* (5.25) *and the controller* (5.38) *has the properties that, for any initial state,*

$$\lim_{t\to\infty} \dot{\hat{\mu}}(t) = 0, \tag{5.47}$$

and, if $d(t)$, $\dot{d}(t)$, $f_a(x_1, t)$, $\frac{\partial f_a(x_1, t)}{\partial t}$, *and* $\frac{\partial f_a(x_1, t)}{\partial x_1}$ *are bounded for all* $t \geq 0$, *then*

$$\lim_{t\to\infty} f_a^{\mathsf{T}}(0, t)\left(\hat{\mu}(t) - \mu\right) = 0. \tag{5.48}$$

Moreover, if $f_a(0, t)$ *is PE, then*

$$\lim_{t\to\infty} \hat{\mu}(t) = \mu. \tag{5.49}$$

Proof In Theorem 5.4, it has been proved that $\lim_{t\to\infty} \operatorname{col}\left(z, \overrightarrow{\mathcal{X}}_r\right) = 0$ which implies $\lim_{t\to\infty} \tau_r\left(\overrightarrow{x}_r(t), \hat{\mu}(t), \hat{b}(t), t\right) = 0$, i.e., (5.47) is satisfied.

To show (5.48), we observe that since both x_1 and $\mathcal{X}_2 = x_2 + f_a^{\mathsf{T}}(x_1, t)\hat{\mu} + x_1\rho_1(x_1)$ approach 0 as $t \to \infty$, so does $x_2 + f_a^{\mathsf{T}}(0, t)\hat{\mu}$, i.e.,

$$\lim_{t\to\infty} \left(x_2(t) + f_a^{\mathsf{T}}(0, t)\hat{\mu}(t)\right) = 0. \tag{5.50}$$

Next, we will show

$$\lim_{t\to\infty} \left(x_2(t) + f_a^{\mathsf{T}}(0, t)\mu\right) = 0. \tag{5.51}$$

In fact, on one hand, x_1 approaches zero as $t \to \infty$. On the other hand, since $d(t)$, $\dot{d}(t)$, $f_a(x_1, t)$, $\dfrac{\partial f_a(x_1, t)}{\partial t}$, and $\dfrac{\partial f_a(x_1, t)}{\partial x_1}$ are bounded, $\ddot{x}_1$ is bounded and hence $\dot{x}_1$ is uniformly continuous. By Barbalat's Lemma, $\dot{x}_1$ approaches zero as $t \to \infty$, which implies (5.51). Comparing (5.50) and (5.51), we have $\lim_{t\to\infty} f_a^{\mathsf{T}}(0, t)\left(\hat{\mu}(t) - \mu\right) = 0$, i.e., (5.48). Finally, by Lemma 2.4, (5.49) holds if $f_a(0, t)$ is PE. □

Example 5.2 Consider a nonlinear system

$$\begin{aligned}\dot{z} &= -z + x_1/2\\ \dot{x}_1 &= d(t)(\sin x_1)z^2 + \mu[\cos(x_1) + a] + x_2\\ \dot{x}_2 &= u\end{aligned} \tag{5.52}$$

where $d(t) \in [-2, 2]$ is an external disturbance and μ is an unknown constant parameter. The parameter a is a known constant. Design a controller u to solve the GASP with the performance output $y = \mathrm{col}(z, x_1, x_2)$.

It is easy to verify the satisfaction of Assumption 5.2. So, the controller can be given in the form (5.38) where the functions ς_2 and τ_2 are given below.

First, we note $|d(\sin x_1)z^2| \le 2|z||z|$. Let $\Delta(z) \ge 1 + 8z^2$. Consider the function $V(z) = z^2 + 4z^4$ whose derivative along the state trajectory of $\dot{z} = -z + x_1/2$ satisfies

$$\begin{aligned}\dot{V}(z) &= 2z(-z + x_1/2) + 16z^3(-z + x_1/2)\\ &\le -z^2 + x_1^2/4 - 8z^4 + 8x_1^4\\ &= -\left(1 + 8z^2\right)z^2 + \left(1/4 + 8x_1^2\right)x_1^2.\end{aligned}$$

Then, we can pick a function $\rho_1(x_1) = 2 + 8x_1^2$ and hence

$$\begin{aligned}\varsigma_1(x_1, \hat{\mu}) &= -[\cos(x_1) + a]\hat{\mu} - x_1\rho_1(x_1)\\ \tau_1(x_1) &= x_1[\cos(x_1) + a].\end{aligned}$$

Define

$$\varsigma_{1x}\left(x_1, \hat{\mu}\right) = \frac{\partial \varsigma_1\left(x_1, \hat{\mu}\right)}{\partial x_1} = \sin(x_1)\hat{\mu} - 2 - 24x_1^2.$$

The remaining functions are given in order:

$$\begin{aligned}\rho_2\left(x_1, \hat{\mu}\right) &= \varsigma_{1x}^2\left(x_1, \hat{\mu}\right)/2 + 9/4\\ \varrho_2\left(x_1, \hat{\mu}\right) &= \varsigma_{1x}\left(x_1, \hat{\mu}\right)[\cos(x_1) + a]\hat{\mu} + \varsigma_{1x}\left(x_1, \hat{\mu}\right)x_2\end{aligned}$$

and

$$\tau_2\left(\vec{x}_2, \hat{\mu}\right) = \tau_1(x_1) - \left[x_2 - \varsigma_1\left(\vec{x}_1, \hat{\mu}\right)\right]\varsigma_{1x}\left(x_1, \hat{\mu}\right)[\cos(x_1) + a]$$
$$\nu_2\left(\vec{x}_2, \hat{\mu}\right) = -[\cos(x_1) + a]\tau_2\left(\vec{x}_2, \hat{\mu}\right).$$

Finally, we have

$$\begin{aligned}\varsigma_2\left(\vec{x}_2, \hat{\mu}\right) &= \varsigma_{1x}\left(x_1, \hat{\mu}\right)[\cos(x_1) + a]\hat{\mu} + \varsigma_{1x}\left(x_1, \hat{\mu}\right)x_2 \\ &\quad - \left[x_2 - \varsigma_1\left(x_1, \hat{\mu}\right)\right]\left[\varsigma_{1x}^2\left(x_1, \hat{\mu}\right)/2 + 9/4\right] \\ &\quad -[\cos(x_1) + a]\tau_2\left(\vec{x}_2, \hat{\mu}\right).\end{aligned}$$

By Theorem 5.4, the GASP is solved by the controller

$$u = \varsigma_2\left(\vec{x}_2, \hat{\mu}\right)$$
$$\dot{\hat{\mu}} = \tau_2\left(\vec{x}_2, \hat{\mu}\right).$$

The performance of the controller is simulated with $\mu = 2$, $z(0) = 5$, $x(0) = [1, 1]^{\mathrm{T}}$ and $\hat{\mu}(0) = 0$, and some results are shown in Figs. 5.3, 5.4, 5.5 and 5.6. Two cases are observed as follows.

Case 1: $a = -1$. In this case, the uncontrolled system has an equilibrium point $\mathrm{col}(z, x) = [0, 0, 0]$. The simulation result is shown in Figs. 5.3 and 5.4. In particular, the state variables z and x of the closed-loop system converge to the origin. However, the function $f_a(0, t) = \cos(0) + a = 0$ is obviously not PE. Therefore, the estimated parameter $\hat{\mu}$ does not converge to its real value of $\mu = 2$.

Case 2: $a = 0$. In this case, the origin $\mathrm{col}(z, x) = [0, 0, 0]$ is not an equilibrium point of the uncontrolled system. The simulation result is shown in Figs. 5.5 and 5.6. We can see that the state vaiables z and x_1 of the closed-loop system converge to 0 but the state x_2 does not. The function $f_a(0, t) = \cos(0) + a = 1$ is obviously PE. Therefore, the estimated parameter $\hat{\mu}$ converges to its real value of $\mu = 2$.

5.4 Adaptive Stabilization of the Duffing Equation

In this section, we consider the controlled Duffing equation (3.33). The external disturbance $v(t) = \gamma\cos(\omega t + \phi)$ can be rewritten as $v(t) = \gamma_1\cos(\omega t) + \gamma_2\sin(\omega t)$ for some constants γ_1 and γ_2. As a result, the model is

$$\begin{aligned}\dot{x}_1 &= x_2 \\ \dot{x}_2 &= -\delta x_2 - \alpha x_1 - \beta x_1^3 + \gamma_1\cos(\omega t) + \gamma_2\sin(\omega t) + u.\end{aligned} \tag{5.53}$$

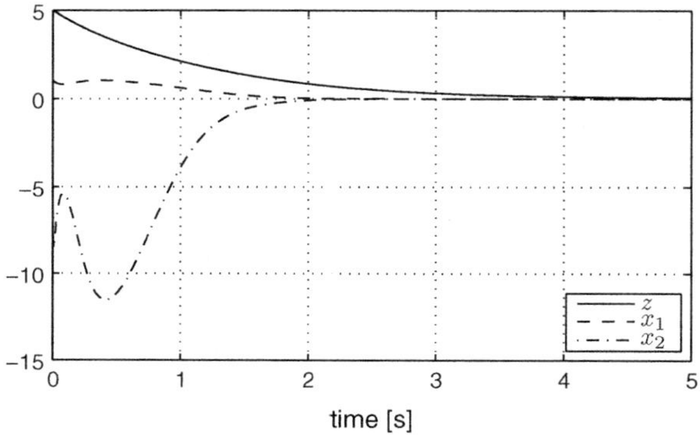

Fig. 5.3 Profile of state trajectories of the closed-loop system in Example 5.2: Case 1

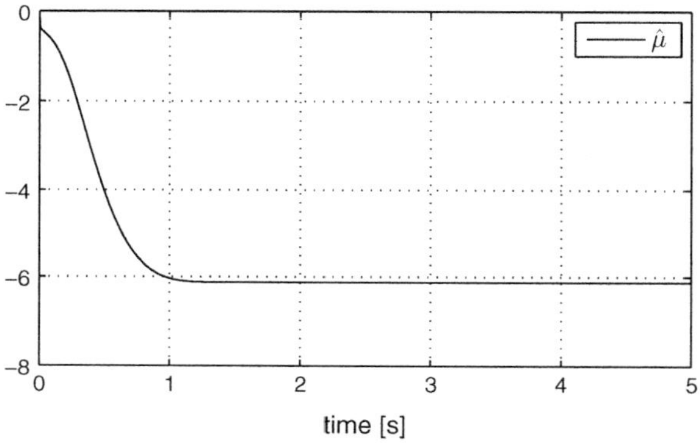

Fig. 5.4 Profile of estimated parameters of the closed-loop system in Example 5.2: Case 1

In (5.53), the parameters δ, α, β, γ_1 and γ_2 are unknown, but the frequency ω of the external signal $v(t)$ is assumed to be known. We will consider the regulation problem with the performance output and the measurement output being $y = y_{\mathrm{m}} = x = [x_1, x_2]^{\mathsf{T}}$.

First, let

$$\mathcal{X}_2 = x_2 + \rho_1 x_1, \quad \rho_1 > 0.$$

As a result, we have

$$\dot{x}_1 = -\rho_1 x_1 + \mathcal{X}_2$$

which is ISS with the state x_1 and the input $\mathcal{X}_2$. The $\mathcal{X}_2$-dynamics become

$$\dot{\mathcal{X}}_2 = f_a^{\mathsf{T}}(x_1, x_2, t)\mu + u \tag{5.54}$$

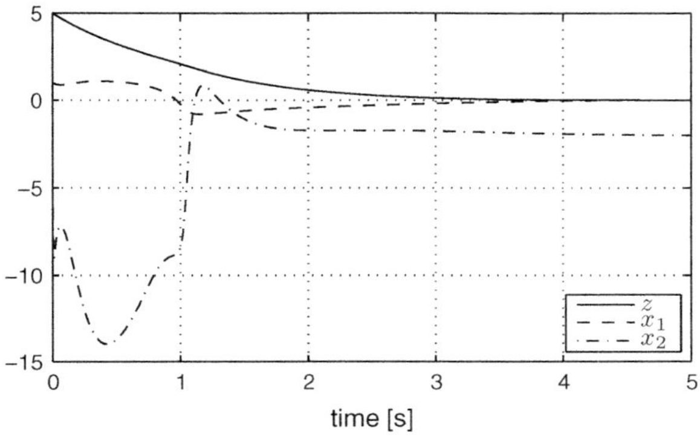

Fig. 5.5 Profile of state trajectories of the closed-loop system in Example 5.2: Case 2

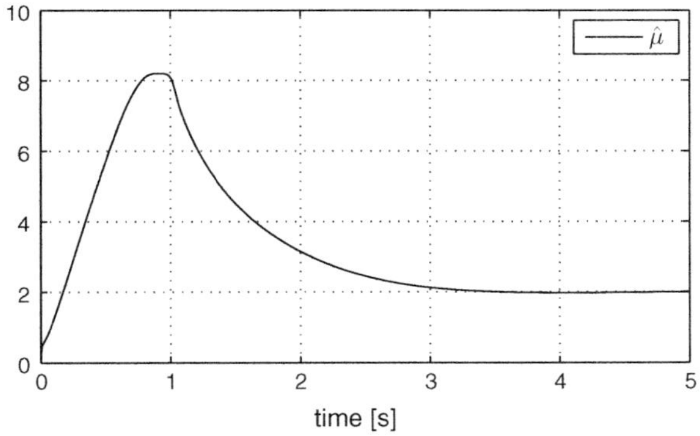

Fig. 5.6 Profile of estimated parameters of the closed-loop system in Example 5.2: Case 2

where

$$f_a(x_1, x_2, t) = \begin{bmatrix} x_2 \\ x_1 \\ x_1^3 \\ \cos(\omega t) \\ \sin(\omega t) \end{bmatrix}, \quad \mu = \begin{bmatrix} -\delta + \rho_1 \\ -\alpha \\ -\beta \\ \gamma_1 \\ \gamma_2 \end{bmatrix}.$$

By Theorem 5.4, an adaptive controller can be designed as follows:

$$\begin{aligned} u &= -\rho_2 \mathcal{X}_2 - f_a^{\mathsf{T}}(x_1, x_2, t)\hat{\mu}, \quad \rho_2 > 0 \\ \dot{\hat{\mu}} &= \Lambda \mathcal{X}_2 f_a(x_1, x_2, t). \end{aligned} \tag{5.55}$$

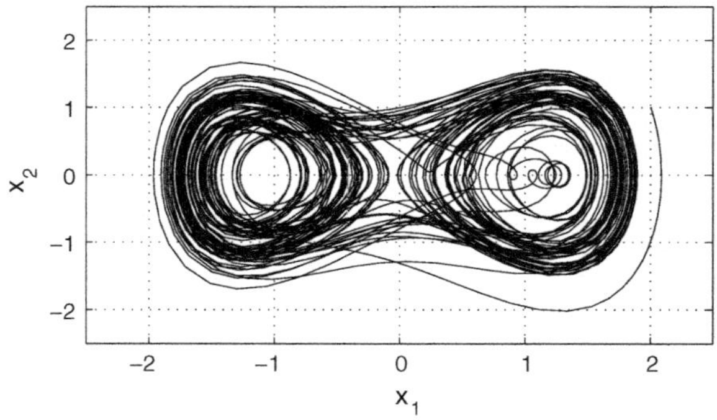

Fig. 5.7 Profile of state trajectories of the uncontrolled Duffing equation

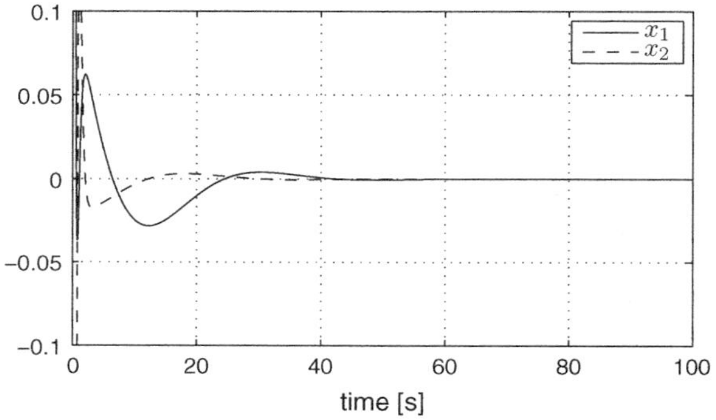

Fig. 5.8 Profile of state trajectories of the controlled Duffing equation

The performance of the controller is simulated with $\delta = 0.03$, $\alpha = -1$, $\beta = 1$, and $v(t) = 0.8\cos(0.2t + 0.1)$. The uncontrolled system trajectories are shown in Fig. 5.7. When the controller is applied with $\rho_1 = 3$ and $\rho_2 = 5$, the results are shown in Figs. 5.8 and 5.9. In particular, it is shown in Fig. 5.8 that the system state asymptotically approaches the equilibrium point $x = 0$. The control signal is shown in Fig. 5.9.

It is noted that the adaptive regulation method developed in this chapter cannot handle the unknown ω because $\cos(\omega t)$ and $\sin(\omega t)$ are not linear in ω. However, the internal model approach to be introduced in Chap. 9 is able to handle the unknown ω.

5.5 Notes and References

The adaptive backstetpping design for time-invariant nonlinear systems was first introduced in [1]. It requires multiple estimates of the same parameter. This over-

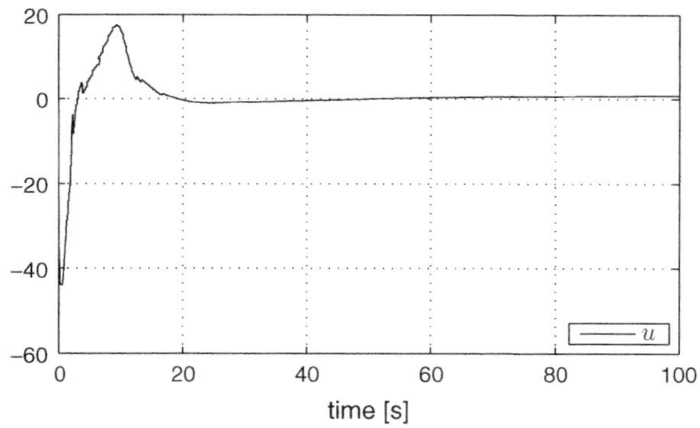

Fig. 5.9 Profile of control signal of the controlled Duffing equation

pamametrization was eliminated in [2] by using the tuning functions design. The adaptive control of time-invariant nonlinear systems is thoroughly covered in the book [3]. In this chapter, we have extended the approach in [1] to time-varying systems so that this approach can be applied to the adaptive output regulation problem studied in Chaps. 9 and 10 and we have also thoroughly addressed the parameter convergence issue. The materials in this chapter can find the references in [4–6]. The example on the Duffing equation can find the reference in [7].

5.6 Problems

Problem 5.1 Solve the GASP for each of the following systems with state x, input u, and an unknown constant parameter $\mu \in \mathbb{R}^2$.

(a) $\dot{x} = [\sin t, x + 1]\mu + u$;
(b) $\dot{x} = [x, x^2]\mu + u$;
(c) $\dot{x} = [\cos x, x/(1 + t)]\mu + u$.

Simulate the closed-loop system for $\mu = [2, 1]^{\mathsf{T}}$ starting from different initial conditions.

Problem 5.2 Use tuning functions design approach to solve the GARP for each of the following systems ($a = 1$) with state x, input u, and unknown constant parameters $\mu_1, \mu_2 \in \mathbb{R}$.

(a) $\dot{x}_1 = \mu_1(x_1^2 + a) + x_2, \dot{x}_2 = \mu_2 x_1 x_2 + u$;
(b) $\dot{x}_1 = \mu_1 x_1^2 + \mu_2 a + x_2, \dot{x}_2 = u$;
(c) $\dot{x}_1 = \mu_1(x_1 + a)/(t + 1) + x_2, \dot{x}_2 = \mu_2 x_2 + u$.

Simulate the closed-loop system for $\mu_1 = 3$ and $\mu_2 = -4$ starting from different initial conditions.

Problem 5.3 Repeat Problem 5.2 for $a = 0$ and show that the controllers indeed solve the GASP for the systems.

Problem 5.4 Solve the GASP for each of the following systems with state x, input u, and an unknown constant parameter $\mu \in \mathbb{R}^2$. The systems contain static uncertainty $-1 \leq d(t) < 1$ and dynamic uncertainty represented by the z-dynamics.

(a) $\dot{z} = -z + 0.5x,\ \dot{x} = d(t)|zx| + [\sin t, x + 1]\mu + u$;
(b) $\dot{z} = -(2 + d(t))z + x,\ \dot{x} = z^2 + 2d(t)x^2 + [x, x^2]\mu + u$;
(c) $\dot{z} = \begin{bmatrix} 0 & 1 \\ -1 & -2 \end{bmatrix} z + \begin{bmatrix} 0 \\ 0.5 \end{bmatrix} x,\ \dot{x} = d(t)\|z\|x + [\cos x, x/(1 + t)]\mu + u$.

Simulate the closed-loop system for $\mu = [2, 1]^{\mathsf{T}}$ starting from different initial conditions.

Problem 5.5 Repeat Problem 5.4 for the systems with an input filter added, i.e.,

(a) $\dot{z} = -z + 0.5x_1,\ \dot{x}_1 = d(t)|zx_1| + [\sin t, x_1 + 1]\mu + x_2,\ \dot{x}_2 = u$;
(b) $\dot{z} = -(2 + d(t))z + x_1,\ \dot{x}_1 = z^2 + 2d(t)x_1^2 + [x_1, x_1^2]\mu + x_2,\ \dot{x}_2 = u$;
(c) $\dot{z} = \begin{bmatrix} 0 & 1 \\ -1 & -2 \end{bmatrix} z + \begin{bmatrix} 0 \\ 0.5 \end{bmatrix} x_1,\ \dot{x}_1 = d(t)\|z\|x_1 + [\cos x_1, x_1/(1 + t)]\mu + x_2,$
$\dot{x}_2 = u$.

Problem 5.6 Determine if the estimation $\hat{\mu}$ of the unknown parameter μ converges to the real value of μ in Problem 5.5. Justify your answer.

References

1. Kanellakopoulos I, okotovi P, Morse AS (1991) Systematic design of adaptive controllers for feedback linearizable systems. IEEE Trans Autom Control 36:1241–1253
2. Krstic M, Kanellakopoulos I, Kokotovi P (1992) Adaptive nonlinear control without overparametrization. Syst control Lett 19:177–185
3. Krstic M, Kanellakopoulos I, Kokotović P (1995) Nonlinear and adaptive control design. Wiley, New York
4. Liu L, Chen Z, Huang J (2009) Parameter convergence and minimal internal model with an adaptive output regulation problem. Automatica 45:1306–1311
5. Liu L, Chen Z, Huang J (2011) Global disturbance rejection of lower triangular systems with an unknown linear exosystem. IEEE Trans Autom Control 56(7):1690–1695
6. Ye XD, Huang J (2003) Decentralized adaptive output regulation for a class of large-scale nonlinear systems. IEEE Trans Autom Control 48:276–281
7. Liu L, Huang J (2008) Asymptotic disturbance rejection of the Duffing system by adaptive output feedback control. IEEE Trans Circuit Syst 55:1030–1066

Chapter 6
Universal Adaptive Stabilization

In Chap. 4, we studied the GRSP for uncertain nonlinear systems containing both dynamic uncertainty and static uncertainty for the case where the boundary of the static uncertainty is known, i.e., $d(t) \in \mathbb{D}$ for a known compact set $\mathbb{D}$. Since the boundary of the static uncertainty is known, it is possible to find a sufficiently high gain to dominate the uncertainty $d(t)$ as we did in Chap. 4. However, as we pointed out in Sect. 2.6, in practice, the static uncertainty can be any bounded unknown time function or arbitrarily large real numbers. Thus, the boundary of $\mathbb{D}$ may be unknown. We recall that we have handled such a case for unity relative degree systems in Theorem 2.9 using the dynamic high gain technique. In this chapter, using the same dynamic high gain technique, we first extend Theorem 2.9 to general output feedback systems in Sect. 6.1. Then, in Sect. 6.2, we further consider lower triangular systems. In Sect. 6.3, we introduce another type of dynamic gain technique called the Nussbaum gain technique. This technique can not only handle the case where the boundary of the static uncertainty is unknown, but also handle the case where the control direction is unknown. In Sect. 6.4, we illustrate the dynamic high gain design method using the hyperchaotic Lorenz system. The notes and references are given in Sect. 6.5.

6.1 Output Feedback Systems

In this section, we will study the global stabilization problem for output feedback nonlinear systems of the following form:

$$\begin{aligned}
\dot{z} &= q(z, x_1, d)\\
\dot{x}_i &= f_i(z, x_1, d) + b_i x_{i+1}, \quad i = 1, \ldots, r-1\\
\dot{x}_r &= f_r(z, x_1, d) + b_r u\\
y &= x_1.
\end{aligned} \tag{6.1}$$

Z. Chen and J. Huang, *Stabilization and Regulation of Nonlinear Systems*,
Advanced Textbooks in Control and Signal Processing,
DOI 10.1007/978-3-319-08834-1_6

As explained in Sect. 4.2, to study the GRSP for the system (6.1), it suffices to study the GRSP for the so-called filter extended system (4.16), i.e.,

$$\begin{aligned} \dot{z} &= q(z, x_1, d) \\ \dot{x}_1 &= f_1(z, x_1, d) + bx_2, \ b > 0 \\ \dot{x}_i &= f_i\left(\overrightarrow{x}_i\right) + x_{i+1}, i = 2, \ldots, r \end{aligned} \tag{6.2}$$

where $\overrightarrow{x}_i = \mathrm{col}(x_1, \ldots, x_i)$ with $x_i \in \mathbb{R}$, $i = 1, \ldots, r$, are the state variables and $u := x_{r+1}$ is the input. The functions q and f_i in the system are sufficiently smooth with $q(0, 0, d) = 0$, $f_1(0, 0, d) = 0$ for all $d(t) \in \mathbb{D}$. The functions $f_i(\overrightarrow{x}_i)$ are sufficiently smooth vanishing at the origin.

In this section, we assume b is known to simplify the controller design. The more general case where b is uncertain can be regarded as a special lower triangular system which will be discussed in the next section. When $d(t)$ belongs to a compact set $\mathbb{D}$ whose boundary is known, the GRSP for the system (6.2) has been studied in Sect. 4.3 in detail. What makes this section distinct from what we have done in Sect. 4.3 is that we will not assume the boundary of $\mathbb{D}$ is known. In other words, we allow the component of $d(t)$ to be any bounded unknown time function or arbitrarily large real number.

The result of this section is an extension of Theorem 2.9 which handles a unity relative degree system. Therefore, an assumption similar to Assumption 2.5 is needed as follows.

Assumption 6.1 The subsystem $\dot{z} = q(z, x_1, d)$ has an ISS Lyapunov function $V(z)$, i.e.,

$$V(z) \sim \{\underline{\alpha}, \bar{\alpha}, \alpha, \hat{\sigma} \mid \dot{z} = q(z, x_1, d)\}$$

and

$$\limsup_{s \to 0^+} \frac{s^2}{\alpha(s)} < \infty, \quad \limsup_{s \to 0^+} \frac{\hat{\sigma}(s)}{s^2} < \infty.$$

Moreover, the functions $\underline{\alpha}$, $\bar{\alpha}$, and α are known and the function $\hat{\sigma}$ is known up to a constant factor in the sense that there exist an unknown constant p and a known function σ such that $\hat{\sigma} = p\sigma$.

As we did in Sect. 2.6 for $r = 1$, there are known smooth functions m_1 and m_2 and a positive real number c not necessarily known such that

$$|f_1(z, x_1, d)| \le cm_1(z)\|z\| + cm_2(x_1)|x_1|. \tag{6.3}$$

Let

$$\Delta(z) \geq 1 + rm_1^2(z). \tag{6.4}$$

By Corollary 2.3, for any sufficiently smooth function $\Delta(z) > 0$, there exists a continuously differentiable function $V'(z)$ satisfying $\underline{\alpha}'(\|z\|) \leq V'(z) \leq \bar{\alpha}'(\|z\|)$ for some class $\mathcal{K}_\infty$ functions $\underline{\alpha}'$ and $\bar{\alpha}'$, such that, along the trajectory of $\dot{z} = \mathcal{Q}(z, x_1, d)$,

$$\dot{V}'(z) \leq -\Delta(z)\|z\|^2 + p'\varkappa(x_1)x_1^2 \tag{6.5}$$

for some unknown positive number p' and a known smooth function $\varkappa$. Let

$$\begin{aligned} \mathcal{S}_1(k, x_1) &= -k\rho_1(x_1)x_1 \\ \rho_1(x_1) &\geq \max\{\varkappa(x_1), m_2(x_1), m_2^2(x_1), 1\}, \end{aligned} \tag{6.6}$$

and, for $i = 2, \ldots, r$, let

$$\mathcal{S}_i(k, \vec{x}_i) = -[x_i - \mathcal{S}_{i-1}(k, \vec{x}_{i-1})]\rho_i(k, \vec{x}_i) + \nu_i(k, \vec{x}_i) \tag{6.7}$$

where the functions ρ_i and ν_i are given as follows

$$\begin{aligned} \rho_i(k, \vec{x}_i) &\geq \left(\frac{\partial \mathcal{S}_{i-1}(k, \vec{x}_{i-1})}{\partial x_1}\right)^2 /2 + 9/4 \\ \nu_i(k, \vec{x}_i) &= -f_i(\vec{x}_i) + \frac{\partial \mathcal{S}_{i-1}(k, \vec{x}_{i-1})}{\partial x_1} bx_2 \\ &\quad + \sum_{j=2}^{i-1} \frac{\partial \mathcal{S}_{i-1}(k, \vec{x}_{i-1})}{\partial x_j}[f_j(\vec{x}_j) + x_{j+1}] \\ &\quad + \frac{\partial \mathcal{S}_{i-1}(k, \vec{x}_{i-1})}{\partial k}\lambda\rho_1(x_1)x_1^2. \end{aligned} \tag{6.8}$$

With these functions defined, we can state the main result as follows.

Theorem 6.1 *Consider the system* (6.2) *for any unknown compact set* $\mathbb{D}$. *Under Assumption* 6.1, the GASP is solved by the following controller:0

$$\begin{aligned} u &= \mathcal{S}_r(k, \vec{x}_r) \\ \dot{k} &= \tau(x_1) = \lambda\rho_1(x_1)x_1^2, \ \lambda > 0 \end{aligned} \tag{6.9}$$

where the functions ρ_1 and $\mathcal{S}_r$ are given in (6.6) and (6.7), respectively, or are obtained in Algorithm 6.1.

Proof Consider the following coordinate transformation

$$\mathcal{X}_1 = x_1$$
$$\mathcal{X}_{i+1} = x_{i+1} - \mathcal{S}_i(k, \overrightarrow{x}_i), \quad i = 1, \ldots, r.$$

For $i = 1, \ldots, r$, let

$$W_i(z, \overrightarrow{\mathcal{X}}_i, \hat{k}) = c^2 V'(z) + \sum_{j=1}^{i} \mathcal{X}_j^2/2 + b\hat{k}^2/(2\lambda),$$

where $\hat{k} = k - k^*$ for some constant $k^* > 0$ to be specified later. Without loss of generality, it is assumed that $c \geq 1$. We will prove the following inequality:

$$\begin{aligned} \dot{W}_i(z, \overrightarrow{\mathcal{X}}_i, \hat{k}) \leq & -(r-i)c^2 m_1^2(z)\|z\|^2 - (r-i)c^2 m_2^2(\mathcal{X}_1)\mathcal{X}_1^2 \\ & - \|z\|^2 - \sum_{j=1}^{i} \mathcal{X}_j^2 + \mathcal{X}_{i+1}^2 \end{aligned} \tag{6.10}$$

for $i = 1, \ldots, r$.

For $i = 1$, direct calculation shows that the derivative of $W_1(z, \mathcal{X}_1, \hat{k})$ along the trajectory of the closed loop system composed of (6.2) and (6.9) satisfies:

$$\begin{aligned} \dot{W}_1(z, \mathcal{X}_1, \hat{k}) \leq & -c^2\Delta(z)\|z\|^2 + c^2 p' \varkappa(\mathcal{X}_1)\mathcal{X}_1^2 + \mathcal{X}_1(f_1(z, \mathcal{X}_1, d) \\ & + b(\mathcal{X}_2 - k\rho_1(\mathcal{X}_1)\mathcal{X}_1)) + b(k - k^*)\rho_1(\mathcal{X}_1)\mathcal{X}_1^2 \\ \leq & -c^2\Delta(z)\|z\|^2 + c^2 m_1^2(z)\|z\|^2 + \mathcal{X}_1^2[c^2 p' \varkappa(\mathcal{X}_1) \\ & + 1/4 + cm_2(\mathcal{X}_1) + b^2/4 - bk^*\rho_1(\mathcal{X}_1)] + \mathcal{X}_2^2. \end{aligned}$$

Having $\Delta(z)$ and $\rho_1(\mathcal{X}_1)$ satisfy (6.4) and (6.6), respectively, and letting

$$k^* = [1 + c^2 p' + 1/4 + c + b^2/4 + (r-1)c^2]/b$$

gives

$$\dot{W}_1(z, \mathcal{X}_1, \hat{k}) \leq -(r-1)c^2 m_1^2(z)\|z\|^2 - (r-1)c^2 m_2^2(\mathcal{X}_1)\mathcal{X}_1^2 - \|z\|^2 - \mathcal{X}_1^2 + \mathcal{X}_2^2.$$

So, the inequality (6.10) with $i = 1$ is satisfied.

Next, we will show that the inequality (6.10) is true for i, $1 < i \leq r$, if it is true for $(i-1)$ by using the relationship

$$\dot{W}_i(z, \overrightarrow{\mathcal{X}}_i, \hat{k}) = \dot{W}_{i-1}(z, \overrightarrow{\mathcal{X}}_{i-1}, \hat{k}) + \mathcal{X}_i\dot{\mathcal{X}}_i.$$

We first note that

$$
\begin{aligned}
\mathcal{X}_i\dot{\mathcal{X}}_i =& \mathcal{X}_i[f_i(\overrightarrow{x}_i) + \mathcal{X}_{i+1} + \mathcal{S}_i(k, \overrightarrow{x}_i) - \dot{\mathcal{S}}_{i-1}(k, \overrightarrow{x}_{i-1})] \\
\le& \mathcal{X}_i[f_i(\overrightarrow{x}_i) + \mathcal{S}_i(k, \overrightarrow{x}_i)] + \mathcal{X}_i^2/4 + \left[\mathcal{X}_i \frac{\partial \mathcal{S}_{i-1}(k, \overrightarrow{x}_{i-1})}{\partial x_1}\right]^2 /2 \\
& - \mathcal{X}_i \frac{\partial \mathcal{S}_{i-1}(k, \overrightarrow{x}_{i-1})}{\partial x_1} b x_2 - \mathcal{X}_i \sum_{j=2}^{i-1} \frac{\partial \mathcal{S}_{i-1}(k, \overrightarrow{x}_{i-1})}{\partial x_j}[f_j(\overrightarrow{x}_j) + x_{j+1}] \\
& - \mathcal{X}_i \frac{\partial \mathcal{S}_{i-1}(k, \overrightarrow{x}_{i-1})}{\partial k} \lambda \rho_1(x_1) x_1^2 + c^2 m_1^2(z)\|z\|^2 + c^2 m_2^2(x_1) x_1^2 + \mathcal{X}_{i+1}^2 \\
\le& -2\mathcal{X}_i^2 + c^2 m_1^2(z)\|z\|^2 + c^2 m_2^2(\mathcal{X}_1)\mathcal{X}_1^2 + \mathcal{X}_{i+1}^2
\end{aligned}
$$

where the function $\mathcal{S}_i(k, \overrightarrow{x}_i)$ is defined in (6.7) and $\dot{\mathcal{S}}_{i-1}(k, \overrightarrow{x}_{i-1})$ is calculated as follows:

$$
\begin{aligned}
\dot{\mathcal{S}}_{i-1}(k, \overrightarrow{x}_{i-1}) =& \frac{\partial \mathcal{S}_{i-1}(k, \overrightarrow{x}_{i-1})}{\partial x_1}[f_1(z, x_1, d) + b x_2] \\
& + \sum_{j=2}^{i-1} \frac{\partial \mathcal{S}_{i-1}(k, \overrightarrow{x}_{i-1})}{\partial x_j}[f_j(\overrightarrow{x}_j) + x_{j+1}] \\
& + \frac{\partial \mathcal{S}_{i-1}(k, \overrightarrow{x}_{i-1})}{\partial k} \lambda \rho_1(x_1) x_1^2.
\end{aligned}
$$

By mathematical induction, the inequality (6.10) is true for $i = 1, \ldots, r$.

Finally, with $\mathcal{X}_{r+1} = 0$, the inequality (6.10) with $i = r$ becomes

$$
\dot{W}_r(z, \overrightarrow{\mathcal{X}}_r, \hat{k}) \le -\|z\|^2 - \sum_{j=1}^{r} \mathcal{X}_j^2. \tag{6.11}
$$

That is, the GASP for the system (6.2) is solved. □

Algorithm 6.1

INPUT: $\underline{\alpha}, \bar{\alpha}, \alpha, \sigma, f_i, i = 1, \ldots, r$

OUTPUT: $\rho_1, \mathcal{S}_i, i = 1, \ldots, r$

STEP 1: For a given f_1, find the functions m_1 and m_2 from (6.3).

STEP 2: Pick the function Δ from (6.4) and call

$$
(\underline{\alpha}', \bar{\alpha}', \varkappa) = \text{ALGORITHM } 2.3(\underline{\alpha}, \bar{\alpha}, \alpha, \sigma, \Delta).
$$

STEP 3: Let $i = 1$; find the functions $\mathcal{S}_1$ and ρ_1 from (6.6).

STEP 4: IF $i = r$, GO TO STEP 6, ELSE GO TO STEP 5.

STEP 5: Let $i = i + 1$; find the function $\mathcal{S}_i$ from (6.7); GO TO STEP 4
STEP 6: END

Example 6.1 Consider the following nonlinear system

$$\begin{aligned}\dot{z} &= -z + w_3 x_1\\ \dot{x}_1 &= w_1 z \sin x_1 + w_2 x_1^3 + x_2\\ \dot{x}_2 &= u\end{aligned}$$

where w_1, w_2, and w_3 are unknown parameters.

First, we verify that the derivative of $V(z) = z^2$ along the inverse dynamics satisfies

$$\dot{V}(z) \le -z^2 + p x_1^2$$

for any unknown constant $p \ge w_3^2$. That is, Assumption 6.1 is satisfied.

Next, we note

$$|w_1 z \sin x_1 + w_2 x_1^3| \le c(|z| + x_1^2 |x_1|)$$

for some $c \ge \max\{|w_1|, |w_2|\}$, that is, (6.3) is verified for $m_1 = 2$ and $m_2(x_1) = x_1^2$. Let $\Delta(z) = 1 + r m_1^2 = 9$. Since Δ is constant, by Corollary 2.3 and Remark 2.11, (6.5) is satisfied with $V'(z) = 9z^2$ and $\varkappa = 9$. Let $\rho_1(x_1) = x_1^4 + x_1^2 + 9 \ge \max\{9, x_1^2, x_1^4, 1\}$ which gives $\mathcal{S}_1(k, x_1) = -k(x_1^5 + x_1^3 + 9x_1)$. Denote

$$\mathcal{S}_{1x}(k, x_1) = \frac{\partial \mathcal{S}_1(k, x_1)}{\partial x_1} = -k(5x_1^4 + 3x_1^2 + 9).$$

Using Algorithm 6.1, we have

$$\mathcal{S}_2(k, x_1, x_2) = -[x_2 - \mathcal{S}_1(k, x_1)]\rho_2(k, x_1, x_2) + \nu_2(k, x_1, x_2)$$

where

$$\begin{aligned}\rho_2(k, x_1, x_2) &= \mathcal{S}_{1x}^2(k, x_1)/2 + 9/4\\ \nu_2(k, x_1, x_2) &= \mathcal{S}_{1x}(k, x_1)x_2 - \lambda \rho_1^2(x_1) x_1^3.\end{aligned}$$

The performance of the controller is simulated with $\lambda = 0.1$, $w_1 = 1.8$, $w_2 = 2$ and $w_3 = 1$, and the initial state values $z(0) = 20$, $x_1(0) = 1$, $x_2(0) = -10$, and $k(0) = 0$. The result is shown in Figs. 6.1 and 6.2. It can be observed that all the state variables of the plant approach the origin asymptotically while the dynamic gain k approaches a finite value asymptotically.

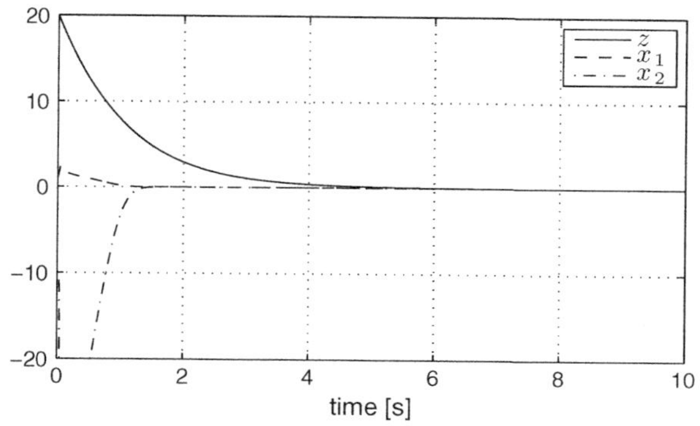

Fig. 6.1 Profile of state trajectories of the closed-loop system in Example 6.1

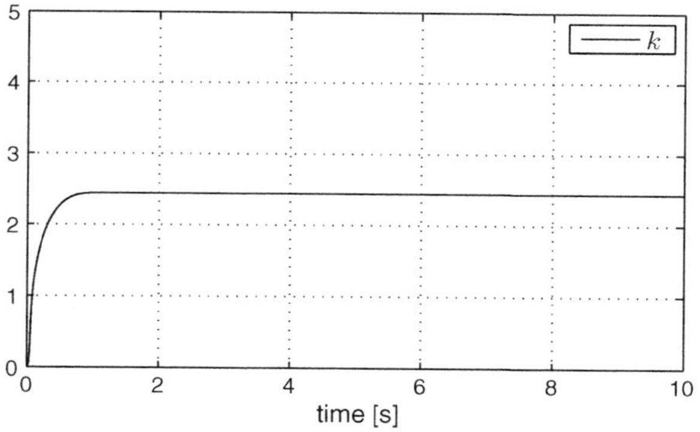

Fig. 6.2 Profile of dynamic gain of the closed-loop system in Example 6.1

6.2 Lower Triangular Systems

In this section, we will further consider the adaptive stabilization problem for lower triangular systems of the form (4.1), i.e.,

$$\begin{aligned} \dot{z}_i &= q_i(\overrightarrow{z}_i, \overrightarrow{x}_i, d(t)) \\ \dot{x}_i &= f_i(\overrightarrow{z}_i, \overrightarrow{x}_i, d(t)) + b_i x_{i+1},\ b_i > 0,\ i = 1, \ldots, r, \end{aligned} \tag{6.12}$$

where $\overrightarrow{z}_i = \text{col}(z_1, \ldots, z_i)$ with $z_i \in \mathbb{R}^{n_i}$ and $\overrightarrow{x}_i = \text{col}(x_1, \ldots, x_i)$ with $x_i \in \mathbb{R}$ are the state variables, $u := x_{r+1} \in \mathbb{R}$ is the input, $d(t) \in \mathbb{D}$ represents system uncertainties with $\mathbb{D}$ an arbitrarily unknown compact set. In Sect. 4.3, it is known that for all $d \in \mathbb{D}$, one has $\bar{b}_i \geq b_i(d) \geq \underline{b}_i$ for some known positive numbers $\bar{b}_i$ and $\underline{b}_i$. In this section, we assume b_i is an arbitrary unknown positive constant. It is assumed that all functions in the system (6.12) are sufficiently smooth satisfying $q_i(0, 0, d) = 0$ and $f_i(0, 0, d) = 0$. Also, we assume that the measurement output and the performance output of (6.12) are $y_m = \vec{x}_r$, and $y = \text{col}(\vec{z}_r, \vec{x}_r)$, respectively. That is, the control law will rely on y_m and will make the state of the closed-loop system bounded and $\lim_{t\to\infty} y(t) = 0$.

Again, we will use the universal adaptive approach as we did in Sect. 2.6 and 6.1. However, what makes the current task more challenging is that we need to design multiple dynamic gains recursively. Thus, we need to extend the changing supply function technique as shown in the following section.

6.2.1 Parameterized Changing Supply Function

Recall that, in Theorem 2.9, for a unity relative degree system, we have derived the closed-loop system as follows

$$\begin{aligned}\dot{z} &= q(z, x, d)\\ \dot{x} &= f(z, x, d) + b(-k\rho(x)x + \bar{u})\\ \dot{k} &= \lambda\rho(x)x^2, \ \lambda > 0.\end{aligned} \tag{6.13}$$

Let $\zeta = \text{col}(z, x, k)$. Then (6.13) can be put in the following form

$$\dot{\zeta} = \varphi(\zeta, \bar{u}, d). \tag{6.14}$$

Also, we have constructed an ISS Lyapunov function $W(\hat{\zeta})$ with $\hat{\zeta} = \text{col}(z, x, \hat{k})$ as follows

$$W(\hat{\zeta}) = V'(z) + x^2/2 + b\hat{k}^2/(2\lambda)$$

which is such that

$$\begin{aligned}\underline{\beta}(\|\hat{\zeta}\|) \le W(\hat{\zeta}, t) &\le \bar{\beta}(\|\hat{\zeta}\|)\\ \dot{W}(\hat{\zeta}) &\le -\|z\|^2 - \|x\|^2 + \|\bar{u}\|^2.\end{aligned}$$

However, the class $\mathcal{K}_\infty$ functions $\underline{\beta}$ and $\bar{\beta}$ constructed from (2.70) are not exactly known as they depend on the unknown constant b. It does not cause any problem in Theorem 2.9 as $\underline{\beta}$ or $\bar{\beta}$ is not used for the controller design. However, it does incur difficulty if $r > 1$ since the information of $\underline{\beta}$ and $\bar{\beta}$ will be used in the subsequent steps in a recursive procedure. To overcome this difficulty, we introduce the concept of the so-called parameterized class $\mathcal{K}_\infty$ function.

Definition 6.1 A continuous function $\alpha : [0, \infty) \times (0, \infty) \mapsto [0, \infty)$ is said to be a *parameterized class* $\mathcal{K}_\infty$ function if, for each fixed $s > 0$, the function $\alpha(\cdot, s)$ is a class $\mathcal{K}_\infty$ function, and for each fixed $r > 0$, the function $\alpha(r, \cdot)$ is nondecreasing.

In Theorem 2.9, if we multiply the function $W(\hat{\zeta})$ by a factor $(1 + 1/b)$, then we have

$$W(\hat{\zeta}) = (1 + 1/b)\left[V'(z) + x^2/2 + b\hat{k}^2/(2\lambda)\right].$$

It can be verified that we still have

$$\dot{W}(\hat{\zeta}) \leq -\|z\|^2 - \|x\|^2 + \bar{u}^2$$

by slightly modifying the proof. Now, we have

$$\underline{\alpha}'(\|z\|) + x^2/2 + \hat{k}^2/(2\lambda) \leq W(\hat{\zeta}) \leq s[\bar{\alpha}'(\|z\|) + x^2/2 + \hat{k}^2/(2\lambda)]$$

for any parameter $s \geq \max\{1+1/b, 1+b\}$, which is not necessarily known. It is easy to pick two known class $\mathcal{K}_\infty$ functions $\underline{\beta}$ and $\bar{\beta}_0$, using Lemma 11.3 of the Appendix so that

$$\begin{aligned}\underline{\beta}(\|\text{col}(z, x, \hat{k})\|) &\leq \underline{\alpha}'(\|z\|) + x^2/2 + \hat{k}^2/(2\lambda)\\ \bar{\beta}_0(\|\text{col}(z, x, \hat{k})\|) &\geq \bar{\alpha}'(\|z\|) + x^2/2 + \hat{k}^2/(2\lambda).\end{aligned}$$

Clearly, $\bar{\beta}(\|\hat{\zeta}\|, s) = s\bar{\beta}_0(\|\hat{\zeta}\|)$ is a parameterized class $\mathcal{K}_\infty$ function. As a result, we have

$$\underline{\beta}(\|\hat{\zeta}\|) \leq W(\hat{\zeta}) \leq \bar{\beta}(\|\hat{\zeta}\|, s).$$

From the above manipulation, we see that the term $1 + 1/b$ is added so that a known class $\mathcal{K}_\infty$ function $\underline{\beta}$ exists as the lower bound of $W(\hat{\zeta})$. But $W(\hat{\zeta})$ has to be bounded from the above by a parameterized class $\mathcal{K}_\infty$ function $\bar{\beta}$.

We now further modify the changing supply function technique so that it can accommodate parameterized class $\mathcal{K}_\infty$ functions.

Lemma 6.1 (Parameterized Changing Supply Function) *Consider a system* $\dot{x} = f(x, u, d)$ *with* $x = col(x_1, x_2) \in \mathbb{R}^n$ *and* $u \in \mathbb{R}^m$. *Suppose there exists a supply function* $V(x)$ *satisfying*

$$\begin{aligned}\underline{\alpha}(\|x\|) &\leq V(x) \leq \bar{\alpha}(\|x\|, s)\\ \dot{V}(x) &\leq -\alpha(\|x_1\|) + p\sigma(\|u\|)\end{aligned} \tag{6.15}$$

for class $\mathcal{K}_\infty$ *functions* α *and* $\underline{\alpha}$, *a class* $\mathcal{K}$ *function* σ, *a parameterized class* $\mathcal{K}_\infty$ *function* $\bar{\alpha}$, *and positive numbers* p *and* s. *Then, for any smooth function*

$\Delta : \mathbb{R}^n \mapsto [0, \infty)$ and positive number k, there exists a continuously differentiable function $V'(x)$ satisfying

$$\begin{aligned} \underline{\alpha}'(\|x\|) &\le V'(x) \le \bar{\alpha}'(\|x\|, s') \\ \dot{V}'(x) &\le -k\Delta(x)\alpha(\|x_1\|) + p'\varkappa(x_2, u)\sigma(\|u\|). \end{aligned} \tag{6.16}$$

for a class $\mathcal{K}_\infty$ function $\underline{\alpha}'$, a parameterized class $\mathcal{K}_\infty$ function $\bar{\alpha}'$, a smooth positive function $\varkappa$, and positive numbers p' and s'. Moreover, if the functions $\alpha, \underline{\alpha}, \bar{\alpha}, \sigma$ and Δ are known, so are the functions $\underline{\alpha}', \bar{\alpha}'$ and $\varkappa$, as given by Algorithm 6.2. The positive numbers s, p, k, s' and p' are not necessarily known.

Proof Let

$$\underline{\alpha}'(\|x\|) = \int_0^{\underline{\alpha}(\|x\|)} \rho(\tau)d\tau, \quad \bar{\alpha}'(\|x\|, s') = s' \int_0^{\bar{\alpha}(\|x\|, s')} \rho(\tau)d\tau \tag{6.17}$$

with $s' = \max\{s, \hat{k}\}$ and $\hat{k} = \max\{k, 1\}$, and let

$$V'(x) = \int_0^{V(x)} \hat{k}\rho(\tau)d\tau$$

where $\rho : [0, \infty) \mapsto [0, \infty)$ is an $\mathcal{SN}$ function. Clearly, $\underline{\alpha}'$ is a class $\mathcal{K}_\infty$ function, $\bar{\alpha}'$ is a parameterized class $\mathcal{K}_\infty$ function, and $V'(x)$ is continuously differentiable and satisfies

$$\underline{\alpha}'(\|x\|) \le V'(x) \le \bar{\alpha}'(\|x\|, s')$$

and

$$\begin{aligned} \dot{V}'(x) &= \hat{k}\rho(V(x))[-\alpha(\|x_1\|) + p\sigma(\|u\|)] \\ &\le -\hat{k}\rho(V(x))\alpha(\|x_1\|) + p\hat{k}\rho(V(x))\sigma(\|u\|). \end{aligned} \tag{6.18}$$

We claim that (6.18) implies

$$\begin{aligned} \dot{V}'(x) \le& -\hat{k}\frac{1}{2}\rho\left[\underline{\alpha}(\|x\|)\right]\alpha(\|x_1\|) \\ &+ p\hat{k}\rho\left[\bar{\alpha}\left(\|x_2\| + \|\alpha^{-1}(2p\sigma(\|u\|))\|, s\right)\right]\sigma(\|u\|). \end{aligned} \tag{6.19}$$

In fact, consider the following two cases.

(i) $\alpha(\|x_1\|)/2 \geq p\sigma(\|u\|)$: In this case, the claim follows from the fact that the right-hand side of (6.18) is bounded from above by $-\hat{k}\rho\,[V(x)]\,\alpha(\|x_1\|)/2$, and hence bounded from above by $-\hat{k}\rho\left[\underline{\alpha}(\|x\|)\right]\alpha(\|x_1\|)/2$.

(ii) $\alpha(\|x_1\|)/2 < p\sigma(\|u\|)$: In this case, it suffices to prove the following two inequalities

$$\begin{aligned} -\rho(V(x)) &\leq -\rho\left[\underline{\alpha}(\|x\|)\right]/2 \\ \rho(V(x)) &\leq \rho\left[\bar{\alpha}\left(\|x_2\| + \|\alpha^{-1}(2p\sigma(\|u\|))\|, s\right)\right]. \end{aligned} \tag{6.20}$$

The first inequality holds since $\underline{\alpha}(\|x\|) \leq V(x)$, and the second one holds since

$$V(x) \leq \bar{\alpha}(\|x\|, s) \leq \bar{\alpha}(\|x_1\| + \|x_2\|, s) \text{ and } \|x_1\| \leq \alpha^{-1}(2p\sigma(\|u\|)).$$

So far, we have shown that (6.19) holds for any continuously differentiable nondecreasing function ρ satisfying $\rho(\tau) > 0, \forall \tau > 0$. We will further show that ρ can be chosen such that (6.16) holds. To this end, for any smooth function Δ, letting

$$\rho_0(\tau) = 2\bar{\Delta}\left(\underline{\alpha}^{-1}(\tau)\right), \quad \bar{\Delta}(\tau) = \max_{\|x\|=\tau}\{\Delta(x)\} \tag{6.21}$$

gives

$$\frac{1}{2}\rho_0[\underline{\alpha}(\|x\|)] = \bar{\Delta}\left(\underline{\alpha}^{-1}(\underline{\alpha}(\|x\|))\right) = \bar{\Delta}(\|x\|) \geq \Delta(x).$$

By construction, ρ_0 is continuous. Let $\rho(\tau) \geq \rho_0(\tau), \ \forall \tau \geq 0$. As a result,

$$-\hat{k}\frac{1}{2}\rho\left[\underline{\alpha}(\|x\|)\right]\alpha(\|x_1\|) \leq -k\Delta(x)\alpha(\|x_1\|).$$

By (11.7), we note

$$p\hat{k}\rho\left[\bar{\alpha}\left(\|x_2\| + \|\alpha^{-1}(2p\sigma(\|u\|))\|, s\right)\right] \leq c(p, \hat{k}, s)\varkappa(x_2, u) \tag{6.22}$$

for some smooth functions $c(p, \hat{k}, s) \geq 1$ and $\varkappa(x_2, u) \geq 1$. Letting $p' = c(p, \hat{k}, s)$ gives

$$p\hat{k}\rho\left[\bar{\alpha}\left(\|x_2\| + \|\alpha^{-1}(2p\sigma(\|u\|))\|, s\right)\right] \leq p'\varkappa(x_2, u).$$

□

Algorithm 6.2

INPUT: $\underline{\alpha}, \bar{\alpha}, \alpha, \sigma, \Delta$
OUTPUT: $\underline{\alpha}', \bar{\alpha}', \varkappa$
STEP 1: Pick an $\mathcal{SN}$ function $\rho(\tau) \geq \rho_0(\tau), \forall \tau \geq 0$ for ρ_0 given by (6.21).
STEP 2: Find the functions $\underline{\alpha}'$ and $\bar{\alpha}'$ from (6.17).
STEP 3: Find the function $\varkappa$ from (6.22).
STEP 4: END

Remark 6.1 Consider a special case of Lemma 6.1 when $k = 1$ and $x = x_1$ (hence $x_2 \in \mathbb{R}^0$), we can rewrite the inequality (6.16) as follows

$$\dot{V}'(x) \leq -\Delta(x)\alpha(\|x\|) + p'\varkappa(u)\sigma(\|u\|). \tag{6.23}$$

Further, if s and p are known, so are s' and p'. Then, Lemma 6.1 reduces to Corollary 2.1.

6.2.2 A Recursive Procedure

In contrast with Theorem 2.9 which deals with unity relative degree systems, when $r > 1$, the approach is recursive. We will repeatedly encounter the stabilization problem of the following inter-connected system:

$$\begin{aligned} \dot{\zeta} &= \varphi(\zeta, \mathcal{X}, d) \\ \dot{z} &= \mathcal{Q}(z, \epsilon(\zeta, \mathcal{X}), d) \\ \dot{\mathcal{X}} &= \mathcal{F}(\zeta, z, \mathcal{X}, d) + bu, \ \ b > 0 \end{aligned} \tag{6.24}$$

where $\zeta \in \mathbb{R}^{n_1}$, $z \in \mathbb{R}^{n_2}$, $\mathcal{X} \in \mathbb{R}$, and $u \in \mathbb{R}$. In the setup, all functions are sufficiently smooth. The top ζ-subsystem is motivated from (6.13), the z-subsystem represents the dynamic uncertainty, and the $\mathcal{X}$-subsystem produces the measured state $\mathcal{X}$. To introduce the performance output and measurement output for (6.24), we partition $\zeta = \text{col}(\varsigma, \xi)$ and define

$$y = h(\zeta, z, \mathcal{X}) = \text{col}(\varsigma, z, \mathcal{X}), \quad \bar{y} = \bar{h}(\zeta, z, \mathcal{X}) = \text{col}(\xi, \mathcal{X}). \tag{6.25}$$

We list the following two assumptions on (6.24).

Assumption 6.2 The subsystem $\dot{z} = \mathcal{Q}(z, v, d)$ has an ISS Lyapunov function $V(z)$, i.e.,

$$V(z) \sim \{\underline{\alpha}, \bar{\alpha}, \alpha, \hat{\sigma} \mid \dot{z} = \mathcal{Q}(z, v, d)\}$$

and

$$\limsup_{s\to 0^+}\frac{s^2}{\alpha(s)}<\infty,\quad \limsup_{s\to 0^+}\frac{\hat{\sigma}(s)}{s^2}<\infty.$$

Moreover, the functions $\underline{\alpha}$, $\bar{\alpha}$, and α are known and the function $\hat{\sigma}$ is known up to a constant factor in the sense that there exist an unknown constant p and a known function σ such that $\hat{\sigma}=p\sigma$.

Assumption 6.3 There exists a continuously differentiable function $W(\hat{\zeta})$ where $\hat{\zeta}=\mathrm{col}(\varsigma,\hat{\xi})$ and $\hat{\xi}=\xi-\xi^*$ for some positive constant vector ξ^* such that

$$\underline{\beta}(\|\hat{\zeta}\|)\leq W(\hat{\zeta})\leq\bar{\beta}(\|\hat{\zeta}\|,s)$$

for a class $\mathcal{K}_\infty$ function $\underline{\beta}$, a parameterized class $\mathcal{K}_\infty$ function $\bar{\beta}$, and a positive constant s which is not necessarily known, and along the trajectory of $\dot{\zeta}=\varphi(\zeta,\mathcal{X},d)$,

$$\dot{W}(\hat{\zeta})\leq-\|\varsigma\|^2+\mathcal{X}^2.$$

We are now ready to develop a recursive control design approach based on the system (6.24).

Lemma 6.2 *Suppose the system* (6.24) *satisfies Assumptions* 6.2 *and* 6.3 *and the following two conditions:*

(i) $\epsilon(\zeta,\mathcal{X})=G(\zeta)[\varsigma^{\mathrm{T}},\mathcal{X}]^{\mathrm{T}}$ *for a matrix* $G(\zeta)$*;*
(ii) $|\mathcal{F}(\zeta,z,\mathcal{X},d)|\leq c\,[m_0(\varsigma,\xi)\|\varsigma\|+m_1(z)\|z\|+m_2(\xi,\mathcal{X})|\mathcal{X}|]$ *for some sufficiently smooth functions* m_0, m_1, *and* m_2, *and a positive number* c *which is not necessarily known.*

Then, there exist a sufficiently smooth positive function $\rho(\xi,\mathcal{X})\geq 1$ *which defines the control law*

$$\begin{aligned}u&=-k\rho(\xi,\mathcal{X})\mathcal{X}+\mathcal{U}\\ \dot{k}&=\lambda\rho(\xi,\mathcal{X})\mathcal{X}^2,\ \lambda>0\end{aligned}\tag{6.26}$$

and a continuously differentiable function $W'(\hat{\zeta},z,\mathcal{X},\hat{k})$ *where* $\hat{k}=k-k^*$ *for a positive number* k^* *such that*

$$\underline{\beta}'(\|col(\hat{\zeta},z,\mathcal{X},\hat{k})\|)\leq W'(\hat{\zeta},z,\mathcal{X},\hat{k})\leq\bar{\beta}'(\|col(\hat{\zeta},z,\mathcal{X},\hat{k})\|,s')\tag{6.27}$$

for a class $\mathcal{K}_\infty$ *function* $\underline{\beta}'$, *a parameterized class* $\mathcal{K}_\infty$ *function* $\bar{\beta}'$, *and a positive number* s' *which is not necessarily known, and along the trajectory of the closed-loop system,*

$$\dot{W}'(\hat{\zeta},z,\mathcal{X},\hat{k})\leq-\|col(\varsigma,z,\mathcal{X})\|^2+\mathcal{U}^2.\tag{6.28}$$

In particular, the construction of ρ, $\underline{\beta}'$, *and* $\bar{\beta}'$ *are summarized in Algorithm* 6.3.

Proof First, under Assumption 6.3, applying Lemma 6.1 shows that, for any smooth function $\Delta_0(\hat{\zeta}) \geq 0$, any positive number k_0 which is not necessarily known, there exists a continuously differentiable function $W_0(\hat{\zeta})$ satisfying

$$\underline{\beta}_0(\|\hat{\zeta}\|) \leq W_0(\hat{\zeta}) \leq \bar{\beta}_0(\|\hat{\zeta}\|, s_0) \tag{6.29}$$

for a class $\mathcal{K}_\infty$ function $\underline{\beta}_0$, a parameterized class $\mathcal{K}_\infty$ function $\bar{\beta}_0$, and a positive number s_0, which is not necessarily known, and along the trajectory of $\dot{\zeta} = \varphi(\zeta, \mathcal{X}, d)$,

$$\dot{W}_0(\hat{\zeta}) \leq -k_0\Delta_0(\hat{\zeta})\|\varsigma\|^2 + p_0'\varkappa_0(\hat{\xi}, \mathcal{X})\|\mathcal{X}\|^2. \tag{6.30}$$

for a smooth function $\varkappa_0(\hat{\xi}, \mathcal{X})$ and a positive number p_0' which is not necessarily known.

Under Assumption 6.2, by Corollary 2.3, for a smooth function $\Delta(z)$, there exists a continuously differentiable function $V'(z)$, satisfying

$$\underline{\alpha}'(\|z\|) \leq V'(z) \leq \bar{\alpha}'(\|z\|)$$

for two class $\mathcal{K}_\infty$ functions $\underline{\alpha}'$ and $\bar{\alpha}'$, such that, along the trajectory of $\dot{z} = \mathcal{Q}(z, v, d)$,

$$\dot{V}'(z) \leq -\Delta(z)\|z\|^2 + p'\varkappa(v)\|v\|^2 \tag{6.31}$$

for a smooth function $\varkappa$ and a positive number p' which is not necessarily known. In the system (6.24), we note $v = \epsilon(\zeta) = G(\zeta)[\varsigma^\top, \mathcal{X}]^\top$ under the condition (i). Then,

$$\varkappa(v)\|v\|^2 = \varkappa(\epsilon(\zeta))\|\epsilon(\zeta)\|^2 \leq \varpi(\varsigma, \xi)\|\varsigma\|^2 + \pi(\xi, \mathcal{X})\|\mathcal{X}\|^2 \tag{6.32}$$

for some smooth functions $\varpi(\varsigma, \xi) \geq 0$ and $\pi(\xi, \mathcal{X}) \geq 0$. As a result,

$$\dot{V}'(z) \leq -\Delta(z)\|z\|^2 + p'\varpi(\varsigma, \xi)\|\varsigma\|^2 + p'\pi(\xi, \mathcal{X})\|\mathcal{X}\|^2 \tag{6.33}$$

The third step is to consider the $\mathcal{X}$-subsystem under the condition (ii). Along the trajectory of the $\mathcal{X}$-subsystem of (6.24), the derivative of $\mathcal{X}^2/2$ satisfies

$$\begin{aligned}
\mathcal{X}\dot{\mathcal{X}} &= \mathcal{X}\left[\mathcal{F}(\zeta, z, \mathcal{X}, d) + b(-k\rho(\xi, \mathcal{X})\mathcal{X} + \mathcal{U})\right] \\
&\leq m_0^2(\varsigma, \xi)\|\varsigma\|^2 + m_1^2(z)\|z\|^2 \\
&\quad + \mathcal{X}^2\left[c^2/2 + cm_2(\xi, \mathcal{X}) + (1 + 1/b)b^2/4 - bk^*\rho(\xi, \mathcal{X})\right] \\
&\quad - b(k - k^*)\rho(\xi, \mathcal{X})\mathcal{X}^2 + (1 + 1/b)^{-1}\mathcal{U}^2.
\end{aligned} \tag{6.34}$$

Next, we note that the derivative of $b\hat{k}^2/(2\lambda)$ along the k-dynamics (6.26) is

$$b\hat{k}\dot{\hat{k}}/\lambda = b(k - k^*)\rho(\xi, \mathcal{X})\mathcal{X}^2. \tag{6.35}$$

With (6.30), (6.33), (6.34), and (6.35) in hand, we choose

$$W'(\hat{\zeta}, z, \mathcal{X}, \hat{k}) = W_0(\hat{\zeta}) + (1 + 1/b)\left[V'(z) + \mathcal{X}^2/2 + b\hat{k}^2/(2\lambda)\right]$$

whose derivative along the trajectory of the closed-loop system satisfies (6.28) if the following inequalities are satisfied

$$\Delta(z) \geq 1 + m_1^2(z) \tag{6.36}$$

$$k_0\Delta_0(\hat{\zeta}) \geq 1 + (1 + 1/b)p'\varpi(\varsigma, \xi) + (1 + 1/b)m_0^2(\varsigma, \xi) \tag{6.37}$$

$$\begin{aligned} k^*\rho(\xi, \mathcal{X}) \geq & \Big[p_0'\varkappa_0(\hat{\xi}, \mathcal{X})(1 + 1/b)^{-1} + p'\pi(\xi, \mathcal{X}) + c^2/2 \\ & + cm_2(\xi, \mathcal{X}) + (1 + 1/b)b^2/4 + 1\Big]/b. \end{aligned} \tag{6.38}$$

We have the following relationship

$$\begin{aligned} \varpi(\varsigma, \xi) &= \varpi(\varsigma, \hat{\xi} + \xi^*) \leq \bar{\varpi}(\varsigma, \hat{\xi})\hat{\varpi}(\xi^*) \\ m_0^2(\varsigma, \xi) &= m_0^2(\varsigma, \hat{\xi} + \xi^*) \leq \bar{m}(\varsigma, \hat{\xi})\hat{m}(\xi^*) \\ \varkappa_0(\hat{\xi}, \mathcal{X}) &= \varkappa_0(\xi - \xi^*, \mathcal{X}) \leq \bar{\varkappa}(\xi, \mathcal{X})\hat{\varkappa}(\xi^*) \end{aligned} \tag{6.39}$$

for some sufficiently smooth functions $\bar{\varpi}, \hat{\varpi}, \bar{m}, \hat{m}, \bar{\varkappa}$, and $\hat{\varkappa}$. Now, (6.36) is satisfied by itself, (6.37) by

$$k_0 \geq 1 + (1 + 1/b)p'\hat{\varpi}(\xi^*) + (1 + 1/b)\hat{m}(\xi^*) \tag{6.40}$$

$$\Delta_0(\hat{\zeta}) \geq \max\{1, \bar{\varpi}(\hat{\zeta}), \bar{m}(\hat{\zeta})\}, \tag{6.41}$$

and (6.38) by

$$\begin{aligned} k^* \geq & \Big[p_0'\hat{\varkappa}(\xi^*)(1 + 1/b)^{-1} + p' + c^2/2 \\ & + c + (1 + 1/b)b^2/4 + 1\Big]/b \end{aligned} \tag{6.42}$$

$$\rho(\xi, \mathcal{X}) \geq \max\{\bar{\varkappa}(\xi, \mathcal{X}), \pi(\xi, \mathcal{X}), m_2(\xi, \mathcal{X}), 1\}. \tag{6.43}$$

Finally, by the construction of $W'(\hat{\zeta}, z, \mathcal{X}, \hat{k})$, we have

$$\begin{aligned} \underline{\beta}'(\|\text{col}(\hat{\zeta}, z, \mathcal{X}, \hat{k})\|) &\leq \underline{\beta}_0(\|\hat{\zeta}\|) + \underline{\alpha}'(\|z\|) + \mathcal{X}^2/2 + \hat{k}^2/(2\lambda) \\ &\leq \underline{\beta}_0(\|\hat{\zeta}\|) + (1 + 1/b)\left(\underline{\alpha}'(\|z\|) + \mathcal{X}^2/2 + b\hat{k}^2/(2\lambda)\right) \\ &\leq W'(\hat{\zeta}, z, \mathcal{X}, \hat{k}) \leq \bar{\beta}_0(\|\hat{\zeta}\|) \\ &\quad + (1 + 1/b)\left(\bar{\alpha}'(\|z\|) + \mathcal{X}^2/2 + b\hat{k}^2/(2\lambda)\right) \\ &\leq \bar{\beta}'(\|\text{col}(\hat{\zeta}, z, \mathcal{X}, \hat{k})\|, s') \end{aligned} \tag{6.44}$$

for some class $\mathcal{K}_\infty$ function $\underline{\beta}'$, some parameterized class $\mathcal{K}_\infty$ function $\bar{\beta}'$, and some positive number s' that is not necessarily known. The proof is thus completed. □

Algorithm 6.3

INPUT: $\epsilon, m_0, m_1, m_2, \underline{\alpha}, \bar{\alpha}, \alpha, \sigma, \underline{\beta}, \bar{\beta}$
OUTPUT: $\rho, \underline{\beta}', \bar{\beta}'$
STEP 1: Pick the function Δ from (6.36) and call

$$(\underline{\alpha}', \bar{\alpha}', \varkappa) = \text{ALGORITHM } 2.5(\underline{\alpha}, \bar{\alpha}, \alpha, \sigma, \Delta).$$

STEP 2: Find the functions ϖ and π from (6.32), and $\bar{\varpi}$ and $\bar{m}$ from (6.39).
STEP 3: Pick the function Δ_0 from (6.41) and call

$$(\underline{\beta}_0, \bar{\beta}_0, \varkappa_0) = \text{ALGORITHM } 6.2(\underline{\beta}, \bar{\beta}, \alpha, \sigma, \Delta_0), \ \alpha(s) = \sigma(s) = s^2.$$

STEP 4: Find the functions $\bar{\varkappa}$ from (6.39), ρ from (6.43), and $\underline{\beta}'$ and $\bar{\beta}'$ from (6.44), in order.
STEP 5: END

6.2.3 Controller Synthesis

In this section, Lemma 6.2 will be iteratively applied to give the solution of the GASP for the lower triangular system (6.12). First, we list the following assumption.

Assumption 6.4 For $i = 1, \ldots, r$, the subsystem $\dot{z}_i = q_i(\vec{z}_i, \vec{x}_i, d)$ has an ISS Lyapunov function $V_i(z_i)$, i.e.,

$$V_i(z_i) \sim \{\underline{\alpha}_i, \bar{\alpha}_i, \alpha_i, \hat{\sigma}_i \mid \dot{z}_i = q_i(\vec{z}_i, \vec{x}_i, d)\}$$

and

$$\limsup_{s\to 0^+} \frac{s^2}{\alpha_i(s)} < \infty, \quad \limsup_{s\to 0^+} \frac{\hat{\sigma}_i(s)}{s^2} < \infty.$$

Moreover, the functions $\underline{\alpha}_i$, $\bar{\alpha}_i$, and α_i are known and the function $\hat{\sigma}_i$ is known up to a constant factor in the sense that there exist an unknown constant p_i and a known function σ_i such that $\hat{\sigma}_i = p_i\sigma_i$.

Next, we solve the GASP for the system (6.12) through a dynamic controller of the form (1.11). As motivated by Lemma 6.2, we consider the following coordinate transformation

$$\mathcal{X}_1 = x_1$$
$$\mathcal{X}_{i+1} = x_{i+1} + k_i \rho_i(\xi_{i-1}, \mathcal{X}_i)\mathcal{X}_i, \quad i = 1, \ldots, r \tag{6.45}$$

where $\rho_i(\xi_{i-1}, \mathcal{X}_i)$ is a sufficiently smooth positive function, $\xi_i = \vec{k}_i$, $\vec{k}_i = \text{col}(k_1, \ldots, k_i)$, and $k_i \in \mathbb{R}$ is governed by

$$\dot{k}_i = \lambda_i \rho_i(\xi_{i-1}, \mathcal{X}_i)\mathcal{X}_i^2, \ \lambda_i > 0. \tag{6.46}$$

For the completeness of notation, we denote $\xi_0 \in \mathbb{R}^0$, $\zeta_i = \text{col}(\varsigma_i, \xi_i)$, and $\varsigma_i = \text{col}(z_1, \mathcal{X}_1, \ldots, z_i, \mathcal{X}_i)$. Based on the coordinate transformation (6.45), we have

$$\begin{bmatrix} x_1 \\ x_2 \\ \vdots \\ x_{i-1} \\ x_i \end{bmatrix} = \begin{bmatrix} 1 & 0 & 0 & \cdots & 0 \\ -k_1\rho_1(\mathcal{X}_1) & 1 & 0 & \ddots & 0 \\ 0 & -k_2\rho_2(\xi_1, \mathcal{X}_2) & 1 & \ddots & 0 \\ \vdots & \ddots & \ddots & \ddots & \vdots \\ 0 & \cdots & 0 & -k_{i-1}\rho_{i-1}(\xi_{i-2}, \mathcal{X}_{i-1}) & 1 \end{bmatrix} \begin{bmatrix} \mathcal{X}_1 \\ \mathcal{X}_2 \\ \vdots \\ \mathcal{X}_{i-1} \\ \mathcal{X}_i \end{bmatrix}.$$

Therefore, it is straightforward to see

$$\text{col}(\vec{z}_{i-1}, \vec{x}_i) = \epsilon_i(\zeta_{i-1}, \mathcal{X}_i) = G_i(\zeta_{i-1})[\varsigma_{i-1}^{\mathsf{T}}, \mathcal{X}_i]^{\mathsf{T}} \tag{6.47}$$

for a sufficiently smooth function ϵ_i and a matrix function G_i.

Under the transformation (6.45), the $\text{col}(\vec{z}_i, \vec{x}_i)$-subsystem (6.12) together with the ξ_{i-1}-dynamics (6.46) can be put into the following form

$$\begin{aligned} \dot{\zeta}_{i-1} &= \varphi_{i-1}(\zeta_{i-1}, \mathcal{X}_i, d) \\ \dot{z}_i &= \mathcal{Q}_i(z_i, \epsilon_i(\zeta_{i-1}, \mathcal{X}_i), d) \\ \dot{\mathcal{X}}_i &= \mathcal{F}_i(\zeta_{i-1}, z_i, \mathcal{X}_i, d) + b_i x_{i+1}, \ b_i > 0. \end{aligned} \tag{6.48}$$

Various functions in (6.48) are defined recursively as follows:

$$\varphi_i(\zeta_i, \mathcal{X}_{i+1}, d) = \begin{bmatrix} \psi_i(\zeta_i, \mathcal{X}_{i+1}, d) \\ \phi_i(\zeta_i, \mathcal{X}_{i+1}, d) \end{bmatrix}$$
$$\psi_i(\zeta_i, \mathcal{X}_{i+1}, d) = \begin{bmatrix} \psi_{i-1}(\zeta_{i-1}, \mathcal{X}_i, d) \\ \mathcal{Q}_i(z_i, \epsilon_i(\zeta_{i-1}, \mathcal{X}_i), d) \\ \mathcal{F}_i(\zeta_{i-1}, z_i, \mathcal{X}_i, d) + b_i(d)(\mathcal{X}_{i+1} - k_i\rho_i(\xi_{i-1}, \mathcal{X}_i)\mathcal{X}_i) \end{bmatrix},$$
$$\phi_i(\zeta_i, \mathcal{X}_{i+1}, d) = \begin{bmatrix} \phi_{i-1}(\zeta_{i-1}, \mathcal{X}_i, d) \\ \lambda_i\rho_i(\xi_{i-1}, \mathcal{X}_i)\mathcal{X}_i^2 \end{bmatrix},$$

for $i = 1, \ldots, r$, and $\psi_0, \phi_0 \in \mathbb{R}^0$ for completeness, and

$$\begin{aligned} \mathcal{Q}_i(z_i, \epsilon_i(\zeta_{i-1}, \mathcal{X}_i), d) &= q_i\left(\vec{z}_i, \vec{x}_i, d\right), \quad i = 1, \ldots, r \\ \mathcal{F}_1(z_1, \mathcal{X}_1, d) &= f_1(z_1, x_1, d) \\ \mathcal{F}_i(\zeta_{i-1}, z_i, \mathcal{X}_i, d) &= f_i\left(\vec{z}_i, \vec{x}_i, d\right) + \frac{d\left(k_{i-1}\rho_{i-1}(\xi_{i-2}, \mathcal{X}_{i-1})\mathcal{X}_{i-1}\right)}{dt}, \\ & \quad i = 2, \ldots, r. \end{aligned} \tag{6.49}$$

In the calculation, it is noted that

$$\frac{d\left(k_i\rho_i(\xi_{i-1}, \mathcal{X}_i)\mathcal{X}_i\right)}{dt} = \frac{\partial\left(k_i\rho_i(\xi_{i-1}, \mathcal{X}_i)\mathcal{X}_i\right)}{\partial \mathrm{col}(\mathcal{X}_i, \xi_i)} \times \begin{bmatrix} \mathcal{F}_i(\zeta_{i-1}, z_i, \mathcal{X}_i, d) + b_i(\mathcal{X}_{i+1} - k_i\rho_i(\xi_{i-1}, \mathcal{X}_i)\mathcal{X}_i) \\ \phi_i(\zeta_i, \mathcal{X}_{i+1}, d) \end{bmatrix}$$

is a function of $\zeta_i, z_{i+1}, \mathcal{X}_{i+1}$ and d.

Clearly, at each step i of the recursive procedure, the system (6.48) is of the form (6.24). Therefore, Lemma 6.2 can be applied at each step to achieve the main result which is summarized in the following theorem.

Theorem 6.2 *Under Assumption* 6.4, *the GASP for the system* (6.12) *with any unknown compact set* $\mathbb{D}$ *is solved by the following controller*

$$\begin{aligned} u &= \mathcal{S}_r(\vec{k}_r, \vec{x}_r) = -k_r\rho_r(\vec{k}_{r-1}, \mathcal{X}_r)\mathcal{X}_r \\ \dot{k}_i &= \tau_i(\vec{k}_{i-1}, \vec{x}_i) = \lambda_i\rho_i(\vec{k}_{i-1}, \mathcal{X}_i)\mathcal{X}_i^2, \ \lambda_i > 0, \\ & \quad i = 1, \ldots, r \end{aligned} \tag{6.50}$$

where

$$\begin{aligned} \mathcal{X}_i &= x_i + k_{i-1}\rho_{i-1}(\vec{k}_{i-2}, \mathcal{X}_{i-1})\mathcal{X}_{i-1}, \ i = r, \ldots, 2 \\ \mathcal{X}_1 &= x_1 \end{aligned}$$

for some sufficiently smooth positive functions $\rho_i(\vec{k}_{i-1}, \mathcal{X}_i), \quad i = 1, \ldots, r$, *as given by Algorithm* 6.4.

Proof The essence of the proof is to show that the following statement is true for $1 \leq i \leq r$.

There exists a continuously differentiable function $W_i(\hat{\zeta}_i)$ where $\hat{\zeta}_i = \mathrm{col}(\varsigma_i, \hat{\xi}_i)$ with $\hat{\xi}_i = \xi_i - \xi_i^*$ for a positive vector ξ_i^* such that

$$\underline{\beta}_i(\|\hat{\zeta}_i\|) \leq W_i(\hat{\zeta}_i) \leq \bar{\beta}_i(\|\hat{\zeta}_i\|, s_i)$$

for a class $\mathcal{K}_\infty$ function $\underline{\beta}_i$, a parameterized class $\mathcal{K}_\infty$ function $\bar{\beta}_i$, and a positive number s_i which is not necessarily known, and along the trajectory of

$\dot{\zeta}_i = \varphi_i\,(\zeta_i, \mathcal{X}_{i+1}, d)$,

$$\dot{W}_i(\hat{\zeta}_i) \le -\|\varsigma_i\|^2 + \mathcal{X}_{i+1}^2.$$

It is noted that the system $\dot{\zeta}_i = \varphi_i\,(\zeta_i, \mathcal{X}_{i+1}, d)$ can be decomposed in the form of (6.48), i.e.,

$$\begin{aligned}\dot{\zeta}_{i-1} &= \varphi_{i-1}\,(\zeta_{i-1}, \mathcal{X}_i, d)\\ \dot{z}_i &= \mathcal{Q}_i(z_i, \epsilon_i(\zeta_{i-1}, \mathcal{X}_i), d)\\ \dot{\mathcal{X}}_i &= \mathcal{F}_i(\zeta_{i-1}, z_i, \mathcal{X}_i, d) + b_i x_{i+1},\ \ b_i > 0\end{aligned} \tag{6.51}$$

and

$$\begin{aligned}x_{i+1} &= -k_i\rho_i(\xi_{i-1}, \mathcal{X}_i)\mathcal{X}_i + \mathcal{X}_{i+1}\\ \dot{k}_i &= \lambda_i\rho_i(\xi_{i-1}, \mathcal{X}_i)\mathcal{X}_i^2,\ \ \lambda_i > 0.\end{aligned} \tag{6.52}$$

Before applying Lemma 6.2 to complete the proof of the aforementioned statement, we need to verify two technical conditions. The condition (i) has been verified during the construction of the system $\dot{\zeta}_i = \varphi_i\,(\zeta_i, \mathcal{X}_{i+1}, d)$. Thus, we only need to verify the condition (ii). From the definition of $\mathcal{F}_i(\zeta_{i-1}, z_i, \mathcal{X}_i, d)$, we have

$$\begin{aligned}|\mathcal{F}_i(\zeta_{i-1}, z_i, \mathcal{X}_i, d)| \le c_i[&m_{0i}(\varsigma_{i-1}, \xi_{i-1})\|\varsigma_{i-1}\|\\ &+m_{1i}(z_i)\|z_i\| + m_{2i}(\xi_{i-1}, \mathcal{X}_i)|\mathcal{X}_i|]\end{aligned} \tag{6.53}$$

for some sufficiently smooth functions m_{0i}, m_{1i}, and m_{2i} and a positive number c_i which is not necessarily known.

Now, it is ready to prove the statement by using mathematical induction. When $i = 1$, the ζ_0-subsystem in (6.51) disappears since $\zeta_0 \in \mathbb{R}^0$. Assumption 6.4 is equivalent to Assumption 6.2 of Lemma 6.2. By applying Lemma 6.2, there exist a controller (6.52) and a function $W_1(\hat{\zeta}_1)$ satisfying the statement for $i = 1$.

Next, we assume the statement is true for $i-1$, that is, the existence of $W_{i-1}(\hat{\zeta}_{i-1})$ for the system $\dot{\zeta}_{i-1} = \varphi_{i-1}\,(\zeta_{i-1}, \mathcal{X}_i, d)$, which is exactly the top subsystem of (6.51). And this inductive assumption actually is Assumption 6.3 of Lemma 6.2. Again, Assumption 6.4 is equivalent to Assumption 6.2 of Lemma 6.2. By applying Lemma 6.2, there exist a controller (6.52) and a function $W_i(\hat{\zeta}_i)$ satisfying the statement.

Thus, we have proven the statement for all $1 \le i \le r$. In particular, when $i = r$, we obtain a continuously differentiable function $W_r(\hat{\zeta}_r)$ satisfying

$$\dot{\hat{W}}_r(\zeta_r) \le -\|\varsigma_r\|^2$$

with $\mathcal{X}_{r+1} = 0$. Thus, by Theorem 2.5, for any initial condition, the trajectory of the close loop is bounded, and

$$\lim_{t\to\infty} \varsigma_r(t) = 0$$

which implies

$$\lim_{t\to\infty} (\vec{z}_r(t), \vec{x}_r(t)) = 0.$$

The proof is thus completed. □

Algorithm 6.4

INPUT: $f_i, \underline{\alpha}_i, \bar{\alpha}_i, \alpha_i, \sigma_i, i = 1, \ldots, r$
OUTPUT: $\rho_i, i = 1, \ldots, r$
STEP 1: Let $i = 1$, $\underline{\beta}_0 = 0$ and $\bar{\beta}_0 = 0$.
STEP 2: Calculate the function ϵ_i from (6.47), $\mathcal{F}_i$ from (6.49), and hence m_{0i}, m_{1i}, and m_{2i} from (6.53).
STEP 3: Call

$$(\rho_i, \underline{\beta}_i, \bar{\beta}_i) = \text{ALGORITHM 6.3}(\epsilon_i, m_{0i}, m_{1i}, m_{2i}, \underline{\alpha}_i, \bar{\alpha}_i, \alpha_i, \sigma_i, \underline{\beta}_{i-1}, \bar{\beta}_{i-1});$$

STEP 4: If $i = r$ GO TO STEP 5, ELSE let $i = i + 1$; GO TO STEP 2.
STEP 5: END

Example 6.2 Consider the following nonlinear system

$$\begin{aligned} \dot{x}_1 &= w_1 \sin x_1 + x_2 \\ \dot{z} &= -z + w_2 x_1^2 + w_3 x_2 \\ \dot{x}_2 &= w_4 x_1 x_2 \sin x_2 + z + u \end{aligned} \tag{6.54}$$

where the state variables x_1 and x_2 are measurable for feedback while the state z represents the dynamic uncertainty, and $\text{col}(w_1, w_2, w_3, w_4)$ is an unknown parameter vector without any prior knowledge. The objective is to design a controller to solve the GASP with the performance output $y = \text{col}(z, x_1, x_2)$.

It is easy to verify Assumption 6.4 for the system (6.54). By Theorem 6.2, a dynamic controller of the form

$$\begin{aligned} u &= -k_2 \rho_2(k_1, \mathcal{X}_2)\mathcal{X}_2 \\ \mathcal{X}_1 &= x_1 \\ \mathcal{X}_2 &= x_2 + k_1 \rho_1(\mathcal{X}_1)\mathcal{X}_1 \\ \dot{k}_1 &= \lambda_1 \rho_1(\mathcal{X}_1)\mathcal{X}_1^2, \ \lambda_1 > 0 \\ \dot{k}_2 &= \lambda_2 \rho_2(k_1, \mathcal{X}_2)\mathcal{X}_2^2, \ \lambda_2 > 0 \end{aligned} \tag{6.55}$$

solves the GASP of (6.54). By Algorithm 6.4, the functions ρ_1 and ρ_2 can be constructed as $\rho_1(\mathcal{X}_1) = 1$ and $\rho_2(k_1, \mathcal{X}_2) = \mathcal{X}_2^4 + k_1^4 + 1$.

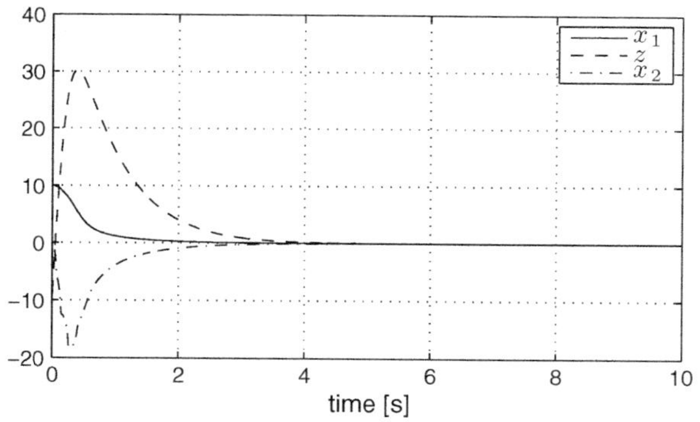

Fig. 6.3 Profile of state trajectories of the closed-loop system in Example 6.2

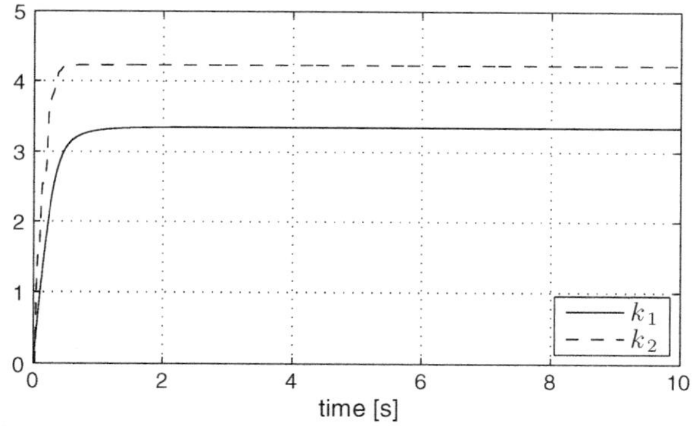

Fig. 6.4 Profile of dynamic gains of the closed-loop system in Example 6.2

The performance of the controller is simulated with parameters $w_1 = -2$, $w_2 = 2$, $w_3 = 2$, and $w_4 = 2$, and initial conditions $x_1(0) = 10$, $z(0) = -10$, $x_2(0) = 0$, and $k_1(0) = k_2(0) = 0$. Let $\lambda_1 = \lambda_2 = 0.1$. The results are shown in Figs. 6.3 and 6.4. The plant state asymptotically converges to the equilibrium point $[x_1, x_2, z]^\top = 0$, while the two dynamic gains asymptotically approach some constants, respectively.

6.3 Unknown Control Direction

So far, we have assumed that the control direction, i.e., the sign of b, is known. In this section, we will study the scenario where the control direction is unknown using the Nussbaum gain technique. For this purpose, we first introduce some technical tools.

For any function $v : \mathbb{R} \mapsto \mathbb{R}$, denote its positive and negative truncated functions by $v^+(s)$ and $v^-(s)$, i.e.,

$$v^+(s) = \max\{0, v(s)\}, \quad v^-(s) = \min\{0, v(s)\}.$$

Obviously, the truncated functions satisfy the following properties

$$\begin{aligned} v^+(s) &\geq 0 \\ v^-(s) &\leq 0 \\ v(s) &= v^+(s) + v^-(s). \end{aligned}$$

Definition 6.2 A continuous function $v : \mathbb{R} \mapsto \mathbb{R}$ is called a class $\mathcal{N}$ function, denoted by $v \in \mathcal{N}$, if

$$\liminf_{k\to\infty} \frac{k - \int_0^k v^-(s)ds}{\int_0^k v^+(s)ds} = 0, \tag{6.56}$$

$$\liminf_{k\to\infty} \frac{k + \int_0^k v^+(s)ds}{-\int_0^k v^-(s)ds} = 0. \tag{6.57}$$

Lemma 6.3 *If $v \in \mathcal{N}$, then*

$$\limsup_{k\to\infty} \frac{1}{k} \int_0^k v(s)ds = +\infty, \tag{6.58}$$

$$\liminf_{k\to\infty} \frac{1}{k} \int_0^k v(s)ds = -\infty. \tag{6.59}$$

Proof From the property (6.56), there exists a sequence $k_1 < k_2 < \ldots$, with $\lim_{i\to\infty} k_i = +\infty$, such that,

$$\lim_{i\to\infty} \frac{k_i - \int_0^{k_i} v^-(s)ds}{\int_0^{k_i} v^+(s)ds} = 0.$$

which is equivalent to

$$\lim_{i\to\infty}\frac{1}{k_i}\int_0^{k_i} v^+(s)ds = +\infty \tag{6.60}$$

and

$$\lim_{i\to\infty}\frac{-\int_0^{k_i} v^-(s)ds}{\int_0^{k_i} v^+(s)ds} = 0. \tag{6.61}$$

From (6.61), one has

$$\lim_{i\to\infty}\frac{\int_0^{k_i} v(s)ds}{\int_0^{k_i} v^+(s)ds} = \lim_{i\to\infty}\frac{\int_0^{k_i} v^+(s)ds + \int_0^{k_i} v^-(s)ds}{\int_0^{k_i} v^+(s)ds} = 1,$$

which, together with (6.60), implies

$$\lim_{i\to\infty}\frac{1}{k_i}\int_0^{k_i} v(s)ds = +\infty. \tag{6.62}$$

Hence, the equation (6.58) is proved. The equation (6.59) can be proved similarly and is left for readers. □

Remark 6.2 A function satisfying the properties (6.58) and (6.59) is called a *Nussbaum function*. Therefore, a class $\mathcal{N}$ function is a type of Nussbaum function.

Remark 6.3 From the proof of Lemma 6.3, we can see that (6.56) and (6.57) are equivalent to

$$\limsup_{k\to\infty}\frac{1}{k}\int_0^{k} v^+(s)ds = +\infty, \quad \liminf_{k\to\infty}\frac{-\int_0^{k} v^-(s)ds}{\int_0^{k} v^+(s)ds} = 0. \tag{6.63}$$

and, respectively,

$$\limsup_{k\to\infty} \frac{-1}{k}\int_0^k v^-(s)ds = +\infty, \quad \liminf_{k\to\infty} \frac{\int_0^k v^+(s)ds}{-\int_0^k v^-(s)ds} = 0. \tag{6.64}$$

Lemma 6.4 *For $v \in \mathcal{N}$, let $\hat{v}(s) = av^+(s) + bv^-(s)$ for two constants a and b satisfying $ab > 0$. Then, $\hat{v} \in \mathcal{N}$.*

Proof We only prove the case with $a, b > 0$. Denote $\hat{v}(s) = \hat{v}^+(s) + \hat{v}^-(s)$ where $\hat{v}^+(s) = av^+(s)$ and $\hat{v}^-(s) = bv^-(s)$. Clearly, $v(s)$ satisfies (6.63) and (6.64) if and only if $\hat{v}(s)$ satisfies (6.63) and (6.64). By Remark 6.3, $v \in \mathcal{N}$ if and only if $\hat{v} \in \mathcal{N}$. □

Example 6.3 The function

$$v(s) = \sin(as)\exp(bs^2), \quad a, b > 0$$

is a class $\mathcal{N}$ function. The verification is given below.

Denote $k_i = i\pi/a$ for an integer i. Let

$$P_i^+ = \int_{k_{2i-2}}^{k_{2i-1}} \sin(as)\exp(bs^2)ds$$

$$P_i^- = \int_{k_{2i-1}}^{k_{2i}} \sin(as)\exp(bs^2)ds.$$

It is noted that

$$\int_0^{k_{2i-1}} v^+(s)ds = P_i^+ + \cdots + P_1^+$$

$$\int_0^{k_{2i-1}} v^-(s)ds = P_{i-1}^- + \cdots + P_1^-.$$

Consider the sequence k_{2i-1}, $i = 1, 2, \ldots$. One has

$$
\begin{aligned}
&\lim_{i\to\infty} \frac{-\int_0^{k_{2i-1}} v^-(s)ds}{\int_0^{k_{2i-1}} v^+(s)ds} \\
&= \lim_{i\to\infty} \frac{-P_{i-1}^- - \cdots - P_1^-}{P_i^+ + \cdots + P_1^+} \le \lim_{i\to\infty} \frac{-(i-1)P_{i-1}^-}{P_i^+} \\
&= \lim_{i\to\infty} \frac{-(i-1)\int_{k_{2i-3}}^{k_{2i-2}} \sin(as)\exp(bs^2)ds}{\int_{k_{2i-2}}^{k_{2i-1}} \sin(as)\exp(bs^2)ds} \\
&= \lim_{i\to\infty} \frac{(i-1)\int_0^{\pi/a} \sin(as)\exp(b(s+k_{2i-3})^2)ds}{\int_0^{\pi/a} \sin(as)\exp(b[(s+k_{2i-3})^2 + 2(s+k_{2i-3})\pi/a + (\pi/a)^2])ds} \\
&\le \lim_{i\to\infty} \frac{i-1}{\exp(2bk_{2i-3}\pi/a)} \\
&= \lim_{i\to\infty} \frac{i-1}{\exp((2i-3)2b(\pi/a)^2)} = 0
\end{aligned}
$$

and

$$
\begin{aligned}
&\lim_{i\to\infty} \frac{k_{2i-1}}{\int_0^{k_{2i-1}} v^+(s)ds} \le \lim_{i\to\infty} \frac{(2i-1)\pi/a}{P_i^+} \\
&\le \lim_{i\to\infty} \frac{(2i-1)\pi/a}{\int_0^{\pi/a} \sin(as)\exp(bs^2)ds \exp((2i-3)2b(\pi/a)^2)} = 0.
\end{aligned}
$$

As a result,

$$
\lim_{i\to\infty} \frac{k_{2i-1} - \int_0^{k_{2i-1}} v^-(s)ds}{\int_0^{k_{2i-1}} v^+(s)ds} = 0
$$

which implies (6.56). The equation (6.57) can be verified in a similar way and is left for readers.

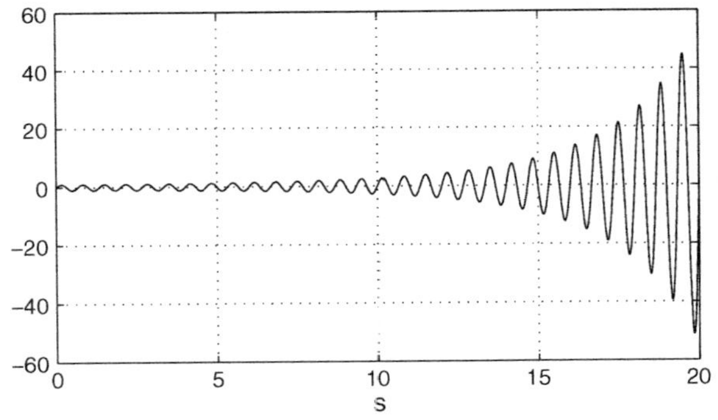

Fig. 6.5 Profile of the class $\mathcal{N}$ function $v(s) = \sin(3\pi s)\exp(0.01s^2)$

The profile of the function $v(s)$ is depicted in Fig. 6.5 with $a = 3\pi$ and $b = 0.01$.

Example 6.4 The function

$$v(s) = \sin(as)s^2, \ a > 0$$

is not a class $\mathcal{N}$ function, but it still satisfies (6.58) and (6.59). The verification is given below.

As in Example 6.3, denote $k_i = i\pi/a$ for an integer i. Using the identity

$$\int \sin(as)s^2 ds = \frac{(2 - a^2s^2)\cos as}{a^3} + \frac{2s\sin as}{a^2},$$

one has

$$P_i^+ = \int_{k_{2i-2}}^{k_{2i-1}} \sin(as)s^2 ds = \frac{-4 + [(2i-1)^2 + (2i-2)^2]\pi^2}{a^3}$$

$$P_i^- = \int_{k_{2i-1}}^{k_{2i}} \sin(as)s^2 ds = \frac{4 - [(2i)^2 + (2i-1)^2]\pi^2}{a^3}.$$

It is noted that

$$\int_0^{k_{2i-1}} v^+(s)ds = P_i^+ + \cdots + P_1^+$$

$$\int_0^{k_{2i-1}} v^-(s)ds = P_{i-1}^- + \cdots + P_1^-.$$

Consider the sequence k_{2i-1}, $i = 1, 2, \ldots$. One has

$$\lim_{i\to\infty} \frac{-\int_0^{k_{2i-1}} v^-(s)ds}{\int_0^{k_{2i-1}} v^+(s)ds} = \lim_{i\to\infty} \frac{-P_{i-1}^- - \cdots - P_1^-}{P_i^+ + \cdots + P_1^+} = 1.$$

It is easy to see that

$$\begin{aligned}
\liminf_{k\to\infty} \frac{k - \int_0^k v^-(s)ds}{\int_0^k v^+(s)ds} &\geq \liminf_{k\to\infty} \frac{-\int_0^k v^-(s)ds}{\int_0^k v^+(s)ds} \\
&= \lim_{i\to\infty} \frac{-\int_0^{k_{2i-1}} v^-(s)ds}{\int_0^{k_{2i-1}} v^+(s)ds} = 1.
\end{aligned}$$

So, $v(s)$ is not a class $\mathcal{N}$ function.

On the other hand, it is noted that

$$\lim_{i\to\infty} \frac{1}{k_{2i-1}} \int_0^{k_{2i-1}} v(s)ds = \lim_{i\to\infty} \frac{(P_i^+ + \cdots + P_1^+) + (P_{i-1}^- + \cdots + P_1^-)}{(2i-1)\pi/a} = +\infty,$$

which implies (6.58). The verification of (6.59) is similar and left for readers.

Lemma 6.5 *Consider two continuously differentiable functions* $V : [0, \infty) \mapsto \mathbb{R}^+$, $k : [0, \infty) \mapsto \mathbb{R}$. *Let* $b : [0, \infty) \mapsto [\underline{b}, \bar{b}]$ *for two constants* $\underline{b}$ *and* $\bar{b}$. *If* $0 \notin [\underline{b}, \bar{b}]$ *and*

$$\begin{aligned}
&\dot{V}(t) \leq (b(t)v(k(t)) + v^*)\dot{k}(t), \\
&\dot{k}(t) \geq 0, \quad \forall t \geq 0
\end{aligned} \tag{6.65}$$

for a constant v^* *and a function* $v \in \mathcal{N}$, *then* $V(t)$ *and* $k(t)$ *are bounded over* $[0, \infty)$.

Proof Let

$$\hat{v}(s) = \bar{b}v^{+}(s) + \underline{b}v^{-}(s).$$

For $\bar{b}\underline{b} > 0$, by Lemma 6.4, $\hat{v}(s) \in \mathcal{N}$. It is noted that, for all $\tau \geq 0$,

$$\begin{aligned} b(\tau)v(k(\tau)) &= b(\tau)v^{+}(k(\tau)) + b(\tau)v^{-}(k(\tau)) \\ &\leq \bar{b}v^{+}(k(\tau)) + \underline{b}v^{-}(k(\tau)) = \hat{v}(k(\tau)). \end{aligned}$$

Integrating the first inequality of (6.65) gives, for all $t \geq 0$,

$$\begin{aligned} 0 \leq V(t) &\leq \int_0^t (b(\tau)v(k(\tau)) + v^*)\dot{k}(\tau)d\tau + V(0) \\ &= \int_0^t b(\tau)v(k(\tau))\dot{k}(\tau)d\tau + \int_0^t v^*\dot{k}(\tau)d\tau + V(0) \\ &\leq \int_{k(0)}^{k(t)} \hat{v}(s)ds + \int_0^t v^*\dot{k}(\tau)d\tau + V(0) \\ &= \int_0^{k(t)} \hat{v}(s)ds - \int_0^{k(0)} \hat{v}(s)ds + v^*k(t) - v^*k(0) + V(0) \end{aligned} \tag{6.66}$$

Denote a constant $c(0) = \int_0^{k(0)} \hat{v}(s)ds + v^*k(0) - V(0)$, one has

$$\int_0^{k(t)} \hat{v}(s)ds + v^*k(t) \geq c(0). \tag{6.67}$$

As $\hat{v} \in \mathcal{N}$, by (6.59) of Lemma 6.3, there exists $k^* > 1$ such that

$$\frac{1}{k^*}\int_0^{k^*} \hat{v}(s)ds < -|c(0)| - v^*.$$

If $k(t)$ is not bounded over $[0, \infty)$, then there exists $t^* > 0$ such that $k(t^*) = k^*$. Thus,

$$\int_0^{k(t^*)} \hat{v}(s)ds < -|c(0)|k(t^*) - v^*k(t^*) < c(0) - v^*k(t^*)$$

which contradicts (6.67).

As $k(t)$ is bounded over $[0, \infty)$, so is $V(t)$ by (6.66). □

Remark 6.4 If b is a constant, $v \in \mathcal{N}$ in Lemma 6.5 can be replaced by a Nussbaum function v satisfying (6.58) and (6.59). In fact, for a constant b, $\hat{v}(s) = bv(s)$ is also a Nussbaum function satisfying (6.58) and (6.59). Then, the proof of Lemma 6.5 simply follows.

Next, we will show how a class $\mathcal{N}$ function can be used to deal with control systems with unknown control direction. For convenience, we use the system (2.46) of relative degree one as a case study. Because the control direction is unknown, Assumption 2.2 used in Theorem 2.8 will be weakened to the following one.

Assumption 6.5 The function $b(d)$ is away from zero, i.e., $b(d) \neq 0, \ \forall d \in \mathbb{D}$.

Since $b(d)$ is continuous in d, we can assume that $b(d) \in [\underline{b}, \bar{b}], \ \forall d \in \mathbb{D}$ for two unknown constants $\underline{b}$ and $\bar{b}$ and $0 \notin [\underline{b}, \bar{b}]$.

Theorem 6.3 *Consider the system* (2.46) *with any unknown compact set* $\mathbb{D}$. *Under Assumptions* 6.5 *and* 2.5, *let* v *be a continuously differentiable class* $\mathcal{N}$ *function, then there exists a controller*

$$\begin{aligned} u &= v(k)\rho(x)x \\ \dot{k} &= \lambda\rho(x)x^2, \ \lambda > 0, \end{aligned} \tag{6.68}$$

that solves the GASP of the system (2.46). *In particular, the function* ρ *is given in Algorithm* 6.5.

Proof By Corollary 2.3, for any sufficient smooth function $\Delta(z) > 0$, there exists a continuously differentiable function $V'(z)$ satisfying $\underline{\alpha}'(\|z\|) \leq V'(z) \leq \bar{\alpha}'(\|z\|)$ for some class $\mathcal{K}_\infty$ functions $\underline{\alpha}'$ and $\bar{\alpha}'$, such that, along the trajectory of $\dot{z} = q(z, x, d)$,

$$\dot{V}'(z) \leq -\Delta(z)\|z\|^2 + p'\varkappa(x)x^2 \tag{6.69}$$

for some unknown constant p' and some known smooth function $\varkappa(x) \geq 1$.

Also, note that

$$|f(z, x, d)| \leq cm_1(z)\|z\| + cm_2(x)|x|, \ \forall d \in \mathbb{D}. \tag{6.70}$$

for some unknown real number $c > 0$ and two known smooth functions $m_1(z)$ and $m_2(x)$.

Let

$$U(z, x) = V'(z) + x^2/2.$$

Then, along the trajectory of the closed-loop system,

$$\begin{aligned}\dot{U}(z,x) = & -\Delta(z)\|z\|^2 + p'\varkappa(x)x^2 + x[f(z,x,d) + b(d)u] \\ \le & -\Delta(z)\|z\|^2 + m_1^2(z)\|z\|^2 + x^2[p'\varkappa(x) + c^2/4 + cm_2(x) \\ & + bv(k)\rho(x)].\end{aligned} \tag{6.71}$$

In (6.71), let

$$\Delta(z) \ge 1 + m_1^2(z) \tag{6.72}$$

$$\rho(x) \ge \max\{\varkappa(x), m_2(x), 1\} \tag{6.73}$$

and

$$v^* \ge p' + c^2/4 + c.$$

One has

$$\dot{U}(z,x) \le (bv(k) + v^*)\rho(x)x^2 = (bv(k) + v^*)\dot{k}/\lambda.$$

From Lemma 6.5, $k(t)$ and $U(z(t), x(t))$ are bounded over $[0, \infty)$, so are $z(t)$ and $x(t)$. Also, it can be seen that $\ddot{k}(t)$ is bounded. Thus, $\dot{k}(t)$ is uniformly continuous over $[0, \infty)$. By Barbalat's Lemma, one has $\lim_{t\to\infty} \dot{k}(t) = 0$ and hence $\lim_{t\to\infty} x(t) = 0$ and $\lim_{t\to\infty} z(t) = 0$. □

Algorithm 6.5

INPUT: $f, \underline{\alpha}, \bar{\alpha}, \alpha, \sigma$
OUTPUT: ρ
STEP1: Find the functions m_1 and m_2 from (6.70).
STEP2: Pick the function Δ from (6.72) and call

$$(\underline{\alpha}', \bar{\alpha}', \varkappa) = \text{ALGORITHM } 2.5(\underline{\alpha}, \bar{\alpha}, \alpha, \sigma, \Delta).$$

STEP3: Calculate the function ρ from (6.73).
STEP4: END

Example 6.5 Consider a second order nonlinear system

$$\begin{aligned}\dot{z} &= -z + w_3 x \\ \dot{x} &= w_1 z \sin x + w_2 x^3 + bu\end{aligned}$$

which was studied in Example 2.18 with $b = 1$. The controller was designed based on the knowledge that b is positive. When b is negative, e.g., $b = -1$, the controller fails as shown in Fig. 6.6.

If the sign of b is unknown, we can apply Theorem 6.3 to design a controller of the form (6.68) for the same system. In particular, with the following class $\mathcal{N}$ function

$$v(s) = \sin(3\pi s)\exp(0.01s^2),$$

the performance of the controller (6.68) is shown in Fig. 6.7 for $b = -1$. All the parameters and initial state values are the same as those used in Example 2.18. The gain k increases to a sufficiently large finite number, with which the state of the plant asymptotically approaches the origin. As the gain k and the state variable x approach their desired values very fast, the details are given in Fig. 6.8 in a smaller time range.

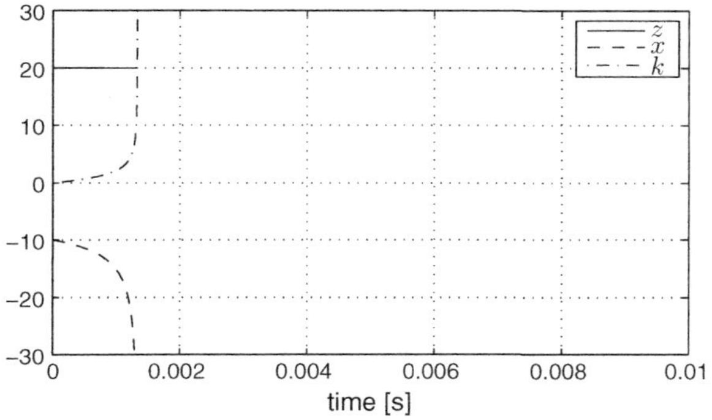

Fig. 6.6 Profile of state trajectories of the closed-loop system in Example 2.18 with the control direction inverted

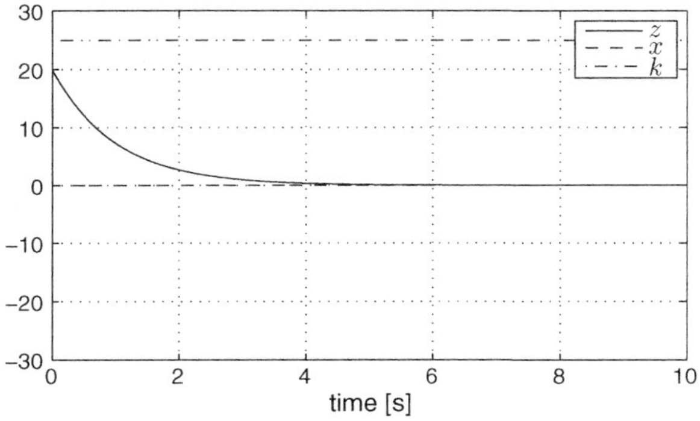

Fig. 6.7 Profile of state trajectories of the closed-loop system in Example 6.5 for $t \in [0, 10]$

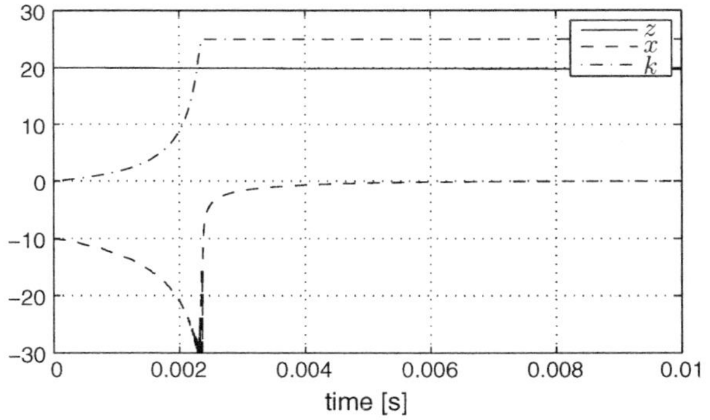

Fig. 6.8 Profile of state trajectories of the closed-loop system in Example 6.5 for $t \in [0, 0.01]$

6.4 Adaptive Stabilization of the Hyperchaotic Lorenz System

In this section, we study the global stabilization problem of the hyperchaotic Lorenz system of the form (3.36). It is known that through the dynamic extension (3.38), the system (3.36) can be put in the form (3.39). By selecting $\lambda_1 = \lambda_2 = 1$, and letting $w = \text{col}(\sigma, \beta, \rho)$, (3.39) takes the following specific form

$$\begin{aligned} \dot{z} &= q(z, x_1, w) \\ \dot{x}_1 &= z_1(\rho - z_2) - x_1 + z_4 + x_2 \\ \dot{x}_2 &= \hat{u} \end{aligned} \tag{6.74}$$

where

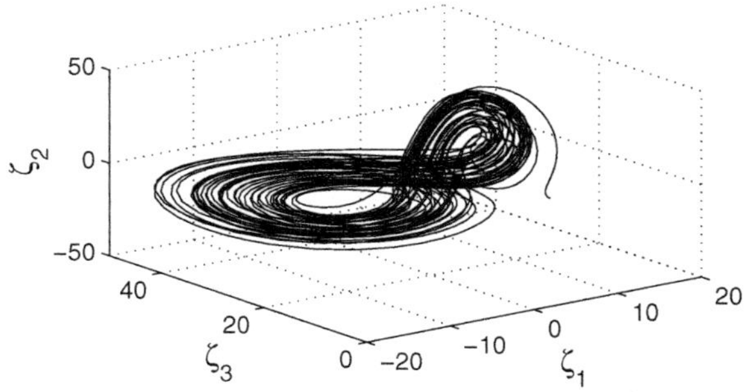

Fig. 6.9 Profile of state trajectories of the uncontrolled hyperchaotic Lorenz system

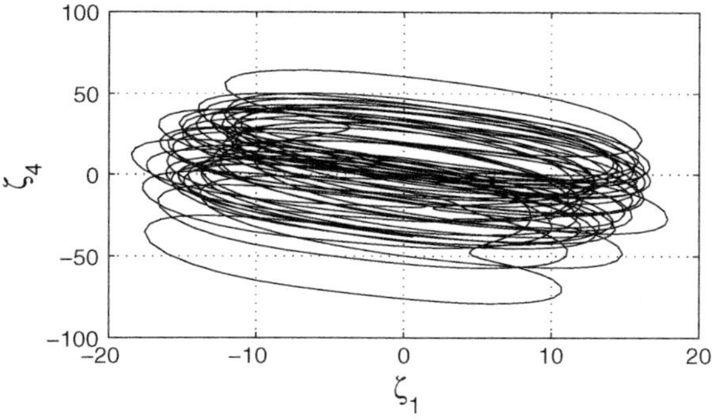

Fig. 6.10 Profile of state trajectories of the uncontrolled hyperchaotic Lorenz system

$$q(z, x_1, w) = \begin{bmatrix} \sigma(x_1 - z_1) \\ z_1 x_1 - \beta z_2 \\ z_1(\rho - z_2) + z_4 - z_3 \\ \beta z_1 + \gamma x_1 z_2 + x_1 - z_3 \end{bmatrix}$$

which is in the form (6.2).

For the uncontrolled system with $u = 0$, let $\xi_1(0) = 0$ and $\xi_2(0) = 0$, i.e., $z_3(0) = x_1(0)$ and $x_2(0) = 0$. Then, we have $\xi_1(t) = 0$ and $\xi_2(t) = 0$. As a result, for all time, $z_1 = \zeta_1, z_2 = \zeta_3, z_3 = \zeta_2, z_4 = \zeta_4, x_1 = \zeta_2$, and $x_2 = 0$ and $\hat{u} = 0$. For $\sigma = 10, \beta = 8/3, \rho = 28, \gamma = -0.5$, and the initial condition $z(0) = [5, 5, 1, 1]^{\mathrm{T}}$ and $x(0) = [1, 0]^{\mathrm{T}}$, the simulation is shown in Figs. 6.9 and 6.10.

We will consider designing a controller u to solve the GASP of (6.74) under the assumption that the three unknown parameters σ, β, ρ are all positive.

To apply Theorem 6.1, we need to verify Assumption 6.1. In particular, we need to find an ISS Lyapunov function for the z-subsystem. Since the z-subsystem is nonlinear and its dimension is high, it is somehow tedious to construct the ISS Lyapunov function, and we will detail the construction as follows.

Let

$$V_1(z_1, z_2) = \epsilon_1(z_1^2/2 + z_1^8/8) + \epsilon_2(z_2^2/2 + z_2^4/4)$$

for two positive numbers ϵ_1 and ϵ_2. The derivative of $V_1(z_1, z_2)$ along the trajectory of the z-subsystem satisfies

$$\begin{aligned}\dot{V}_1(z_1,z_2) &= \epsilon_1\sigma(z_1+z_1^7)(x_1-z_1)+\epsilon_2(z_2+z_2^3)(z_1x_1-\beta z_2)\\ &\le -\epsilon_1\sigma(z_1^2+z_1^8)/2+\epsilon_1\sigma(x_1^2/2+2^7x_1^8)\\ &\quad -\epsilon_2\beta(z_2^2+z_2^4)/2+\epsilon_2(z_1^4+x_1^4)/(4\beta)+\epsilon_2(z_1^8+x_1^8)/(16\beta^3)\\ &\le \left[-\epsilon_1\sigma(z_1^2+z_1^8)/2+\epsilon_2z_1^4/(4\beta)+\epsilon_2z_1^8/(16\beta^3)\right]-\epsilon_2\beta(z_2^2+z_2^4)/2\\ &\quad +\epsilon_1\sigma(x_1^2/2+2^7x_1^8)+\epsilon_2x_1^4/(4\beta)+\epsilon_2x_1^8/(16\beta^3).\end{aligned}$$

Next, we define

$$V_2(z_3,z_4)=2z_3^2+(z_3-z_4)^2/2$$

whose derivative along the trajectory of the z-subsystem satisfies

$$\begin{aligned}\dot{V}_2(z_3,z_4) &= (5z_3-z_4)[z_1(\rho-z_2)+z_4-z_3]\\ &\quad -(z_3-z_4)[\beta z_1+\gamma x_1z_2+x_1-z_3]\\ &\le \left[-z_3^2/4-z_4^2/4+2\epsilon_3(5z_3-z_4)^2+3\epsilon_3(z_3-z_4)^2\right]\\ &\quad +(\rho^2z_1^2+z_1^4/4+z_2^4)/(4\epsilon_3)+(\beta^2z_1^2+\gamma^2x_1^4/4+\gamma^2z_2^4+x_1^2)/(4\epsilon_3)\\ &\le \left[-z_3^2/4-z_4^2/4+\epsilon_3(5z_3-z_4)^2+3\epsilon_3(z_3-z_4)^2\right]\\ &\quad +(\rho^2z_1^2+z_1^4/4+\beta^2z_1^2)/(4\epsilon_3)+(1+\gamma^2)z_2^4/(4\epsilon_3)\\ &\quad +(\gamma^2x_1^4/4+x_1^2)/(4\epsilon_3)\end{aligned}$$

for any positive number ϵ_3.

From the degrees of the polynomials on the right hand sides of the following four inequalities, we can see that there exist a sufficiently small ϵ_3, sufficiently large ϵ_1 and ϵ_2, and a sufficiently large p, such that

$$\begin{aligned}&-z_3^2/4-z_4^2/4+\epsilon_3(5z_3-z_4)^2+\epsilon_3(z_3-z_4)^2\le -z_3^2/8-z_4^2/8,\\ &-\epsilon_2\beta(z_2^2+z_2^4)/2+(1+\gamma^2)z_2^4/(4\epsilon_3)\le -(z_2^2+z_2^4),\\ &\left[-\epsilon_1\sigma(z_1^2+z_1^8)/2+\epsilon_2z_1^4/(4\beta)+\epsilon_2z_1^8/(16\beta^3)\right]+(\rho^2z_1^2+z_1^4/4+\beta^2z_1^2)/(4\epsilon_3)\\ &\le -(\rho^2z_1^2+z_1^4),\end{aligned}$$

and

$$\begin{aligned}p(x_1^2+x_1^8)\ge{}&\epsilon_1\sigma(x_1^2/2+2^7x_1^8)+\epsilon_2x_1^4/(4\beta)+\epsilon_2x_1^8/(16\beta^3)\\ &+(\gamma^2x_1^4/4+x_1^2)/(4\epsilon_3).\end{aligned}$$

It should be noted that it suffices to guarantee the existence of the parameters ϵ_1, ϵ_2, ϵ_3 and p and it is not necessary to explicitly find them. Now, let

$$V(z) = V_1(z_1, z_2) + V_2(z_3, z_4).$$

Then, along the trajectory of the z-subsystem,

$$\dot{V}(z) \leq -\alpha(z) + p(x_1^2 + x_1^8)$$

where

$$\alpha(z) = (\rho^2 z_1^2 + z_1^4) + (z_2^2 + z_2^4) + z_3^2/8 + z_4^2/8.$$

Thus, Assumption 6.1 is verified with $V(z)$ as an ISS Lyapunov function for the z-subsystem.

By Theorem 6.1, we can construct a controller of the form

$$\begin{aligned} \hat{u} &= \nu(k, x_1, x_2) - \rho_2(x_1)\mathcal{X}_2 \\ \mathcal{X}_2 &= x_2 + k\rho_1(x_1)x_1 \\ \dot{k} &= \lambda\rho_1(x_1)x_1^2, \ \lambda > 0 \end{aligned} \tag{6.75}$$

where the functions ρ_1, ρ_2, and ν are explicitly calculated as follows.

At the first step, we define

$$W_1(z, x_1, \hat{k}) = 3V(z) + x_1^2/2 + \hat{k}^2/(2\lambda)$$

where $\hat{k} = k - k^*$ for a constant k^*. Direct calculation shows

$$\begin{aligned} x_1\dot{x}_1 &= x_1(z_1(\rho - z_2) - x_1 + z_4 + x_2) \\ &\leq \rho^2 z_1^2 + z_1^4/2 + z_2^4/2 + z_4^2/8 + 2x_1^2 + \mathcal{X}_2^2 - k\rho_1(x_1)x_1^2 \\ &\leq \alpha(z) + 2x_1^2 + \mathcal{X}_2^2 - k\rho_1(x_1)x_1^2. \end{aligned}$$

Thus letting

$$\begin{aligned} \rho_1(x_1) &= 10 + x_1^6 \\ k^* &\geq 3p + 3 \end{aligned}$$

gives

$$\dot{W}_1(z, x_1, \hat{k}) \leq -2\alpha(z) - 2x_1^2 + \mathcal{X}_2^2.$$

At the second step, let $\mathcal{S}_1(k, x_1) = -k(10 + x_1^6)x_1$, $\mathcal{S}_{1x}(k, x_1) = -k(10 + 7x_1^6)$, and $\mathcal{X}_2 = x_2 - \mathcal{S}_1(k, x_1)$. Then

$$\dot{\mathcal{X}}_2 = \hat{u} + \lambda\rho_1^2(x_1)x_1^3 - \mathcal{S}_{1x}(k, x_1)x_2 - \mathcal{S}_{1x}(k, x_1)(z_1(\rho - z_2) - x_1 + z_4).$$

Let

$$W_2(z, x_1, x_2, \hat{k}) = W_1(z, x_1, \hat{k}) + \mathcal{X}_2^2/2.$$

Then

$$\dot{W}_2(z, x_1, x_2, \hat{k}) = \dot{W}_1(z, x_1, \hat{k}) + \mathcal{X}_2\dot{\mathcal{X}}_2.$$

Direct calculation shows

$$\begin{aligned}\mathcal{X}_2\dot{\mathcal{X}}_2 &= \mathcal{X}_2[\hat{u} + \lambda\rho_1^2(x_1)x_1^3 - \mathcal{S}_{1x}(k, x_1)x_2] - \mathcal{X}_2\mathcal{S}_{1x}(k, x_1)(z_1(\rho - z_2) - x_1 + z_4)\\ &\leq \mathcal{X}_2[\hat{u} + \lambda\rho_1^2(x_1)x_1^3 - \mathcal{S}_{1x}(k, x_1)x_2] + 3\mathcal{X}_2^2\mathcal{S}_{1x}^2(k, x_1) + \alpha(z) + x_1^2.\end{aligned}$$

Letting $\hat{u} = \nu(k, x_1, x_2) - \rho_2(x_1)\mathcal{X}_2$ with

$$\begin{aligned}\nu(k, x_1, x_2) &= -\lambda\rho_1^2(x_1)x_1^3 + \mathcal{S}_{1x}(k, x_1)x_2\\ \rho_2(x_1) &= 3\mathcal{S}_{1x}^2(k, x_1) + 2\end{aligned}$$

gives

$$\mathcal{X}_2\dot{\mathcal{X}}_2 \leq -2\mathcal{X}_2^2 + \alpha(z) + x_1^2.$$

Thus,

$$\dot{W}_2(z, x_1, x_2, \hat{k}) \leq -\alpha(z) - x_1^2 - \mathcal{X}_2^2.$$

By Theorem 2.5, for any initial condition, the state trajectory of the closed-loop system is bounded for $t \geq 0$, and

$$\lim_{t\to\infty} \text{col}(z(t), x_1(t), \mathcal{X}_2(t)) = 0$$

which simply implies

$$\lim_{t\to\infty} \text{col}(z(t), x_1(t), x_2(t)) = 0.$$

Thus, the GASP of (6.74) is solved.

The performance of the controller is simulated with $\sigma = 10$, $\beta = 8/3$, $\rho = 28$, $\gamma = -0.5$, and $\lambda = 1$, and the initial state values $z(0) = [5, 5, 1, 1]^\mathsf{T}$, $x(0) = [1, 0]^\mathsf{T}$,

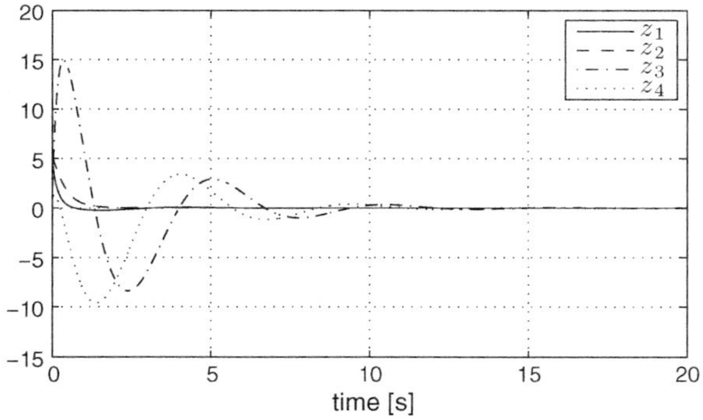

Fig. 6.11 Profile of state trajectories of the controlled hyperchaotic Lorenz system (part 1)

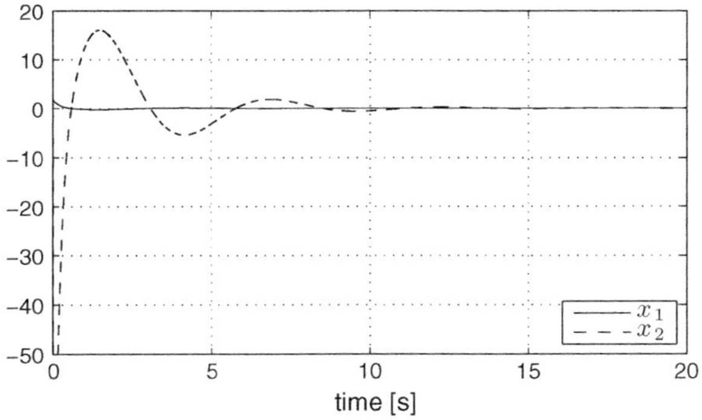

Fig. 6.12 Profile of state trajectories of the controlled hyperchaotic Lorenz system (part 2)

and $k(0) = 0$. It can be seen that the state of the plant asymptotically approaches the origin $z = 0$ and $x = 0$ as illustrated in Figs. 6.11 and 6.12. The dynamic gain asymptotically approaches a finite constant as shown in Fig. 6.13.

6.5 Notes and References

The universal adaptive control technique has been used for handling static uncertainty with unknown boundary in [1–5], etc. Sect. 6.1 is adapted from [6] and Sect. 6.2 is mainly based on [7, 8]. The Nussbaum gain technique has been used for handling the unknown control direction in many papers. The main references for Sect.6.3 are [9–13]. The example on the Hyperchaotic Lorenz system can find its reference in [8].

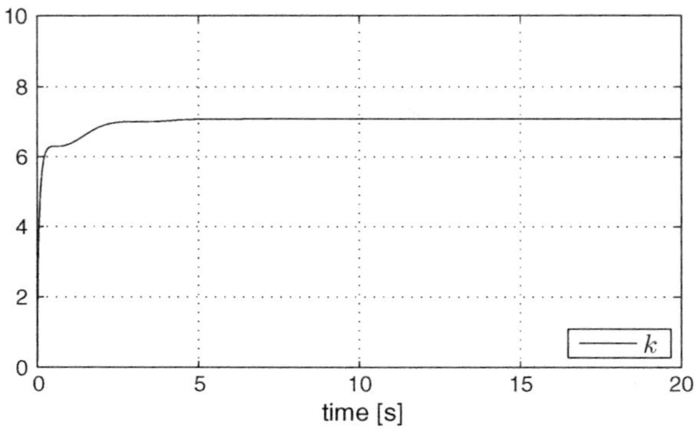

Fig. 6.13 Profile of dynamic gain of the controlled hyperchaotic Lorenz system

6.6 Problems

Problem 6.1 Find a controller u with partial state feedback (x_1, x_2, x_3) that solves the GASP for each of the following systems where w_1 and w_2 are unknown parameters.

(a) $\dot{x}_1 = w_1(\sin x_1 + x_1) + x_2$
$\dot{x}_2 = u$;

(b) $\dot{z} = -z + w_1 \sin x_1$
$\dot{x}_1 = w_2 z x_1 + x_2$
$\dot{x}_2 = u$;

(c) $\dot{z} = -z + w_1 x_1$
$\dot{x}_1 = w_2 z + x_2$
$\dot{x}_2 = x_2^2 + x_3$
$\dot{x}_3 = u$.

Problem 6.2 Find a controller u with partial state feedback (x_1, x_2) that solves the GASP for each of the following systems where w_1, w_2, and w_3 are unknown parameters.

(a) $\dot{z} = -z + w_1 x_1$
$\dot{x}_1 = w_2 x_1 + z + x_2$
$\dot{x}_2 = w_3 x_1^2 + u$;

(b) $\dot{x}_1 = w_1 x_1 + x_2$
$\dot{z} = -z + w_2 x_2$
$\dot{x}_2 = w_3 x_1^2 + z + u$;

(c) $\dot{z} = -z + w_1 x_1$
$\dot{x}_1 = z + (2 + w_2^2) x_2$
$\dot{x}_2 = w_3 x_1^2 + z + u$.

Problem 6.3 Find a controller u with partial state feedback x that solves the GASP for each of the following systems where w and $b \neq 0$ are unknown parameters.

(a) $\dot{z} = -z + w \sin x$
$\dot{x} = z + bu;$

(b) $\dot{z} = \begin{bmatrix} 0 & 1 \\ -1 & -2 \end{bmatrix} z + \begin{bmatrix} x \\ wx \end{bmatrix}$
$\dot{x} = \|z\|x + bu.$

Problem 6.4 The GASP for the system (5.26) has been solved in Chap. 5 for a known compact set $\mathbb{D}$. Using the universal adaptive technique discussed in Chap. 6 to solve the same problem for an unknown compact set $\mathbb{D}$.

Problem 6.5 Repeat Problem 6.4 for the system (5.25).

Problem 6.6 The GASP for the system (6.2) has been solved in Chap. 6.1 for a known parameter $b > 0$. Extend the result to the case that the parameter $b > 0$ is unknown (Note: the extension is a special case of (6.12).)

Problem 6.7 Repeat Problem 6.6 with $b > 0$ replaced by any $b(d)$ satisfying Assumption 6.5.

Problem 6.8 Use the result developed in Problem 6.7 to solve the problem in Sect. 6.4, i.e., the GASP of the following system

$$\begin{aligned} \dot{z} &= q(z, x_1, w) \\ \dot{x}_1 &= z_1(\rho - z_2) - x_1 + z_4 + b(d)x_2 \\ \dot{x}_2 &= u \end{aligned}$$

for any $b(d)$ satisfying Assumption 6.5.

References

1. Ilchmann A (1993) Non-identifier-based high-gain adaptive control. Springer-Verlag, Berlin
2. Ilchmann A, Ryan EP (1994) Universal λ-tracking for nonlinearly perturbed systems in the presence of noise. Automatica 30:337–346
3. Mareels I, Polderman JW (1996) Adaptive systems: an introduction. Birkhäauser, Boston
4. Ryan EP (1994) A nonlinear universal servomechanism. IEEE Trans Autom Control 39: 753–761
5. Ye XD, Huang J (2003) Decentralized adaptive output regulation for a class of large-scale nonlinear systems. IEEE Trans Autom Control 48:276–281
6. Xu D, Huang J (2010) Output regulation for a class of nonlinear systems using the observer based on output feedback control. Dyn Contin Discret Impul Syst Ser A: Math Anal 17:789–807
7. Chen Z, Huang J (2004) Dissipativity, stabilization, and regulation of cascade-connected systems. IEEE Trans Autom Control 49:635–650
8. Xu D, Huang J (2010) Global output regulation for output feedback systems with an uncertain exosystem and its application. Int J Robust Nonlinear Control 20:1678–1691

9. Jiang ZP, Mareels I, Hills DJ, Huang J (2004) A unifying framework for global regulation via a nonlinear output feedback: from iss to integral iss. IEEE Trans Autom Control 49(4):549–562
10. Liu L, Huang J (2006) Global robust output regulation of output feedback systems with unknown high-frequency gain sign. IEEE Trans Autom Control 51:625–631
11. Nussbaum RD (1983) Some remarks on a conjecture in parameter adaptive control. Syst Control Lett 3:243–246
12. Xu D, Huang J (2010) Output regulation design for a class of nonlinear systems with an unknown control direction. ASME J Dyn Syst Meas Contr 132:014503
13. Ye X (1999) Asymptotic regulation of time-varying uncertain nonlinear systems with unknown control directions. Automatica 35:929–935

Chapter 7
Robust Output Regulation: A Framework

Having studied the stabilization problem from Chaps. 2–6, we now turn to the output regulation problem, or alternatively, servomechanism problem. As roughly described in Chap. 1, the output regulation problem is to achieve, via a feedback control law, asymptotic tracking and disturbance rejection for a class of reference inputs and disturbances while maintaining the stability of the closed-loop system. Thus it poses a more challenging problem than the stabilization problem. The output regulation problem can be handled by two approaches: feedforward design approach and internal model approach. Both approaches try to convert the output regulation problem of a given plant to the stabilization problem of some auxiliary system. However, the feedforward design approach can only handle systems without model uncertainty. Therefore, in this book, we will focus on the internal model design approach. This chapter aims to give a precise formulation of the output regulation problem and establish a general framework for handling the output regulation problem via the internal model approach. The chapter is organized as follows. In Sect. 7.1, we present the definition for two versions of the output regulation problem: local robust output regulation problem and global robust output regulation problem. We also introduce a set of partial differential equations and algebraic equations that define the steady state of a system under the constraint that the error output is equal to zero. This set of equations are called regulator equations whose solvability is instrumental for the solvability of the output regulation problem. In Sect. 7.2, we introduce the concept of the steady-state generator that is an autonomous system independent of unknown parameters of the plant and can reproduce the solution or partial solution of the regulator equations. However, the state of the steady-state generator cannot be directly used for feedback control. Therefore, in Sect. 7.3, we further introduce the concept of internal model which can be viewed as an asymptotic observer of the steady-state generator. Then we construct various internal models corresponding to various steady-state generators. In Sect. 7.4, we complete the establishment of the general framework for handling the output regulation problem via the internal model approach. An internal model can be viewed as a dynamic compensator which together with the given plant constitutes a so-called augmented system. The internal model is conceived in such a way that certain invariant manifold of the augmented system is

Z. Chen and J. Huang, *Stabilization and Regulation of Nonlinear Systems*,
Advanced Textbooks in Control and Signal Processing,
DOI 10.1007/978-3-319-08834-1_7

stablizable, and the stabilizability of this invariant manifold of the augmented system implies the solvability of the output regulation problem of the original plant. Thus, the key of solving the output regulation problem of a given plant is to find an appropriate internal model for this plant based on the steady-state generator. In Sect. 7.5, we apply this framework to solve the robust output regulation problem for linear systems. The framework will be further applied to solve the robust output regulation problem for various classes of nonlinear systems in the subsequent chapters. This chapter ends with some notes and references in Sect. 7.6.

7.1 Problem Description

This chapter is concerned with a nonlinear control system of the following form:

$$\begin{aligned} \dot{x} &= f(x, u, v(t), w) \\ e &= h_{\mathrm{e}}(x, u, v(t), w) \\ y_{\mathrm{m}} &= h_{\mathrm{m}}(x, u, v(t), w) \end{aligned} \tag{7.1}$$

where $x \in \mathbb{R}^n$ is the state, $u \in \mathbb{R}^m$ is the input, $e \in \mathbb{R}^p$ is the performance output, $y_{\mathrm{m}} \in \mathbb{R}^q$ is the measurement output, $v(t) \in \mathbb{R}^{l_1}$ represents external disturbances and/or reference inputs, and $w \in \mathbb{R}^{l_2}$ represents an unknown constant vector whose nominal value is zero. Here, we denote the performance output by e instead of y in (2.7)–(2.9) to indicate that the tracking error is the performance output in the output regulation problem. A typical tracking error is defined as $e = y - y_{\mathrm{d}}$ where $y_{\mathrm{d}} = h_{\mathrm{d}}(v, w)$ is the reference trajectory to be tracked. We also assume that the tracking error e is always measurable, i.e., e is a component of y_{m}.

What distinguishes the output regulation problem from other tracking or disturbance rejection problems is that v is not a given time-varying signal, but a class of signals generated by an autonomous system of the following form:

$$\dot{v} = a(v, \sigma), \quad v(0) = v_0 \tag{7.2}$$

where $\sigma \in \mathbb{R}^{l_3}$ is an unknown constant parameter vector. In what follows, we call the system (7.2) an *exosystem*, and the signal $v(t)$ an exogenous signal. If the function a does not contain any uncertain parameter σ, or what is the same, $\sigma \in \mathbb{R}^0$, then we say the exosystem is known exactly. A special case of (7.2) is when the function a is linear, i.e., there exists a matrix $A_1(\sigma) \in \mathbb{R}^{l_1 \times l_1}$ such that $a(v, \sigma) = A_1(\sigma)v$.

Example 7.1 Let $v(t) \in \mathbb{R}^2$, and

$$a(v, \sigma) = A_1(\sigma)v \text{ with } A_1(\sigma) = \begin{bmatrix} 0 & -\sigma \\ \sigma & 0 \end{bmatrix}. \tag{7.3}$$

Then the solution of the exosystem (7.3) is given by

$$v(t) = \alpha \begin{bmatrix} \cos(\sigma t + \beta) \\ \sin(\sigma t + \beta) \end{bmatrix}, \quad \alpha \begin{bmatrix} \cos \beta \\ \sin \beta \end{bmatrix} = v_0 = [v_{01}, v_{02}]^{\mathrm{T}}.$$

Thus, the system (7.3) can produce sinusoidal disturbance and/or reference input with arbitrary unknown amplitude $\alpha = \|v_0\|$ and initial phase $\beta = \tan^{-1}\left(\frac{v_{02}}{v_{01}}\right)$ determined by the initial condition v_0 and arbitrary unknown frequency determined by the unknown parameter σ.

We will consider a control law of the following form:

$$\begin{aligned} u &= \kappa_1(\nu, y_{\mathrm{m}}) \\ \dot{\nu} &= \kappa_2(\nu, y_{\mathrm{m}}) \end{aligned} \tag{7.4}$$

for some functions κ_1 and κ_2. For simplicity, we assume all the functions in (7.1), (7.2), and (7.4) are globally defined, vanishing at the respective origins, sufficiently smooth, and satisfy, for all $w \in \mathbb{R}^{l_2}$, $f(0, 0, 0, w) = 0$, $h_{\mathrm{e}}(0, 0, 0, w) = 0$, $h_{\mathrm{m}}(0, 0, 0, w) = 0$, and, for all $\sigma \in \mathbb{R}^{l_3}$, $a(0, \sigma) = 0$.

To guarantee the implementability of the controller, we assume the function h_{m} does not depend on u explicitly, i.e., $y_{\mathrm{m}} = h_{\mathrm{m}}(x, v, w)$, in the sequel. Let $x_{\mathrm{c}} = \mathrm{col}(x, \nu)$. Then the closed-loop system composed of the plant and the control law is described as follows:

$$\begin{aligned} \dot{x}_{\mathrm{c}} &= f_{\mathrm{c}}(x_{\mathrm{c}}, v, w) \\ e &= h_{\mathrm{c}}(x_c, v, w) \end{aligned} \tag{7.5}$$

where

$$\begin{aligned} f_{\mathrm{c}}(x_{\mathrm{c}}, v, w) &= \begin{bmatrix} f(x, \kappa_1(\nu, h_{\mathrm{m}}(x, v, w)), v, w) \\ \kappa_2(\nu, h_{\mathrm{m}}(x, v, w)) \end{bmatrix} \\ h_{\mathrm{c}}(x_{\mathrm{c}}, v, w) &= h_{\mathrm{e}}(x, \kappa_1(\nu, h_{\mathrm{m}}(x, v, w)), v, w). \end{aligned}$$

It can be seen that, for all $w \in \mathbb{R}^{l_2}$, $f_{\mathrm{c}}(0, 0, w) = 0$ and $h_{\mathrm{c}}(0, 0, w) = 0$. Roughly, the robust output regulation problem is concerned with designing a control law of the form (7.4) such that the closed-loop system (7.5) satisfies two requirements: internal stability, and asymptotic tracking and disturbance rejection. Depending on the characterization of internal stability, several versions of the robust output regulation problem can be defined. In this chapter, we only introduce the local and global versions of the robust output regulation problem.

For convenience, we denote $d = \mathrm{col}(v, w, \sigma) \in \mathbb{R}^l$ with $l = l_1 + l_2 + l_3$. We use the notation $\mathbb{V}$, $\mathbb{W}$, and $\mathbb{S}$ to denote a generic compact subset of $\mathbb{R}^{l_i}$, $i = 1, 2, 3$, respectively, such that $0 \in \mathbb{V}$, $0 \in \mathbb{W}$, and $\sigma_0 \in \mathbb{S}$ with σ_0 being the nominal value of σ. Finally, let $\mathbb{D} = \mathbb{V} \times \mathbb{W} \times \mathbb{S}$.

Robust Output Regulation Problem(RORP): *Given the plant* (7.1), *and the exosystem* (7.2) *with* $\sigma \in \mathbb{R}^0$, *find a controller of the form* (7.4) *such that, for all sufficiently* *small* $x_c(0)$, v_0, *and* w, *the closed-loop system* (7.5) *satisfies the following two properties:*

(i) *The trajectory of the closed-loop system* (7.5) *exists and is bounded for all* $t > 0$.
(ii) *The trajectory described in (i) satisfies* $\lim_{t\to\infty} e(t) = 0$.

Global Robust Output Regulation Problem (GRORP) *on a compact set* $\mathbb{D}$:*Given the plant* (7.1), *the exosystem* (7.2), *and some compact set* $\mathbb{D}$, *find a control law of the form* (7.4) *such that, for any* $d(t) \in \mathbb{D}$, *and any initial condition* $x_c(0)$, *the closed-loop system* (7.5) *satisfies the following two properties:*

(i) *The trajectory of the closed-loop system* (7.5) *exists and is bounded for all* $t > 0$.
(ii) *The trajectory described in (i) satisfies* $\lim_{t\to\infty} e(t) = 0$.

Remark 7.1 If for any compact set $\mathbb{D}$ containing $(0, 0, \sigma_0)$, the GRORP on $\mathbb{D}$ is solvable, then we say the GRORP is solvable.

Remark 7.2 In Chaps. 4, 5, and 7, three problems, GRSP, GARP (GASP) and GRORP, are defined, respectively. The difference among them lies on the quantities to be regulated to zero. In GRSP, the whole state vector of the closed-loop system is regulated to zero; in GARP (GASP), the performance output (typically part of the state vector of the closed-loop system) is regulated to zero; and, in GRORP, the specified tracking error is regulated to zero and the exogenous signal is generated by an exosystem.

To have the problem well posed, let us first state the following standard assumptions which guarantee the boundedness of disturbances and/or reference trajectories.

Assumption 7.1 There exists a class $\mathcal{K}$ function γ such that $\|v(t)\| \le \gamma(\|v(0)\|)$, $\forall t \ge 0$, for all $\sigma \in \mathbb{S}$ and all $v(0) \in \mathbb{R}^{l_1}$.

Under Assumption 7.1, for any compact set $\mathbb{V} \subset \mathbb{R}^{l_1}$ containing the origin of $\mathbb{R}^{l_1}$, there exists a compact set $\mathbb{V}_0$ such that $v(t) \in \mathbb{V}$, $\forall t \ge 0$ if $v(0) \in \mathbb{V}_0$. A special case of Assumption 7.1 is as follows.

Assumption 7.2 The exosystem is of the form

$$\dot{v} = A_1(\sigma)v, \quad v(0) = v_0, \tag{7.6}$$

and, for any $\sigma \in \mathbb{S}$, all eigenvalues of $A_1(\sigma)$ are simple with zero real parts.

In particular, when the matrix A_1 is exactly known, Assumption 7.2 reduces to the following one.

Assumption 7.3 The exosystem is of the form

$$\dot{v} = A_1 v, \quad v(0) = v_0, \tag{7.7}$$

and all eigenvalues of A_1 are simple with zero real parts.

Next we will introduce a set of important equations. For this purpose, define a so-called composite system as follows:

$$\begin{aligned} \dot{x} &= f(x, u, v, w) \\ \dot{v} &= a(v, \sigma) \\ e &= h_e(x, u, v, w). \end{aligned} \tag{7.8}$$

Associated with the composite system is the following set of partial differential and algebraic equations:

$$\begin{aligned} \frac{\partial \mathbf{x}(v, w, \sigma)}{\partial v} a(v, \sigma) &= f(\mathbf{x}(v, w, \sigma), \mathbf{u}(v, w, \sigma), v, w) \\ 0 &= h_e(\mathbf{x}(v, w, \sigma), \mathbf{u}(v, w, \sigma), v, w) \end{aligned} \tag{7.9}$$

where $\mathbf{x}(v, w, \sigma)$ and $\mathbf{u}(v, w, \sigma)$ are two sufficiently smooth functions satisfying $\mathbf{x}(0, 0, 0) = 0$ and $\mathbf{u}(0, 0, 0) = 0$. We call (7.9) the *regulator equations* associated with (7.8).

Example 7.2 Consider a nonlinear system

$$\begin{aligned} \dot{x}_1 &= x_2 \\ \dot{x}_2 &= w \sin x_1^2 + u \\ e &= x_1 - v_1 \end{aligned}$$

and a linear exosystem

$$\dot{v} = A_1(\sigma)v, \ A_1(\sigma) = \begin{bmatrix} 0 & -\sigma \\ \sigma & 0 \end{bmatrix}.$$

The regulator equations are

$$\begin{aligned} \frac{\partial \mathbf{x}_1(v, w, \sigma)}{\partial v} A_1(\sigma)v &= \mathbf{x}_2(v, w, \sigma) \\ \frac{\partial \mathbf{x}_2(v, w, \sigma)}{\partial v} A_1(\sigma)v &= w \sin \mathbf{x}_1^2(v, w, \sigma) + \mathbf{u}(v, w, \sigma) \\ 0 &= \mathbf{x}_1(v, w, \sigma) - v_1 \end{aligned}$$

which have a solution

$$\begin{aligned} \mathbf{x}_1(v, w, \sigma) &= v_1 \\ \mathbf{x}_2(v, w, \sigma) &= -\sigma v_2 \\ \mathbf{u}(v, w, \sigma) &= -\sigma^2 v_1 - w \sin v_1^2. \end{aligned}$$

Let $\bar{x} = x - \mathbf{x}(v, w, \sigma)$ and $\bar{u} = u - \mathbf{u}(v, w, \sigma)$. Then (7.9) implies that $\bar{x}$ satisfies

$$\begin{aligned} \dot{\bar{x}} &= \bar{f}(\bar{x}, \bar{u}, v, w, \sigma) \\ e &= \bar{h}_e(\bar{x}, \bar{u}, v, w, \sigma) \end{aligned} \tag{7.10}$$

where

$$\begin{aligned} \bar{f}(\bar{x}, \bar{u}, v, w, \sigma) &= f(\bar{x} + \mathbf{x}(v, w, \sigma), \bar{u} + \mathbf{u}(v, w, \sigma), v, w) \\ &\quad - f(\mathbf{x}(v, w, \sigma), \mathbf{u}(v, w, \sigma), v, w) \\ \bar{h}_e(\bar{x}, \bar{u}, v, w, \sigma) &= h_e(\bar{x} + \mathbf{x}(v, w, \sigma), \bar{u} + \mathbf{u}(v, w, \sigma), v, w). \end{aligned}$$

It can be verified that $\bar{f}(0, 0, v, w, \sigma) = 0$ and $\bar{h}(0, 0, v, w, \sigma) = 0$ for all v, w, σ.

In the sequel, we call the system (7.10) the *error system*. Under Assumption 7.1, for any given compact set $\mathbb{D}$, if there exists a control law of the form $\bar{u} = k(\bar{x})$ vanishing at the origin such that $\bar{x} = 0$ is globally asymptotically stable for any $d(t) \in \mathbb{D}$, then, for any $\bar{x}(0)$,

$$\lim_{t \to \infty} \bar{x}(t) = \lim_{t \to \infty} (x(t) - \mathbf{x}(v(t), w, \sigma)) = 0 \tag{7.11}$$

and

$$\lim_{t \to \infty} \bar{u}(t) = \lim_{t \to \infty} (u(t) - \mathbf{u}(v(t), w, \sigma)) = \lim_{t \to \infty} k(\bar{x}(t)) = k(0) = 0. \tag{7.12}$$

As a result,

$$\lim_{t \to \infty} e(t) = \lim_{t \to \infty} \bar{h}_e(\bar{x}(t), \bar{u}(t), v(t), w, \sigma) = \bar{h}_e(0, 0, v(t), w, \sigma) = 0.$$

That is, the output regulation problem of the original plant is solved by the following control law globally

$$u = \mathbf{u}(v, w, \sigma) + k(x - \mathbf{x}(v, w, \sigma)). \tag{7.13}$$

From this observation, we can see that the solution of the regulator equations provides the necessary feedforward information for designing a control law to convert the output regulation problem of the given system (7.1) to the stabilization problem of the well defined error system (7.10). A control law of the form (7.13) is called feedforward control law. It can be seen that the feedforward control law works only if neither the given plant nor the exosystem contains unknown parameters. In what follows, we will introduce the so-called internal model design approach which is capable of handling the uncertain parameters w and σ by making use of a well defined dynamic compensator called internal model, and the robust and adaptive control techniques developed from Chaps. 2–6.

Remark 7.3 From (7.11) and (7.12), we can interpret the solution of the regulator equations $\mathbf{x}(v, w, \sigma)$ and $\mathbf{u}(v, w, \sigma)$ as the *zero-error constrained state* and the *zero-error constrained input*, respectively.

Remark 7.4 We can also give a geometric interpretation to the solution of the regulator equations. The solvability of the regulator equations means that the composite system (7.8) has an output zeroing manifold characterized by

$$\mathbb{M} = \{(x, v) \in \mathbb{R}^n \times \mathbb{R}^{l_1} \mid x = \mathbf{x}(v, w, \sigma) \}. \tag{7.14}$$

In fact, the first equation of (7.9) means that $\mathbb{M}$ is a control invariant manifold of the composite system (7.8) rendered by the state feedback control $u = \mathbf{u}(v, w, \sigma)$, and the second equation of (7.9) means that this manifold is contained in the kernel of the error output mapping $h_e(x, \mathbf{u}(v, w, \sigma), v, w)$.

Under the control law (7.13), the closed-loop system can be put in the form (7.5) with $x_c = x$, and

$$\begin{aligned} f_c(x_c, v, w, \sigma) &= f(x, \mathbf{u}(v, w, \sigma) + k(x - \mathbf{x}(v, w, \sigma)), v, w) \\ h_c(x_c, v, w, \sigma) &= h_e(x, \mathbf{u}(v, w, \sigma) + k(x - \mathbf{x}(v, w, \sigma)), v, w). \end{aligned}$$

Moreover, the solvability of the regulator equations implies that $\mathbf{x}(v, w, \sigma)$ satisfies the following partial differential equation

$$\frac{\partial \mathbf{x}(v, w, \sigma)}{\partial v} a(v, \sigma) = f_c(\mathbf{x}(v, w, \sigma), v, w). \tag{7.15}$$

Thus, the manifold defined in (7.14) can also be interpreted as a stable invariant manifold of the closed-loop system under the control law (7.13). By Theorem 11.4 of the Appendix, if all the eigenvalues of the Jacobian matrix

$$\frac{\partial a}{\partial v}(0, \sigma)$$

have zero real part and all the eigenvalues of the Jacobian matrix

$$\frac{\partial f_c}{\partial x_c}(0, 0, w, \sigma)$$

have negative real part, then the manifold defined in (7.14) is a center manifold of the composite system (7.8). In this case, the equation (7.15) is called a center manifold equation associated with the composite system (7.8). If, in addition, the equilibrium point of the exosystem at the origin is Lyapunov stable, then the local output regulation problem is solvable if the regulator equations (7.9) are solvable. Thus, in what follows, we make the following assumption.

Assumption 7.4 There exist sufficiently smooth functions $\mathbf{x}(v, w, \sigma)$ and $\mathbf{u}(v, w, \sigma)$ with $\mathbf{x}(0, 0, 0) = 0$ and $\mathbf{u}(0, 0, 0) = 0$ that satisfy the regulator equations (7.9) for all $d \in \mathbb{R}^l$.

Example 7.3 For linear systems, that is, the systems described by the following equations:

$$\begin{aligned} \dot{x} &= A(w)x + B(w)u + E(w)v \\ e &= C(w)x + D(w)u + F(w)v \end{aligned} \tag{7.16}$$

where $x \in \mathbb{R}^n$, $u, e \in \mathbb{R}^m$, and all the matrix functions $A(w)$, $B(w)$, $C(w)$, $D(w)$, $E(w)$, and $F(w)$ are continuous in w. The regulator equations associated with (7.16) and the exosystem satisfying Assumption 7.3 take the following form

$$\begin{aligned} X(w)A_1 v &= (A(w)X(w) + B(w)U(w) + E(w))v \\ 0 &= (C(w)X(w) + D(w)U(w) + F(w))v \end{aligned} \tag{7.17}$$

which reduces to the following linear matrix equations by removing v on both sides of (7.17):

$$\begin{aligned} X(w)A_1 &= A(w)X(w) + B(w)U(w) + E(w) \\ 0 &= C(w)X(w) + D(w)U(w) + F(w). \end{aligned} \tag{7.18}$$

Equation (7.18) is a type of Sylvester equations and the solvability of (7.18) has been well studied. For example, by Theorem 1.9 of [1], for any $w \in \mathbb{W}$, any $E(w)$ and $F(w)$, Eq. (7.18) admits a unique solution pair $(X(w), U(w))$ if and only if the following condition is satisfied:

For any $w \in \mathbb{W}$, rank $\begin{bmatrix} A(w) - \lambda I & B(w) \\ C(w) & D(w) \end{bmatrix} = n + m$ for all $\lambda \in \text{eig}(A_1)$, where $\text{eig}(A_1) = \{\lambda \mid \det(A_1 - \lambda I) = 0\}$.

The solution to the regulator equations (7.17) is then given by $\mathbf{x}(v, w) = X(w)v$ and $\mathbf{u}(v, w) = U(w)v$.

For general nonlinear systems, the solvability of the regulator equations is more complicated, and we will provide sufficient conditions for the solvability of the regulator equations associated with a few classes of nonlinear systems in the subsequent chapters.

7.2 Steady-State Generator

As we have seen from the previous section, the output regulation problem can be viewed as a stabilization problem about an invariant manifold defined by the solution to the regulator equations. This stabilization problem can always be converted to a conventional stabilization problem about an equilibrium point after a coordinate and input transformation defined by the solution to the regulator equations. Unfortunately, since the solution of the regulator equations depends on the unknown parameters w and/or σ, the new coordinates cannot be used for feedback. Therefore, this idea can be implemented only if we can reproduce the solution of the regulator equations by some dynamic compensator that does not rely on the unknown parameters. This idea motivates the so-called internal model design approach. Roughly, an internal model is a dynamic compensator independent of the uncertain parameters w and σ and can asymptotically reproduce the solution of the regulator equations. The composition of the internal model and the given plant constitutes a so-called augmented system. The internal model will be conceived such that the augmented system is stabilizable and the control law that stabilizes the augmented system together with the internal model will be the overall control law that solves the output regulation problem of the given plant. To ascertain the existence of an internal model, we need to first introduce the concept of the steady-state generator for the composite system (7.8) as follows.

Definition 7.1 Let $c : \mathbb{R}^{n+m} \mapsto \mathbb{R}^r$ be a mapping for some positive integer $m \le r \le n+m$. Under Assumption 7.4, the composite system (7.8) is said to have a *steady-state generator* with output $c(x, u) \in \mathbb{R}^r$ if there exists a triplet $\{\theta, \phi, \psi\}$ where, for some integer ℓ, $\theta : \mathbb{R}^l \mapsto \mathbb{R}^\ell$, $\phi : \mathbb{R}^{\ell+l_1+l_3} \mapsto \mathbb{R}^\ell$, and $\psi : \mathbb{R}^{\ell+l_3} \mapsto \mathbb{R}^r$ are sufficiently smooth functions vanishing at the origin, such that, for any trajectory $v(t) \in \mathbb{R}^{l_1}$ of the exosystem, $w \in \mathbb{R}^{l_2}$, and $\sigma \in \mathbb{R}^{l_3}$,

$$
\begin{aligned}
\dot{\theta}(v, w, \sigma) &= \phi(\theta(v, w, \sigma), v, \sigma) \\
\mathbf{c}(v, w, \sigma) &= \psi(\theta(v, w, \sigma)), \\
&\quad \mathbf{c}(v, w, \sigma) := c(\mathbf{x}(v, w, \sigma), \mathbf{u}(v, w, \sigma)).
\end{aligned}
\tag{7.19}
$$

Remark 7.5 Define an autonomous system as follows:

$$
\dot{v} = a(v, \sigma), \quad y = \mathbf{c}(v, w, \sigma). \tag{7.20}
$$

Then, that the composite system admits a steady-state generator with output $c(x, u) \in \mathbb{R}^r$ means that, for any $\sigma \in \mathbb{R}^{l_3}$, the function $\mathbf{c}(v, w, \sigma)$ can be reproduced by the following system:

$$
\dot{\theta} = \phi(\theta, v, \sigma), \quad y = \psi(\theta) \tag{7.21}
$$

which is independent of the unknown parameter w. If the exosystem does not contain the unknown parameter σ, then the system (7.21) is also independent of σ. We say that the system (7.20) is immersed into the system (7.21). If $c(x, u) = \text{col}(x, u)$, then the solution of the regulator equations can be reproduced by the system (7.21). In many cases, we assume that $c(x, u) = \left[x_{i_1}, x_{i_2}, \ldots, x_{i_{\bar{n}}}, u\right]^{\mathrm{T}}$ where $1 \le i_1 < i_2 < \cdots < i_{\bar{n}} \le n$ for some integer $\bar{n}$ satisfying $0 \le \bar{n} \le n$. With $c(x, u)$ thus defined, a steady-state generator can be viewed as a dynamic system which reproduces the partial ($\bar{n} < n$) or whole ($\bar{n} = n$) solution of the regulator equations. In the special case where $\bar{n} = 0$, only the control part of the solution of the regulator equations can be reproduced. Thus, we call (7.19) a partial (full) *steady-state state generator* if $0 < \bar{n} < n$ ($\bar{n} = n$) or a *steady-state input generator* if $\bar{n} = 0$. Without loss of generality, we always assume $i_j = j$ for $j = 1, \ldots, \bar{n}$ since the index of state variables can be relabeled to have this assumption satisfied.

Example 7.4 For a linear system of the form (7.16) with the exosystem given by (7.7), if the Sylvester equations (7.18) have a solution pair $(X(w), U(w))$, then the composite system (7.8) always admits a steady-state generator with the output u. In fact, let $p(\lambda) = \lambda^\ell - \phi_1 - \phi_2\lambda - \cdots - \phi_\ell\lambda^{\ell-1}$ be the minimum polynomial of A_1, and

$$\Phi = \begin{bmatrix} 0 & 1 & 0 & \cdots & 0 \\ 0 & 0 & 1 & \cdots & 0 \\ \vdots & \vdots & \vdots & \ddots & \vdots \\ 0 & 0 & 0 & \cdots & 1 \\ \phi_1 & \phi_2 & \phi_3 & \cdots & \phi_\ell \end{bmatrix}, \quad \Psi = \begin{bmatrix} 1 \\ 0 \\ \vdots \\ 0 \\ 0 \end{bmatrix}^{\mathrm{T}}. \tag{7.22}$$

For $i = 1, \ldots, m$, let

$$\theta_i(v, w) = \begin{bmatrix} U_i(w)v \\ U_i(w)A_1v \\ U_i(w)A_1^2v \\ \vdots \\ U_i(w)A_1^{\ell-1}v \end{bmatrix}$$

where $U_i(w)$ is the ith row of $U(w)$. Then it can be verified that the triplet $(\theta_i(v, w), \Phi, \Psi)$ forms a steady-state generator with output u_i, i.e., it is such that

$$\dot{\theta}_i(v, w) = \Phi\theta_i(v, w), \ \mathbf{u}_i(v, w) = \Psi\theta_i(v, w), i = 1, \ldots, m. \tag{7.23}$$

In what follows, we will study various conditions under which the steady-state generator of (7.8) exists. For convenience, we focus on the case where $r = 1$, i.e., $c(x, u)$ is a scalar function. The general case where $r > 1$ can be similarly handled as shown by Lemma 7.6 later.

7.2.1 Linear Immersion Assumption

In this subsection, we first consider a simple case where the exosystem with the output $\mathbf{c}(v, w, \sigma)$ can be immersed to a linear system. For this purpose, let

$$L_a\mathbf{c}(v, w, \sigma) := \frac{\partial \mathbf{c}(v, w, \sigma)}{\partial v}a(v, \sigma)$$

$$L_a^i \mathbf{c}(v, w, \sigma) := \frac{\partial L_a^{i-1}\mathbf{c}(v, w, \sigma)}{\partial v} a(v, \sigma), i = 2, 3, \ldots. \tag{7.24}$$

Assumption 7.5 There exist a positive integer ℓ and real scalars $\phi_i(\sigma), i = 1, \ldots, \ell$, such that

$$\begin{aligned} &L_a^{\ell}\mathbf{c}(v, w, \sigma) - \phi_1(\sigma)\mathbf{c}(v, w, \sigma) - \phi_2(\sigma)L_a\mathbf{c}(v, w, \sigma) - \cdots \\ &\quad -\phi_\ell(\sigma)L_a^{\ell-1}\mathbf{c}(v, w, \sigma) = 0, \ \forall \sigma \in \mathbb{S}, v \in \mathbb{R}^{l_1}, w \in \mathbb{R}^{l_2}. \end{aligned} \tag{7.25}$$

Lemma 7.1 *Under Assumption 7.5, the composite system (7.8) admits a linear steady-state generator with output $c(x, u)$:*

$$\dot{\theta}(v, w, \sigma) = \Phi(\sigma)\theta(v, w, \sigma), \quad \mathbf{c}(v, w, \sigma) = \Psi\theta(v, w, \sigma). \tag{7.26}$$

where $\theta(v, w, \sigma) = col(\mathbf{c}(v, w, \sigma), L_a\mathbf{c}(v, w, \sigma), \ldots, L_a^{(\ell-1)}\mathbf{c}(v, w, \sigma)$, and

$$\Phi(\sigma) = \begin{bmatrix} 0 & 1 & 0 & \cdots & 0 \\ 0 & 0 & 1 & \cdots & 0 \\ \vdots & \vdots & \vdots & \ddots & \vdots \\ 0 & 0 & 0 & \cdots & 1 \\ \phi_1(\sigma) & \phi_2(\sigma) & \phi_3(\sigma) & \cdots & \phi_\ell(\sigma) \end{bmatrix}, \quad \Psi = \begin{bmatrix} 1 \\ 0 \\ \vdots \\ 0 \\ 0 \end{bmatrix}^{\mathrm{T}}. \tag{7.27}$$

Moreover, the pair $(\Psi, \Phi(\sigma))$ is observable.

Proof It is straightforward to verify that the triplet $(\theta, \Phi(\sigma), \Psi)$ satisfies (7.26), and the pair $(\Psi, \Phi(\sigma))$ is observable for all σ. □

Next we will further show that Assumption 7.5 always holds if $\mathbf{c}(v, w, \sigma)$ is a polynomial in v and Assumption 7.2 is satisfied. For this purpose, let us first introduce some notation. For any $v \in \mathbb{R}^{l_1}$, let

$$\begin{aligned} v^{[0]} &= 1 \\ v^{[1]} &= v = \mathrm{col}(v_1, \ldots, v_{l_1}) \\ v^{[i]} &= \mathrm{col}(v_1^i, v_1^{i-1}v_2, \ldots, v_1^{i-1}v_{l_1}, v_1^{i-2}v_2^2, \ldots, v_1^{i-2}v_2 v_{l_1}, \ldots, v_{l_1}^i), \\ &\quad i = 2, 3, \ldots. \end{aligned} \tag{7.28}$$

It can be seen that the dimension of $v^{[i]}$ is given by the binomial coefficient

$$C_{l_1+i-1}^i = \frac{(l_1 + i - 1)!}{(l_1 - 1)!i!}.$$

Let P_i be a row vector with dimension $C_{l_1+i-1}^i$ whose entries are real numbers. Then we call $P_i v^{[i]}$ a homogeneous polynomial of degree i. We call $\mathbf{c}(v, w, \sigma) \in \mathbb{R}$ a polynomial in v of degree $\delta > 0$ if there exist constant row vectors $P_i(w, \sigma)$ which may depend on (w, σ), $i = 0, 1, \ldots, \delta$, such that

$$\mathbf{c}(v, w, \sigma) = \sum_{i=0}^{\delta} P_i(w, \sigma) v^{[i]} \tag{7.29}$$

where $P_\delta \neq 0$.

Lemma 7.2 *Under Assumption 7.2, assume the function $\mathbf{c}(v, w, \sigma)$ is a polynomial in v of degree δ. Then,*

(i) For some positive integer $\ell > 0$, the function $\mathbf{c}(v(t), w, \sigma)$ is a trigonometric polynomial of the form

$$\mathbf{c}(v(t), w, \sigma) = \sum_{i=1}^{\ell} C_i(v_0, w, \sigma) e^{j\hat{\omega}_i t}, \quad \forall \sigma \in \mathbb{S}, v_0 \in \mathbb{R}^{l_1}, w \in \mathbb{R}^{l_2} \tag{7.30}$$

where $\hat{\omega}_i = -\hat{\omega}_{1+\ell-i}$, $C_i = C^*_{1+\ell-i}$ (C^* is the complex conjugate of C), and $\hat{\omega}_i \neq \hat{\omega}_j$ for $i \neq j$. Moreover, $C_i(v_0, w, \sigma)$ is not identically zero for all $v_0 \in \mathbb{R}^{l_1}$, $w \in \mathbb{R}^{l_2}$, $\sigma \in \mathbb{R}^{l_3}$.

(ii) There exist a positive integer ℓ and real scalars $\phi_1(\sigma), \phi_2(\sigma), \ldots, \phi_\ell(\sigma)$ such that (7.25) is satisfied, i.e., Assumption 7.5 is satisfied.

Proof Part (i). For any $\sigma \in \mathbb{S}$, assume the nonzero eigenvalues of $A_1(\sigma)$ are $\pm j\omega_1, \ldots, \pm j\omega_k$ where $k = l_1/2$ if l_1 is even and $k = (l_1 - 1)/2$ if l_1 is odd (l_1 is the dimension of the exosystem, i.e., $v \in \mathbb{R}^{l_1}$). Let

$$\Omega = \{m_1\omega_1 + \cdots + m_k\omega_k, m_1, \ldots, m_k = 0, \pm 1, \ldots, \pm\delta\}.$$

Since $\mathbf{c}(v, w, \sigma)$ is a polynomial function of v and $v(t)$ is the linear combination of sinusoids of frequencies ω_i, we have

$$\mathbf{c}(v(t), w, \sigma) = \sum_{i=1}^{\ell} C_i(v_0, w, \sigma) e^{j\hat{\omega}_i t}, \ \forall v_0 \in \mathbb{R}^{l_1}, w \in \mathbb{R}^{l_2}, \sigma \in \mathbb{R}^{l_3} \tag{7.31}$$

where $\hat{\omega}_i \in \Omega$ and $C_i \in \mathbb{C}$ for some integer ℓ. Also, $\hat{\omega}_1, \ldots, \hat{\omega}_\ell$ are distinct and, for $i = 1, \ldots, \ell$, $\hat{\omega}_i = -\hat{\omega}_{1+\ell-i}$, $C_i = C^*_{1+\ell-i}$, and $C_i(v_0, w, \sigma)$ is not identically zero for $v_0 \in \mathbb{R}^{l_1}$, $w \in \mathbb{R}^{l_2}$, $\sigma \in \mathbb{R}^{l_3}$.

Part (ii). Define a polynomial function

$$p(\lambda, \sigma) = \prod_{i=1}^{\ell} (\lambda - j\hat{\omega}_i) = \lambda^\ell - \phi_1(\sigma) - \phi_2(\sigma)\lambda - \cdots - \phi_\ell(\sigma)\lambda^{\ell-1} \tag{7.32}$$

where $\phi_i(\sigma)$ may depend on σ since $\hat{\omega}_i$ may depend on σ. Using (7.31) gives

$$\frac{d^k \mathbf{c}(v(t), w, \sigma)}{dt^k} = \sum_{i=1}^{\ell} (j\hat{\omega}_i)^k C_i(v_0, w, \sigma) e^{j\hat{\omega}_i t}, \quad k = 1, 2, \ldots, \ell. \tag{7.33}$$

Thus,

$$\begin{aligned}
&\frac{d^\ell \mathbf{c}(v(t), w, \sigma)}{dt^\ell} - \phi_1(\sigma)\mathbf{c}(v(t), w, \sigma) - \cdots - \phi_\ell(\sigma)\frac{d^{(\ell-1)}\mathbf{c}(v(t), w, \sigma)}{dt^{(\ell-1)}} \\
&= \sum_{i=1}^{\ell} ((j\hat{\omega}_i)^\ell - \phi_1(\sigma) - \phi_2(\sigma)(j\hat{\omega}_i) - \cdots - \phi_\ell(\sigma)(j\hat{\omega}_i)^{\ell-1}) C_i(v_0, w, \sigma) e^{j\hat{\omega}_i t} \\
&= \sum_{i=1}^{\ell} p(j\hat{\omega}_i, \sigma) C_i(v_0, w, \sigma) e^{j\hat{\omega}_i t} = 0.
\end{aligned} \tag{7.34}$$

Note that

$$\begin{aligned}
L_a \mathbf{c}(v, w, \sigma) &= \frac{\partial \mathbf{c}(v, w, \sigma)}{\partial v} A_1(\sigma) v = \frac{d\mathbf{c}(v(t), w, \sigma)}{dt} \\
L_a^i \mathbf{c}(v, w, \sigma) &= \frac{\partial L_a^{i-1} \mathbf{c}(v, w, \sigma)}{\partial v} A_1(\sigma) v = \frac{d^i \mathbf{c}(v(t), w, \sigma)}{dt^i}, \quad i = 2, 3, \ldots.
\end{aligned}$$

Then (7.34) implies (7.25). □

Remark 7.6 From the proof of Lemma 7.2, if the matrix A_1 is known exactly, i.e., $\sigma \in \mathbb{R}^0$, then the coefficients ϕ_i, $i = 1, \ldots, \ell$, are independent of σ. Hence, the steady-state generator (7.26) is also independent of σ.

Remark 7.7 Let $p(\lambda, \sigma) = \lambda^\ell - \phi_1(\sigma) - \phi_2(\sigma)\lambda - \cdots - \phi_\ell(\sigma)\lambda^{\ell-1}$ be a monic polynomial in λ. If a function $\mathbf{c}(v, w, \sigma)$ satisfies (7.25) for all $\sigma \in \mathbb{S}$, we call $p(\lambda, \sigma)$ a *zeroing polynomial* of $\mathbf{c}(v, w, \sigma)$ on $\mathbb{S}$, and the matrix $\Phi(\sigma)$ described in (7.27) a *companion matrix* of the polynomial $p(\lambda, \sigma)$. We call $p(\lambda, \sigma)$ a *minimal zeroing polynomial* of $\mathbf{c}(v, w, \sigma)$ on $\mathbb{S}$ if it is a zeroing polynomial of $\mathbf{c}(v, w, \sigma)$ on $\mathbb{S}$ of least degree. Clearly, the polynomial constructed in (7.32) is the minimal zeroing polynomial of $\mathbf{c}(v, w, \sigma)$ on $\mathbb{S}$. It can be seen from the proof of Lemma 7.2 that, if all the eigenvalues of the matrix $A_1(\sigma)$ have zero real part, so do the eigenvalues of the matrix $\Phi(\sigma)$.

Example 7.5 Suppose

$$\mathbf{c}(v, w, \sigma) = w v_1 + v_1 v_2,$$

where

$$\dot{v}_1 = \sigma v_2, \ \dot{v}_2 = -\sigma v_1.$$

Recursive calculation shows that

$$
\begin{aligned}
L_a\mathbf{c}(v, w, \sigma) &= w\sigma v_2 - \sigma v_1^2 + \sigma v_2^2 \\
L_a^2\mathbf{c}(v, w, \sigma) &= -w\sigma^2 v_1 - 4\sigma^2 v_1 v_2 \\
L_a^3\mathbf{c}(v, w, \sigma) &= -w\sigma^3 v_2 + 4\sigma^3 v_1^2 - 4\sigma^3 v_2^2 \\
L_a^4\mathbf{c}(v, w, \sigma) &= w\sigma^4 v_1 + 16\sigma^4 v_1 v_2.
\end{aligned}
$$

It is easy to verify

$$L_a^4\mathbf{c}(v, w, \sigma) + 4\sigma^4\mathbf{c}(v, w, \sigma) + 5\sigma^2 L_a^2\mathbf{c}(v, w, \sigma) = 0.$$

Therefore, Assumption 7.5 is satisfied with $\ell = 4$, and

$$\phi_1(\sigma) = -4\sigma^4, \quad \phi_2(\sigma) = 0, \quad \phi_3(\sigma) = -5\sigma^2, \ \phi_4(\sigma) = 0.$$

7.2.2 Nonlinear Immersion Assumption

The validity of Lemma 7.2 relies on two assumptions, i.e., the function $\mathbf{c}(v, w, \sigma)$ is a polynomial in v, and the exosystem is linear. We will consider relaxing these two assumptions in this subsection and the next subsection, respectively. For convenience, in this subsection, we will assume that the exosystem is linear and is known exactly.

With the operator L_a defined in (7.24), let

$$\mathcal{L}_a^i\pi(v, w) := \text{col}\left(\pi(v, w), L_a\pi(v, w), \ldots, L_a^{i-1}\pi(v, w)\right), \quad i = 1, 2, \ldots,$$

for any sufficiently smooth scalar function $\pi(v, w)$.

Assumption 7.6 There exist polynomials $\pi^1(v, w), \ldots, \pi^p(v, w)$ in v for some integer $p > 0$, and a sufficiently smooth function ψ vanishing at the origin such that

$$\mathbf{c}(v, w) = \psi(\mathcal{L}_a^{\ell_1}\pi^1(v, w), \ldots, \mathcal{L}_a^{\ell_p}\pi^p(v, w)), \ \forall v \in \mathbb{R}^{l_1}, w \in \mathbb{R}^{l_2}$$

where, for $i = 1, \ldots, p$, ℓ_i are the degree of the minimal zero polynomials of $\pi^i(v, w)$.

Lemma 7.3 *Under Assumption 7.3, if $\mathbf{c}(v, w)$ satisfies Assumption 7.6, then the composite system (7.8) admits a steady-state generator of the following form:*

$$\dot{\theta}(v, w) = \Phi\theta(v, w), \quad \mathbf{c}(v, w) = \psi(\theta(v, w)) \tag{7.35}$$

where

$$\theta = col(\theta^1, \ldots, \theta^p), \ \Phi = diag(\Phi^1, \ldots, \Phi^p)$$

with $\theta^i(v,w) = \mathcal{L}_a^{\ell_i}\pi^i(v,w)$ *and* $\Phi^i \in \mathbb{R}^{\ell_i \times \ell_i}$ *being the companion matrix of* $\pi^i(v,w)$. *Let* $\Psi = [\Psi^1, \ldots, \Psi^p]$ *with* $\Psi^i \in \mathbb{R}^{1\times\ell_i}$ *being the Jacobian matrix of* ψ *at the origin. The steady-state generator (7.35) is linearly observable if the minimal zeroing polynomials of* $\pi^i(v,w)$*'s are pairwise coprime and the pairs* (Ψ^i, Φ^i), $i = 1, \ldots, p$, *are observable.*

Proof From Lemma 7.1, it is clear that (7.35) is a steady-state generator of the composite system (7.8). To show that (Ψ, Φ) is observable, it suffices to show, by PBH (Popov-Belevitch-Hautus) test, that, for any λ,

$$\begin{aligned}
&\operatorname{rank}\begin{bmatrix} \lambda I - \Phi \\ \Psi \end{bmatrix} \\
&= \operatorname{rank}\begin{bmatrix} \lambda I - \Phi^1 & 0 & \cdots & 0 \\ 0 & \lambda I - \Phi^2 & \cdots & 0 \\ \vdots & \vdots & \vdots & \vdots \\ 0 & 0 & 0 & \lambda I - \Phi^p \\ \Psi^1 & \Psi^2 & \cdots & \Psi^p \end{bmatrix} \\
&= \ell_1 + \cdots + \ell_p.
\end{aligned} \tag{7.36}$$

It is clear that (7.36) holds for any $\lambda \notin \operatorname{eig}(\Phi)$. For any $\lambda \in \operatorname{eig}(\Phi)$, there exists $1 \le i \le p$ such that $\lambda \in \operatorname{eig}(\Phi^i)$ and $\lambda \notin \operatorname{eig}(\Phi^j)$, $j \neq i$ and $1 \le j \le p$, since, for any $j \neq i$, $P_j(\lambda)$ and $P_i(\lambda)$ are coprime. Thus

$$\operatorname{rank}\begin{bmatrix} \lambda I - \Phi \\ \Psi \end{bmatrix} = \operatorname{rank}\begin{bmatrix} \lambda I - \Phi^i \\ \Psi^i \end{bmatrix} + \sum_{j=1\&j\neq i}^{p} \operatorname{rank}\begin{bmatrix} \lambda I - \Phi^j \end{bmatrix} = \ell_1 + \cdots + \ell_p$$

since (Ψ^i, Φ^i) is observable. The proof is thus completed. □

Example 7.6 Being an *(almost)* arbitrary nonlinear function of polynomial functions, the class of functions satisfying Assumption 7.6 is much larger than the class of the polynomial or trigonometric polynomial functions. Suppose the solution to the regulator equations of the system (7.8) is such that

$$\mathbf{c}(v,w) = \sin(wv_1 + v_2) + wv_3,$$

where

$$\begin{aligned}
\dot v_1 &= v_2,\ \ \dot v_2 = -v_1 \\
\dot v_3 &= 2v_4,\ \ \dot v_4 = -2v_3.
\end{aligned}$$

Letting $\pi_1(v,w) = wv_1 + v_2$ and $\pi_2(v,w) = wv_3$ gives

$$\mathbf{c}(v,w) = \psi\left(\pi_1(v,w), \dot\pi_1(v,w), \pi_2(v,w), \dot\pi_2(v,w)\right) = \sin(\pi_1(v,w)) + \pi_2(v,w)$$

with $\Psi = [1, 0, 1, 0]$. Thus, this system admits a linearly observable steady-state generator as given in Lemma 7.3.

7.2.3 Generalized Linear Immersion Assumption

Both Lemmas 7.2 and 7.3 require the exosystem to be a linear system. To relax this assumption, we will extend Assumption 7.5 to the so-called generalized linear immersion assumption which has two equivalent versions. For convenience, again we assume that the exosystem is known exactly, or, $\sigma \in \mathbb{R}^0$.

Assumption 7.7 There exist a positive integer ℓ and sufficiently smooth scalar functions $\phi_i(v)$, $i = 1, \ldots, \ell$, such that

$$\begin{aligned} L_a^\ell \mathbf{c}(v, w) - \phi_1(v)\mathbf{c}(v, w) - \phi_2(v)L_a\mathbf{c}(v, w) - \cdots \\ -\phi_\ell(v)L_a^{\ell-1}\mathbf{c}(v, w) = 0, \forall v \in \mathbb{R}^{l_1}, w \in \mathbb{R}^{l_2}. \end{aligned} \tag{7.37}$$

Assumption 7.8 There exist a positive integer ℓ and sufficiently smooth scalar functions $\varphi_i(v)$, $i = 1, \ldots, \ell$, such that

$$\begin{aligned} L_a^\ell \mathbf{c}(v, w) - \varphi_1(v)\mathbf{c}(v, w) - L_a[\varphi_2(v)\mathbf{c}(v, w)] - \ldots \\ -L_a^{\ell-1}[\varphi_\ell(v)\mathbf{c}(v, w)] = 0, \ \forall v \in \mathbb{R}^{l_1}, w \in \mathbb{R}^{l_2}. \end{aligned} \tag{7.38}$$

Lemma 7.4 *A function* $\mathbf{c}(v, w)$ *satisfies Assumption 7.7 if and only if it satisfies Assumption 7.8.*

Proof First, we note that, for $i = 1, \ldots, \ell$,

$$L_a^{i-1}[\varphi_i(v)\mathbf{c}(v, w)] = \sum_{j=0}^{i-1} C_{i-1}^j L_a^j \varphi_i(v) L_a^{i-1-j}\mathbf{c}(v, w)$$

where C_{i-1}^j is the binomial coefficient. Then, the equation (7.38) becomes

$$\begin{aligned} L_a^\ell \mathbf{c}(v, w) &= \sum_{i=1}^{\ell} L_a^{i-1}[\varphi_i(v)\mathbf{c}(v, w)] \\ &= \sum_{i=1}^{\ell}\sum_{j=0}^{i-1} C_{i-1}^j L_a^j \varphi_i(v) L_a^{i-1-j}\mathbf{c}(v, w) \\ &\overset{(k=i-j)}{=} \sum_{i=1}^{\ell}\sum_{k=1}^{i} C_{i-1}^{i-k} L_a^{i-k}\varphi_i(v) L_a^{k-1}\mathbf{c}(v, w) \end{aligned}$$

$$= \sum_{k=1}^{\ell} \left(\sum_{i=k}^{\ell} C_{i-1}^{i-k} L_a^{i-k} \varphi_i(v) \right) L_a^{k-1} \mathbf{c}(v, w).$$

Comparing this equation with (7.37) gives

$$\phi_k(v) = \sum_{i=k}^{\ell} C_{i-1}^{i-k} L_a^{i-k} \varphi_i(v), \quad k = 1, \ldots, \ell. \tag{7.39}$$

Thus, the if part is proved by (7.39). To prove the only if part, it suffices to show that, for any given $\phi_1, \ldots, \phi_\ell$,

$$\varphi_i(v) = \sum_{j=0}^{\ell-i} (-1)^j C_{i+j-1}^j L_a^j \phi_{i+j}(v), \quad i = 1, \ldots, \ell \tag{7.40}$$

is the solution to (7.39). For this purpose, we substitute (7.40) into (7.39) as follows:

$$\begin{aligned}
\phi_k(v) &= \sum_{i=k}^{\ell} C_{i-1}^{i-k} L_a^{i-k} \varphi_i(v) \\
&= \sum_{i=k}^{\ell} \sum_{j=0}^{\ell-i} (-1)^j C_{i-1}^{i-k} C_{i+j-1}^j L_a^{i+j-k} \phi_{i+j}(v) \\
&\overset{(p=i+j)}{=} \sum_{i=k}^{\ell} \sum_{p=i}^{\ell} (-1)^{p-i} C_{i-1}^{i-k} C_{p-1}^{p-i} L_a^{p-k} \phi_p(v) \\
&= \sum_{p=k}^{\ell} \left(\sum_{i=k}^{p} (-1)^{p-i} C_{i-1}^{i-k} C_{p-1}^{p-i} \right) L_a^{p-k} \phi_p(v) = \phi_k(v).
\end{aligned}$$

In the last step, we have used the fact that

$$\begin{aligned}
\sum_{i=k}^{p} (-1)^{p-i} C_{i-1}^{i-k} C_{p-1}^{p-i} &= \left(\sum_{i=k}^{p} (-1)^{p-i} C_{p-k}^{p-i} \right) C_{p-1}^{k-1} \\
&\overset{(j=p-i)}{=} \left(\sum_{j=0}^{p-k} (-1)^j C_{p-k}^j \right) C_{p-1}^{k-1} = \begin{cases} 0 & p > k \\ 1 & p = k \end{cases}.
\end{aligned}$$

The proof is thus completed. □

Lemma 7.5 *If* $\mathbf{c}(v, w)$ *satisfies Assumption 7.7 (or equivalently Assumption 7.8), then the composite system (7.8) admits a steady-state generator along the exosystem* $\dot{v} = a(v)$

$$\dot{\theta}(v, w) = \Phi(v)\theta(v, w), \quad \mathbf{c}(v, w) = \Psi\theta(v, w). \tag{7.41}$$

Moreover, Eq. (7.41) is observable in the sense that the pair $(\Psi, \Phi(v))$ *is observable for any* v.

Proof Let $\theta(v, w) = \mathcal{L}_a^\ell \mathbf{c}(v, w)$, and

$$\Phi(v) = \begin{bmatrix} 0 & 1 & 0 & \cdots & 0 \\ 0 & 0 & 1 & \cdots & 0 \\ \vdots & \vdots & \vdots & \ddots & \vdots \\ 0 & 0 & 0 & \cdots & 1 \\ \phi_1(v) & \phi_2(v) & \phi_3(v) & \cdots & \phi_\ell(v) \end{bmatrix}, \quad \Psi = \begin{bmatrix} 1 \\ 0 \\ \vdots \\ 0 \\ 0 \end{bmatrix}^{\mathsf{T}}. \tag{7.42}$$

Then, Assumption 7.7 implies (7.41), and it is obvious that the pair $(\Psi, \Phi(v))$ is observable. □

Under Assumption 7.8, the steady-state generator (7.41) can be constructed in the following alternative structure:

$$\Phi(v) = \begin{bmatrix} \varphi_\ell(v) & 1 & 0 & \cdots & 0 \\ \varphi_{\ell-1}(v) & 0 & 1 & \cdots & 0 \\ \vdots & \vdots & \vdots & \ddots & \vdots \\ \varphi_2(v) & 0 & 0 & \cdots & 1 \\ \varphi_1(v) & 0 & 0 & \cdots & 0 \end{bmatrix}, \quad \Psi = \begin{bmatrix} 1 \\ 0 \\ \vdots \\ 0 \\ 0 \end{bmatrix}^{\mathsf{T}}. \tag{7.43}$$

Let $\theta(v, w) = \mathrm{col}(\theta_1(v, w), \ldots, \theta_\ell(v, w))$. Then, Eq. (7.41) becomes

$$\begin{aligned} \theta_1(v, w) &= \mathbf{c}(v, w) \\ \theta_{i+1}(v, w) &= \dot{\theta}_i(v, w) - \varphi_{\ell-i+1}(v)\theta_1(v, w), \ i = 1, \ldots, \ell - 1 \\ 0 &= \dot{\theta}_\ell(v, w) - \varphi_1(v)\theta_1(v, w). \end{aligned} \tag{7.44}$$

From the top $\ell + 1$ equations of (7.44), we have

$$\begin{aligned} \theta_1(v, w) &= \mathbf{c}(v, w) \\ \theta_i(v, w) &= L_a^{i-1}\mathbf{c}(v, w) - \sum_{j=0}^{i-2} L_{a(v)}^{i-2-j}[\varphi_{\ell-j}(v)\mathbf{c}(v, w)], \ i = 2, \ldots, \ell. \end{aligned}$$

These ℓ functions must satisfy the last equation of (7.44), i.e.,

$$0 = L_a^\ell \mathbf{c}(v, w) - \sum_{j=0}^{\ell-2} L_a^{\ell-1-j}[\varphi_{\ell-j}(v)\mathbf{c}(v, w)] - \varphi_1(v)\mathbf{c}(v, w)$$

which is guaranteed by Assumption 7.8.

Example 7.7 Consider a nonlinear exosystem:

$$\dot{v} = a(v) = A_1 v + A_2 v v_3$$

where $v = [v_1, v_2, v_3, v_4]^{\mathrm{T}}$,

$$A_1 = \begin{bmatrix} 0 & 0 & 0 & 0 \\ 0 & 0 & 0 & 0 \\ 0 & 0 & 0 & 1 \\ 0 & 0 & -1 & 0 \end{bmatrix}, \quad A_2 = \begin{bmatrix} 0 & 1 & 0 & 0 \\ -1 & 0 & 0 & 0 \\ 0 & 0 & 0 & 0 \\ 0 & 0 & 0 & 0 \end{bmatrix}.$$

Consider a polynomial function $\mathbf{c}(v, w) = w v_1$. It can be verified that Assumption 7.7 holds with

$$L_a^3 \mathbf{c}(v, w) - (-3 v_3 v_4)\mathbf{c}(v, w) - (-1 - v_3^2) L_a \mathbf{c}(v, w) = 0.$$

By Lemma 7.5, a steady-state generator of the form (7.41) exists.

7.3 Internal Model

The internal model principle was developed in the 1970s for solving the output regulation problem for uncertain linear systems and is one of the most remarkable methods for linear control system design. The well-known PID (proportion-integral-derivative) control can be viewed as a special application of the internal model principle to the case where both the reference and the disturbance are constant. For the class of linear systems, an internal model is a dynamic compensator determined by the exosystem. The internal model together with the given plant forms a so-called augmented system. As pointed out in the previous section, the essential feature of an internal model for linear systems is that it leads to a stabilizable augmented system and the stabilization solution of the augmented system leads to the solution of the robust output regulation problem of the given plant. For a general nonlinear system, the global stabilizability problem is untractable. Therefore, we will define the concept of internal model in two steps. First, we introduce the concept of internal model candidate which is such that the stabilizibility of the augmented system implies the solvability of the robust output regulation problem of the given plant. It will be seen shortly that a steady-state generator itself is an internal model candidate. However, this internal model candidate does not lead to a stabilizable augmented system. If, for a particular class of nonlinear systems, an internal model candidate leads to a stabilizable augmented system, then this internal model candidate is further called an internal model for this class of nonlinear systems.

Below, we first define an internal model candidate corresponding to an existing steady-state generator.

Definition 7.2 Under Assumption 7.4, suppose the composite system (7.8) admits a steady-state generator (7.19). Let $\gamma : \mathbb{R}^{\ell+r+l_1} \mapsto \mathbb{R}^{\ell}$ be a sufficiently smooth function vanishing at the origin. Then we call the following system

$$\dot{\eta} = \gamma(\eta, c(x, u), v) \tag{7.45}$$

an *internal model candidate* corresponding to (7.19) on a mapping τ if there exists a diffeomorphic mapping $\tau(\cdot, \sigma) : \mathbb{R}^{\ell} \mapsto \mathbb{R}^{\ell}$ such that

$$\frac{\partial \tau(\theta(v, w, \sigma), \sigma)}{\partial \theta} \phi(\theta(v, w, \sigma), v, \sigma) = \gamma(\tau(\theta(v, w, \sigma), \sigma), \mathbf{c}(v, w, \sigma), v) \tag{7.46}$$

for all $d \in \mathbb{R}^l$.

Remark 7.8 Under Assumption 7.4, suppose the composite system (7.8) admits a steady-state generator (7.19) and a corresponding internal model candidate (7.45) on a mapping $\tau(\theta, \sigma)$. Define a new vector function

$$\theta'(v, w, \sigma) = \tau(\theta(v, w, \sigma), \sigma).$$

Then, the system has an alternative steady-stage generator

$$\begin{aligned} \dot{\theta}'(v, w, \sigma) &= \phi'(\theta'(v, w, \sigma), v, \sigma) \\ \mathbf{c}(v, w, \sigma) &= \psi'(\theta'(v, w, \sigma)) \end{aligned} \tag{7.47}$$

where

$$\begin{aligned} \phi'(\theta', v, \sigma) &= \frac{\partial \tau(\theta, \sigma)}{\partial \theta} \phi(\theta, v, \sigma)|_{\theta=\tau^{-1}(\theta', \sigma)} \\ \psi'(\theta') &= \psi(\tau^{-1}(\theta', \sigma)). \end{aligned}$$

Here $\tau^{-1}(\cdot, \sigma)$ is the inverse mapping of $\tau(\cdot, \sigma)$, i.e, $\tau^{-1}(\tau(\theta, \sigma), \sigma) = \theta$. As a result, the equation (7.46) implies

$$\phi'(\theta'(v, w, \sigma), v, \sigma) = \gamma(\theta'(v, w, \sigma), \mathbf{c}(v, w, \sigma), v).$$

In other words, Eq. (7.45) is also an internal model candidate corresponding to the steady-state generator (7.47) on an identity mapping. In particular, if $\tau(\theta, \sigma) = T(\sigma)\theta$ for some nonsingular matrix $T(\sigma)$, then the steady-state generator (7.47) is defined by

$$\begin{aligned} \phi'(\theta', v, \sigma) &= T(\sigma)\phi(T^{-1}(\sigma)\theta', v, \sigma) \\ \psi'(\theta') &= \psi(T^{-1}(\sigma)\theta'). \end{aligned}$$

If the composite system (7.8) admits a steady-state generator with output $c(x, u)$ and $\sigma \in \mathbb{R}^0$, then the steady-state generator itself is an internal model candidate on the identity mapping τ upon defining $\gamma(\eta, c(x, u), v) = \phi(\eta, v)$. However, this particular internal model candidate is uncoupled with the given plant, and cannot lead to a stabilizable augmented system. That is why we need to give a more general characterization of the internal model candidate in Definition 7.2 which allows the internal model to be coupled to the given plant through the function $c(x, u)$.

Remark 7.9 It will be seen in Sect. 7.4 that an internal model together with the given plant will define a so-called augmented system of the form (7.54). If the augmented system is stabilizable in the sense to be described in Sect. 7.4, then $\lim_{t\to\infty}(\eta - \tau(\theta(v, w, \sigma))) = 0$. In other words, an internal model can be viewed as an asymptotic observer of the corresponding steady-state generator.

We will now construct some classes of internal model candidates corresponding to the steady-state generators (7.26), (7.35), and (7.41) presented in Sect. 7.2 under various assumptions.

Example 7.8 Under Assumption 7.5, the composite system (7.8) admits a steady-state generator of the form (7.26) with the pair $(\Psi, \Phi(\sigma))$ observable. Under the additional Assumption 7.2, by Remark 7.7, all the eigenvalues of $\Phi(\sigma)$ have zero real parts. From Problem 7.8, for any controllable pair (M, N) where $M \in \mathbb{R}^{\ell\times\ell}$ with M Hurwitz and $N \in \mathbb{R}^{\ell\times 1}$, the Sylvester equation

$$T(\sigma)\Phi(\sigma) - MT(\sigma) = N\Psi \tag{7.48}$$

has a unique nonsingular solution $T(\sigma)$. It can be verified that the following system

$$\dot{\eta} = M\eta + Nc(x, u), \tag{7.49}$$

is an internal model candidate on $\tau(\theta, \sigma) = T(\sigma)\theta$ corresponding to (7.26).

Example 7.9 Consider the steady-state generator of the form (7.35) and assume the pair (Φ, Ψ) is observable. Pick $M \in \mathbb{R}^{\ell\times\ell}$ and $N \in \mathbb{R}^{\ell\times 1}$ such that M is Hurwitz and (M, N) is controllable. Then the Sylvester equation $T\Phi - MT = N\Psi$ has a unique nonsingular solution T. The system

$$\dot{\eta} = M\eta + N(c(x, u) - \psi(T^{-1}\eta) + \Psi T^{-1}\eta) \tag{7.50}$$

is an internal model candidate on the mapping $\tau(\theta, \sigma) = T\theta$ corresponding to (7.35) since

$$\begin{aligned} &T^{-1}[MT\theta(v, w) + N(\mathbf{c}(v, w) - \psi(\theta(v, w)) + \Psi\theta(v, w))] \\ &\quad = T^{-1}MT\theta(v, w) + T^{-1}N\Psi\theta(v, w) = \Phi\theta(v, w). \end{aligned}$$

Example 7.10 Consider the steady-state generator (7.41) with $\Phi(v)$, Ψ being given by (7.43). Let $\Phi = \Phi(0)$. We note that the matrix $\Phi(v)$ in (7.43) can be written in the form $\Phi(v) = \Phi + \bar{\varphi}(v)\Psi$ with $\bar{\varphi}(0) = 0$ where $\bar{\varphi}(v) = \text{col}(\varphi_\ell(v) - \varphi_\ell(0), \ldots, \varphi_1(v) - \varphi_1(0))$. The pair (Ψ, Φ) is observable. Pick $M \in \mathbb{R}^{\ell\times\ell}$ and $N \in \mathbb{R}^{\ell\times 1}$ such that M is Hurwitz, M and Φ have disjoint spectrums, and (M, N) is controllable. Then the Sylvester equation $T\Phi - MT = N\Psi$ has a unique nonsingular solution T. Let $N(v) = N + T\bar{\varphi}(v)$. Then we have

$$T\Phi(v) - MT = N(v)\Psi.$$

Thus, the following system

$$\dot{\eta} = M\eta + N(v)c(x, u) \tag{7.51}$$

is an internal model candidate on the mapping $\tau(\theta, \sigma) = T\theta$ corresponding to the steady-state generator (7.41) since

$$\begin{aligned} T^{-1}[MT\theta(v, w) + N(v)\mathbf{c}(v, w)] &= T^{-1}[MT + N\Psi + T\bar{\varphi}(v)\Psi]\theta(v, w) \\ &= [\Phi + \bar{\varphi}(v)\Psi]\theta(v, w) = \Phi(v)\theta(v, w). \end{aligned}$$

In the above examples, various steady-state generators and the corresponding internal model candidates are constructed assuming $c(x, u) \in \mathbb{R}$. The examples are summarized in Table 7.1. In fact, all these results can be used to handle the more general case with $c(x, u) \in \mathbb{R}^r$, $r > 1$. In particular, the steady-state generator and the corresponding internal model candidate can be constructed separately for each $c_i(x, u)$ as shown in the following lemma.

Lemma 7.6 *Under Assumption 7.4, for $i = 1, \ldots, r$, assume the composite system (7.8) admits a steady-state generator with output $c_i(x, u) \in \mathbb{R}$, respectively, of the*

Table 7.1 Steady-state generators and internal models under different assumptions

Exosystem	Steady-state generator	Internal model	Mapping
Assumption 7.5 (Linear immersion)			
$\dot{v} = A_1(\sigma)v$	$\dot{\theta}(v, w, \sigma) = \Phi(\sigma)\theta(v, w, \sigma)$ $\mathbf{c}(v, w, \sigma) = \Psi\theta(v, w, \sigma)$	$\dot{\eta} = M\eta + Nc(x, u)$	$\tau(\theta, \sigma) = T(\sigma)\theta$ $T(\sigma)\Phi(\sigma)$ $-MT(\sigma) = N\Psi$
Assumption 7.6 (Nonlinear immersion)			
$\dot{v} = A_1 v$	$\dot{\theta}(v, w) = \Phi\theta(v, w)$ $\mathbf{c}(v, w) = \psi(\theta(v, w))$	$\dot{\eta} = M\eta + Nc(x, u)$ $-N(\psi(T^{-1}\eta) - \Psi T^{-1}\eta)$	$\tau(\theta, \sigma) = T\theta$ $T\Phi - MT = N\Psi$
Assumption 7.7 or 7.8 (Generalized linear immersion)			
$\dot{v} = a(v)$	$\dot{\theta}(v, w) = (\Phi + \bar{\varphi}(v)\Psi)\theta(v, w)$ $\mathbf{c}(v, w) = \Psi\theta(v, w)$	$\dot{\eta} = M\eta + N(v)c(x, u)$ $N(v) = N + T\bar{\varphi}(v)$	$\tau(\theta, \sigma) = T\theta$ $T\Phi - MT = N\Psi$

following form:

$$\begin{aligned}\dot{\theta}_i(v, w, \sigma) &= \phi_i(\theta_i(v, w, \sigma), v, \sigma)\\ \mathbf{c}_i(v, w, \sigma) &= \psi_i(\theta_i(v, w, \sigma)), \quad \mathbf{c}_i(v, w, \sigma) := c_i(\mathbf{x}(v, w, \sigma), \mathbf{u}(v, w, \sigma)).\end{aligned}$$

Also, assume there exists an internal model candidate

$$\dot{\eta}_i = \gamma_i(\eta_i, c_i(x, u), v)$$

on a mapping $\tau_i(\theta_i, \sigma)$ *corresponding to the steady-state generator. Then, the system has a steady-state generator (7.19) with output*

$$c(x, u) = col(c_1(x, u), \ldots, c_r(x, u))$$

and a corresponding internal model candidate (7.45) on a mapping $\tau(\theta, \sigma)$ *where*

$$\begin{aligned}\theta &= col(\theta_1, \ldots, \theta_r)\\ \eta &= col(\eta_1, \ldots, \eta_r)\\ \phi(\theta, v, \sigma) &= col(\phi_1(\theta_1, v, \sigma), \ldots, \phi_r(\theta_r, v, \sigma))\\ \psi(\theta) &= col(\psi_1(\theta_1), \ldots, \psi_r(\theta_r))\\ \tau(\theta, \sigma) &= col(\tau_1(\theta_1, \sigma), \ldots, \tau_r(\theta_r, \sigma))\\ \gamma(\eta, c(x, u), v) &= col(\gamma_1(\eta_1, c_1(x, u), v), \ldots, \gamma_r(\eta_r, c_r(x, u), v)).\end{aligned}$$

Example 7.11 Consider the linear system described by (7.16) with the exosystem given by (7.7). It is known from Remark 7.4 that, if the Sylvester equations (7.18) have a solution pair $(X(w), U(w))$, then the composite system always admits a steady-state generator with the output u_i as follows

$$\dot{\theta}_i(v, w) = \Phi_i\theta_i(v, w), \ \mathbf{u}_i(v, w) = \Psi_i\theta_i(v, w), \ i = 1, \ldots, m. \tag{7.52}$$

Let $M_i \in \mathbb{R}^{\ell\times\ell}$ and $N_i \in \mathbb{R}^{\ell\times 1}$ be any controllable pair such that M_i is Hurwitz. It can be verified that, for each $i = 1, \ldots, m$, (Ψ_i, Φ_i) is observable and all the eigenvalues of Φ_i are the same as those of A_1 (that is, M_i and Φ_i have disjoint spectrum). Then, from Problem 7.8, the Sylvester equation

$$T_i\Phi_i - M_iT_i = N_i\Psi_i$$

admits a unique nonsingular solution T_i. Thus, the system has an internal model candidate on the mapping $\tau_i(\theta_i) = T_i\theta_i$ with output u_i as follows:

$$\dot{\eta}_i = M_i\eta_i + N_iu_i, i = 1, \ldots, m.$$

By Lemma 7.6, the system has an internal model candidate on the mapping $\tau(\theta) = \text{col}(\tau_1(\theta_1), \ldots, \tau_m(\theta_m))$ with output u as follows:

$$\dot{\eta} = M\eta + Nu \tag{7.53}$$

where

$$\begin{aligned} \eta &= \text{col}(\eta_1, \ldots, \eta_m) \\ M &= \text{diag}(M_1, \ldots, M_m) \\ N &= \text{diag}(N_1, \ldots, N_m). \end{aligned}$$

7.4 From Output Regulation to Stabilization

Having introduced the concept of internal model candidate, we will further define, in this section, the augmented composite system composed of the composite system and the internal model candidate as follows:

$$\begin{aligned} \dot{x} &= f(x, u, v, w) \\ \dot{\eta} &= \gamma(\eta, c(x, u), v) \\ \dot{v} &= a(v, \sigma) \\ e &= h_e(x, v, w) \\ y_m &= h_m(x, u, v, w). \end{aligned} \tag{7.54}$$

Recall that the internal model candidate is defined such that the augmented composite system has an output zeroing invariant manifold $\mathbb{M} = \{(x, \eta, v) \mid x = \mathbf{x}(v, w, \sigma), \eta = \tau(\theta(v, w, \sigma), \sigma), v \in \mathbb{R}^{l_1}\}$ under the control $\mathbf{u}(v, w, \sigma)$ in the sense that

$$\begin{aligned} \frac{\partial \mathbf{x}(v, w, \sigma)}{\partial v} a(v, \sigma) &= f(\mathbf{x}(v, w, \sigma), \mathbf{u}(v, w, \sigma), v, w) \\ \frac{\partial \tau(\theta(v, w, \sigma), \sigma)}{\partial v} a(v, \sigma) &= \gamma(\tau(\theta(v, w, \sigma), \sigma), c(\mathbf{x}(v, w, \sigma), \mathbf{u}(v, w, \sigma)), v) \\ 0 &= h_e(\mathbf{x}(v, w, \sigma), \mathbf{u}(v, w, \sigma), v, w). \end{aligned} \tag{7.55}$$

Consider a control law of the form

$$\begin{aligned} u &= \kappa_1(\xi, y_m) \\ \dot{\xi} &= \kappa_2(\xi, y_m) \end{aligned} \tag{7.56}$$

where the functions κ_1 and κ_2 are sufficiently smooth satisfying, for some sufficiently smooth function $\boldsymbol{\xi}(v, w, \sigma)$,

$$\begin{aligned}\mathbf{u}(v, w, \sigma) &= \kappa_1\left(\boldsymbol{\xi}(v, w, \sigma), h_{\mathrm{m}}(\mathbf{x}(v, w, \sigma), v, w)\right)\\ \frac{\partial \boldsymbol{\xi}(v, w, \sigma)}{\partial v} a(v, \sigma) &= \kappa_2\left(\boldsymbol{\xi}(v, w, \sigma), h_{\mathrm{m}}(\mathbf{x}(v, w, \sigma), v, w)\right). \end{aligned} \tag{7.57}$$

Let $x_{\mathrm{c}} = \mathrm{col}(x, \eta, \xi)$ be the state of the closed-loop system composed of (7.54) and (7.56), $\mathbf{x}_{\mathrm{c}}(v, w, \sigma) = \mathrm{col}(\mathbf{x}(v, w, \sigma), \tau(\theta(v, w, \sigma), \sigma), \boldsymbol{\xi}(v, w, \sigma))$, and $\bar{x}_{\mathrm{c}} = x_{\mathrm{c}} - \mathbf{x}_{\mathrm{c}}(v, w, \sigma)$. Denote the closed-loop system by the following equations:

$$\dot{\bar{x}}_{\mathrm{c}} = f_{\mathrm{c}}(\bar{x}_{\mathrm{c}}, v, w, \sigma), \; e = h_{\mathrm{c}}(\bar{x}_{\mathrm{c}}, v, w, \sigma). \tag{7.58}$$

Then, using (7.55) and (7.57), it can be verified that the closed-loop system has the property that $f_{\mathrm{c}}(0, v, w, \sigma) = 0$ and $h_{\mathrm{c}}(0, v, w, \sigma) = 0$ for all $v \in \mathbb{R}^{l_1}$, $w \in \mathbb{R}^{l_2}$, and $\sigma \in \mathbb{R}^{l_3}$.

The manifold $\mathbb{M}$ is said to be globally stabilizable if, for arbitrarily large compact subsets $\mathbb{V} \in \mathbb{R}^{l_1}$, $\mathbb{W} \in \mathbb{R}^{l_2}$, and $\mathbb{S} \in \mathbb{R}^{l_3}$, there exists a control law of the form (7.56) satisfying (7.57) such that, for all $v \in \mathbb{V}$, all $w \in \mathbb{W}$, and all $\sigma \in \mathbb{S}$, the solution of (7.58) is bounded for all $t \geq 0$ and $\lim_{t\to\infty} \bar{x}_c(t) = 0$. As a result, $\lim_{t\to\infty} e(t) = \lim_{t\to\infty} = h_{\mathrm{c}}(\bar{x}_c(t), v, w, \sigma) = h_{\mathrm{c}}(0, v, w, \sigma) = 0$.

If the manifold $\mathbb{M}$ is globally stabilizable by a control law of the form (7.56) satisfying (7.57), then the internal model candidate is further called a (global) *internal model* of the composite system (7.8). Thus an internal model can be interpreted as a dynamic compensator such that the invariant manifold $\mathbb{M}$ of the augmented system (7.61) is globally stabilizable by a control law of the form (7.56) satisfying (7.57).

Next, we will further elaborate the notion of the internal model for the special case where $\sigma \in \mathbb{R}^0$. In this case, it is possible to perform a coordinate and input transformation on the augmented system so that the invariant manifold $\mathbb{M}$ can be transformed to the equilibrium point at the origin of an transformed augmented system. For convenience, suppose the state vector x has the decomposition $x = \mathrm{col}(x_{\mathrm{o}}, x_{\mathrm{m}})$ such that $y_{\mathrm{m}} = \mathrm{col}(x_{\mathrm{m}}, e)$ (recall that we assume e is always measurable) and the output function $c(x, u)$ has the following structure:

$$c(x, u) = \mathrm{col}(x_{\mathrm{m}}, u). \tag{7.59}$$

Let $c_{\mathrm{o}}(x) = x_{\mathrm{o}}$. Performing on (7.54) the following coordinate and input transformation

$$\begin{aligned}\bar{\eta} &= \eta - \tau(\theta(v, w, \sigma), \sigma)\\ \begin{bmatrix}\bar{x}_{\mathrm{o}}\\ \bar{x}_{\mathrm{m}}\end{bmatrix} &= \begin{bmatrix} x_{\mathrm{o}} - c_{\mathrm{o}}(\mathbf{x}(v, w, \sigma))\\ x_{\mathrm{m}} - \psi_{\mathrm{x}}(\tau^{-1}(\eta, \sigma))\end{bmatrix}\\ \bar{u} &= u - \psi_{\mathrm{u}}(\tau^{-1}(\eta, \sigma))\end{aligned} \tag{7.60}$$

with $\psi = \mathrm{col}(\psi_{\mathrm{x}}, \psi_{\mathrm{u}})$ and $\bar{x} = \mathrm{col}(\bar{x}_{\mathrm{o}}, \bar{x}_{\mathrm{m}})$, gives a new system denoted by

$$\begin{aligned}\dot{\bar{x}} &= \bar{f}(\bar{x}, \bar{\eta}, \bar{u}, d)\\ \dot{\bar{\eta}} &= \bar{\gamma}(\bar{x}, \bar{\eta}, \bar{u}, d)\\ e &= \bar{h}_{\mathrm{e}}(\bar{x}, \bar{\eta}, \bar{u}, d)\end{aligned} \tag{7.61}$$

where

$$\begin{aligned}\bar{f}_{\mathrm{o}}(\bar{x}, \bar{\eta}, \bar{u}, d) &= f_{\mathrm{o}}(x, u, v, w) - f_{\mathrm{o}}(\mathbf{x}(v, w, \sigma), \mathbf{u}(v, w, \sigma), v, w)\\ \bar{f}_{\mathrm{m}}(\bar{x}, \bar{\eta}, \bar{u}, d) &= f_{\mathrm{m}}(x, u, v, w) - \frac{\partial \psi_{\mathrm{x}}(\tau^{-1}(\eta, \sigma))}{\partial \eta}\gamma(\eta, c(x, u), v)\\ \bar{\gamma}(\bar{x}, \bar{\eta}, \bar{u}, d) &= \gamma(\eta, c(x, u), v) - \gamma(\tau(\theta(v, w, \sigma), c(\mathbf{x}(v, w, \sigma), \mathbf{u}(v, w, \sigma)), v)\\ \bar{h}_{\mathrm{e}}(\bar{x}, \bar{\eta}, \bar{u}, d) &= h_{\mathrm{e}}(x, u, v, w)\end{aligned}$$

for

$$\bar{f}(\bar{x}, \bar{\eta}, \bar{u}, d) = \begin{bmatrix} \bar{f}_{\mathrm{o}}(\bar{x}, \bar{\eta}, \bar{u}, d) \\ \bar{f}_{\mathrm{m}}(\bar{x}, \bar{\eta}, \bar{u}, d) \end{bmatrix}, \quad f(x, u, v, w) = \begin{bmatrix} f_{\mathrm{o}}(x, u, v, w) \\ f_{\mathrm{m}}(x, u, v, w) \end{bmatrix}.$$

We call (7.61) the *transformed augmented system.* This system has the following property.

Lemma 7.7 *Under Assumption 7.4, suppose the composite system (7.8) admits a steady-state generator (7.19) with output $c(x, u)$ and a corresponding internal model candidate (7.45) on a mapping τ. Then the augmented system (7.61) has the property that, for all $d \in \mathbb{R}^l$,*

$$\begin{aligned}\bar{f}(0, 0, 0, d) &= 0\\ \bar{\gamma}(0, 0, 0, d) &= 0\\ \bar{h}_{\mathrm{e}}(0, 0, 0, d) &= 0.\end{aligned} \tag{7.62}$$

Proof Recall that, for any d, under the control $\mathbf{u}(v, w, \sigma)$, the manifold $\mathbb{M} = \{(x, \eta, v) \mid x = \mathbf{x}(v, w, \sigma), \eta = \tau(\theta(v, w, \sigma), \sigma), v \in \mathbb{R}^{l_1}\}$ is the output zeroing invariant manifold of the augmented composite system (7.54) in the sense of (7.55). Thus, under the state and input transformation (7.60), this output zeroing manifold is represented by $\mathbb{M} = \{(\bar{x}, \bar{\eta}, v) \mid \bar{x} = 0, \ \bar{\eta} = 0, v \in \mathbb{R}^{l_1}\}$ with the feedback control given by $\bar{u} = 0$. This is to say that the origin $(\bar{x}, \bar{\eta}) = (0, 0)$ is the equilibrium point of the unforced transformed augmented system and the error output equation is identically zero at $(\bar{x}, \bar{\eta}, \bar{u}) = (0, 0, 0)$ for all d. Thus the proof is completed. □

By Lemma 7.7, the augmented system (7.61) has an equilibrium point at the origin when $\bar{u} = 0$ for all d, and, at the origin, the error output e is identically equal to zero regardless of the value of d. Thus, if the equilibrium point at the origin can be locally or globally stabilized by a control $\bar{u}$ of the following form

$$\begin{aligned} \bar{u} &= k_1(\xi, \bar{x}_{\mathrm{m}}, e) \\ \dot{\xi} &= k_2(\xi, \bar{x}_{\mathrm{m}}, e) \end{aligned} \tag{7.63}$$

where k_1 and k_2 are sufficiently smooth functions vanishing at the origin, then, from the transformation (7.60), the RORP/GRORP is solved by a control law relying on x_{m} and e only provided that $\sigma \in \mathbb{R}^0$. The above observation will further lead to the various versions of the concept of internal model. Two such versions will be given as follows.

Definition 7.3 Under Assumptions 7.1 and 7.4, suppose the composite system (7.8) with $\sigma \in \mathbb{R}^0$ admits a steady-state generator (7.19) with output $c(x, u)$, a corresponding internal model candidate (7.45) on a mapping τ. If the equilibrium point at the origin of the transformed augmented system (7.61) with $d = 0$ is exponentially stabilizable by the control law of the form (7.63), then the internal model candidate is called a *(local) internal model*. If the GRSP/GASP of the transformed augmented system (7.61) for any compact set $\mathbb{D}$ with $\sigma \in \mathbb{R}^0$ is solvable by the control law of the form (7.63), then the internal model candidate is called a *(global) internal model*.

By this definition, we immediately have the following results.

Proposition 7.1 *Suppose the equilibrium point at the origin of the exosystem (7.2) is stable and all the eigenvalues of the Jacobian matrix*

$$\frac{\partial a}{\partial v}(0, 0)$$

have zero real part, and suppose the composite system (7.8) with $\sigma \in \mathbb{R}^0$ admits a steady-state generator (7.19) with output $c(x, u)$, a corresponding local internal model (7.45) on a mapping τ, and an augmented system (7.61). If the closed-loop system composed of (7.61) and (7.63) has a Hurwitz Jacobian matrix at $\mathrm{col}(\bar{x}, \bar{\eta}, \xi) = 0$ with $v(t) = 0$ and $w = 0$, then the controller

$$\begin{aligned} u &= \psi_{\mathrm{u}}(\tau^{-1}(\eta)) + k_1(\xi, x_{\mathrm{m}} - \psi_{\mathrm{x}}(\tau^{-1}(\eta)), e) \\ \dot{\eta} &= \gamma(\eta, c(x, u), v) \\ \dot{\xi} &= k_2(\xi, x_{\mathrm{m}} - \psi_{\mathrm{x}}(\tau^{-1}(\eta)), e). \end{aligned} \tag{7.64}$$

solves the RORP for the system (7.8).

Proof Denote the state of the closed-loop system composed of the plant (7.61) and the controller (7.63) by $\bar{x}_{\mathrm{c}} = \mathrm{col}(\bar{x}, \bar{\eta}, \xi)$. By Theorem 11.4 in the Appendix, for all sufficiently small $\bar{x}_{\mathrm{c}}(0)$, w, $v(0)$, $\bar{x}_{\mathrm{c}}(t)$ and $v(t)$ exist and are bounded for all $t > 0$, and $\lim_{t\to\infty} \bar{x}_{\mathrm{c}}(t) = 0$. Thus, by Lemma 7.7, we have $\lim_{t\to\infty} e(t) = 0$.

Denote the state of the closed-loop system composed of the plant (7.1) and the controller (7.64) by $x_{\mathrm{c}} = \mathrm{col}(x, \eta, \xi)$. Then

$$x_c = \bar{x}_c + \mathrm{col}(c_o(\mathbf{x}(v, w, \sigma)),\ \psi_x(\tau^{-1}(\bar{\eta} + \tau(\theta(v, w, \sigma)))),\ \tau(\theta(v, w, \sigma)), 0).$$

Then, for all sufficiently small $x_c(0)$, w, $v(0)$, $x_c(t)$ exists and is bounded for all $t > 0$. Therefore, the closed-loop system composed of (7.1) and (7.64) satisfies both properties (i) and (ii) of RORP. □

Next, we will present the global version of Proposition 7.1.

Proposition 7.2 *Under Assumption 7.4, suppose the composite system (7.8) with $\sigma \in \mathbb{R}^0$ admits a steady-state generator (7.19) with output $c(x, u)$, and a corresponding internal model (7.45) on a mapping τ. If the GRSP/GASP of the transformed augmented system (7.61) for any compact set $\mathbb{D}$ is solved by the controller (7.63), then the controller (7.64) solves the GRORP for the system (7.1).*

Proof The proof is straightforward by examining the definitions of GRSP/GASP and GRORP. □

Remark 7.10 So far we have established a general framework to handle the output regulation problem for the composite system (7.8). This framework consists of two steps. The first step is to convert the robust output regulation problem for a given plant into a robust stabilization problem of the augmented system composed of the given plant and the internal model, and the second one is to robustly stabilize the augmented system. We have shown that the first step can always be accomplished if the solution of the regulator equations satisfies various immersion assumptions introduced in Sect. 7.2. The second step poses a more challenging task than the stabilization problem of the given plant because the augmented plant is much more complicated than the given plant due to the attachment of the internal model. Clearly, the accomplishment of the second step has to rely on the progress of the stabilization techniques for nonlinear systems. In the next few chapters, we will consider a few typical nonlinear systems whose GRORP can be converted to GRSP/GASP of nonlinear systems studied in Chaps. 3–6.

7.5 Linear Robust Output Regulation

In this section we will apply the framework for handling the robust output regulation to the linear system described by (7.16) with the exosystem satisfying Assumption 7.3. It is also assumed that the system satisfies the following assumptions.

Assumption 7.9 For any $w \in \mathbb{W}$, rank $\begin{bmatrix} A(w) - \lambda I & B(w) \\ C(w) & D(w) \end{bmatrix} = n + m$ for all $\lambda \in \mathrm{eig}(A_1)$, where $\mathrm{eig}(A_1) = \{\lambda \mid \det(A_1 - \lambda I) = 0\}$.

Assumption 7.10 For any $w \in \mathbb{W}$, the pair $(A(w), B(w))$ is stabilizable, and the pair $(C(w), A(w))$ is detectable.

From Example 7.10, under Assumption 7.9, the linear composite system admits a linear internal model on the mapping $T\theta$ with output $c(x, u) = u$ given by (7.53). The augmented system (7.54) now consists of the plant (7.16) and the internal model (7.53), and the transformation (7.60) with $c_o(x) = x_o = x$ takes the following form

$$\begin{aligned}\bar{\eta} &= \eta - T\theta(v, w)\\ \bar{x} &= x - \mathbf{x}(v, w)\\ \bar{u} &= u - \Psi T^{-1}\eta\end{aligned} \tag{7.65}$$

where

$$\begin{aligned}\theta &= \text{col}(\theta_1, \ldots, \theta_m)\\ T &= \text{diag}(T_1, \ldots, T_m)\\ \Psi &= \text{diag}(\Psi_1, \ldots, \Psi_m).\end{aligned}$$

Applying the transformation (7.65) to the augmented system consisting of (7.16) and (7.53) gives the transformed augmented system as follows:

$$\begin{aligned}\dot{\bar{x}} &= A(w)\bar{x} + B(w)\bar{u} + B(w)\Psi T^{-1}\bar{\eta}\\ \dot{\bar{\eta}} &= (M + N\Psi T^{-1})\bar{\eta} + N\bar{u}\\ e &= C(w)\bar{x} + D(w)\bar{u} + D(w)\Psi T^{-1}\bar{\eta}.\end{aligned} \tag{7.66}$$

By Proposition 7.1, it suffices to exponentially stabilize the equilibrium point at the origin of (7.66). Under Assumptions 7.9 and 7.10, using the PBH test, for any $w \in \mathbb{W}$, the augmented system (7.66) is both stabilizable and detectable (See Problem 7.11). Therefore, it is always possible to find a control law

$$\begin{aligned}\bar{u} &= -K\xi\\ \dot{\xi} &= L\xi + Qe.\end{aligned} \tag{7.67}$$

that stabilizes (7.66) for all sufficiently small w. Thus, the following control law

$$\begin{aligned}u &= \Psi T^{-1}\eta - K\xi\\ \dot{\eta} &= M\eta + Nu\\ \dot{\xi} &= L\xi + Qe\end{aligned}$$

solves the robust output regulation problem of the linear system (7.16) for all sufficiently small w.

Next we will consider the robust output regulation problem of system (7.16) where $\mathbb{W}$ is an arbitrarily prescribed compact set. For this purpose, we need to solve the robust stabilization problem for the augmented system (7.66). Nevertheless, the robust stabilization problem for the system of the form (7.66) is in general untractable.

We will thus focus on a subclass of the linear systems (7.16) satisfying the following additional assumption.

Assumption 7.11 (i) The system is single input single output with $u, e \in \mathbb{R}$ and $D(w) = 0$.
(ii) The system has the relative degree r with input u and output $y = C(w)x$ in the sense that, for all $w \in \mathbb{W}$, $C(w)A^k(w)B(w) = 0$, for $k = 0, 1, \ldots, r-2$, and $C(w)A^{r-1}(w)B(w) \neq 0$.
(iii) For $k = 0, 1, \ldots, r-2$, $C(w)A^k(w)E(w) = 0$ for all $w \in \mathbb{W}$.
(iv) The system (7.16) is minimum phase for any $w \in \mathbb{W}$.

Parts (i)–(iii) of Assumption 7.11 mean that the relative degree of the system (7.16) from u to y is not greater than the relative degree of the same system from v to y. Under (i)–(iii), there exists a matrix $T(w)$ such that under the linear transformation (See Problem 7.12)

$$T(w)x = \operatorname{col}(\mathcal{X}_0, \mathcal{X}_1, \ldots, \mathcal{X}_r), \tag{7.68}$$

where $\mathcal{X}_0 \in \mathbb{R}^{n-r}$, $\mathcal{X}_k \in \mathbb{R}$, $k = 1, \ldots, r$, the system (7.16) is equivalent to

$$\begin{aligned}
\dot{\mathcal{X}}_0 &= A_1(w)\mathcal{X}_0 + A_2(w)\mathcal{X}_1 + E_0(w)v \\
\dot{\mathcal{X}}_k &= \mathcal{X}_{(k+1)}, \ k = 1, \ldots, r-1 \\
\dot{\mathcal{X}}_r &= A_3(w)\mathcal{X}_0 + \sum_{k=1}^{r} c_k(w)\mathcal{X}_k + E_r(w)v + b(w)u \\
e &= \mathcal{X}_1 + F(w)v,
\end{aligned} \tag{7.69}$$

where $b(w) = C(w)A(w)^{r-1}B(w) \neq 0$ for any $w \in \mathbb{W}$. Moreover, under part (iv) of Assumption 7.11, $A_1(w)$ is Hurwitz for all $w \in \mathbb{W}$. Since $\mathbb{W}$ is a compact set, by the continuity of $b(w)$, it has the same sign for all $w \in \mathbb{W}$. Without loss of generality, from now on, we assume that, for all $w \in \mathbb{W}$, $b(w) > 0$.

Similarly, under the linear transformation

$$T(w)\bar{x} = \operatorname{col}(\bar{\mathcal{X}}_0, \bar{\mathcal{X}}_1, \ldots, \bar{\mathcal{X}}_r), \tag{7.70}$$

the augmented system (7.66) becomes

$$\begin{aligned}
\dot{\bar{\mathcal{X}}}_0 &= A_1(w)\bar{\mathcal{X}}_0 + A_2(w)\bar{\mathcal{X}}_1 \\
\dot{\bar{\mathcal{X}}}_k &= \bar{\mathcal{X}}_{(k+1)}, \ k = 1, \ldots, r-1 \\
\dot{\bar{\mathcal{X}}}_r &= A_3(w)\bar{\mathcal{X}}_0 + \sum_{k=1}^{r} c_k(w)\bar{\mathcal{X}}_k + b(w)\bar{u} + b(w)\Psi T^{-1}\bar{\eta} \\
\dot{\bar{\eta}} &= (M + N\Psi T^{-1})\bar{\eta} + N\bar{u} \\
e &= \bar{\mathcal{X}}_1.
\end{aligned} \tag{7.71}$$

We will use the so-called high-gain feedback control method to robustly stabilize the system (7.71). For this purpose, we further perform another coordinate transformation on (7.71),

$$\begin{aligned}\zeta &= \gamma_0\bar{x}_1 + \gamma_1\bar{x}_2 + \cdots + \gamma_{r-2}\bar{x}_{(r-1)} + \bar{x}_r,\\ \tilde{\eta} &= \bar{\eta} - \frac{1}{b(w)}N\zeta,\end{aligned} \tag{7.72}$$

where the coefficients γ_k, $k = 0, 1, \ldots, r-2$, are such that the polynomial $s^{r-1} + \gamma_{r-2}s^{r-2} + \cdots + \gamma_1 s + \gamma_0$ is stable. Then the augmented system (7.71) is transformed to

$$\begin{aligned}\dot{\bar{x}}_0 &= A_1(w)\bar{x}_0 + A_2(w)D\bar{x}\\ \dot{\bar{x}} &= \Lambda\bar{x} + G\zeta\\ \dot{\tilde{\eta}} &= M\tilde{\eta} + \hat{A}_3(w)\bar{x}_0 + \hat{c}(w)\bar{x} + \hat{c}_r(w)\zeta\\ \dot{\zeta} &= A_3(w)\bar{x}_0 + \tilde{c}(w)\bar{x} + \tilde{c}_r(w)\zeta + b(w)\Psi T^{-1}\tilde{\eta} + b(w)\bar{u}\end{aligned} \tag{7.73}$$

where $\bar{x} = \mathrm{col}(\bar{x}_1, \ldots, \bar{x}_{(r-1)})$, $\tilde{c}(w) = [\tilde{c}_1(w), \ldots, \tilde{c}_{(r-1)}(w)]$, and $\hat{c}(w) = [\hat{c}_1(w), \ldots, \hat{c}_{(r-1)}(w)]$. The other parameters are defined as follows,

$$\begin{aligned}\tilde{c}_1(w) &= c_1(w) - (c_r(w) + \gamma_{r-2})\gamma_0,\\ \tilde{c}_k(w) &= c_k(w) + \gamma_{k-2} - (c_r(w) + \gamma_{r-2})\gamma_{k-1},\ k = 2, \ldots, r-1,\\ \tilde{c}_r(w) &= c_r(w) + \gamma_{r-2} + \Psi T^{-1}N,\\ \hat{c}_k(w) &= -\frac{\tilde{c}_k(w)}{b(w)}N,\ k = 1, \ldots, r-1,\\ \hat{c}_r(w) &= \frac{1}{b(w)}(MN + N\Psi T^{-1}N - \tilde{c}_r(w)N),\\ \hat{A}_3(w) &= -\frac{1}{b(w)}NA_3(w),\end{aligned}$$

and

$$D = [1, 0, \ldots, 0],\ G = [0, \ldots, 0, 1]^{\mathsf{T}},\ \Lambda = \begin{bmatrix} 0 & 1 & \cdots & 0\\ \vdots & \vdots & \ddots & \vdots\\ 0 & 0 & \cdots & 1\\ -\gamma_0 & -\gamma_1 & \cdots & -\gamma_{r-2}\end{bmatrix}.$$

Note that e and its k-th derivatives are

$$\begin{aligned}e &= \bar{x}_1,\\ e^{(k)} &= \bar{x}_{(k+1)},\quad k = 1, \ldots, r-1.\end{aligned}$$

It now suffices to design a controller of the form (7.67) to robustly stabilize the system (7.73). For this purpose, we first synthesize a static state feedback controller using e and its derivatives, and then further design a dynamic output feedback controller of the form (7.67) which relies on e only. Before we do so, we need to establish the following technical lemma.

Lemma 7.8 *Consider the linear uncertain system*

$$\begin{bmatrix} \dot{x} \\ \dot{y} \end{bmatrix} = \begin{bmatrix} M_1(w) & N_1(w, \nu) \\ N_2(w, \nu) & N_3(w, \nu) + \nu M_2(w) \end{bmatrix} \begin{bmatrix} x \\ y \end{bmatrix}, \tag{7.74}$$

where $\nu \in \mathbb{R}$, $w \in \mathbb{W}$, $M_1(w) \in \mathbb{R}^{n\times n}$, $M_2(w) \in \mathbb{R}^{m\times m}$ are Hurwitz for all $w \in \mathbb{W}$. Suppose for some $\varepsilon > 0$, $\nu_0 > 0$, $\|N_i(w, \nu)\| \le \varepsilon$, $i = 1, 2, 3$, for all $\nu \ge \nu_0$, and all $w \in \mathbb{W}$. Then, there exists $\nu^* > 0$ such that, for all $\nu \ge \nu^*$, the origin of (7.74) is asymptotically stable for every $w \in \mathbb{W}$.

Proof Since $M_1(w)$ and $M_2(w)$ are Hurwitz for any $w \in \mathbb{W}$, there exist symmetric positive definite matrices $P_1(w)$ and $P_2(w)$ such that, for all $w \in \mathbb{W}$

$$\begin{aligned} M_1(w)^{\mathsf T} P_1(w) + P_1(w) M_1(w) &\le -I_n, \\ M_2(w)^{\mathsf T} P_2(w) + P_2(w) M_2(w) &\le -I_m. \end{aligned} \tag{7.75}$$

Define the positive definite function

$$V(x, y) = \mu x^{\mathsf T} P_1(w) x + y^{\mathsf T} P_2(w) y,$$

where $\mu > 2$. Then $\beta_1 \|\mathrm{col}(x, y)\|^2 \le V(x, y) \le \beta_2 \|\mathrm{col}(x, y)\|^2$, where

$$\begin{aligned} \beta_1 &= \min_{w\in\mathbb{W}} \{\mu \lambda_{\min}(P_1(w)), \lambda_{\min}(P_2(w))\} > 0, \\ \beta_2 &= \max_{w\in\mathbb{W}} \{\mu \lambda_{\max}(P_1(w)), \lambda_{\max}(P_2(w))\} > 0. \end{aligned}$$

Here $\lambda_{\min}$ and $\lambda_{\max}$ represent the minimal and maximal eigenvalues, respectively. Furthermore, the derivative of $V(x, y)$ along the system (7.74) is given by

$$\begin{aligned} \dot{V}(x, y) &= \mu x^{\mathsf T} (M_1(w)^{\mathsf T} P_1(w) + P_1(w) M_1(w)) x \\ &\quad + \nu y^{\mathsf T} (M_2(w)^{\mathsf T} P_2(w) + P_2(w) M_2(w)) y \\ &\quad + y^{\mathsf T} (N_3(w, \nu)^{\mathsf T} P_2(w) + P_2(w) N_3(w, \nu)) y \\ &\quad + 2\mu y^{\mathsf T} N_1(w, \nu)^{\mathsf T} P_1(w) x + 2 x^{\mathsf T} N_2(w, \nu)^{\mathsf T} P_2(w) y. \end{aligned} \tag{7.76}$$

By mean square inequality, we have when $\nu \ge \nu_0$,

$$\begin{aligned} & 2\mu y^{\mathsf T} N_1(w, \nu)^{\mathsf T} P_1(w) x \\ & \le \frac{\mu}{2} \|x\|^2 + 2\mu \|N_1(w, \nu)\|^2 \|P_1(w)\|^2 \|y\|^2 \end{aligned}$$

$$
\begin{aligned}
&\le \frac{\mu}{2}x^{\mathsf T}x + 2\mu\varepsilon^2\|P_1(w)\|^2\|y\|^2,\\
&2x^{\mathsf T}N_2(w,\nu)^{\mathsf T}P_2(w)y\\
&\quad\le x^{\mathsf T}x + \|N_2(w,\nu)\|^2\|P_2(w)\|^2\|y\|^2\\
&\quad\le x^{\mathsf T}x + \varepsilon^2\|P_2(w)\|^2\|y\|^2,\\
&y^{\mathsf T}(N_3(w,\nu)^{\mathsf T}P_2(w) + P_2(w)N_3(w,\nu))y\\
&\quad\le \|N_3^{\mathsf T}(w,\nu)P_2(w) + P_2(w)N_3(w,\nu)\|\,\|y\|^2\\
&\quad\le 2\varepsilon\|P_2(w)\|\,\|y\|^2.
\end{aligned}
\tag{7.77}
$$

Hence, by (7.75) to (7.77), we have

$$
\begin{aligned}
\dot V(x,y) \le &-\left(\frac{\mu}{2}-1\right)\|x\|^2\\
&-\left(\nu - 2\mu\varepsilon^2\|P_1(w)\|^2 - \varepsilon^2\|P_2(w)\|^2 - 2\varepsilon\|P_2(w)\|\right)\|y\|^2.
\end{aligned}
$$

Let $\nu_1(w) = \max\{\nu_0, 2\mu\varepsilon^2\|P_1(w)\|^2 + \varepsilon^2\|P_2(w)\|^2 + 2\varepsilon\|P_2(w)\|\}$, and $\nu^* = \max_{w\in\mathbb{W}}\{\nu_1(w)\}$. Then, for any $\nu > \nu^*$, $\dot V(x,y) < -\beta_3\|\mathrm{col}(x,y)\|^2$ where

$$
\beta_3 = \min_{w\in\mathbb{W}}\left\{\frac{\mu}{2} - 1,\ \nu - 2\mu\varepsilon^2\|P_1(w)\|^2 - \varepsilon^2\|P_2(w)\|^2 - 2\varepsilon\|P_2(w)\|\right\}.
$$

Thus, the origin of linear uncertain system (7.74) is asymptotically stable for every $w \in \mathbb{W}$. □

Lemma 7.9 *Consider the system (7.73) where $b(w) > 0$, A_1 and M are Hurwitz, and the polynomial $s^{r-1} + \gamma_{r-2}s^{r-2} + \cdots + \gamma_1 s + \gamma_0$ is stable. There exists $K^* > 0$ such that, for any $K > K^*$, the static state feedback control law*

$$
\begin{aligned}
\bar u &= -K\zeta,\\
\zeta &= \gamma_0 e + \gamma_1 e^{(1)} + \cdots + \gamma_{r-2}e^{(r-2)} + e^{(r-1)}
\end{aligned}
\tag{7.78}
$$

stabilizes the system for every $w \in \mathbb{W}$.

Proof The closed-loop system can be put to the following compact form

$$
\begin{bmatrix} \dot{\bar{x}} \\ \dot{\bar{x}}_0 \\ \dot{\tilde\eta} \\ \dot\zeta \end{bmatrix} = A(w,K)\begin{bmatrix} \bar{x} \\ \bar{x}_0 \\ \tilde\eta \\ \zeta \end{bmatrix},
\tag{7.79}
$$

where

$$A(w,K)=\begin{bmatrix} \Lambda & 0 & 0 & G \\ A_2(w)D & A_1(w) & 0 & 0 \\ \hat{c}(w) & \hat{A}_3(w) & M & \hat{c}_r(w) \\ \tilde{c}(w) & A_3(w) & b(w)\Psi T^{-1} & \tilde{c}_r(w)-b(w)K \end{bmatrix}. \tag{7.80}$$

In other words, $A(w, K)$ can be put in the form

$$A(w,K)=\begin{bmatrix} M_1(w) & N_1(w) \\ N_2(w) & N_3(w)+KM_2(w) \end{bmatrix}$$

where

$$M_1(w)=\begin{bmatrix} \Lambda & 0 & 0 \\ A_2(w)D & A_1(w) & 0 \\ \hat{c}(w) & \hat{A}_3(w) & M \end{bmatrix},\ N_1(w)=\begin{bmatrix} G \\ 0 \\ \hat{c}_r(w) \end{bmatrix},$$

$$N_2(w)=\begin{bmatrix} \tilde{c}(w) & A_3(w) & b(w)\Psi T^{-1} \end{bmatrix},\ N_3(w)=\tilde{c}_r(w),\ M_2(w)=-b(w).$$

Under the conditions of the lemma, it is easy to see $M_1(w)$ and $M_2(w)$ are Hurwitz for all $w \in \mathbb{W}$. Since $N_i(w)$, $i = 1, 2, 3$, are independent of K, all the conditions in Lemma 7.8 are satisfied. Thus, by Lemma 7.8, there exists a positive number K^* such that, for any $K > K^*$, the origin of system (7.79) is asymptotically stable for every value of $w \in \mathbb{W}$. □

The controller (7.78) depends on the derivatives of e which in turn depend on the state variables $\bar{x}_k$, $k = 1, \ldots, r-1$. Next, we will further consider synthesizing an output feedback controller that incorporates a high gain observer. The main result is summarized below.

Lemma 7.10 *Consider the system (7.71) where $b(w) > 0$, A_1 and M are Hurwitz, and the polynomial $s^{r-1} + \gamma_{r-2}s^{r-2} + \cdots + \gamma_1 s + \gamma_0$ is stable. There exist positive numbers K^* and h^* such that, for any $K > K^*$ and $h > h^*$, the following dynamic output feedback control law*

$$\begin{aligned} \bar{u} &= -K\hat{\zeta}, \\ \dot{\xi} &= A_0(h)\xi + B_0(h)e, \\ \hat{\zeta} &= \gamma_0\xi_1 + \gamma_1\xi_2 + \cdots + \gamma_{r-2}\xi_{(r-1)} + \xi_r, \end{aligned} \tag{7.81}$$

where $\xi = col(\xi_1, \ldots, \xi_r) \in \mathbb{R}^r$,

$$A_0(h)=\begin{bmatrix} -h\delta_{r-1} & 1 & 0 & \cdots & 0 \\ -h^2\delta_{r-2} & 0 & 1 & \cdots & 0 \\ \vdots & \vdots & \vdots & \vdots & \vdots \\ -h^{r-1}\delta_1 & 0 & 0 & \cdots & 1 \\ -h^r\delta_0 & 0 & 0 & \cdots & 0 \end{bmatrix},\quad B_0(h)=\begin{bmatrix} h\delta_{r-1} \\ h^2\delta_{r-2} \\ \vdots \\ h^{r-1}\delta_1 \\ h^r\delta_0 \end{bmatrix},$$

and δ_j, $j = 0, 1, \ldots, r-1$, *are such that the polynomial* $s^r + \delta_{r-1}s^{r-1} + \cdots + \delta_1 s + \delta_0$ *is stable, stabilizes the system for every* $w \in \mathbb{W}$.

Proof Let $\tilde{\bar{x}} = \mathrm{col}(\bar{x}_1, \ldots, \bar{x}_r)$ and $\Gamma = [\gamma_0, \gamma_1, \ldots, \gamma_{r-2}, 1]$. Then the closed-loop system consisting of the augmented system (7.71) and the control law (7.81) is

$$\begin{aligned}
\dot{\bar{x}}_0 &= A_1(w)\bar{x}_0 + A_2(w)\bar{x}_1 \\
\dot{\tilde{\bar{x}}} &= A_4(w)\bar{x}_0 + A_5(w)\tilde{\bar{x}} + B(w)\Psi T^{-1}\bar{\eta} - B(w)K\Gamma\xi \\
\dot{\bar{\eta}} &= (M + N\Psi T^{-1})\bar{\eta} - NK\Gamma\xi \\
\dot{\xi} &= A_0(h)\xi + B_0(h)e
\end{aligned} \tag{7.82}$$

where various matrices are defined as follows:

$$A_4(w) = \begin{bmatrix} 0_{(r-1)\times(n-r)} \\ A_3(w) \end{bmatrix}, \quad A_5(w) = \begin{bmatrix} 0 & 1 & 0 & \cdots & 0 \\ 0 & 0 & 1 & \cdots & 0 \\ \vdots & \vdots & \vdots & \ddots & \vdots \\ 0 & 0 & 0 & \cdots & 1 \\ c_1(w) & c_2(w) & c_3(w) & \cdots & c_r(w) \end{bmatrix},$$

$$B(w) = \begin{bmatrix} 0 \; 0 \cdots 0 \; b(w) \end{bmatrix}^{\mathsf{T}}.$$

Let $\psi_k = h^{r-k}(e^{(k-1)} - \xi_k)$, $k = 1, \ldots, r$. Then

$$\begin{aligned}
\dot{\psi}_k &= h^{r-k}(e^{(k)} - \dot{\xi}_k) = h^{r-k}(e^{(k)} - (-h^k\delta_{r-k}\xi_1 + \xi_{(k+1)} + h^k\delta_{r-k}e)) \\
&= h^{r-k}((e^{(k)} - \xi_{(k+1)}) - h^k\delta_{r-k}(e - \xi_1)) = h(\psi_{(k+1)} - \delta_{r-k}\psi_1), \\
\dot{\psi}_r &= e^{(r)} - \dot{\xi}_r = -h^r\delta_0(e - \xi_1) + e^{(r)} = -h\delta_0\psi_1 + e^{(r)} \\
&= -h\delta_0\psi_1 + \dot{\bar{x}}_r,
\end{aligned}$$

and

$$\begin{aligned}
\dot{\bar{x}}_r &= A_3(w)\bar{x}_0 + \sum_{k=1}^{r} c_k(w)\bar{x}_k + b(w)\Psi T^{-1}\bar{\eta} - b(w)K\zeta + b(w)K(\zeta - \hat{\zeta}), \\
\dot{\tilde{\bar{x}}} &= A_4(w)\bar{x}_0 + A_5(w)\tilde{\bar{x}} + B(w)\Psi T^{-1}\bar{\eta} - B(w)K\zeta + B(w)K(\zeta - \hat{\zeta}), \\
\dot{\bar{\eta}} &= (M + N\Psi T^{-1})\bar{\eta} - NK\zeta + NK(\zeta - \hat{\zeta}).
\end{aligned}$$

Here ζ is given by (7.78), and moreover,

$$\begin{aligned}
\zeta - \hat{\zeta} &= \gamma_0(e - \xi_1) + \gamma_1(e^{(1)} - \xi_2) + \cdots \\
&\quad + \gamma_{r-2}\,(e^{(r-2)} - \xi_{(r-1)}) + (e^{(r-1)} - \xi_r) \\
&= \Gamma D_h^{-1}\psi
\end{aligned}$$

where $\psi = \mathrm{col}(\psi_1, \ldots, \psi_r)$ and $D_h = \mathrm{diag}(h^{r-1}, h^{r-2}, \ldots, 1)$.

Let $\phi = \text{col}(\bar{x}_0, \tilde{x}, \bar{\eta})$. Then the closed-loop system (7.82) can be put to the following compact form

$$\begin{bmatrix} \dot{\phi} \\ \dot{\psi} \end{bmatrix} = \begin{bmatrix} \bar{A}(w, K) & Z_1(w, K, h) \\ Z_2(w, K) & hA_0(1) + Z_3(w, K, h) \end{bmatrix} \begin{bmatrix} \phi \\ \psi \end{bmatrix}, \tag{7.83}$$

where $\bar{A}(w, K)$ is similar to $A(w, K)$ given by (7.80) through the linear transformation (7.72), and

$$\begin{aligned} Z_1(w, K, h) &= \begin{bmatrix} 0_{(n-r)\times r} \\ B(w)K\Gamma D_h^{-1} \\ NK\Gamma D_h^{-1}, \end{bmatrix}, \\ Z_2(w, K) &= \begin{bmatrix} 0_{(r-1)\times(n+\ell)} \\ \Theta(w, K) \end{bmatrix} + \left[\begin{array}{c|c|c} 0_{(r-1)\times(n-r)} & 0_{(r-1)\times(r)} & 0_{(r-1)\times\ell} \\ \hline 0_{1\times(n-r)} & -b(w)K\Gamma & 0_{1\times\ell} \end{array}\right], \\ Z_3(w, K, h) &= B(w)K\Gamma D_h^{-1}, \\ \Theta(w, K) &= \left[A_3(w), c_1(w), \ldots, c_r(w), b(w)\Psi T^{-1}\right]. \end{aligned}$$

From the proof of Lemma 7.9, there exists a positive number K^* such that $A(w, K)$ and hence $\bar{A}(w, K)$ are Hurwitz for any $K > K^*$ and any $w \in \mathbb{W}$. Since the polynomial $s^r + \delta_{r-1}s^{r-1} + \cdots + \delta_1 s + \delta_0$ is stable, $A_0(1)$ is Hurwitz. Further, (7.83) takes the form of (7.74) with $v = h$, and $M_1(w) = \bar{A}(w, K)$, $N_1(w, v) = Z_1(w, K, h)$, $N_2(w, v) = Z_2(w, K)$, $N_3(w, v) = Z_3(w, K, h)$, $M_2(w) = A_0(1)$, and $\|D_h^{-1}\| \leq \|D_1^{-1}\|$ for any $h \geq 1$. Therefore, for any $h \geq 1$, there exists $\varepsilon > 0$, independent of h, such that $\|N_i(w, v)\| \leq \varepsilon$, $i = 1, 2, 3$. Thus, by Lemma 7.8, there exists $h^* > 0$ such that, for any $h > h^*$, the origin of the closed-loop system (7.82) is asymptotically stable for every $w \in \mathbb{W}$. □

Finally, the solvability of the GRORP of a linear system is summarized in the following theorem.

Theorem 7.1 *Consider the class of linear systems (7.16) with the exosystem (7.7). Under Assumptions 7.3, 7.9, and 7.11, the GRORP is solved on a prescribed compact set $\mathbb{D}$ by a controller of the form*

$$\begin{aligned} u &= \Psi T^{-1}\eta - K\hat{\zeta}, \\ \dot{\eta} &= M\eta + Nu, \\ \dot{\xi} &= A_0(h)\xi + B_0(h)e, \\ \hat{\zeta} &= \gamma_0\xi_1 + \gamma_1\xi_2 + \cdots + \gamma_{r-2}\xi_{(r-1)} + \xi_r. \end{aligned} \tag{7.84}$$

7.6 Notes and References

The output regulation problem for the class of linear systems was thoroughly studied in the 1970s in [2–4], among others. A salient outcome of this research is the internal model principle which enables the conversion of the output regulation problem into an eigenvalue placement problem for an augmented linear system. For the class of nonlinear systems, the same problem was first treated for the special case in which the exogenous signals are constant [4–6]. The nonlinear output regulation problem with time-varying exogenous signals was first studied in 1990 by Isidori and Byrnes without considering the parameter uncertainty [7]. One of the key contributions of [7] is to link the solvability of the regulator equations to that of the output regulation problem. The solvability of the regulator equations has been extensively studied in the literature [7, 8] using the center manifold theory which can be found in [9, 10]. The robust version of the same problem was pursued by quite a few people in [11–17]. Various solvability conditions have been given which impose assumptions on the solution of the regulator equations. Later, some authors have also addressed the semiglobal or global output regulation problem for nonlinear systems with special structures [16, 18–22]. Based on these works, a systematic framework was proposed in [23], which lay the foundation of this chapter. The problem formulation in Sect. 7.1 is quite standard and is mainly adapted from [24]. Section 7.2.1 handles the existence of the steady-state generator for the case where the solution of the regulator equations satisfies certain polynomial or immersion assumption. The polynomial and the linear immersion assumptions were first used in [11, 12]. It was proved in [25] that both the polynomial and the linear immersion assumptions are equivalent to the trigonometric polynomial assumption when the exosystem is linear. The nonlinear immersion assumption was proposed in [23] which is satisfied for more general nonlinear systems (Sect. 7.2.2). The output regulation problem for nonlinear exosystems was studied in [26] for the local case, and in [27] for the global case. Section 7.2.3 is mainly from [27]. Section 7.4 is mainly from [23, 24]. The exposition of Sect. 7.5 is adapted from [28]. References on the construction of internal model based on observer theory can be found in [29–31].

7.7 Problems

Problem 7.1 Find an appropriate exosystem to generate each of the following signals.

(a) $v(t) = 3 + \sin 2t$;
(b) $v(t) = 2\cos t + \sin 2t$;
(c) $v(t) = \sin 2t \cos t$.

Problem 7.2 Consider a nonlinear system

$$\dot{x}_1 = x_1^2 + x_2$$
$$\dot{x}_2 = w \cos x_1 \sin x_2 + u$$
$$e = x_1 - v_1$$

and a linear exosystem with $v = [v_1\ v_2]^{\mathrm{T}}$,

$$\dot{v} = A_1(\sigma)v, \ A_1(\sigma) = \begin{bmatrix} 0 & -\sigma \\ \sigma & 0 \end{bmatrix}.$$

Write down the regulator equations and find the solution.

Problem 7.3 Repeat Problem 7.2 for an nonlinear exosystem

$$\dot{v}_1 = v_2$$
$$\dot{v}_2 = -av_1 + b(1 - v_1^2)v_2, \ a > 0, \ b > 0.$$

Problem 7.4 Given the following exosystem

$$\dot{v}_1 = \sigma v_2, \ \dot{v}_2 = -\sigma v_1,$$

for each of the following $\mathbf{c}(v, w, \sigma)$, verify Assumption 7.5 and find a corresponding linearly observable steady-state generator.

(a) $\mathbf{c}(v, w, \sigma) = wv_1 + v_2$;
(b) $\mathbf{c}(v, w, \sigma) = wv_1v_2$;
(c) $\mathbf{c}(v, w, \sigma) = wv_1 + \sigma v_2^2$.

Problem 7.5 For a nonlinear function $a(v)$ in the following form

$$a(v) = A_1 v + \sum_{i=2}^{k} A_i v a_i(v)$$

for some integer $k \geq 2$ and some matrices $A_i \in \mathbb{R}^{l_1 \times l_1}$. The functions $a_k : \mathbb{R}^{l_1} \mapsto \mathbb{R}$, $i = 2, \ldots, k$ are sufficiently smooth and satisfy $a_i(0) = 0$. Let $\mathbf{c}(v, w)$ be a polynomial in v.

(a) Find a triplet $(\theta(v, w), \Phi, \Psi)$ such that, along the exosystem $\dot{v} = A_1 v$,

$$\dot{\theta}(v, w) = \Phi\theta(v, w), \ \mathbf{c}(v, w) = \Psi\theta(v, w).$$

(b) Suppose there exist some matrices Δ_i, $i = 2, \ldots, k$, satisfying

$$\frac{\partial \theta(v, w)}{\partial v} A_i v = \Delta_i \theta(v, w).$$

Verify that a steady-state generator exists, along the exosystem $\dot{v} = a(v)$.

Problem 7.6 Let

$$\mathbf{c}(v, w) = \sin(w_1 v_1) + w_2 v_3^2,$$

where

$$\begin{aligned}\dot{v}_1 &= v_2,\ \ \dot{v}_2 = -v_1\\ \dot{v}_3 &= 3v_4,\ \ \dot{v}_4 = -3v_3.\end{aligned}$$

Verify Assumption 7.6 and find a corresponding linearly observable steady-state generator.

Problem 7.7 Let $\mathbf{c}(v, w) = wv_1$, where

$$\begin{aligned}\dot{v}_1 &= v_2\\ \dot{v}_2 &= -2v_1 + (1 - v_1^2)v_2.\end{aligned}$$

Verify Assumptions 7.7 and 7.8 and find a corresponding linearly observable steady-state generator.

Problem 7.8 Suppose all the eigenvalues of $\Phi(\sigma)$ have zero real parts and the pair $(\Psi, \Phi(\sigma))$ is observable. Also, M is Hurwitz and the pair (M, N) is controllable. Show that the Sylvester equation

$$T(\sigma)\Phi(\sigma) - MT(\sigma) = N\Psi$$

has a unique nonsingular solution $T(\sigma)$ (see Appendix A of [32]).

Problem 7.9 For each steady-state generator found in Problems 7.4, 7.5, 7.6, and 7.7, construct a corresponding internal model candidate.

Problem 7.10 Let $\mathbf{c}(v, w) = c(\mathbf{x}(v, w), \mathbf{u}(v, w)) \in \mathbb{R}$. Assume there exist an integer ℓ and a sufficiently smooth function ϕ vanishing at the origin such that

$$L_a^{\ell}\mathbf{c}(v, w) + \phi(\mathbf{c}(v, w), L_a\mathbf{c}(v, w), \ldots, L_a^{\ell-1}\mathbf{c}(v, w)) = 0.$$

Let $\theta(v, w) = [\theta_1(v, w), \ldots, \theta_\ell(v, w)]^{\mathrm{T}} = [\mathbf{c}(v, w), L_a\mathbf{c}(v, w), \ldots, L_a^{\ell-1}\mathbf{c}(v, w)]^{\mathrm{T}}$ and

$$\alpha(\theta) = \begin{bmatrix} \theta_2 \\ \theta_3 \\ \vdots \\ -\phi(\theta_1, \theta_2, \ldots, \theta_\ell) \end{bmatrix}, \quad \Psi = \begin{bmatrix} 1 \\ 0 \\ \vdots \\ 0 \end{bmatrix}^{\mathrm{T}}.$$

(a) Verify that

$$\dot{\theta}(v, w) = \alpha(\theta(v, w)), \ \mathbf{c}(v, w) = \Psi\theta(v, w). \tag{7.85}$$

Thus, Eq. (7.85) is a nonlinear steady-state generator on the identity mapping with output $c(x, u)$.

(b) Show that $\dot{\eta} = \alpha(\eta) + N(c(x, u) - \Psi\eta)$, where $N \in \mathbb{R}^{\ell}$ is a constant vector, is an internal model candidate corresponding to the steady-state generator (7.85).

Problem 7.11 Under Assumptions 7.9 and 7.10, using the PBH test to show that, for any $w \in \mathbb{W}$, the augmented system (7.66) is both stabilizable and detectable.

Problem 7.12 Find the linear transformation matrix $T(w)$ in (7.68) such that the system (7.16) is equivalent to (7.69).

Problem 7.13 Consider the following linear system

$$\begin{aligned}\dot{x}_0 &= -x_0 + wx_1 + v_1 \\ \dot{x}_1 &= x_2 \\ \dot{x}_2 &= w(x_0 + x_1 + x_2) + u \\ e &= x_1 - v_1\end{aligned}$$

where

$$\dot{v}_1 = v_2, \ \dot{v}_2 = -v_1.$$

Solve the robust output regulation problem for

(a) sufficiently small w;
(b) $-2 \leq w \leq 2$.

References

1. Huang J (2004) Nonlinear output regulation problem: theory and applications. SIAM, Philadelphia
2. Davison EJ (1976) The robust control of a servomechanism problem for linear time-invariant multivariable systems. IEEE Trans Autom Control 21:25–34
3. Francis BA (1977) The linear multivariable regulator problem. SIAM J Control Optim 15:486–505
4. Francis BA, Wonham WM (1976) The internal model principle of control theory. Automatica 12:457–465
5. Desoer C, Lin CA (1985) Tracking and disturbance rejection of mimo nonlinear systems with PI controller. IEEE Trans Autom Control 30:861–867
6. Huang J, Rugh WJ (1990) On a nonlinear multivariable servomechanism problem. Automatica 26:963–972

7. Isidori A, Byrnes CI (1990) Output regulation of nonlinear systems. IEEE Trans Autom Control 35:131–140
8. Huang J (2003) On the solvability of the regulator equations for a class of nonlinear systems. IEEE Trans Autom Control 48:880–885
9. Carr J (1981) Applications of the center manifold theory. Springer, New York
10. Isidori A (1995) Nonlinear control systems, 3rd edn. Springer, New York
11. Byrnes CI, Priscoli FD, Isidori A, Kang W (1997) Structurally stable output regulation of nonlinear systems. Automatica 33:369–385
12. Huang J (1995) Asymptotic tracking and disturbance rejection in uncertain nonlinear systems. IEEE Trans Autom Control 40:1118–1122
13. Huang J (1995) Output regulation of nonlinear systems with nonhyperbolic zero dynamics. IEEE Trans Autom Control 40:1497–1500
14. Huang J, Lin C-F (1994) On a robust nonlinear servomechanism problem. IEEE Trans Autom Control 39:1510–1513
15. Huang J, Lin C-F (1993) Internal model principle and robust control of nonlinear systems. In: Proceedings of the 32nd IEEE conference on decision and control, pp 1501–1513
16. Khalil H (1994) Robust servomechanism output feedback controllers for feedback linearizable systems. Automatica 30:1587–1589
17. Priscoli FD (1993) Robust tracking for polynomial plants. In: Proceedings of the European control conference, pp 369–373
18. Isidori A (1997) A remark on the problem of semiglobal nonlinear output regulation. IEEE Trans Autom Control 42:1734–1738
19. Khalil H (2000) On the design of robust servomechanisms for minimum phase nonlinear systems. Int J Robust Nonlinear Control 10:339–361
20. Serrani A, Isidori A (2000) Global robust output regulation for a class of nonlinear systems. Syst Control Lett 39:133–139
21. Serrani A, Isidori A, Marconi L (2000) Semiglobal robust output regulation of minimum-phase nonlinear systems. Int J Robust Nonlinear Control 10:379–396
22. Serrani A, Isidori A, Marconi L (2001) Semiglobal nonlinear output regulation with adaptive internal model. IEEE Trans Autom Control 46:1178–1194
23. Huang J, Chen Z (2004) A general framework for tackling the output regulation problem. IEEE Trans Autom Control 49:2203–2218
24. Huang J (2011) An overview of the output regulation problem. J Syst Sci Math Sci 31(9):1055–1081
25. Huang J (2001) Remarks on robust output regulation problem for nonlinear systems. IEEE Trans Autom Control 46:2028–2031
26. Chen Z, Huang J (2005) Robust output regulation with nonlinear exosystems. Automatica 41(8):1447–1454
27. Yang X, Huang J (2012) New results on robust output regulation of nonlinear systems with a nonlinear exosystem. Int J Robust Nonlinear Control 22:1703–1719
28. Su Y, Huang J (2013) Cooperative robust output regulation of a class of heterogeneous linear uncertain multi-agent systems, Int J Robust Nonlinear Control, doi:10.1002/rnc.3027
29. Marconi L, Praly L (2008) Uniform practical nonlinear output regulation. IEEE Trans Autom Control 53(5):1184–1202
30. Marconi L, Praly L, Isidori A (2007) Output stabilization via nonlinear Luenberger observers. SIAM J Control Optim 45(6):2277–2298
31. Marconi L, Praly L, Isidori A (2010) Robust asymtotic stabilization of nonlinear systems with non-hyperbolic zero dynamics. IEEE Trans Autom Control 55:907–921
32. Huang J (2004) Nonlinear output regulation problem: theory and Applications. SIAM, Philadelphia

Chapter 8
Global Robust Output Regulation

We already investigated in Chap. 7 that, under some suitable assumptions, the GRORP for a given nonlinear system can be converted into a GRSP/GASP for an augmented system. However, the GRSP/GASP itself is a challenging task. So far, the results are mainly limited to nonlinear systems with special structures such as those studied from Chaps. 2–6. In this chapter, using the framework established in Chap. 7, we will first study the GRORP for both output feedback systems and lower triangular systems for the case where $v(t) \in \mathbb{V}$ and $w \in \mathbb{W}$ for two compact sets $\mathbb{V}$ and $\mathbb{W}$ whose boundary is known, and the exosystem is known exactly. It turns out that, for this case, the GRORP of the output feedback systems and lower triangular systems can be converted to the GRSP of the systems studied in Chap. 4. Then, we will further consider the GRORP for the case where the boundary of $\mathbb{V}$ and $\mathbb{W}$ is unknown. The chapter is organized as follows. In Sects. 8.1–8.3, we study systems with relative degree one, output feedback systems, and lower triangular systems, respectively, under the assumption that the solution of the regulator equations satisfies the nonlinear immersion condition. In Sect. 8.4, we further consider the GRORP under the assumption that the solution of the regulator equations satisfies the generalized linear immersion condition which allows the exosystem to be nonlinear. In Sect. 8.5, we consider the GRORP with the boundary of $\mathbb{V}$ and $\mathbb{W}$ unknown, or what is the same, both the unknown parameter w and the initial state of the exosystem can be arbitrary large. The approach is illustrated via the Lorenz system in Sect. 8.6. Finally, the notes and references are given in Sect. 8.7.

8.1 Systems with Relative Degree One

We first consider the following composite system

$$\begin{aligned} \dot{x}_0 &= f_0(x_0, x, v(t), w) \\ \dot{x} &= f(x_0, x, v(t), w) + b(w)u \end{aligned}$$

Z. Chen and J. Huang, *Stabilization and Regulation of Nonlinear Systems*,
Advanced Textbooks in Control and Signal Processing,
DOI 10.1007/978-3-319-08834-1_8

$$\begin{aligned}\dot{v} &= A_1 v\\ y &= x\end{aligned} \tag{8.1}$$

where $x_0 \in \mathbb{R}^{n-1}$ and $x \in \mathbb{R}$ are the state variables, $u \in \mathbb{R}$ is the input, $v(t) \in \mathbb{R}^{l_1}$ represents external disturbances and/or reference inputs, and $w \in \mathbb{R}^{l_2}$ represents unknown parameters. It is assumed that all functions in system (8.1) are sufficiently smooth with $f_0(0, 0, 0, w) = 0$ and $f(0, 0, 0, w) = 0$, and A_1 is known exactly. The system (8.1) is composed of a nonlinear system with relative degree one and an exosystem satisfying Assumption 7.3.

Let the tracking error be

$$e = y - y_d, \ \ y_d = h_d(v, w) \tag{8.2}$$

where $h_d(v, w)$ is a sufficiently smooth function satisfying $h_d(0, w) = 0$ for all w.

Let $d(t) = \text{col}(v(t), w) \in \mathbb{D}$ where $\mathbb{D} = \mathbb{V} \times \mathbb{W} \subset \mathbb{R}^l$ with $l = l_1 + l_2$ and is assumed to have a known boundary. We are now ready to consider the GRORP for the composite system (8.1) on $\mathbb{D}$. First, we need two assumptions.

Assumption 8.1 *The function $b(w)$ is away from zero, e.g., $b(w) > 0, \forall w \in \mathbb{W}$.*

Assumption 8.2 *There exists a sufficiently smooth function $\mathbf{x}_0(v, w)$ with $\mathbf{x}_0(0, 0) = 0$ satisfying*

$$\frac{\partial \mathbf{x}_0(v, w)}{\partial v} A_1 v = f_0(\mathbf{x}_0(v, w), h_d(v, w), v, w), \ \forall d \in \mathbb{D}. \tag{8.3}$$

Example 8.1 Equation (8.3) is a type of center manifold equation. Consider an example with

$$f_0(x_0, x, v, w) = F(w)x_0 + G(x, v, w)x + D(v, w).$$

If none of the eigenvalues of the matrix $F(w)$ coincide with any λ given by $\{\, \lambda \mid \lambda = i_1\lambda_1 + \cdots + i_{l_1}\lambda_{l_1}, i_1 + \cdots + i_{l_1} = i, \ i_1, \ldots, i_{l_1} \geq 0, \ i = 1, 2, \ldots, \}$ where $\lambda_1, \ldots, \lambda_{l_1}$ are the eigenvalues of A_1, then (8.3) has a power series solution of the form

$$\mathbf{x}_0(v, w) = \sum_{i \geq 1} X_i(w) v^{[i]} \tag{8.4}$$

where $v^{[i]}$ is defined in (7.28). Using the property of the Kronecker product, it can be shown that there exists a constant matrix $A^{[i]}$ such that (Problem 8.1)

$$\frac{\partial v^{[i]}}{\partial v} A_1 v = A^{[i]} v^{[i]}.$$

Then, $X_i(w)$ satisfies the Sylvester equation of the form

$$X_i(w)A^{[i]} = F(w)X_i(w) + G_i(w) \tag{8.5}$$

where $G_i(w)$ is such that

$$G(h_{\rm d}(v, w), v, w)h_{\rm d}(v, w) + D(v, w) = \sum_{i \geq 1} G_i(w)v^{[i]}.$$

In particular, when $h_{\rm d}(v, w)$ and $D(v, w)$ are polynomials in v, and $G(x, v, w)$ is a polynomial in x and v, then, for some integer p,

$$G(h_{\rm d}(v, w), v, w)h_{\rm d}(v, w) + D(v, w)$$

is a polynomial in v of degree p. In this case, Eq. (8.3) has a unique globally defined solution which is a polynomial of degree p in v.

Under Assumptions 8.1 and 8.2, the solution to the regulator equations is

$$\text{col}(\mathbf{x}_0(v, w), \mathbf{x}(v, w), \mathbf{u}(v, w))$$

where

$$\begin{aligned}
\mathbf{x}(v, w) &= h_{\rm d}(v, w) \\
\mathbf{u}(v, w) &= b^{-1}(w)\left[\frac{\partial h_{\rm d}(v, w)}{\partial v}A_1 v - f(\mathbf{x}_0(v, w), h_{\rm d}(v, w), v, w)\right].
\end{aligned}$$

We will study the GRORP under the nonlinear immersion assumption, i.e., Assumption 7.6 with $\mathbf{c}(v, w) = \mathbf{u}(v, w)$. For convenience, let $\mathbf{x}_2(v, w) = \mathbf{u}(v, w)$.

Assumption 8.3 *The composite system* (8.1) *admits a steady-state generator of the following form*

$$\dot{\theta}(v, w) = \Phi\theta(v, w), \ \ \mathbf{x}_2(v, w) = \psi(\theta(v, w)) \tag{8.6}$$

where, for some integer $\ell > 0$, $\theta : \mathbb{R}^l \mapsto \mathbb{R}^\ell$ *and* $\psi : \mathbb{R}^\ell \mapsto \mathbb{R}$ *are sufficiently smooth functions vanishing at the origin, and* $\Phi \in \mathbb{R}^{\ell \times \ell}$ *is a constant matrix. Moreover, the pair* (Ψ, Φ) *is observable where* Ψ *is the Jacobian matrix of the function* ψ *at the origin.*

Remark 8.1 Sufficient conditions for the satisfaction of Assumption 8.3 was given in Lemma 7.3 in terms of $\mathbf{u}(v, w)$. In particular, if $\mathbf{u}(v, w)$ satisfies Assumption 7.5, then Assumption 8.3 always holds with ψ a linear function of θ.

By Example 7.9, under Assumption 8.3, for any controllable pair (M, N) with $M \in \mathbb{R}^{\ell \times \ell}$ Hurwitz and $N \in \mathbb{R}^{\ell \times 1}$, there is a unique nonsingular matrix T

satisfying the Sylvester equation $T\Phi - MT = N\Psi$. Then, associated with the steady-state generator (8.6), we can construct an internal model candidate on the mapping $\tau(\theta) = T\theta$ as follows:

$$\dot{\eta} = M\eta + N(u - \psi(T^{-1}\eta) + \Psi T^{-1}\eta) \tag{8.7}$$

For convenience, we denote

$$\delta(\eta) = \psi(T^{-1}\eta) - \Psi T^{-1}\eta \tag{8.8}$$

which is the nonlinear part of the function $\psi(T^{-1}\eta)$.

Attaching the internal model candidate (8.7) to the given plant yields an augmented system of the form (7.54). The coordinate and input transformation (7.60) takes the following specific form:

$$\begin{aligned}
\bar{\eta} &= \eta - T\theta(v, w)\\
\bar{x}_0 &= x_0 - \mathbf{x}_0(v, w)\\
\bar{x} &= x - \mathbf{x}(v, w)\\
\bar{u} &= u - \psi(T^{-1}\eta)
\end{aligned} \tag{8.9}$$

Applying the transformation (8.9) to the augmented system (7.54) gives the transformed augmented system (7.61) with the following specific form:

$$\begin{aligned}
\dot{\bar{x}}_0 &= f_0(x_0, x, v, w) - f_0(\mathbf{x}_0(v, w), \mathbf{x}(v, w), v, w)\\
\dot{\bar{\eta}} &= (M + N\Psi T^{-1})\bar{\eta} + N\bar{u}\\
\dot{\bar{x}} &= f(x_0, x, v, w) - f(\mathbf{x}_0(v, w), \mathbf{x}(v, w), v, w)\\
&\quad - b(w)\psi(\theta(v, w)) + b(w)\psi(T^{-1}\eta) + b(w)\bar{u}.
\end{aligned} \tag{8.10}$$

By Proposition 7.2, all we need to do is to globally stabilize the augmented system (8.10). However, this system is not in the familiar lower triangular form as studied in Chap. 4. To eliminate the term $\bar{u}$ in the second equation of (8.10), let us perform on the $\bar{\eta}$-subsystem of (8.10) another coordinate transformation

$$\tilde{\eta} = \bar{\eta} - Nb^{-1}(w)\bar{x} \tag{8.11}$$

which gives

$$\begin{aligned}
\dot{\bar{x}}_0 &= \bar{f}_0(\bar{x}_0, \bar{x}, d)\\
\dot{\tilde{\eta}} &= M\tilde{\eta} - N[\delta(\tilde{\eta} + \mu) - \delta(\mu)] + \rho(\bar{x}_0, \bar{x}, d)\\
\dot{\bar{x}} &= \bar{f}(\bar{x}_0, \tilde{\eta}, \bar{x}, d) + b(w)\bar{u}
\end{aligned} \tag{8.12}$$

where various quantities are described as follows:

$$\begin{aligned}
\mu &= T\theta(v, w) + Nb^{-1}(w)\bar{x} \\
\rho(\bar{x}_0, \bar{x}, d) &= -N[\delta(T\theta(v, w) + Nb^{-1}(w)\bar{x}) - \delta(T\theta(v, w))] \\
&\quad + b^{-1}(w)MN\bar{x} \\
&\quad - Nb^{-1}(w)[f(x_0, x, v, w) - f(\mathbf{x}_0(v, w), \mathbf{x}(v, w), v, w)] \\
\bar{f}_0(\bar{x}_0, \bar{x}, d) &= f_0(\bar{x}_0 + \mathbf{x}_0(v, w), \bar{x} + \mathbf{x}(v, w), v, w) - f_0(\mathbf{x}_0(v, w), \mathbf{x}(v, w), v, w) \\
\bar{f}(\bar{x}_0, \bar{\eta}, \bar{x}, d) &= f(\bar{x}_0 + \mathbf{x}_0(v, w), \bar{x} + \mathbf{x}(v, w), v, w) - f(\mathbf{x}_0(v, w), \mathbf{x}(v, w), v, w) \\
&\quad - b(w)\psi(\theta(v, w)) + b(w)\psi(\theta(v, w) + T^{-1}\bar{\eta} + T^{-1}Nb^{-1}(w)\bar{x}).
\end{aligned}$$

By Lemma 7.7, or by direct inspection, we have, for all $d \in \mathbb{D}$, $\bar{f}_0(0, 0, d) = 0$, $\rho(0, 0, d) = 0$, and $\bar{f}(0, 0, 0, d) = 0$.

Let $z = \text{col}(\bar{x}_0, \bar{\eta})$. Then, the system (8.12) can be put in the form

$$\begin{aligned}
\dot{z} &= q(z, \bar{x}, d) \\
\dot{\bar{x}} &= \bar{f}(z, \bar{x}, d) + b(w)\bar{u}
\end{aligned} \tag{8.13}$$

which takes the same form as (2.46).

The global robust stabilization problem of the system of the form (8.13) is well studied in Chap. 2. In particular, by Theorem 2.8, it suffices to require the z-subsystem to satisfy Assumption 2.3 in order to stabilize (8.13). For this purpose, we need two more assumptions.

Assumption 8.4 *The subsystem* $\dot{\bar{x}}_0 = \bar{f}_0(\bar{x}_0, \bar{x}, d)$ *has an ISS Lyapunov function* $V(\bar{x}_0)$, *i.e.,*

$$V(\bar{x}_0) \sim \{\underline{\alpha}, \bar{\alpha}, \alpha, \sigma \mid \dot{\bar{x}}_0 = \bar{f}_0(\bar{x}_0, \bar{x}, d)\}$$

and

$$\limsup_{s \to 0^+} \frac{s^2}{\alpha(s)} < \infty, \quad \limsup_{s \to 0^+} \frac{\sigma(s)}{s^2} < \infty.$$

Example 8.2 Suppose the x_0-dynamics are of the form

$$\dot{x}_0 = f_0(x_0, x, v, w) = F(w)x_0 + G(x, v, w)x + D(v, w).$$

Let $\mathbf{x}(v, w) = h_d(v, w)$ and $\mathbf{x}_0(v, w)$ be the solution to the equation

$$\frac{\partial \mathbf{x}_0(v, w)}{\partial v} A_1 v = F(w)\mathbf{x}_0(v, w) + G(\mathbf{x}(v, w), v, w)\mathbf{x}(v, w) + D(v, w).$$

The coordinate transformation $\bar{x}_0 = x_0 - \mathbf{x}_0(v, w)$, $\bar{x} = x - \mathbf{x}(v, w)$ gives

$$\begin{aligned}\dot{\bar{x}}_0 &= \bar{f}_0(\bar{x}_0, \bar{x}, d) \\ &= F(w)\bar{x}_0 + G(\bar{x} + \mathbf{x}(v, w), v, w)(\bar{x} + \mathbf{x}(v, w)) - G(\mathbf{x}(v, w), v, w)\mathbf{x}(v, w).\end{aligned}$$

Using the approach of Example 2.11, it is ready to verify that Assumption 8.4 is satisfied if $F(w)$ is a Hurwitz matrix.

Assumption 8.5 *Let P be a symmetric positive definite matrix such that $PM + M^\top P = -I$. The function δ in (8.8) satisfies*

$$-2\tilde{\eta}^\top PN\left(\delta(\tilde{\eta} + \mu) - \delta(\mu)\right) \le (1 - \epsilon)\|\tilde{\eta}\|^2, \quad \forall \tilde{\eta}, \mu \tag{8.14}$$

for a constant $0 < \epsilon < 1$.

Remark 8.2 Assumption 8.5 is to restrict the growth of the nonlinear part of the function ψ. The inequality (8.14) is satisfied in some interesting cases. In fact, (8.14) holds for some $0 < \epsilon < 1$ if

$$|\delta(\tilde{\eta} + \mu) - \delta(\mu)| < \frac{1 - \epsilon}{2\|PN\|}\|\tilde{\eta}\|.$$

Thus, (8.14) holds if δ is globally Lipschitz, i.e.,

$$|\delta(\tilde{\eta} + \mu) - \delta(\mu)| \le L\|\tilde{\eta}\|$$

for some positive number L, and moreover L satisfies

$$L < \frac{1 - \epsilon}{2\|PN\|}.$$

In the special case where the function δ in (8.8) is identically equal to zero, Assumption 8.5 is satisfied automatically. Clearly, a linear internal model candidate

$$\dot{\eta} = M\eta + Nu$$

is an example of this special case.

Lemma 8.1 *Under Assumption 8.5, the system $\dot{\tilde{\eta}} = M\tilde{\eta} - N[\delta(\tilde{\eta} + \mu) - \delta(\mu)] + \rho(\bar{x}_0, \bar{x}, d)$ with $(\bar{x}_0, \bar{x})$ as input has an ISS Lyapunov function $V_2(\tilde{\eta})$, i.e.,*

$$V_2(\tilde{\eta}) \sim \{\underline{\alpha}_2, \bar{\alpha}_2, \alpha_2, (\zeta, \sigma_2) \mid \dot{\tilde{\eta}} = M\tilde{\eta} - N[\delta(\tilde{\eta} + \mu) - \delta(\mu)] + \rho(\bar{x}_0, \bar{x}, d)\}$$

and

$$\limsup_{s \to 0^+} \frac{s^2}{\alpha_2(s)} < \infty, \ \limsup_{s \to 0^+} \frac{\sigma_2(s)}{s^2} < \infty, \ \limsup_{s \to 0^+} \frac{\zeta(s)}{s^2} < \infty.$$

Proof Let

$$V_2(\tilde{\eta}) = \frac{2}{\epsilon}\tilde{\eta}^{\mathrm{T}} P \tilde{\eta}.$$

Then

$$\frac{2}{\epsilon}\lambda_{\min}\|\tilde{\eta}\|^2 \leq V_2(\tilde{\eta}) \leq \frac{2}{\epsilon}\lambda_{\max}\|\tilde{\eta}\|^2$$

where $\lambda_{\min}$ and $\lambda_{\max}$ are the minimal and maximal eigenvalues of P, respectively. And the derivative of $V_2(\tilde{\eta})$ along the trajectory of the $\tilde{\eta}$-subsystem satisfies

$$\begin{aligned}
&\frac{\partial V_2(\tilde{\eta})}{\partial \tilde{\eta}}\left(M\tilde{\eta} - N\left(\delta(\tilde{\eta}+\mu) - \delta(\mu)\right) + \rho(\bar{x}_0, \bar{x}, d)\right) \\
&= \frac{2}{\epsilon}\left[2\tilde{\eta}^T P M \tilde{\eta} - 2\tilde{\eta}^T P N\left(\delta(\tilde{\eta}+\mu) - \delta(\mu)\right) + 2\tilde{\eta}^T P\rho(\bar{x}_0, \bar{x}, d)\right] \\
&= \frac{2}{\epsilon}\left[-\tilde{\eta}^T \tilde{\eta} - 2\tilde{\eta}^T P N\left(\delta(\tilde{\eta}+\mu) - \delta(\mu)\right) + 2\tilde{\eta}^T P\rho(\bar{x}_0, \bar{x}, d)\right] \\
&\leq \frac{2}{\epsilon}\left[-\epsilon\tilde{\eta}^{\mathrm{T}}\tilde{\eta} + 2\tilde{\eta}^{\mathrm{T}} P\rho(\bar{x}_0, \bar{x}, d)\right] \\
&\leq \frac{2}{\epsilon}\left[-\frac{\epsilon}{2}\|\tilde{\eta}\|^2 + \frac{2}{\epsilon}\|P\rho(\bar{x}_0, \bar{x}, d)\|^2\right] \\
&\leq -\|\tilde{\eta}\|^2 + \left\|\frac{2}{\epsilon}P\rho(\bar{x}_0, \bar{x}, d)\right\|^2.
\end{aligned} \tag{8.15}$$

Noting the function $\rho(\bar{x}_0, \bar{x}, d)$ is continuously differentiable satisfying $\rho(0, 0, d) = 0$ for $d \in \mathbb{D}$ with $\mathbb{D}$ a compact set, we have

$$\left\|\frac{2}{\epsilon}P\rho(\bar{x}_0, \bar{x}, d)\right\|^2 \leq a_0(\bar{x}_0)\|\bar{x}_0\|^2 + a_1(\bar{x})\bar{x}^2$$

for some smooth function $a_0(\bar{x}_0) \geq 0$ and $a_1(\bar{x}) \geq 0$. As a result, we have

$$\dot{V}_2(\tilde{\eta}) \leq -\|\tilde{\eta}\|^2 + a_0(\bar{x}_0)\|\bar{x}_0\|^2 + a_1(\bar{x})\bar{x}^2.$$

Pick two non-decreasing continuous functions c_0 and c_1 such that $c_0(|\bar{x}_0|) \geq a_0(\bar{x}_0)$ and $c_1(|\bar{x}|) \geq a_1(\bar{x})$. The proof is completed by defining $\underline{\alpha}_2(s) = (2/\epsilon)\lambda_{\min}s^2$, $\bar{\alpha}_2(s) = (2/\epsilon)\lambda_{\max}s^2$, $\alpha_2(s) = s^2$, $\zeta(s) = c_0(s)s^2$, and $\sigma_2(s) = c_1(s)s^2$. □

Using Theorem 2.11 and Lemma 8.1 concludes that the z-subsystem of (8.13) satisfies Assumption 2.3 as summarized below.

Lemma 8.2 *Under Assumptions* 8.4 *and* 8.5, *the* z*-subsystem* $\dot{z} = q(z, \bar{x}, d)$ *of* (8.13) *has an ISS Lyapunov function* $U(z)$, *i.e.,*

$$U(z) \sim \{\underline{\beta}, \bar{\beta}, \beta, \varsigma \mid \dot{z} = q(z, \bar{x}, d)\}$$

and

$$\limsup_{s\to 0^+} \frac{s^2}{\beta(s)} < \infty, \quad \limsup_{s\to 0^+} \frac{\varsigma(s)}{s^2} < \infty.$$

Now, we are ready to state the following theorem.

Theorem 8.1 *Consider the composite system* (8.1). *Under Assumptions* 7.3, 8.1–8.5, *the GRORP is solved on the prescribed compact set* $\mathbb{D}$ *by a controller of the form*

$$\begin{aligned} u &= \psi(T^{-1}\eta) + s(e) \\ \dot{\eta} &= M\eta + N(u - \psi(T^{-1}\eta) + \Psi T^{-1}\eta). \end{aligned} \tag{8.16}$$

Proof Observe that the system (8.13) is of the form (2.46). With Lemma 8.2 and Theorem 2.8, the GRSP for the system (8.13) is solved by a static controller $\bar{u} = s(\bar{x})$. Then, by Proposition 7.2, the GRORP for the system (8.1) is solved on $\mathbb{D}$ by the controller (8.16). □

Example 8.3 Consider the following system

$$\begin{aligned} \dot{x}_0 &= -x_0 + 0.2v_2x + v_1^2 \\ \dot{x} &= x_0 - 0.1v_1x - \sin^2(0.1wx) + 10wv_1 + 10v_2 + 10u \\ y &= x \\ e &= y - 10v_1 \end{aligned} \tag{8.17}$$

where the external disturbance $v = \mathrm{col}(v_1, v_2)$ is governed by the exosystem

$$\dot{v} = A_1v, \; A_1 = \begin{bmatrix} 0 & 1 \\ -1 & 0 \end{bmatrix}.$$

It is assumed that $v(t) \in \mathbb{V} = \{v_1^2 + v_2^2 \leq 1\}$ and $-1 \leq w \leq 1$.

It can be verified that Assumption 8.1 is trivially true. Also, the system satisfies Assumptions 8.2 and 8.3. In particular, the solution to the regulator equations exists globally and is given by

$$\begin{aligned} \mathbf{x}_0(v, w) &= v_1^2 \\ \mathbf{x}(v, w) &= 10v_1 \\ \mathbf{u}(v, w) &= -wv_1 + 0.1\sin^2(wv_1). \end{aligned}$$

Let $\pi(v, w) = wv_1$. Then the minimal zeroing polynomial of $\pi(v, w)$ is $\lambda^2 + 1$, and

$$\mathbf{u}(v, w) = \psi(\mathcal{L}^2_{A_1 v}\pi(v, w)) = -\pi(v, w) + 0.1\sin^2\pi(v, w).$$

With $\theta(v, w) = \mathcal{L}^2_{A_1 v}\pi(v, w) = [wv_1 \ wv_2]^{\mathsf{T}}$, the system has a steady-state generator with output u:

$$\dot{\theta}(v, w) = \Phi\theta(v, w), \ \mathbf{u}(v, w) = \psi(\theta(v, w)).$$

Let Ψ be the Jacobian matrix of ψ at the origin. Then it can be verified that the following pair

$$\Phi = \begin{bmatrix} 0 & 1 \\ -1 & 0 \end{bmatrix}, \ \Psi = \begin{bmatrix} -1 & 0 \end{bmatrix}$$

is observable. Choose a pair of controllable matrices

$$M = \begin{bmatrix} -1 & 0 \\ 0 & -2 \end{bmatrix}, \ N = \begin{bmatrix} 0.2 \\ 0.4 \end{bmatrix}.$$

For this pair of matrices, the solution to the Sylvester equation $MT + N\Psi = T\Phi$ is given by

$$T = \begin{bmatrix} -0.1 & 0.1 \\ -0.16 & 0.08 \end{bmatrix},$$

and hence,

$$T^{-1} = \begin{bmatrix} 10 & -12.5 \\ 20 & -12.5 \end{bmatrix}.$$

Now, we are ready to define

$$\psi(\theta) = -\theta_1 + 0.1\sin^2\theta_1$$

and hence

$$\psi(T^{-1}\theta) = -10\theta_1 + 12.5\theta_2 + 0.1\sin^2(10\theta_1 - 12.5\theta_2).$$

Then, there exists a corresponding internal model

$$\dot{\eta} = M\eta + N(u - \psi(T^{-1}\eta) + \Psi T^{-1}\eta). \tag{8.18}$$

Define the following coordinate and input transformation

$$\begin{aligned}\tilde{\eta} &= \eta - T\theta(v, w) - 0.1N\bar{x}\\ \bar{x}_0 &= x_0 - \mathbf{x}_0(v, w)\\ \bar{x} &= x - \mathbf{x}(v, w)\\ \bar{u} &= u - \psi(T^{-1}\eta).\end{aligned}$$

Performing the coordinate transformation on the system composed of (8.17) and (8.18) gives the following augmented system

$$\begin{aligned}\dot{\bar{x}}_0 &= -\bar{x}_0 + 0.2v_2\bar{x}\\ \dot{\tilde{\eta}} &= M\tilde{\eta} - N(\delta(\tilde{\eta} + \mu) - \delta(\mu)) + \rho(\bar{x}_0, \bar{x}, d)\\ \dot{\bar{x}} &= \bar{x}_0 - \sin^2(0.1w\bar{x} + wv_1) + \sin^2(wv_1) - 0.1v_1\bar{x}\\ &\quad + 10\Psi T^{-1}(\tilde{\eta} + 0.1N\bar{x}) + 10\delta(\tilde{\eta} + 0.1N\bar{x} + T\theta) - 10\delta(T\theta) + 10\bar{u}\end{aligned}$$

where

$$\begin{aligned}\mu &= 0.1N\bar{x} + T\theta\\ \delta(\theta) &= 0.1\sin^2(10\theta_1 - 12.5\theta_2)\\ \rho(\bar{x}_0, \bar{x}, d) &= -N\left(\delta(0.1N\bar{x} + T\theta) - \delta(T\theta)\right) + 0.1MN\bar{x}\\ &\quad -0.1N\left(\bar{x}_0 - \sin^2(0.1w\bar{x} + wv_1) + \sin^2(wv_1) - 0.1v_1\bar{x}\right).\end{aligned}$$

It can be seen that the system $\dot{\bar{x}}_0 = -\bar{x}_0 + 0.2v_2\bar{x}$ satisfies Assumption 8.4. To verify Assumption 8.5, we resort to Remark 8.2. Solving the Lyapunov equation $PM + M^{\mathsf{T}}P = -I$ gives

$$P = \begin{bmatrix} 0.5 & 0\\ 0 & 0.25\end{bmatrix}.$$

Direct calculation gives

$$\begin{aligned}&-2\tilde{\eta}^{\mathsf{T}}\begin{bmatrix} 0.5 & 0\\ 0 & 0.25\end{bmatrix}\begin{bmatrix} 0.2\\ 0.4\end{bmatrix}(\delta(\tilde{\eta} + \mu) - \delta(\mu))\\ &= 0.02(\tilde{\eta}_1 + \tilde{\eta}_2)\left(-\sin^2(10(\tilde{\eta}_1 + \mu_1) - 12.5(\tilde{\eta}_2 + \mu_2)) + \sin^2(10\mu_1 - 12.5\mu_2)\right)\\ &\le 0.02\,|(\tilde{\eta}_1 + \tilde{\eta}_2)(10\tilde{\eta}_1 - 12.5\tilde{\eta}_2)| < 0.28\|\tilde{\eta}\|^2.\end{aligned}$$

Thus, the inequality (8.14) holds for $0 < \epsilon < 0.72$. Therefore, Assumption 8.5 is satisfied. Thus, by Theorem 8.1, the GRORP for the system (8.17) is solvable. The controller can be explicitly synthesized as follows:

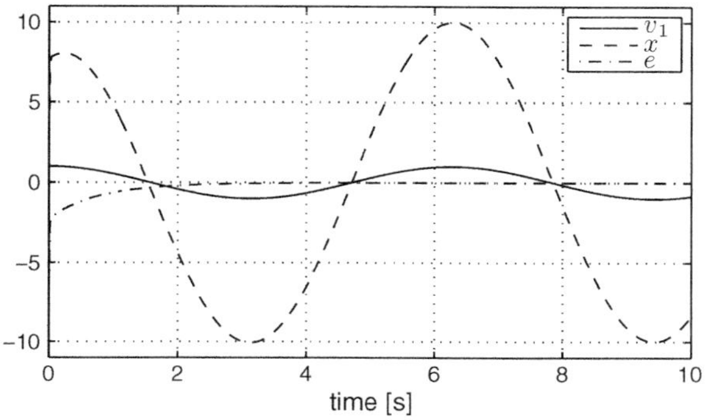

Fig. 8.1 Profile of state trajectories of the closed-loop system in Example 8.3 (part 1)

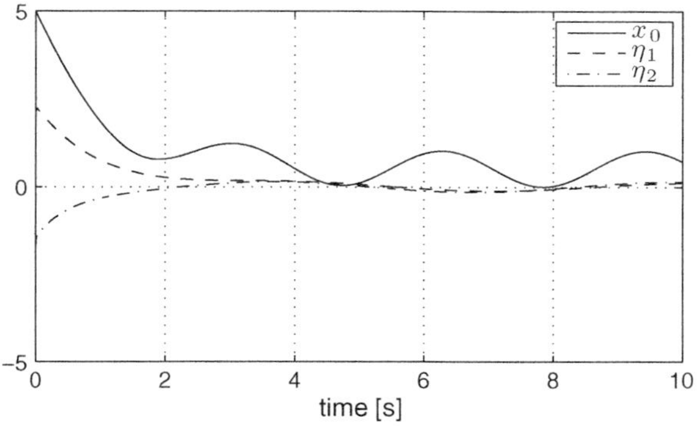

Fig. 8.2 Profile of state trajectories of the closed-loop system in Example 8.3 (part 2)

$$u = \psi(T^{-1}\eta) - 17.3e$$
$$\dot{\eta} = M\eta + N(u - \psi(T^{-1}\eta) + \Psi T^{-1}\eta).$$

The simulation is conducted with $w = 0.8$, $v(0) = [1, 0]^{\mathrm{T}}$, $x_0(0) = 5$, $x(0) = -6$, and $\eta(0) = [2, -2]^{\mathrm{T}}$. The results are shown in Figs. 8.1 and 8.2. It can be seen that the controller achieves $\lim_{t\to\infty} e(t) = 0$ while keeping all state variables of the closed-loop system bounded.

8.2 Output Feedback Systems

In this section, we consider solving, via output feedback control, the global robust output regulation problem for the following class of nonlinear systems

$$\begin{aligned}
\dot{x}_0 &= f_0(x_0, x_1, v(t), w) \\
\dot{x}_i &= f_i(x_0, x_1, v(t), w) + b_i x_{i+1}, \ i = 1, \ldots, r-1 \\
\dot{x}_r &= f_r(x_0, x_1, v(t), w) + b_r u \\
y &= x_1.
\end{aligned} \tag{8.19}$$

Like in Sect. 8.1, the tracking error is

$$e = y - y_{\mathrm{d}}, \ y_{\mathrm{d}} = h_{\mathrm{d}}(v, w), \tag{8.20}$$

and the exosystem is linear and known exactly.

It can be seen that the system (8.19) is of the output feedback form (3.26) with a specific uncertainty $d(t) = \mathrm{col}(v(t), w)$.

As studied in Sect. 3.2, attaching the filter (3.28), i.e.,

$$\dot{\xi} = A\xi + Bu, \ \xi \in \mathbb{R}^r \tag{8.21}$$

to the system (8.19) with $b_i \neq 0, i = 1, \ldots, r$, yields the so-called filter extended system. Performing on the extended system the transformation of the form (3.29) and (3.30), i.e.,

$$\begin{aligned}
\hat{x}_0 &= \mathrm{col}(x_0, \epsilon), \ \epsilon = Dx - \xi, \ x = [x_1, \ldots, x_r]^{\mathrm{T}} \\
\hat{x}_1 &= x_1 \\
\begin{bmatrix} \hat{x}_2 \\ \vdots \\ \hat{x}_r \end{bmatrix} &= \begin{bmatrix} C \\ CA \\ \vdots \\ CA^{r-2} \end{bmatrix} \xi \\
\hat{u} &= CA^{r-1}\xi + u
\end{aligned} \tag{8.22}$$

converts the filter extended system to the form (3.31), i.e.,

$$\begin{aligned}
\dot{\hat{x}}_0 &= \hat{f}_0(\hat{x}_0, \hat{x}_1, v(t), w) \\
\dot{\hat{x}}_1 &= \hat{f}_1(\hat{x}_0, \hat{x}_1, v(t), w) + b\hat{x}_2 \\
\dot{\hat{x}}_i &= \hat{x}_{i+1}, \ i = 2, \ldots, r \\
y &= \hat{x}_1
\end{aligned} \tag{8.23}$$

with $\hat{x}_{r+1} = \hat{u}$. Thus, similarly to what we have already pointed out in Sect. 3.2, if the GRORP for the system (8.23) is solved by a measurement output feedback controller with $y_{\mathrm{m}} = \mathrm{col}(\hat{x}_1, \ldots, \hat{x}_r)$, i.e.,

$$\begin{aligned} \hat{u} &= \kappa_1(\nu, \hat{x}_1, \ldots, \hat{x}_r) \\ \dot{\nu} &= \kappa_2(\nu, \hat{x}_1, \ldots, \hat{x}_r), \end{aligned} \tag{8.24}$$

then the GRORP for the original system (8.19) is solved by an output feedback controller of the following form:

$$\begin{aligned} u &= \kappa_1(\nu, y, C\xi, \ldots, CA^{r-2}\xi) - CA^{r-1}\xi \\ \dot{\nu} &= \kappa_2(\nu, y, C\xi, \ldots, CA^{r-2}\xi) \\ \dot{\xi} &= A\xi + Bu. \end{aligned}$$

To study the GRORP for the system (8.23), we first show that the corresponding regulator equations admit a solution under the same conditions on the original system (8.19). The statement is summarized in the following lemma.

Lemma 8.3 *Under Assumption* 8.2, *and suppose* $b_i \neq 0,\ i = 1, \ldots, r$. *Then,*

(i) The solution to the regulator equations for the system composed of (8.19) *and* (8.20) *exists and is denoted by*

$$col(\mathbf{x}_0(v, w), \mathbf{x}_1(v, w), \ldots, \mathbf{x}_r(v, w), \mathbf{u}(v, w)). \tag{8.25}$$

(ii) The solution to the regulator equations for the system composed of (8.23) *and* (8.20) *also exists and is denoted by*

$$col(\hat{\mathbf{x}}_0(v, w), \hat{\mathbf{x}}_1(v, w), \ldots, \hat{\mathbf{x}}_r(v, w), \hat{\mathbf{u}}(v, w)). \tag{8.26}$$

Proof Part (i). Let

$$\begin{aligned} \mathbf{x}_1(v, w) &= h_{\mathrm{d}}(v, w) \\ \mathbf{x}_{i+1}(v, w) &= b_i^{-1}\left[\frac{\partial \mathbf{x}_i(v, w)}{\partial v} A_1 v - f_i(\mathbf{x}_0(v, w), h_{\mathrm{d}}(v, w), v, w)\right], \quad i = 1, \ldots, r, \end{aligned}$$

and $\mathbf{u}(v, w) = \mathbf{x}_{r+1}(v, w)$. Then it is straightforward to verify that the functions in (8.25) satisfy $\mathbf{x}_i(0, 0) = 0, i = 0, \ldots, r + 1$, and constitute the global solution to the regulator equations for the composite system (8.19) and (8.20).

Part (ii). From the structure of the system (8.23), it suffices to show that there exists a sufficiently smooth function $\hat{\mathbf{x}}_0(v, w)$ with $\hat{\mathbf{x}}_0(0, 0) = 0$ such that

$$\frac{\partial \hat{\mathbf{x}}_0(v, w)}{\partial v} A_1 v = \hat{f}_0(\hat{\mathbf{x}}_0(v, w), h_{\mathrm{d}}(v, w), v, w), \ \forall d \in \mathbb{D}. \tag{8.27}$$

For this purpose, let $\phi_v(t, \nu)$ denote the flow of the exosystem $\dot{v} = A_1 v$ satisfying $\phi_v(0, \nu) = \nu$. Under Assumption 7.3, $\phi_v(t, \nu)$ is bounded over $t \in (-\infty, \infty)$. In particular, if $\nu = v(0)$, one has $\phi_v(t, v(0)) = v(t)$; if $\nu = v(t)$, one has $\phi_v(s, v(t)) = v(s + t)$.

Let

$$\boldsymbol{\xi}(v, w) = \int_{-\infty}^{0} e^{-As} B\mathbf{u}(\phi_v(s, v), w)ds. \tag{8.28}$$

Since A is Hurwitz, and $\mathbf{u}(v, w)$ exists globally and is sufficiently smooth, $\boldsymbol{\xi}(v, w)$ also exists globally and is sufficiently smooth. Clearly, $\boldsymbol{\xi}(0, 0) = 0$. We now verify that the function $\boldsymbol{\xi}(v, w)$ satisfies the following equation

$$\frac{\partial \boldsymbol{\xi}(v, w)}{\partial v} A_1 v = A\boldsymbol{\xi}(v, w) + B\mathbf{u}(v, w), \ \forall d \in \mathbb{D}. \tag{8.29}$$

It suffices to show that $\xi(t) = \boldsymbol{\xi}(v(t), w)$ is the solution to (8.21) with $u = \mathbf{u}(v(t), w)$ from the initial solution $\xi(0) = \boldsymbol{\xi}(v(0), w)$. In fact, the solution can be explicitly calculated as follows.

First we note

$$\begin{aligned}
\xi(0) = \boldsymbol{\xi}(v(0), w) &= \int_{-\infty}^{0} e^{-As} B\mathbf{u}(\phi_v(s, v(0)), w)ds \\
&= \int_{-\infty}^{0} e^{-As} B\mathbf{u}(v(s), w)ds.
\end{aligned}$$

Thus,

$$\begin{aligned}
\xi(t) &= e^{At}\xi(0) + \int_{0}^{t} e^{A(t-s)} B\mathbf{u}(v(s), w)ds \\
&= e^{At} \int_{-\infty}^{0} e^{-As} B\mathbf{u}(v(s), w)ds + \int_{0}^{t} e^{A(t-s)} B\mathbf{u}(v(s), w)ds \\
&= \int_{-\infty}^{t} e^{A(t-s)} B\mathbf{u}(v(s), w)ds \\
&= \int_{-\infty}^{0} e^{-As} B\mathbf{u}(v(s+t), w)ds \\
&= \int_{-\infty}^{0} e^{-As} B\mathbf{u}(\phi_v(s, v(t)), w)ds = \boldsymbol{\xi}(v(t), w).
\end{aligned}$$

From above, the solution to the regulator equations for the composite system (8.19), (8.20), and (8.21), is denoted by

$$\mathrm{col}(\mathbf{x}_0(v, w), \mathbf{x}(v, w), \boldsymbol{\xi}(v, w), \mathbf{u}(v, w))$$

with $\mathbf{x}(v, w) = [\mathbf{x}_1(v, w), \ldots, \mathbf{x}_r(v, w)]^{\mathrm{T}}$. It is known that the system (8.23) is converted from (8.19) and (8.21) through the coordinate transformation (8.22), in particular,

$$\hat{x}_0 = \mathrm{col}(x_0, \epsilon),\ \epsilon = Dx - \xi,\ x = [x_1, \ldots, x_r]^{\mathrm{T}}.$$

Then, there exists a solution to the regulator equations for the composite system (8.23) and (8.20). So, there exists a sufficiently smooth function $\hat{\mathbf{x}}_0(v, w)$ with $\hat{\mathbf{x}}_0(0, 0) = 0$ satisfying (8.27). Specifically, the function $\hat{\mathbf{x}}_0(v, w)$ is

$$\hat{\mathbf{x}}_0(v, w) = \mathrm{col}(\mathbf{x}_0(v, w), \boldsymbol{\epsilon}(v, w)),\ \boldsymbol{\epsilon}(v, w) = D\mathbf{x}(v, w) - \boldsymbol{\xi}(v, w). \qquad \square$$

We now focus on the GRORP for the system (8.23) by the partial state feedback control law (8.24) assuming $b > 0$. In this section, we assume b is known to simplify the controller design. The more general case where b is uncertain can be regarded as a special case of lower triangular systems to be discussed in the next section. To save notation, we consider the following system:

$$\begin{aligned}
\dot{x}_0 &= f_0(x_0, x_1, v(t), w)\\
\dot{x}_1 &= f(x_0, x_1, v(t), w) + bx_2,\ b > 0\\
\dot{x}_i &= -\lambda_{i-1}x_i + x_{i+1},\ i = 2, \ldots, r\\
y &= x_1
\end{aligned} \tag{8.30}$$

where $x_0 \in \mathbb{R}^{n-r}$ and $x_i \in \mathbb{R},\ i = 1, \ldots, r$, are the state variables, $u = x_{r+1}$ is the input, and $\lambda_i,\ i = 2, \ldots, r$, are known positive constants. It is also assumed that all functions in the system (8.30) are sufficiently smooth with $f_0(0, 0, 0, w) = 0$ and $f(0, 0, 0, w) = 0$. All the other notation has the same meaning as that used in (8.1). This system includes (8.23) as a special case by letting $\lambda_i = 0$.

The assumption of the existence of $\hat{\mathbf{x}}_0(v, w)$ satisfying (8.27) becomes Assumption 8.2. We are now ready to consider the GRORP for the system (8.30) under the same assumptions used in the previous section.

Under Assumption 8.2, the solution to the regulator equations for the system composed of (8.30) and (8.20) is

$$\mathrm{col}(\mathbf{x}_0(v, w), \mathbf{x}_1(v, w), \ldots, \mathbf{x}_{r+1}(v, w))$$

where $\mathbf{u}(v, w) = \mathbf{x}_{r+1}(v, w)$ and

$$\mathbf{x}_1(v, w) = h_{\mathrm{d}}(v, w)$$

$$\mathbf{x}_2(v, w) = b^{-1}\left[\frac{\partial h_d(v, w)}{\partial v}A_1 v - f(\mathbf{x}_0(v, w), h_d(v, w), v, w)\right]$$
$$\mathbf{x}_{i+1}(v, w) = \frac{\partial \mathbf{x}_i(v, w)}{\partial v}A_1 v + \lambda_{i-1}\mathbf{x}_i(v, w),\ i = 2, \ldots, r.$$

It is noted that Assumption 8.3 means that the system (8.30) admits a steady-state generator with output $c(x, u) = x_2$. When $r > 1$, we need the system (8.30) to admit a steady-state generator with output $c(x, u) = \mathrm{col}(x_2, \ldots, x_{r+1})$. For this purpose, define the functions $\chi_1, \ldots, \chi_r$ as follows:

$$\chi_1(\theta) = \psi(\theta)$$
$$\chi_i(\theta) = \frac{\partial \chi_{i-1}(\theta)}{\partial \theta}\Phi\theta + \lambda_{i-1}\chi_{i-1}(\theta),\ i = 2, \ldots, r.$$

As a result, we have

$$\mathbf{x}_{i+1}(v, w) = \chi_i(\theta(v, w)),\ i = 1, \ldots, r$$

Thus, Assumption 8.3 also guarantees that the system (8.30) admits a steady-state generator with output $c(x, u) = \mathrm{col}(x_2, \ldots, x_{r+1})$:

$$\dot{\theta}(v, w) = \Phi\theta(v, w), \quad \mathbf{c}(v, w) = \chi(\theta(v, w)), \tag{8.31}$$

where $\chi = \mathrm{col}(\chi_1, \cdots, \chi_r)$.

Remark 8.3 In synthesizing the steady-state generator with output $\mathrm{col}(x_2, \ldots, x_{r+1})$, we have taken the advantage of the fact that the functions $\mathbf{x}_i(v, w)$, $i = 2, \ldots, r + 1$, rely on the same function $\theta(v, w)$. Therefore, the dimension of the steady-state generator with output $\mathrm{col}(x_2, \ldots, x_{r+1})$ is the same as that of the steady-state generator with output x_2.

The internal model corresponding to the steady-state generator (8.31) takes the same form as (8.7) and is repeated as follows:

$$\dot{\eta} = M\eta + N(x_2 - \psi(T^{-1}\eta) + \Psi T^{-1}\eta). \tag{8.32}$$

Attaching the internal model (8.32) to the system (8.30) yields an augmented system of the form (7.54). Performing on the augmented system the following transformation which is the composition of the transformations (7.60) and (8.11)

$$\begin{aligned}
\tilde{\eta} &= \eta - T\theta(v, w) - Nb^{-1}\bar{x}_1\\
\bar{x}_0 &= x_0 - \mathbf{x}_0(v, w)\\
\bar{x}_1 &= x_1 - \mathbf{x}_1(v, w)\\
\bar{x}_i &= x_i - \chi_{i-1}(T^{-1}\eta),\ i = 2, \ldots, r\\
\bar{u} &= u - \chi_r(T^{-1}\eta)
\end{aligned} \tag{8.33}$$

gives the transformed augmented system (7.61) with the following specific form:

$$
\begin{aligned}
\dot{\bar{x}}_0 &= \bar{f}_0(\bar{x}_0, \bar{x}_1, d) \\
\dot{\tilde{\eta}} &= M\tilde{\eta} - N[\delta(\tilde{\eta}+\mu) - \delta(\mu)] + \rho(\bar{x}_0, \bar{x}_1, d) \\
\dot{\bar{x}}_1 &= f_1(\bar{x}_0, \tilde{\eta}, \bar{x}_1, d) + b\bar{x}_2 \\
\dot{\bar{x}}_i &= f_i(\bar{x}_0, \tilde{\eta}, \bar{x}_1, \ldots, \bar{x}_i, d) + \bar{x}_{i+1}, \; i = 2, \ldots, r
\end{aligned} \tag{8.34}
$$

where the quantities $\bar{f}_0$, $f_1 = \bar{f}$, δ, ρ, and μ are defined as in (8.12), and

$$
f_i(\bar{x}_0, \tilde{\eta}, \bar{x}_1, \ldots, \bar{x}_i, d) = -\lambda_{i-1}\bar{x}_i - \left.\frac{\partial \chi_{i-1}(T^{-1}\eta)}{\partial \eta}\right|_{\eta=\tilde{\eta}+T\theta(v,w)+Nb^{-1}\bar{x}_1} N\bar{x}_2.
$$

In deriving the last equation of (8.34), we have made use of the following identity:

$$
-\lambda_{i-1}\chi_{i-1}(T^{-1}\eta) + \chi_i(T^{-1}\eta) - \frac{\partial \chi_{i-1}(T^{-1}\eta)}{\partial \eta}[M\eta + N\Psi T^{-1}\eta] = 0.
$$

Like in the previous section, we can put the system (8.34) in the following compact form

$$
\begin{aligned}
\dot{z} &= q(z, \bar{x}_1, d) \\
\dot{\bar{x}}_1 &= f_1(z, \bar{x}_1, d) + b\bar{x}_2 \\
\dot{\bar{x}}_i &= f_i(z, \bar{x}_1, \ldots, \bar{x}_i, d) + \bar{x}_{i+1}, \; i = 2, \ldots, r
\end{aligned} \tag{8.35}
$$

where $z = \mathrm{col}(\bar{x}_0, \tilde{\eta})$. The global robust stabilization of a system of the form (8.35) is well studied in Chap. 4.

By Lemma 8.2, under Assumptions 8.4 and 8.5, the z-dynamics satisfy Assumption 4.1. Thus, by Theorem 4.2, the GRSP for the system (8.35) is solved by a controller of the form $\bar{u} = \mathcal{S}_r(\bar{x}_1, \ldots, \bar{x}_r)$. Then, by Proposition 7.2, the GRORP for the system (8.30) on $\mathbb{D}$ is solved by a controller of the following form:

$$
\begin{aligned}
u &= \chi_{\mathrm{u}}(T^{-1}\eta) + \mathcal{S}_r(e, x_{\mathrm{m}} - \chi_{\mathrm{x}}(T^{-1}\eta)) \\
\dot{\eta} &= M\eta + N(x_2 - \psi(T^{-1}\eta) + \Psi T^{-1}\eta)
\end{aligned} \tag{8.36}
$$

with $x_{\mathrm{m}} = \mathrm{col}(x_2, \ldots, x_r)$, $\chi_{\mathrm{x}} = \mathrm{col}(\chi_1, \ldots, \chi_{r-1})$, and $\chi_{\mathrm{u}} = \chi_r$. As a result, we have the following result.

Theorem 8.2 *Consider the system* (8.30) *and the exosystem* (7.7). *Under Assumptions* 7.3, 8.2–8.5, *the GRORP is solved on the prescribed compact set* $\mathbb{D}$ *by a controller of the form* (8.36).

Example 8.4 For the following system

$$\begin{aligned}
\dot{x}_1 &= -(0.1x_1)^2 + wx_1 + 10v_2 + x_2 \\
\dot{x}_2 &= (0.2v_2 + 0.5w)x_1 - 5wv_2 + u \\
y &= x_1 \\
e &= y - 10v_1,
\end{aligned} \tag{8.37}$$

where $v = \text{col}(v_1, v_2)$ is governed by the exosystem

$$\dot{v} = A_1 v, \ A_1 = \begin{bmatrix} 0 & 1 \\ -1 & 0 \end{bmatrix}.$$

We will design an output feedback controller to solve the GRORP for $v(t) \in \mathbb{V} = \{v_1^2 + v_2^2 \leq 1\}$ and $-1 \leq w \leq 1$.

Attaching the filter (3.28), i.e.,

$$\begin{aligned}
\dot{\xi}_1 &= -\xi_1 + \xi_2 \\
\dot{\xi}_2 &= -\xi_1 + u
\end{aligned}$$

to the system (8.37) yields the filter extended system. Performing on the extended system the transformation (8.22), i.e.,

$$\begin{aligned}
\hat{x}_0 &= [\hat{x}_{01}, \ \hat{x}_{02}]^{\mathrm{T}} = [x_1 - \xi_1, \ x_2 - \xi_2]^{\mathrm{T}} \\
\hat{x}_1 &= x_1 \\
\hat{x}_2 &= \xi_2 \\
\hat{u} &= -\xi_1 + u
\end{aligned}$$

converts it to the form (8.23), i.e.,

$$\begin{aligned}
\dot{\hat{x}}_{01} &= -\hat{x}_{01} + \hat{x}_{02} - (0.1\hat{x}_1)^2 + w\hat{x}_1 + 10v_2 + \hat{x}_1 \\
\dot{\hat{x}}_{02} &= -\hat{x}_{01} + (0.2v_2 + 0.5w)\hat{x}_1 - 5wv_2 + \hat{x}_1 \\
\dot{\hat{x}}_1 &= -(0.1\hat{x}_1)^2 + w\hat{x}_1 + 10v_2 + \hat{x}_{02} + \hat{x}_2 \\
\dot{\hat{x}}_2 &= \hat{u}.
\end{aligned} \tag{8.38}$$

Now, it suffices to solve the GRORP for the system (8.38) using a partial state (i.e., $(\hat{x}_1, \hat{x}_2)$) feedback controller. To save notation, we put the system (8.38) in the form (8.30), i.e.,

$$\begin{aligned}
\dot{x}_{01} &= -x_{01} + x_{02} - (0.1x_1)^2 + wx_1 + 10v_2 + x_1 \\
\dot{x}_{02} &= -x_{01} + (0.2v_2 + 0.5w)x_1 - 5wv_2 + x_1
\end{aligned}$$

$$\dot{x}_1 = -(0.1x_1)^2 + wx_1 + 10v_2 + x_{02} + x_2$$
$$\dot{x}_2 = u. \tag{8.39}$$

It can be verified that the system (8.39) satisfies Assumptions 8.2 and 8.3. In particular, the solution to the regulator equations exists globally and is given by

$$\begin{aligned}
\mathbf{x}_{01}(v, w) &= (10 + 5w)v_1 - 5wv_2 \\
\mathbf{x}_{02}(v, w) &= v_1^2 \\
\mathbf{x}_1(v, w) &= 10v_1 \\
\mathbf{x}_2(v, w) &= -10wv_1 \\
\mathbf{u}(v, w) &= -10wv_2.
\end{aligned}$$

Let $\pi(v, w) = 10wv_1$. With $\theta(v, w) = \mathcal{L}^2_{A_1 v}\pi(v, w) = [10wv_1 \;\; 10wv_2]^{\mathrm{T}}$, the system has a steady-state generator with output x_2:

$$\dot{\theta}(v, w) = \Phi\theta(v, w), \;\; \mathbf{x}_2(v, w) = \Psi\theta(v, w).$$

Then it can be verified that the following pair

$$\Phi = \begin{bmatrix} 0 & 1 \\ -1 & 0 \end{bmatrix}, \;\; \Psi = \begin{bmatrix} -1 & 0 \end{bmatrix}$$

is observable. Choose a pair of controllable matrices

$$M = \begin{bmatrix} -1 & 0 \\ 0 & -2 \end{bmatrix}, \;\; N = \begin{bmatrix} 0.2 \\ 0.4 \end{bmatrix}.$$

For this pair of matrices, the solution to the Sylvester equation $MT + N\Psi = T\Phi$ is given by

$$T = \begin{bmatrix} -0.1 & 0.1 \\ -0.16 & 0.08 \end{bmatrix},$$

and hence,

$$T^{-1} = \begin{bmatrix} 10 & -12.5 \\ 20 & -12.5 \end{bmatrix}.$$

Now, we are ready to define

$$\begin{aligned}
\chi_1(\theta) &= -\theta_1 \\
\chi_2(\theta) &= -\theta_2
\end{aligned}$$

and hence

$$\chi_1(T^{-1}\theta) = -10\theta_1 + 12.5\theta_2$$
$$\chi_2(T^{-1}\theta) = -20\theta_1 + 12.5\theta_2.$$

Then, there exist a steady-state generator

$$\dot{\theta}(v,w) = \Phi\theta(v,w),\ \mathbf{x}_2(v,w) = \chi_1(\theta(v,w)),\ \mathbf{u}(v,w) = \chi_2(\theta(v,w))$$

with output $\text{col}(x_2, u)$ and a corresponding internal model

$$\dot{\eta} = M\eta + Nx_2. \tag{8.40}$$

Define the following coordinate and input transformation

$$\begin{aligned}\tilde{\eta} &= \eta - T\theta(v,w) - N\bar{x}_1\\ \bar{x}_{01} &= x_{01} - \mathbf{x}_{01}(v,w)\\ \bar{x}_{02} &= x_{02} - \mathbf{x}_{02}(v,w)\\ \bar{x}_1 &= x_1 - \mathbf{x}_1(v,w)\\ \bar{x}_2 &= x_2 - \chi_1(T^{-1}\eta)\\ \bar{u} &= u - \chi_2(T^{-1}\eta).\end{aligned}$$

Performing the coordinate transformation on the system composed of (8.39) and (8.40) gives the following augmented system

$$\begin{aligned}\dot{\bar{x}}_{01} &= -\bar{x}_{01} + \bar{x}_{02} - (0.1\bar{x}_1)^2 + (w + 1 - 0.2v_1)\bar{x}_1\\ \dot{\bar{x}}_{02} &= -\bar{x}_{01} + (0.2v_2 + 0.5w + 1)\bar{x}_1\\ \dot{\tilde{\eta}} &= M\tilde{\eta} + MN\bar{x}_1 - N[-(0.1\bar{x}_1)^2 - 0.2v_1\bar{x}_1 + w\bar{x}_1 + \bar{x}_{02}]\\ \dot{\bar{x}}_1 &= -(0.1\bar{x}_1)^2 - 0.2v_1\bar{x}_1 + w\bar{x}_1 + \Psi T^{-1}N\bar{x}_1 + \bar{x}_{02} + \Psi T^{-1}\tilde{\eta} + \bar{x}_2\\ \dot{\bar{x}}_2 &= -3\bar{x}_2 + \bar{u}.\end{aligned}$$

It can be verified that the system governing $\bar{x}_0 = [\bar{x}_{01},\ \bar{x}_{02}]^{\mathsf{T}}$ satisfies Assumption 8.4. Also Assumption 8.5 is satisfied trivially. Thus, by Theorem 8.2, the GRORP for the system (8.39) is solvable. The controller can be explicitly synthesized as follows:

$$\begin{aligned}u &= \chi_2(T^{-1}\eta) - 130\mathcal{X}_2 - 2\mathcal{X}_2^5\\ \mathcal{X}_2 &= x_2 - \chi_1(T^{-1}\eta) + 17e + 2e^3\\ \dot{\eta} &= M\eta + Nx_2.\end{aligned}$$

The simulation is conducted with $w = 0.8$, $v(0) = [1, 0]^{\mathsf{T}}$, $x_0(0) = [5, 1]^{\mathsf{T}}$, $x_1(0) = -6$, $x_2(0) = 10$, and $\eta(0) = [2, -2]^{\mathsf{T}}$. The results are shown in Figs. 8.3

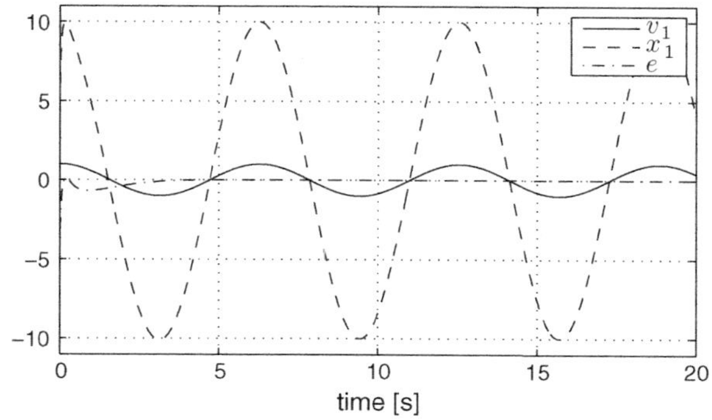

Fig. 8.3 Profile of state trajectories of the closed-loop system in Example 8.4 (part 1)

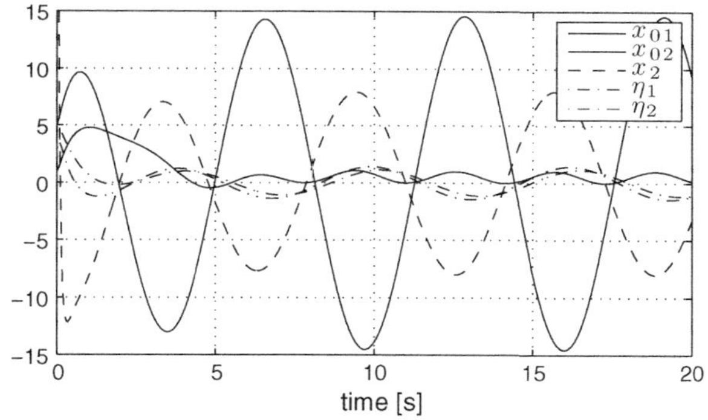

Fig. 8.4 Profile of state trajectories of the closed-loop system in Example 8.4 (part 2)

and 8.4. It can be seen that the controller achieves $\lim_{t\to\infty} e(t) = 0$ while keeping all state variables of the closed-loop system bounded.

8.3 Lower Triangular Systems

In this section, we will consider the GRORP for a class of lower triangular systems of the following form:

$$\begin{aligned}\dot{x}_0 &= f_0(x_0, x_1, v(t), w)\\ \dot{x}_i &= f_i(x_0, \overrightarrow{x}_i, v(t), w) + b_i(w)x_{i+1},\ i = 1, \ldots, r\\ \dot{v} &= A_1 v\\ y &= x_1\end{aligned} \tag{8.41}$$

where $x_0 \in \mathbb{R}^{n-r}$ and $\overrightarrow{x}_i = \text{col}(x_1, \ldots, x_i)$ with $x_i \in \mathbb{R}$, $i = 1, \ldots, r$, are the state variables, and $u := x_{r+1} \in \mathbb{R}$ is the input. It is assumed that all functions in the system (8.41) are sufficiently smooth with $f_0(0, 0, 0, w) = 0$ and $f_i(0, 0, 0, w) = 0$. Like in Sect. 7.2, let the tracking error be

$$e = y - y_d, \ y_d = h_d(v, w) \tag{8.42}$$

and the exosystem satisfy Assumption 7.3. Also, we assume $v(t) \in \mathbb{V}$ and $w \in \mathbb{W}$ for two known compact sets $\mathbb{V}$ and $\mathbb{W}$. So, $d(t) = \text{col}(v(t), w) \in \mathbb{D}$ where $\mathbb{D} = \mathbb{V} \times \mathbb{W}$ has a known boundary. To consider the GRORP for the system (8.41) on $\mathbb{D}$, let us make the following assumptions.

Assumption 8.6 *For $i = 1, \ldots, r$, the function $b_i(w)$ is away from zero, e.g., $b_i(w) > 0$, $\forall w \in \mathbb{W}$.*

Assumption 8.7 *There exists a sufficiently smooth function $\mathbf{x}_0(v, w)$ with $\mathbf{x}_0(0, 0) = 0$ satisfying*

$$\frac{\partial \mathbf{x}_0(v, w)}{\partial v} A_1 v = f_0(\mathbf{x}_0(v, w), h_d(v, w), v, w), \forall d \in \mathbb{D}. \tag{8.43}$$

Under Assumptions 8.6 and 8.7, the solution to the regulator equations of the system (8.41) exists globally and is given as follows:

$$\begin{aligned} \mathbf{x}_1(v, w) &= h_d(v, w) \\ \mathbf{x}_i(v, w) &= b_{i-1}^{-1}(w) \Big(\frac{\partial \mathbf{x}_{i-1}(v, w)}{\partial v} A_1 v \\ &\quad - f_{i-1}(\mathbf{z}(v, w), \mathbf{x}_1(v, w), \ldots, \mathbf{x}_{i-1}(v, w), v, w) \Big), \quad i = 2, \ldots, r+1. \end{aligned}$$

Let $\mathbf{x}(v, w) = \text{col}(\mathbf{x}_1(v, w), \ldots, \mathbf{x}_r(v, w))$ and $\mathbf{u}(v, w) = \mathbf{x}_{r+1}(v, w)$. Then the solution to the regulator equations is given by

$$\text{col}(\mathbf{x}_0(v, w), \mathbf{x}(v, w), \mathbf{u}(v, w)).$$

As in the previous two sections, we need to convert the GRORP of the system (8.41) into the GRSP of an augmented system. For this purpose, we need to make the following assumption which is similar to Assumption 8.3.

Assumption 8.8 *For $i = 1, \ldots, r$, there exist some integers $\ell_i > 0$, and sufficiently smooth functions $\theta_i : \mathbb{R}^l \mapsto \mathbb{R}^{\ell_i}$ and $\psi_i : \mathbb{R}^{\ell_i} \mapsto \mathbb{R}$ vanishing at the origin, and constant matrices $\Phi_i \in \mathbb{R}^{\ell_i \times \ell_i}$, such that*

$$\dot{\theta}_i(v, w) = \Phi_i \theta_i(v, w), \quad \mathbf{x}_{i+1}(v, w) = \psi_i(\theta_i(v, w)). \tag{8.44}$$

Moreover, the pair (Ψ_i, Φ_i) is observable where Ψ_i is the Jacobian matrix of the function ψ_i at the origin.

Let $\theta = \text{col}(\theta_1, \ldots, \theta_r)$, $\Phi = \text{diag}(\Phi_1, \ldots, \Phi_r)$, $\psi(\theta) = \text{col}(\psi_1(\theta_1), \cdots, \psi_r(\theta_r))$, and $c(x, u) = \text{col}(x_2, \ldots, x_r, u)$. Then (8.44) can be put in the following compact form:

$$\dot{\theta}(v, w) = \Phi\theta(v, w), \ \mathbf{c}(v, w) = \psi(\theta(v, w)) \tag{8.45}$$

By Lemma 7.6, the system (8.45) is a steady-state generator of the system (8.41) with output $c(x, u)$.

Remark 8.4 Like Assumption 8.3, sufficient conditions for the satisfaction of Assumption 8.8 was given in Lemma 7.3 in terms of the solution of the regulator equations. In particular, if all components of $\mathbf{x}(v, w)$ and $\mathbf{u}(v, w)$ satisfy Assumption 7.5, then Assumption 8.8 always holds with ψ_i a linear function of θ_i for all $i = 1, \ldots, r$.

By Example 7.9, for each $i = 1, \ldots, r$, and for any controllable pair (M_i, N_i) with $M_i \in \mathbb{R}^{\ell_i \times \ell_i}$ Hurwitz, and $N_i \in \mathbb{R}^{\ell_i \times 1}$, there exists a unique nonsingular matrix T_i satisfying the Sylvester equation $M_i T_i + N_i \Psi_i = T_i \Phi_i$. Thus, corresponding to the steady-state generator (8.45), there exists the following internal model candidate on the mapping $\tau_i = T_i \theta_i$:

$$\dot{\eta}_i = M_i \eta_i + N_i(x_{i+1} - \psi_i(T_i^{-1}\eta_i) + \Psi_i T_i^{-1}\eta_i) \tag{8.46}$$

For convenience, we denote

$$\delta_i(\eta_i) = \psi_i(T_i^{-1}\eta_i) - \Psi_i T_i^{-1}\eta_i$$

which is the nonlinear part of the function $\psi_i(T_i^{-1}\eta)$.

Let $\eta = \text{col}(\eta_1, \ldots, \eta_r)$, and let T, M, N, and Ψ be diagonal matrices with their diagonal elements being T_i, M_i, N_i, and Ψ_i, respectively, for $i = 1, \ldots, r$. Then, Eq. (8.46) can be put in the following compact form:

$$\dot{\eta} = M\eta + N(c(x, u) - \psi(T^{-1}\eta) + \Psi T^{-1}\eta). \tag{8.47}$$

By Lemma 7.6, the system (8.47) is an internal model candidate of (8.41) on the mapping $\tau(\theta) = T\theta$ corresponding to the steady-state generator (8.45).

Next, we apply the coordinate and input transformation according to (7.60) which becomes

$$\begin{aligned}
\bar{\eta}_i &= \eta_i - T_i\theta_i(v, w), \quad i = 1, \ldots, r \\
\bar{x}_0 &= x_0 - \mathbf{x}_0(v, w) \\
\bar{x}_1 &= x_1 - \mathbf{x}_1(v, w) \\
\bar{x}_{i+1} &= x_{i+1} - \psi_i(T_i^{-1}\eta_i), \quad i = 1, \ldots, r
\end{aligned} \tag{8.48}$$

with $\bar{u} = \bar{x}_{r+1}$. Denote $\overrightarrow{\bar{\eta}}_i = \text{col}(\bar{\eta}_1, \ldots, \bar{\eta}_i)$ and $\overrightarrow{\bar{x}}_i = \text{col}(\bar{x}_1, \ldots, \bar{x}_i)$. In particular, $\bar{x} = \overrightarrow{\bar{x}}_r$. The transformation converts the augmented system composed of the original plant (8.41) and the internal model (8.47) into the following form

$$\begin{aligned}
\dot{\bar{x}}_0 &= \hat{f}_0(\bar{x}_0, \bar{x}_1, d) \\
\dot{\bar{\eta}}_i &= \left(M_i + N_i \Psi_i T_i^{-1}\right) \bar{\eta}_i + N_i \bar{x}_{i+1}, \; i = 1, \ldots, r \\
\dot{\bar{x}}_i &= \hat{f}_i\left(\bar{x}_0, \overrightarrow{\bar{\eta}}_i, \overrightarrow{\bar{x}}_i, d\right) + b_i(w)\bar{x}_{i+1}, \; i = 1, \ldots, r
\end{aligned} \tag{8.49}$$

where

$$\begin{aligned}
\hat{f}_0(\bar{x}_0, \bar{x}_1, d) &= f_0(\bar{x}_0 + \mathbf{x}_0(v, w), \bar{x}_1 + \mathbf{x}_1(v, w), v, w) \\
&\quad - f_0(\mathbf{x}_0(v, w), \mathbf{x}_1(v, w), v, w) \\
\hat{f}_1(\bar{x}_0, \bar{\eta}_1, \bar{x}_1, d) &= f_1(\bar{x}_0 + \mathbf{x}_0(v, w), \bar{x}_1 + \mathbf{x}_1(v, w), v, w) \\
&\quad - f_1(\mathbf{x}_0(v, w), \mathbf{x}_1(v, w), v, w) \\
&\quad + b_1(w)\psi_1(T_1^{-1}\bar{\eta}_1 + \theta_1) - b_1(w)\psi_1(\theta_1) \\
\hat{f}_i(\bar{x}_0, \overrightarrow{\bar{\eta}}_i, \overrightarrow{\bar{x}}_i, d) &= f_i(\bar{x}_0 + \mathbf{x}_0(v, w), \bar{x}_1 + \mathbf{x}_1(v, w), \bar{x}_2 + \psi_1(T_1^{-1}\bar{\eta}_1 + \theta_1(v, w)), \\
&\quad \ldots, \bar{x}_i + \psi_{i-1}(T_{i-1}^{-1}\bar{\eta}_{i-1} + \theta_{i-1}(v, w)), v, w) \\
&\quad - f_i(\mathbf{x}_0(v, w), \ldots, \mathbf{x}_i(v, w), v, w) \\
&\quad + b_i(w)\psi_i(T_i^{-1}\bar{\eta}_i + \theta_i) - b_i(w)\psi_i(\theta_i) \\
&\quad - \frac{\partial \psi_{i-1}(\theta_{i-1})}{\partial \theta_{i-1}} T_{i-1}^{-1} \dot{\bar{\eta}}_{i-1} \\
&\quad - \left(\frac{\partial \psi_{i-1}(T_{i-1}^{-1}\eta_{i-1})}{\partial \eta_{i-1}} - \frac{\partial \psi_{i-1}(\theta_{i-1})}{\partial \theta_{i-1}} T_{i-1}^{-1}\right) \dot{\eta}_{i-1}, \\
&\quad i = 2, \ldots, r.
\end{aligned} \tag{8.50}$$

By Proposition 7.2, the GRORP for the system (8.41) is solved if we can make the equilibrium point of system (8.49) at $(\bar{x}_0, \bar{x}, \bar{\eta}) = (0, 0, 0)$ RUGAS for all $d \in \mathbb{D}$. To apply Theorem 4.2 to the system (8.49), we need to convert (8.49) to the standard lower triangular form (4.1). For this purpose, similar to what has been done in Sect. 8.1, we further perform on (8.49) another coordinate transformation

$$\tilde{\eta}_i = \bar{\eta}_i - b_i^{-1}(w) N_i \bar{x}_i, \; i = 1, \ldots, r \tag{8.51}$$

which yields

$$\dot{\tilde{\eta}}_i = M_i \tilde{\eta}_i - N_i \delta_i(\tilde{\eta}_i + \mu_i) + N_i \delta_i(\mu_i) + \rho_i(\bar{x}_0, \overrightarrow{\bar{\eta}}_{i-1}, \overrightarrow{\bar{x}}_i, d)$$

where

$$\begin{aligned}\rho_i(\bar{x}_0, \overrightarrow{\bar{\eta}}_{i-1}, \overrightarrow{\bar{x}}_i, d) = &-N_i\delta_i(b_i^{-1}(w)N_i\bar{x}_i + T_i\theta_i(v, w)) + N_i\delta_i(T_i\theta_i(v, w))\\ &+b_i^{-1}(w)M_iN_i\bar{x}_i - N_i\psi_i(\theta_i(v, w))\\ &-b_i^{-1}(w)N_i\left(\hat{f}_i(\bar{x}_0, \overrightarrow{\bar{\eta}}_i, \overrightarrow{\bar{x}}_i, v, w) - b_i(w)\psi_i(T_i^{-1}\eta_i)\right)\\ \mu_i = &\ b_i^{-1}(w)N_i\bar{x}_i + T_i\theta_i(v, w).\end{aligned}$$

It is important to note that, from the definition,

$$\hat{f}_i(\bar{x}_0, \overrightarrow{\bar{\eta}}_i, \overrightarrow{\bar{x}}_i, v, w) - b_i(w)\psi_i(T_i^{-1}\eta_i)$$

and hence $\rho_i(\bar{x}_0, \overrightarrow{\bar{\eta}}_{i-1}, \overrightarrow{\bar{x}}_i, v, w)$ do not depend on η_i.

Now, let $z_1 = \text{col}(\bar{x}_0, \bar{\eta}_1)$ and $z_i = \bar{\eta}_i$, $i = 2, \ldots, r$. Then, under the transformation (8.51), the system (8.49) can be put into the standard lower triangular form (4.1), i.e.,

$$\begin{aligned}\dot{z}_i &= q_i(\overrightarrow{z}_i, \overrightarrow{\bar{x}}_i, d)\\ \dot{\bar{x}}_i &= \bar{f}_i(\overrightarrow{z}_i, \overrightarrow{\bar{x}}_i, d) + b_i(w)\bar{x}_{i+1}, \ i = 1, \ldots, r.\end{aligned} \tag{8.52}$$

where

$$\begin{aligned}q_1(z_1, \bar{x}_1, d) &= \begin{bmatrix}\hat{f}_0(\bar{x}_0, \bar{x}_1, d)\\ M_1\bar{\eta}_1 - N_1\delta_1(\bar{\eta}_1 + \mu_1) + N_1\delta_1(\mu_1) + \rho_1(\bar{x}_0, \overrightarrow{\bar{x}}_1, d)\end{bmatrix}\\ q_i(\overrightarrow{z}_i, \overrightarrow{\bar{x}}_i, d) &= M_i\bar{\eta}_i - N_i\delta_i(\bar{\eta}_i + \mu_i) + N_i\delta_i(\mu_i)\\ &\quad +\rho_i(\bar{x}_0, \overrightarrow{\bar{\eta}}_{i-1}, \overrightarrow{\bar{x}}_i, d), \ i = 2, \ldots, r\\ \bar{f}_i(\overrightarrow{z}_i, \overrightarrow{\bar{x}}_i, d(t)) &= \hat{f}_i\left(\bar{x}_0, \overrightarrow{\bar{\eta}}_i, \overrightarrow{\bar{x}}_i, d\right), \ i = 1, \ldots, r.\end{aligned}$$

To apply Theorem 4.2 to (8.52), we require all z_i-subsystems in (8.52) to satisfy Assumption 4.3. For this purpose, we need the following two assumptions.

Assumption 8.9 *The subsystem* $\dot{\bar{x}}_0 = \bar{f}_0(\bar{x}_0, \bar{x}_1, d)$ *has an ISS Lyapunov function* $V(\bar{x}_0)$, *i.e.,*

$$V(\bar{x}_0) \sim \{\underline{\alpha}, \bar{\alpha}, \alpha, \sigma \mid \dot{\bar{x}}_0 = \bar{f}_0(\bar{x}_0, \bar{x}_1, d)\}$$

and

$$\limsup_{s\to 0^+} \frac{s^2}{\alpha(s)} < \infty, \quad \limsup_{s\to 0^+} \frac{\sigma(s)}{s^2} < \infty.$$

Assumption 8.10 *Let P_i be a symmetric positive definite matrix such that $P_i M_i + M_i^{\mathrm{T}} P_i = -I$. The function δ_i satisfies*

$$-2\tilde{\eta}^{\mathrm{T}} P_i N_i \left(\delta_i(\tilde{\eta}+\mu) - \delta_i(\mu)\right) \le (1-\epsilon_i)\|\tilde{\eta}\|^2, \quad \forall \tilde{\eta}, \mu \tag{8.53}$$

for a constant $0 < \epsilon_i < 1$.

It can be seen that Assumptions 8.9 and 8.10 are similar to Assumptions 8.4 and 8.5 for output feedback systems, respectively. By Lemmas 8.1 and 8.2, under Assumptions 8.9 and 8.10, for $i = 1, \ldots, r$, the subsystem $\dot{z}_i = q_i(\vec{z}_i, \vec{\bar{x}}_i, d)$ has ISS Lyapunov function $U_i(\zeta)$, i.e.,

$$U_i(\zeta) \sim \{\underline{\alpha}'_i, \bar{\alpha}'_i, \alpha'_i, \sigma'_i \mid \dot{z}_i = q_i(\vec{z}_i, \vec{\bar{x}}_i, d)\}$$

and

$$\limsup_{s\to 0^+} \frac{s^2}{\alpha'_i(s)} < \infty, \quad \limsup_{s\to 0^+} \frac{\sigma'_i(s)}{s^2} < \infty.$$

That is, all z_i subsystems of (8.52) satisfy Assumption 4.3. By Theorem 4.2, the GRSP for the system (8.52) is solved on $\mathbb{D}$ by a static controller $\bar{u} = \mathcal{S}_r(\bar{x}_1, \ldots, \bar{x}_r)$. Then, by Proposition 7.2, the GRORP for the system (8.30) is solved on $\mathbb{D}$ by the controller of the following form:

$$\begin{aligned}
u &= \psi_r(T_r^{-1}\eta_r) + \mathcal{S}_r(e, x_2 - \psi_1(T_1^{-1}\eta_1), \ldots, x_r - \psi_{r-1}(T_{r-1}^{-1}\eta_{r-1})) \\
\dot{\eta} &= M\eta + N(c(x,u) - \psi(T^{-1}\eta) + \Psi T^{-1}\eta).
\end{aligned} \tag{8.54}$$

In summary, we have the following result.

Theorem 8.3 *Consider the composite system* (8.41). *Under Assumptions* 7.3, 8.6–8.9, *the GRORP is solved on the prescribed compact set $\mathbb{D}$ by a controller of the form* (8.54).

Remark 8.5 It can be verified that, under the controller (8.54), the closed-loop system satisfies the following three properties.

(i) There exists a sufficiently smooth function

$$\mathbf{x}_c(v, w) = \mathrm{col}(\mathbf{x}_1(v, w), \ldots, \mathbf{x}_{r+1}(v, w))$$

satisfying the center manifold equation for all $(v, w) \in \mathbb{D}$.

(ii) For all $(v, w) \in \mathbb{D}$, the trajectories $x_c(t) = \text{col}(x_1(t), \ldots, x_r(t), u(t))$ of the closed-loop system starting from any initial state exist and satisfy

$$\|x_c(t) - \mathbf{x}_c(v(t), w)\| \leq \varrho(\|x_c(0) - \mathbf{x}_c(v(0), w)\|, t), \ \forall t \geq 0$$

for some class $\mathcal{KL}$ function ϱ.

(iii) The trajectories described in (ii) also satisfy $\lim_{t\to\infty} e(t) = 0$.

Example 8.5 Consider the following lower triangular system

$$\begin{aligned}
\dot{x}_0 &= -5x_0^3 + w_1 x_0^2 e \\
\dot{x}_1 &= x_0 + 0.1w_1 e + 12.5x_2 \\
\dot{x}_2 &= -0.2v_1 + (0.1x_0 - 0.8w_2v_2)x_1 + \sin^2(w_2v_1x_2) + u \\
y &= x_1 \\
e &= y - 10v_1
\end{aligned} \tag{8.55}$$

with the exosystem

$$\begin{bmatrix} \dot{v}_1 \\ \dot{v}_2 \end{bmatrix} = A_1 \begin{bmatrix} v_1 \\ v_2 \end{bmatrix}, \ A_1 = \begin{bmatrix} 0 & -0.5 \\ 0.5 & 0 \end{bmatrix}. \tag{8.56}$$

These equations formulate the control problem of designing a partial state feedback regulator to make the output y of the system (8.55) asymptotically track a sinusoidal signal of frequency 0.5 with arbitrarily amplitude in the presence of two uncertain parameters w_1 and w_2. Let $\mathbb{V} = \{v_1^2 + v_2^2 \leq 1\}$ and $\mathbb{W} = \{-1 \leq w_i \leq 1, \ i = 1, 2\}$. It can be verified that this system satisfies Assumptions 8.6–8.8. In particular, the regulator equations associated with this system have a globally defined solution as follows:

$$\begin{aligned}
\mathbf{x}_0\,(v, w) &= 0 \\
\mathbf{x}_1\,(v, w) &= 10v_1 \\
\mathbf{x}_2\,(v, w) &= -0.4v_2 \\
\mathbf{u}\,(v, w) &= 8w_2v_1v_2 - \sin^2(0.4w_2v_1v_2).
\end{aligned} \tag{8.57}$$

Let $c(x, u) = \text{col}(x_2, u)$, $\pi_1(v, w) = -0.4v_2$, and $\pi_2(v, w) = 8w_2v_1v_2$. Then, the minimal zeroing polynomials of $\pi_1^1(v, w)$ and $\pi_2^1(v, w)$ are $4\lambda^2 + 1$ and $\lambda^2 + 1$, respectively. There exist two functions ψ_1 and ψ_2 such that

$$\begin{aligned}
\psi_1(\mathcal{L}^2_{A_1 v}\pi_1(v, w)) &= \pi_1(v, w) \\
\psi_2(\mathcal{L}^2_{A_1 v}\pi_2(v, w)) &= \pi_2(v, w) - \sin^2(0.05\pi_2(v, w)).
\end{aligned}$$

Then, with $\theta_i(v, w) = \mathcal{L}^2_{A_1 v}\pi_i(v, w)$, the system has a steady-state generator with output $[x_2, u]^{\mathsf{T}}$, respectively, as follows:

$$\begin{aligned}\dot{\theta}_1(v,w) &= \Phi_1\theta_1(v,w),\ \mathbf{x}_2(v,w) = \psi_1(\theta(v,w))\\ \dot{\theta}_2(v,w) &= \Phi_2\theta_2(v,w),\ \mathbf{u}(v,w) = \psi_2(\theta(v,w))\end{aligned}$$

Let Ψ_i be the Jacobian matrix of ψ_i at the origin. Then, it can be verified that the following pairs of matrices

$$\Phi_1 = \begin{bmatrix} 0 & 1 \\ -0.25 & 0 \end{bmatrix},\ \Phi_2 = \begin{bmatrix} 0 & 1 \\ -1 & 0 \end{bmatrix},\ \Psi_1 = \Psi_2 = [1\ 0]$$

are observable. Choose two pairs of controllable matrices as follows:

$$M_1 = M_2 = \begin{bmatrix} -2 & 0 \\ 0 & -1 \end{bmatrix},\ N_1 = N_2 = \begin{bmatrix} 2 \\ 1 \end{bmatrix}.$$

Solving the pertinent Sylvester equations gives

$$T_1 = \begin{bmatrix} 0.9412 & -0.4706 \\ 0.8 & -0.8 \end{bmatrix} \text{ and } T_2 = \begin{bmatrix} 0.8 & -0.4 \\ 0.5 & -0.5 \end{bmatrix}.$$

Thus,

$$\begin{aligned}\psi_1(T_1^{-1}\theta_1) &= \begin{bmatrix} 2.125 & -1.25 \end{bmatrix}\theta_1\\ \psi_2(T_2^{-1}\theta_2) &= \begin{bmatrix} 2.5 & -2 \end{bmatrix}\theta_2 - \sin^2\left(\begin{bmatrix} 0.125 & -0.1 \end{bmatrix}\theta_2\right).\end{aligned}$$

Then

$$\begin{aligned}\delta_1(\theta_1) &= 0\\ \delta_2(\theta_2) &= -\sin^2\left(\begin{bmatrix} 0.125 & -0.1 \end{bmatrix}\theta_2\right).\end{aligned} \tag{8.58}$$

Now, the internal model can be constructed as follows,

$$\begin{aligned}\dot{\eta}_1 &= M_1\eta_1 + N_1 x_2\\ \dot{\eta}_2 &= M_2\eta_2 + N_2\left(u + \sin^2\left(\begin{bmatrix} 0.125 & -0.1 \end{bmatrix}\eta_2\right)\right).\end{aligned} \tag{8.59}$$

The following coordinate and input transformation

$$\begin{aligned}\bar{\eta}_i &= \eta_i - T_i\theta_i(v,w),\ i = 1,2\\ \bar{x}_0 &= x_0 - \mathbf{x}_0(v,w)\\ \bar{x}_1 &= x_1 - \mathbf{x}_1(v,w) = e\\ \bar{x}_2 &= x_2 - \psi_1(T_1^{-1}\eta_1)\\ \bar{u} &= u - \psi_2(T_2^{-1}\eta_2)\end{aligned}$$

and

$$z_1 = \begin{bmatrix} \bar{x}_0 \\ \tilde{\eta}_1 \end{bmatrix} = \begin{bmatrix} \bar{x}_0 \\ \bar{\eta}_1 - 0.08N_1\bar{x}_1 \end{bmatrix}$$
$$z_2 = \tilde{\eta}_2 = \bar{\eta}_2 - N_2\bar{x}_2$$

puts the augmented system (8.55) and (8.59) into the following form:

$$\begin{aligned}
\dot{z}_1 &= \begin{bmatrix} -5\bar{x}_0^3 + w_1\bar{x}_0^2\bar{x}_1 \\ M_1\tilde{\eta}_1 + 0.08M_1N_1\bar{x}_1 - 0.08N_1(\bar{x}_0 + 0.1w_1\bar{x}_1) \end{bmatrix} \\
\dot{\bar{x}}_1 &= \bar{x}_0 + 0.1w_1\bar{x}_1 + 12.5\psi_1(T_1^{-1}(\tilde{\eta}_1 + 0.08N_1\bar{x}_1)) + 12.5\bar{x}_2 \\
\dot{z}_2 &= M_2z_2 - N_2\left(\delta_2(z_2 + N_2\bar{x}_2 + T_2\theta_2) - \delta_2(N_2\bar{x}_2 + T_2\theta_2)\right) + M_2N_2\bar{x}_2 \\
&\quad -N_2a(z_1, \bar{x}_1, \bar{x}_2, v, w) - N_2\left(\delta_2(N_2\bar{x}_2 + T_2\theta_2) - \delta_2(T_2\theta_2)\right) \\
\dot{\bar{x}}_2 &= a(z_1, \bar{x}_1, \bar{x}_2, v, w) + \psi_2(T_2^{-1}(z_2 + N_2\bar{x}_2 + \theta_2)) \\
&\quad -\psi_2(T_2^{-1}\theta_2) + \bar{u},
\end{aligned} \tag{8.60}$$

where

$$\begin{aligned}
&a(z_1, \bar{x}_1, \bar{x}_2, v, w) \\
&= 0.1\bar{x}_0(\bar{x}_1 + 10v_1) + \sin^2\big(w_2v_1(\bar{x}_2 + \psi_1(T_1^{-1}\bar{\eta}_1) - 0.4v_2)\big) \\
&\quad - \sin^2(0.4w_2v_1v_2) - 0.8w_2v_2\bar{x}_1 - [2.125 \quad - 1.25]((M_1 + N_1\Psi T^{-1})\bar{\eta}_1 \\
&\quad +N_1\bar{x}_2).
\end{aligned}$$

Next, we will verify that all the assumptions needed for applying Theorem 8.3 are satisfied. Let $V(\bar{x}_0) = \bar{x}_0^2/2$. Then the derivative of $V(\bar{x}_0)$ along the first equation of (8.60) is

$$\dot{V}(\bar{x}_0) = \bar{x}_0(-5\bar{x}_0^3 + w_1\bar{x}_0^2\bar{x}_1) \leq -5\bar{x}_0^4 + |\bar{x}_0|^3|\bar{x}_1| \leq -4\bar{x}_0^4 + \bar{x}_1^4$$

which verifies Assumption 8.9.

When $i = 1$, Assumption 8.10 is trivially satisfied by noting $\delta_1(\theta_1) = 0$. When $i = 2$, solving the Lyapunov equation $P_2M_2 + M_2^{\mathrm{T}}P_2 = -I$ gives

$$P_2 = \begin{bmatrix} 0.25 & 0 \\ 0 & 0.5 \end{bmatrix}.$$

Then, the inequality (8.53) becomes

$$-2\tilde{\eta}^{\mathrm{T}}\begin{bmatrix} 0.25 & 0 \\ 0 & 0.5 \end{bmatrix}\begin{bmatrix} 2 \\ 1 \end{bmatrix}(\delta_2(\tilde{\eta} + \mu) - \delta_2(\mu)) \leq (1 - \epsilon_2)\|\tilde{\eta}\|^2. \tag{8.61}$$

Letting $\tilde{\eta} = \mathrm{col}(\tilde{\eta}_1, \tilde{\eta}_2)$ puts (8.61) to

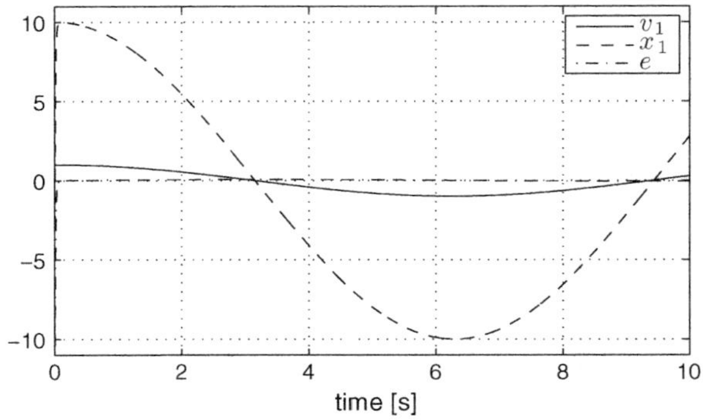

Fig. 8.5 Profile of state trajectories of the closed-loop system in Example 8.5 (part 1)

$$
\begin{aligned}
&(\tilde{\eta}_1+\tilde{\eta}_2)\left(\sin^2([0.125 \quad -0.1](\tilde{\eta}+\mu))-\sin^2([0.125 \quad -0.1]\mu)\right) \\
&\quad \le (1-\epsilon_2)(\tilde{\eta}_1^2+\tilde{\eta}_2^2).
\end{aligned} \tag{8.62}
$$

Simple manipulation shows that (8.62), hence the inequality (8.61), holds for $0 < \epsilon_2 < 0.773$.

By Theorem 8.3, the GRORP of the system (8.60) is solvable. In fact, using the procedure described in Theorem 4.2 shows that, for the given $\mathbb{V}$ and $\mathbb{W}$, the following controller

$$
\begin{aligned}
\bar{u} &= -(k_2\mathcal{X}_2^2+k_3)^2\mathcal{X}_2 \\
\mathcal{X}_2 &= \bar{x}_2+k_1\bar{x}_1
\end{aligned}
$$

where $k_1 = 6.5$, $k_2 = 0.3$, and $k_3 = 50$ globally robustly stabilizes the system (8.60). Therefore, the overall controller for solving the GRORP for the system (8.55) is given by the composition of the internal model (8.59) and

$$
\begin{aligned}
u &= -(k_2\mathcal{X}_2^2+k_3)^2\mathcal{X}_2+\psi_2(T_2^{-1}\eta_2) \\
\mathcal{X}_2 &= x_2+k_1e-\psi_1(T_1^{-1}\eta_1).
\end{aligned}
$$

The simulation is conducted with $w = 0.8$, $v(0) = [1, 0]^{\mathsf{T}}$, $z(0) = 10$, $x(0) = [-6, 2]^{\mathsf{T}}$, $\eta_1(0) = [1, 2]^{\mathsf{T}}$, and $\eta_2(0) = [1, 3]^{\mathsf{T}}$. The results are shown in Figs. 8.5–8.7. It can be seen that the controller achieves $\lim_{t\to\infty} e(t) = 0$ while keeping all state variables of the closed-loop system bounded.

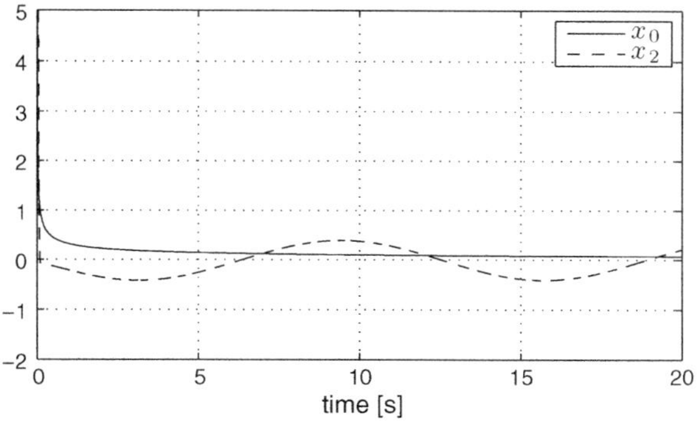

Fig. 8.6 Profile of state trajectories of the closed-loop system in Example 8.5 (part 2)

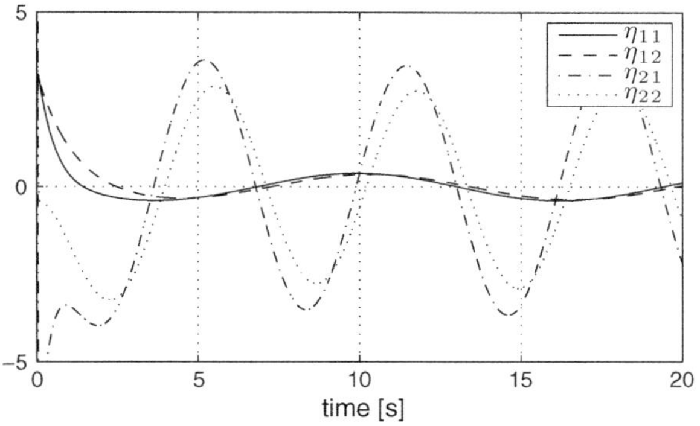

Fig. 8.7 Profile of state trajectories of the closed-loop system in Example 8.5 (part 3)

8.4 Nonlinear Exosystems

The internal model used in the previous sections relies on Assumption 7.3. Thus it can only handle linear exosystems. To handle nonlinear exosystems of the form (7.2) with $\sigma \in \mathbb{R}^0$ satisfying Assumption 7.1, we will consider the internal model of the form (7.51) which corresponds to the steady-state generator of the form (7.41). Since using different internal models only leads to slightly different augmented systems, we will only focus on the system with relative degree one, i.e., the system (8.1), in this section.

As derived in Sect. 8.1, under Assumptions 8.1 and 8.2, the solution to the regulator equations is

$$\mathrm{col}(\mathbf{x}_0(v, w), \mathbf{x}(v, w), \mathbf{u}(v, w)).$$

Now, instead of Assumption 8.3, we assume the following:

Assumption 8.11 *The system composed of* (8.19) *and* (7.2) *admits a steady-state generator with output* x_2 *of the following form:*

$$\dot{\theta}(v, w) = \Phi(v)\theta(v, w), \ \mathbf{u}(v, w) = \Psi\theta(v, w) \tag{8.63}$$

where, for some integer $\ell > 0$, $\theta : \mathbb{R}^l \mapsto \mathbb{R}^\ell$ *is some sufficiently smooth function vanishing at the origin,* $\Phi(v) \in \mathbb{R}^{\ell\times\ell}$ *is a matrix whose elements are sufficiently smooth functions of* v, *and* Ψ *is a constant column vector of dimension* ℓ. *Moreover, the pair* $(\Psi, \Phi(v))$ *is observable for all* v.

Remark 8.6 By Lemma 7.5, Assumption 8.11 is satisfied if there exist an integer ℓ and sufficiently smooth scalar functions $\varphi_i(v)$, $i = 1, \ldots, \ell$, such that $\mathbf{u}(v, w)$ satisfies

$$\begin{aligned} L_a^\ell \mathbf{u}(v, w) - \varphi_1(v)\mathbf{u}(v, w) - L_a[\varphi_2(v)\mathbf{u}(v, w)] - \cdots \\ - L_a^{\ell-1}[\varphi_\ell(v)\mathbf{u}(v, w)] = 0, \ \forall v \in \mathbb{R}^{l_1}, w \in \mathbb{R}^{l_2} \end{aligned} \tag{8.64}$$

or, equivalently, if there exist an integer ℓ and sufficiently smooth scalar functions $\phi_i(v)$, $i = 1, \ldots, \ell$, such that $\mathbf{u}(v, w)$ satisfies

$$\begin{aligned} L_a^\ell \mathbf{u}(v, w) - \phi_1(v)\mathbf{u}(v, w) - \phi_2(v)L_a\mathbf{u}(v, w) - \cdots \\ - \phi_\ell(v)L_a^{\ell-1}\mathbf{u}(v, w) = 0, \ \forall v \in \mathbb{R}^{l_1}, w \in \mathbb{R}^{l_2}. \end{aligned} \tag{8.65}$$

Note that we can always write $\Phi(v)$ as $\Phi(v) = \Phi + \bar{\varphi}(v)\Psi$ with $\bar{\varphi}(0) = 0$. From Example 7.10, under Assumption 8.11, for any controllable pair (M, N) where $M \in \mathbb{R}^{\ell\times\ell}$ is Hurwitz, and $N \in \mathbb{R}^{\ell\times 1}$, there is a unique nonsingular matrix T satisfying the Sylvester equation $T\Phi - MT = N\Psi$. Then an internal model candidate on the mapping $\tau(\theta) = T\theta$ associated with the steady-state generator (8.63) is as follows:

$$\dot{\eta} = M\eta + N(v)u, \ N(v) = N + T\bar{\varphi}(v) \tag{8.66}$$

Attaching the internal model (8.66) to the given plant yields the augmented system. Performing on the augmented system the following coordinate and input transformation:

$$\begin{aligned} \tilde{\eta} &= \eta - T\theta(v, w) - N(v)b^{-1}(w)\bar{x} \\ \bar{x}_0 &= x_0 - \mathbf{x}_0(v, w) \\ \bar{x} &= x - \mathbf{x}(v, w) \\ \bar{u} &= u - \Psi T^{-1}\eta \end{aligned} \tag{8.67}$$

gives

$$\begin{aligned}
\dot{\bar{x}}_0 &= \bar{f}_0(\bar{x}_0, \bar{x}, d) \\
\dot{\tilde{\eta}} &= M\tilde{\eta} + \rho(\bar{x}_0, \bar{x}, d) \\
\dot{\bar{x}} &= \bar{f}(\bar{x}_0, \tilde{\eta}, \bar{x}, d) + b(w)\bar{u}
\end{aligned} \tag{8.68}$$

where

$$\begin{aligned}
\rho(\bar{x}_0, \bar{x}, d) &= b^{-1}(w)MN(v)\bar{x} - N(v)b^{-1}(w) \\
&\quad [f(\bar{x}_0 + \mathbf{x}_0(v,w), \bar{x} + \mathbf{x}(v,w), v, w) \\
&\quad - f(\mathbf{x}_0(v,w), \mathbf{x}(v,w), v, w)] \\
&\quad - \frac{\partial N(v)}{\partial v} a(v) b^{-1}(w)\bar{x} \\
\bar{f}_0(\bar{x}_0, \bar{x}, d) &= f_0(\bar{x}_0 + \mathbf{x}_0(v,w), \bar{x} + \mathbf{x}(v,w), v, w) \\
&\quad - f_0(\mathbf{x}_0(v,w), \mathbf{x}(v,w), v, w) \\
\bar{f}(\bar{x}_0, \tilde{\eta}, \bar{x}, d) &= f(\bar{x}_0 + \mathbf{x}_0(v,w), \bar{x} + \mathbf{x}(v,w), v, w) \\
&\quad - f(\mathbf{x}_0(v,w), \mathbf{x}(v,w), v, w) \\
&\quad + b(w)\Psi T^{-1}(\tilde{\eta} + N(v)b^{-1}(w)\bar{x}).
\end{aligned}$$

The system (8.68) is in the same form as the system (8.12). Moreover, by Remark 8.2, the $\tilde{\eta}$ subsystem satisfies Assumption 8.5 automatically since the function δ in (8.8) is identically equal to zero in the current case. Thus, we can directly obtain a result similar to Theorem 8.1 as follows.

Theorem 8.4 *Consider the composite system* (8.1). *Under Assumptions* 7.1, 8.1, 8.2, 8.4, *and* 8.11, *the GRORP is solved on the prescribed compact set* $\mathbb{D}$ *by a controller of the form*

$$\begin{aligned}
u &= \Psi T^{-1}\eta + s(e) \\
\dot{\eta} &= M\eta + N(v)u.
\end{aligned} \tag{8.69}$$

Example 8.6 Consider the following system

$$\begin{aligned}
\dot{x}_0 &= -x_0 + 0.2v_2 x + v_1^2 \\
\dot{x} &= x_0 - 0.1v_1 x - \sin^2(0.1we) + 10wv_1 + 10v_2 + 10u \\
y &= x \\
e &= y - 10v_1
\end{aligned} \tag{8.70}$$

where $v = [v_1 \ v_2]^\top$ is governed by the following van der Pol oscillator

$$\begin{aligned}\dot{v}_1 &= v_2\\ \dot{v}_2 &= -av_1 + b(1 - v_1^2)v_2.\end{aligned}$$

It is well known that, for all $a > 0$ and $b > 0$, the van der Pol oscillator produces a globally asymptotically stable limit cycle.

The system (8.70) is similar to (8.17) with $\sin^2(0.1wx_1)$ replaced by $\sin^2(0.1we)$. As a result, like (8.17), the system (8.70) satisfies Assumptions 8.1, 8.2, and 8.4. In particular, a polynomial solution to the regulator equations exists globally and is given by

$$\begin{aligned}\mathbf{x}_0(v, w) &= v_1^2\\ \mathbf{x}(v, w) &= 10v_1\\ \mathbf{u}(v, w) &= -wv_1.\end{aligned}$$

Next, we verify that Assumption 8.11 is satisfied by noting

$$L_a^2\mathbf{u}(v, w) - (-a - bv_1v_2)\mathbf{u}(v, w) - bL_a\mathbf{u}(v, w) = 0. \tag{8.71}$$

Then, there exists a steady-state generator with output u with $\theta(v, w) = [-wv_1, -wv_2]^\top$ and

$$\Phi(v) = \begin{bmatrix} b & 1\\ -a - bv_1v_2 & 0\end{bmatrix}, \quad \Psi = \begin{bmatrix} 1 & 0\end{bmatrix}.$$

A corresponding internal model is given by (8.66) with

$$M = \begin{bmatrix} -2 & 1\\ -1 & 0\end{bmatrix}, \quad N(v) = \begin{bmatrix} b + 2\\ 1 - a - bv_1v_2\end{bmatrix}, \quad T = I.$$

By Theorem 8.4, the GRORP is solved by the following controller

$$\begin{aligned}u &= \Psi\eta - ke\\ \dot{\eta} &= M\eta + N(v)x_2\end{aligned}$$

for a sufficiently large k.

With $a = 1$ and $b = 2$, the phase portrait of the van der Pol oscillator is shown in Fig. 8.8. The performance of the controller is simulated with $w = 1$, $k = 6.5$, $v(0) = [1, -3]^\top$, $x_0(0) = 10$, $x(0) = -6$, and $\eta(0) = [1, 2]^\top$. The results are shown in Figs. 8.9 and 8.10. It can be seen that the controller achieves $\lim_{t\to\infty} e(t) = 0$ while keeping all state variables of the closed-loop system bounded.

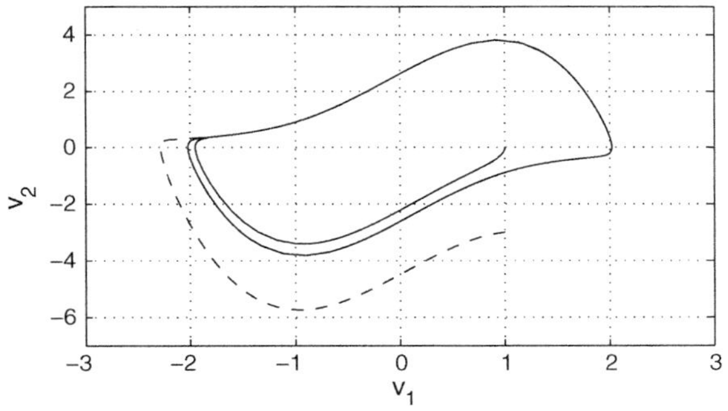

Fig. 8.8 Profile of van der Pol oscillator in Example 8.6

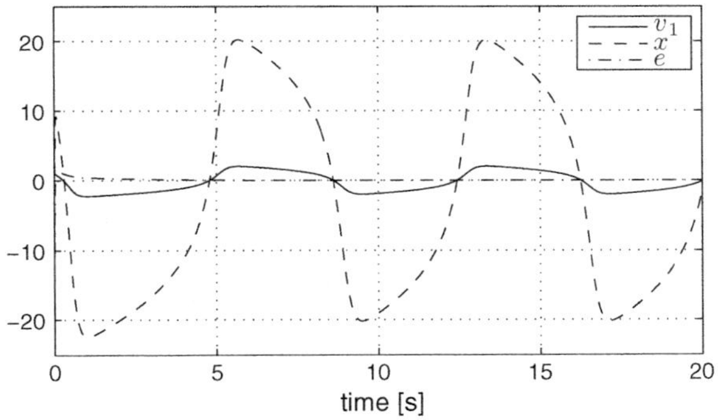

Fig. 8.9 Profile of state trajectories of the closed-loop system in Example 8.6 (part 1)

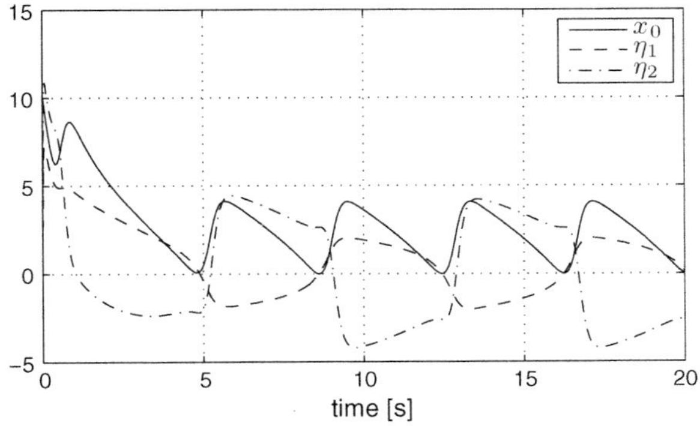

Fig. 8.10 Profile of state trajectories of the closed-loop system in Example 8.6 (part 2)

8.5 Uncertainties with Unknown Boundary

In the previous sections, we have studied the GRORP for various nonlinear systems on a compact $\mathbb{D} = \mathbb{V} \times \mathbb{W}$ with a known boundary. In this section, we will further consider the solution of GRORP for the same systems without the assumption that $\mathbb{V}$ and $\mathbb{W}$ are known. In other words, we allow GRORP to be solved for any unknown parameter w, and any initial state v_0 of the exosystem.

8.5.1 Systems with Relative Degree One

Consider the system (8.1). Under Assumptions 7.3, 8.1–8.3, by performing the same manipulation as in Sect. 7.1, we can obtain the system (8.13). What distinguishes the current task from previous sections is that the compact set $\mathbb{D}$ is unknown. For this purpose, we need to modify Assumption 8.4 to the following.

Assumption 8.12 *The subsystem $\dot{\bar{x}}_0 = \bar{f}_0(\bar{x}_0, \bar{x}, d)$ has an ISS Lyapunov function $V(\bar{x}_0)$, i.e.,*

$$V(\bar{x}_0) \sim \{\underline{\alpha}, \bar{\alpha}, \alpha, \hat{\sigma} \mid \dot{\bar{x}}_0 = \bar{f}_0(\bar{x}_0, \bar{x}, d)\}$$

and

$$\limsup_{s\to 0^+} \frac{s^2}{\alpha(s)} < \infty, \quad \limsup_{s\to 0^+} \frac{\hat{\sigma}(s)}{s^2} < \infty.$$

Moreover, the functions $\underline{\alpha}$, $\bar{\alpha}$, and α are known and the function $\hat{\sigma}$ is known up to a constant factor in the sense that there exist an unknown constant p and a known function σ such that $\hat{\sigma} = p\sigma$.

Now, we are ready to state the following result.

Theorem 8.5 *Consider the composite system* (8.1) *with any unknown compact set $\mathbb{D}$. Under Assumptions* 7.3, 8.1–8.3, 8.5, *and* 8.12, *the GRORP is solved by the following controller:*

$$\begin{aligned} u &= \psi(T^{-1}\eta) - k\rho(e)e \\ \dot{\eta} &= M\eta + N(x_2 - \psi(T^{-1}\eta) + \Psi T^{-1}\eta) \\ \dot{k} &= \lambda\rho(e)e^2, \quad \lambda > 0. \end{aligned} \tag{8.72}$$

Proof It suffices to show the GASP for the system (8.13) is solved under Assumptions 8.5 and 8.12. Since the system (8.13) is in the form of (2.55), we can apply Theorem 2.9 to the system (8.13). Therefore, we will verify that the system (8.13) satisfies Assumption 2.5 under Assumptions 8.5 and 8.12. This is indeed the case by using Lemmas 8.1 and 8.2 with slight modification. Therefore, by Theorem 2.9, the GASP for the system (8.13) is solved by the following controller

$$\begin{aligned}\bar{u} &= -k\rho(\bar{x})\bar{x}\\ \dot{k} &= \lambda\rho(\bar{x})\bar{x}^2, \lambda > 0.\end{aligned}$$

Thus, the GRORP for the system (8.1) is solved by the controller (8.72). □

8.5.2 Output Feedback Systems

Consider the system (8.30) with $\mathbb{D}$ any compact set. Under Assumptions 7.3, 8.2, and 8.3, by conducting the same manipulation as in Sect. 8.2, we can obtain the system (8.35). Then we only need to solve the GASP for the system (8.35) which is of the form (6.12) with $z_2, \ldots, z_r \in \mathbb{R}^0$, $b_1(d) = b$, and $b_i(d) = 1$, $i = 2, \ldots, r$. For this purpose, we need Assumptions 8.5 and 8.12. The main result is summarized as follows.

Theorem 8.6 *Consider the system* (8.30) *with any unknown compact set* $\mathbb{D}$ *and the exosystem* (7.7). *Under Assumptions* 7.3, 8.2, 8.3, 8.5, *and* 8.12, *the GRORP is solved by the following controller*

$$\begin{aligned}u &= \chi_{\mathrm{u}}(T^{-1}\eta) + \mathcal{S}_r(k, e, x_{\mathrm{m}} - \chi_{\mathrm{x}}(T^{-1}\eta))\\ \dot{\eta} &= M\eta + N(u - \psi(T^{-1}\eta) + \Psi T^{-1}\eta)\\ \dot{k} &= \tau(k, e, x_{\mathrm{m}} - \chi_{\mathrm{x}}(T^{-1}\eta)), \quad k = [k_1, \ldots, k_r]^{\mathrm{T}}, \ k_i \in \mathbb{R}, \ i = 1, \ldots, r, \quad (8.73)\end{aligned}$$

with $x_{\mathrm{m}} = col(x_2, \ldots, x_r)$, $\chi_{\mathrm{x}} = col(\chi_1, \ldots, \chi_{r-1})$, *and* $\chi_{\mathrm{u}} = \chi_r$.

Proof It suffices to show the GASP for the system (8.35) is solved under Assumptions 8.5 and 8.12. Since the system (8.35) is in the form of (6.12), we can apply Theorem 6.2 to (8.35). Therefore, we will verify that (8.35) satisfies Assumption 6.4 under Assumptions 8.5 and 8.12. Since $z_2, \ldots, z_r \in \mathbb{R}^0$, we only need to show that Assumption 6.4 is satisfied for $i = 1$. This is true by using Lemmas 8.1 and 8.2 with slight modification. Therefore, by Theorem 6.2, the GASP for the system (8.35) is solved by the following controller:

$$\begin{aligned}\bar{u} &= \mathcal{S}_r(k_1, \ldots, k_r, \bar{x}_1, \ldots, \bar{x}_r)\\ \dot{k}_i &= \tau_i(k_1, \ldots, k_{i-1}, \bar{x}_1, \ldots, \bar{x}_i) \quad i = 1, \ldots, r.\end{aligned}$$

Define

$$\tau(k_1,\ldots,k_r,\bar{x}_1,\ldots,\bar{x}_r) = \begin{bmatrix} \tau_1(\bar{x}_1) \\ \vdots \\ \tau_r(k_1,\ldots,k_{r-1},\bar{x}_1,\ldots,\bar{x}_r) \end{bmatrix}. \tag{8.74}$$

Then, the GRORP for the system (8.30) is solved by the controller (8.73). □

8.5.3 Lower Triangular Systems

Consider the system (8.41) with $\mathbb{D}$ an arbitrarily unknown compact set. Under Assumptions 7.3, 8.6–8.8, by conducting the same manipulation as in Sect. 8.3, we can obtain the system (8.52) which is in the form (6.12). To solve the GASP for the system (8.52), we need Assumption 8.10, and a modified version of Assumption 8.9 as follows.

Assumption 8.13 *The subsystem* $\dot{\bar{x}}_0 = \bar{f}_0(\bar{x}_0, \bar{x}_1, d)$ *has an ISS Lyapunov function* $V(\bar{x}_0)$, *i.e.,*

$$V(\bar{x}_0) \sim \{\underline{\alpha}, \bar{\alpha}, \alpha, \hat{\sigma} \mid \dot{\bar{x}}_0 = \bar{f}_0(\bar{x}_0, \bar{x}_1, d)\}$$

and

$$\limsup_{s\to 0^+} \frac{s^2}{\alpha(s)} < \infty, \quad \limsup_{s\to 0^+} \frac{\hat{\sigma}(s)}{s^2} < \infty.$$

Moreover, the functions $\underline{\alpha}$, $\bar{\alpha}$, *and* α *are known and the function* $\hat{\sigma}$ *is known up to a constant factor in the sense that there exist an unknown constant* p *and a known function* σ *such that* $\hat{\sigma} = p\sigma$.

Now, we are ready to state the following theorem.

Theorem 8.7 *Consider the system* (8.41) *with any unknown compact set* $\mathbb{D}$. *Under Assumptions* 7.3, 8.6–8.8, 8.10, *and* 8.13, *the GRORP is solved by the following controller:*

$$\begin{aligned} u &= \psi_r(T_r^{-1}(\eta_r)) + \mathcal{S}_r(k, e, x_2 - \psi_1(T_1^{-1}\eta_1), \ldots, x_r - \psi_{r-1}(T_{r-1}^{-1}\eta_{r-1})) \\ \dot{\eta} &= M\eta + N(c(x,u) - \psi(T^{-1}\eta) + \Psi T^{-1}\eta) \\ \dot{k} &= \tau(k, e, x_2 - \psi_1(T_1^{-1}\eta_1), \ldots, x_r - \psi_{r-1}(T_{r-1}^{-1}\eta_{r-1})), \\ k &= [k_1, \ldots, k_r]^{\mathsf{T}}, \quad k_i \in \mathbb{R}, \ i = 1, \ldots, r, \end{aligned} \tag{8.75}$$

Proof It suffices to show the GASP for the system (8.52) is solved under Assumptions 8.10 and 8.13. Since the system (8.52) is in the form of (6.12), we can apply

Theorem 6.2 to (8.52). Therefore, we will verify that (8.52) satisfies Assumption 6.4 under Assumptions 8.10 and 8.13. This is indeed the case by using Lemmas 8.1 and 8.2 with slight modification. Therefore, by Theorem 6.2, the GASP for the system (8.52) is solved by the controller

$$\begin{aligned}\bar{u} &= \mathcal{S}_r(k_1, \ldots, k_r, \bar{x}_1, \ldots, \bar{x}_r)\\ \dot{k}_i &= \tau_i(k_1, \ldots, k_{i-1}, \bar{x}_1, \ldots, \bar{x}_i) \qquad i = 1, \ldots, r.\end{aligned}$$

Define $\tau(k_1, \ldots, k_r, \bar{x}_1, \ldots, \bar{x}_r)$ as in (8.74). Then, the GRORP for the system (8.41) for the unknown compact set $\mathbb{D}$ is solved by the controller (8.75). □

8.6 Asymptotic Tracking of the Lorenz System

In this section, we consider a controlled Lorenz system (3.34) or (3.35) which can be put in the form (8.1), i.e.,

$$\begin{aligned}\dot{x}_0 &= \begin{bmatrix} -\alpha & 0 \\ x & -\beta \end{bmatrix} x_0 + \begin{bmatrix} \alpha x \\ 0 \end{bmatrix}\\ \dot{x} &= [1,\ 0]x_0(\rho - [0,\ 1]x_0) - x + u.\end{aligned} \tag{8.76}$$

That is, $x_{01} = \zeta_1, x_{02} = \zeta_3$, and $x = \zeta_2$.

We will design a control law to make the output $y = x$ asymptotically track a sinusoidal reference input v_1 of frequency $\sigma \geq 0$, which is generated by an exosystem

$$\begin{bmatrix} \dot{v}_1 \\ \dot{v}_2 \end{bmatrix} = A_1 \begin{bmatrix} v_1 \\ v_2 \end{bmatrix}, \quad A_1 = \begin{bmatrix} 0 & \sigma \\ -\sigma & 0 \end{bmatrix}. \tag{8.77}$$

The problem can be formulated as the GRORP of (8.76) with the tracking error

$$e = y - v_1.$$

In the formulation, we allow the parameters undergo some perturbation, i.e., $\text{col}(\alpha, \beta, \rho) = \text{col}(\alpha_0, \beta_0, \rho_0) + w$ where $w_0 = \text{col}(\alpha_0, \beta_0, \rho_0)$ is the nominal value and w the perturbation. We assume that $\mathbb{V}$ is any compact set containing the origin, and $\mathbb{W}$ is any compact set containing w_0 such that all the parameters α, β and ρ are positive.

We will first apply Theorem 8.1 to solve the GRORP of (8.76) on $\mathbb{D} = \mathbb{V} \times \mathbb{W}$. For this purpose, we need to verify Assumptions 8.1–8.5. Assumption 8.1 is self verified. To verify Assumption 8.2, we note that, with $\mathbf{x}(v, w) = v_1$, the first equation of (8.76) gives

$$\mathbf{x}_0(v, w) = \begin{bmatrix} r_{11}(w)v_1 + r_{12}(w)v_2 \\ r_{21}(w)v_1^2 + r_{22}(w)v_2^2 + r_{23}(w)v_1 v_2 \end{bmatrix} \tag{8.78}$$

where

$$
\begin{aligned}
r_{11}(w) &= \frac{\alpha^2}{\sigma^2 + \alpha^2} \\
r_{12}(w) &= \frac{-\alpha\sigma}{\sigma^2 + \alpha^2} \\
r_{21}(w) &= \frac{\beta^2 r_{11}(w) + \beta\sigma r_{12}(w) + 2\sigma^2 r_{11}(w)}{\beta(\beta^2 + 4\sigma^2)} \\
r_{22}(w) &= \frac{-\sigma r_{23}(w)}{\beta} \\
r_{23}(w) &= \frac{r_{12}(w)\beta - 2\sigma r_{11}(w)}{\beta^2 + 4\sigma^2}.
\end{aligned}
$$

Assumption 8.2 is thus satisfied. Furthermore, substituting (8.78) into the second equation of (8.76) gives

$$
\begin{aligned}
\mathbf{u}(v, w) &= \sigma v_2 + v_1 - [1\ 0]\mathbf{x}_0(v, w)(\rho - [0\ 1]\mathbf{x}_0(v, w)) \\
&= r_{31}(w)v_1 + r_{32}(w)v_2 + r_{33}(w)v_1^3 + r_{34}(w)v_2^3 \\
&\quad + r_{35}(w)v_1^2 v_2 + r_{36}(w)v_1 v_2^2
\end{aligned} \tag{8.79}
$$

where

$$
\begin{aligned}
r_{31}(w) &= 1 - \rho r_{11}(w)) \\
r_{32}(w) &= \sigma - \rho r_{12}(w)) \\
r_{33}(w) &= r_{11}(w)r_{21}(w) \\
r_{34}(w) &= r_{12}(w)r_{22}(w) \\
r_{35}(w) &= r_{12}(w)r_{21}(w) + r_{11}(w)r_{23}(w) \\
r_{36}(w) &= r_{11}(w)r_{22}(w) + r_{12}(w)r_{23}(w).
\end{aligned}
$$

Next, let $\pi(v, w) = \mathbf{u}(v, w)$ and

$$
\theta(v, w) = \mathcal{L}^4_{A_1 v}\pi(v, w) = \left[\mathbf{u}(v, w), \frac{d\mathbf{u}(v, w)}{dt}, \frac{d^2\mathbf{u}(v, w)}{dt^2}, \frac{d^3\mathbf{u}(v, w)}{dt^3}\right]^{\mathsf{T}}
$$

Then, along the trajectory of the exosystem, we have

$$
\frac{d^4\mathbf{u}(v, w)}{dt^4} + 9\sigma^4\mathbf{u}(v, w) + 10\sigma^2\frac{d^2\mathbf{u}(v, w)}{dt^2} = 0.
$$

Thus, there exists a steady-state generator

$$
\dot{\theta}(v, w) = \Phi\theta(v, w), \quad \mathbf{u}(v, w) = \Psi\theta(v, w)
$$

with output u, where

$$\Phi = \begin{bmatrix} 0 & 1 & 0 & 0 \\ 0 & 0 & 1 & 0 \\ 0 & 0 & 0 & 1 \\ -9\sigma^4 & 0 & -10\sigma^2 & 0 \end{bmatrix}, \quad \Psi = \begin{bmatrix} 1 \\ 0 \\ 0 \\ 0 \end{bmatrix}^{\mathrm{T}}.$$

Assumption 8.3 is thus satisfied.

Pick a pair of controllable matrices

$$M = \begin{bmatrix} 0 & 1 & 0 & 0 \\ 0 & 0 & 1 & 0 \\ 0 & 0 & 0 & 1 \\ -4 & -12 & -13 & -6 \end{bmatrix}, \quad N = \begin{bmatrix} 0 \\ 0 \\ 0 \\ 1 \end{bmatrix}^{\mathrm{T}}$$

where M is Hurwitz. Solving the Sylvester equation $MT + N\Psi = T\Phi$ yields

$$T^{-1} = \begin{bmatrix} 4-9\sigma^4 & 12 & 13-10\sigma^2 & 6 \\ -54\sigma^4 & 4-9\sigma^4 & 12-60\sigma^2 & 13-10\sigma^2 \\ \star_{(1)} & -54\sigma^4 & \star_{(2)} & 12-60\sigma^2 \\ \star_{(3)} & \star_{(1)} & \star_{(4)} & \star_{(2)} \end{bmatrix}$$

$$\star_{(1)} := 9\sigma^4(10\sigma^2 - 13), \quad \star_{(2)} := 91\sigma^4 - 130\sigma^2 + 4$$
$$\star_{(3)} := 108\sigma^4(5\sigma^2 - 1), \quad \star_{(4)} := 6\sigma^2(91\sigma^2 - 20).$$

Thus,

$$\Psi T^{-1} = \left[4 - 9\sigma^4 \; 12 \; 13 - 10\sigma^2 \; 6\right]. \tag{8.80}$$

The matrices M and N constitute an internal model on the mapping $\tau(\theta) = T\theta$ as follows:

$$\dot{\eta} = M\eta + Nu. \tag{8.81}$$

Performing the following coordinate transformation

$$\begin{aligned} \bar{x}_0 &= x_0 - \mathbf{x}_0(v, w) \\ \bar{x} &= x - \mathbf{x}(v, w) \\ \tilde{\eta} &= \eta - T\theta(v, w) - N\bar{x} \\ \bar{u} &= u - \Psi T^{-1}\eta \end{aligned}$$

on the system composed of (8.76) and (8.81) gives the following transformed augmented system:

$$\begin{aligned}
\dot{\bar{x}}_0 &= \begin{bmatrix} -\alpha\bar{x}_{01} + \alpha\bar{x} \\ -\beta\bar{x}_{02} + (\bar{x}_{01} + \mathbf{x}_{01}(v,w))(\bar{x} + v_1) - \mathbf{x}_{01}(v,w)v_1 \end{bmatrix} \\
\dot{\tilde{\eta}} &= M\tilde{\eta} + MN\bar{x} - Na(\bar{x}_0, \bar{x}, v, w) \\
\dot{\bar{x}} &= a(\bar{x}_0, \bar{x}, v, w) + \Psi T^{-1}(\tilde{\eta} + N\bar{x}) + \bar{u}
\end{aligned} \tag{8.82}$$

where

$$\begin{aligned}
a(\bar{x}_0, \bar{x}, v, w) = &-\bar{x} + \rho\bar{x}_{01} - (\bar{x}_{01} + \mathbf{x}_{01}(v,w))(\bar{x}_{02} + \mathbf{x}_{02}(v,w)) \\
&+\mathbf{x}_{01}(v,w)\mathbf{x}_{02}(v,w).
\end{aligned}$$

We now need to verify that the $\bar{x}_0$-subsystem satisfies Assumption 8.4. Let

$$V(\bar{x}_0) = \frac{\hbar}{2}\bar{x}_{01}^2 + \frac{\hbar}{4}\bar{x}_{01}^4 + \frac{1}{2}\bar{x}_{02}^2$$

for some $\hbar > 0$ which satisfies, along the trajectory of the $\bar{x}_0$-subsystem,

$$\begin{aligned}
\dot{V}(\bar{x}_0) = &-\hbar\alpha\bar{x}_{01}^2 + \hbar\alpha\bar{x}_{01}\bar{x} - \hbar\alpha\bar{x}_{01}^4 + \hbar\alpha\bar{x}_{01}^3\bar{x} \\
&-\beta\bar{x}_{02}^2 + \bar{x}_{01}\bar{x}_{02}\bar{x} + v_1\bar{x}_{01}\bar{x}_{02} + \mathbf{x}_{01}(v,w)\bar{x}_{02}\bar{x}.
\end{aligned} \tag{8.83}$$

For any $\varepsilon > 0$, using Young's inequality in (8.83) gives

$$\begin{aligned}
\hbar\alpha\bar{x}_{01}\bar{x} &\le \frac{1}{2}\bar{x}_{01}^2 + \frac{\hbar^2\alpha^2}{2}\bar{x}^2 \\
\hbar\alpha\bar{x}_{01}^3\bar{x} &\le \frac{3}{4}\bar{x}_{01}^4 + \frac{\hbar^4\alpha^4}{4}\bar{x}^4 \\
\bar{x}_{01}\bar{x}_{02}\bar{x} &\le \frac{1}{4}\bar{x}_{01}^4 + \frac{\varepsilon}{2}\bar{x}_{02}^2 + \frac{1}{4\varepsilon^2}\bar{x}^4 \\
v_1\bar{x}_{01}\bar{x}_{02} &\le \frac{1}{2\varepsilon}\bar{x}_{01}^2 + \frac{\varepsilon v_1^2}{2}\bar{x}_{02}^2 \\
\mathbf{x}_{01}(v,w)\bar{x}_{02}\bar{x} &\le \frac{\varepsilon}{2}\bar{x}_{02}^2 + \frac{\mathbf{x}_{01}^2(v,w)}{2\varepsilon}\bar{x}^2.
\end{aligned} \tag{8.84}$$

Substituting (8.84) into (8.83) gives

$$\begin{aligned}
\dot{V}(\bar{x}_0) \le &-\left(\hbar\alpha - \frac{1}{2} - \frac{1}{2\varepsilon}\right)\bar{x}_{01}^2 - (\hbar\alpha - 1)\,\bar{x}_{01}^4 - \left(\beta - \varepsilon - \frac{\varepsilon v_1^2}{2}\right)\bar{x}_{02}^2 \\
&+\left(\frac{\hbar^2\alpha^2}{2} + \frac{\mathbf{x}_{01}^2(v,w)}{2\varepsilon}\right)\bar{x}^2 + \left(\frac{\hbar^4\alpha^4}{4} + \frac{1}{4\varepsilon^2}\right)\bar{x}^4.
\end{aligned} \tag{8.85}$$

Since $\mathbb{D}$ is compact, there exist a sufficiently small $\varepsilon > 0$ and a sufficiently large $\hbar > 0$, such that

$$\hbar\alpha - \frac{1}{2} - \frac{1}{2\varepsilon} > 0, \quad \hbar\alpha - 1 > 0, \quad \beta - \varepsilon - \frac{\varepsilon v_1^2}{2} > 0.$$

Moreover, if the boundary of $\mathbb{D}$ is known, then $\varepsilon > 0$ and $\hbar > 0$ are also known. Thus, Assumption 8.4 is satisfied. Finally Assumption 8.5 is trivially satisfied since $\delta = 0$.

Thus, if $\mathbb{D}$ is a known compact set, by Theorem 8.1, the GRORP for the system (8.76) is solved by an output feedback controller as follows:

$$\begin{aligned} u &= \Psi T^{-1}\eta - (k_1 e^6 + k_2)e \\ \dot{\eta} &= M\eta + Nu \end{aligned} \tag{8.86}$$

where k_1 and k_2 are two sufficiently large real numbers depending on the size of $\mathbb{D}$. For example, we invite the readers to find k_1 and k_2 for the case where $\sigma = 0.8$, $\alpha = 10 \pm 1$, $\beta = 2.5 \pm 0.5$, $\rho = 30 \pm 2$, and $\|v(0)\| \le 1$ (Problem 8.5).

Next, we will consider the case where $\mathbb{D}$ is an arbitrarily unknown compact set. This case can be handled by Theorem 8.5. For this purpose, we need to further verify Assumption 8.13. This is indeed true from the inequality (8.85). Thus, the problem for this case can be solved by the following controller:

$$\begin{aligned} u &= \Psi T^{-1}\eta - k\rho(e)e \\ \dot{\eta} &= M\eta + Nu \\ \dot{k} &= \lambda\rho(e)e^2, \quad \lambda > 0. \end{aligned} \tag{8.87}$$

With the dynamic gain, it is not necessary to specify the gain k. The simulation is conducted with $\rho(e) = e^6 + 10$, $\lambda = 0.1$, $\alpha = 10$, $\beta = 8/3$, and $\rho = 28$. The initial condition of the closed-loop system is $v(0) = [1, 0]^{\mathrm{T}}$, $x_0(0) = [5, 5]^{\mathrm{T}}$, $x(0) = 0$, $\eta(0) = [1, 1, 0, 0]^{\mathrm{T}}$, and $k(0) = 0$. The simulation results are shown in

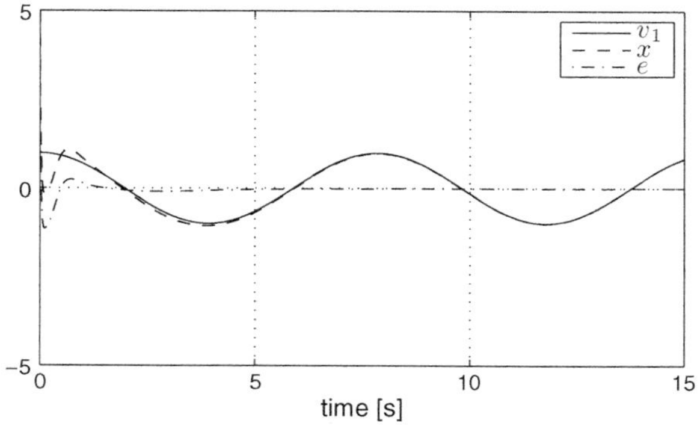

Fig. 8.11 Profile of state trajectories of the controlled Lorenz system with unbounded uncertainties (part 1)

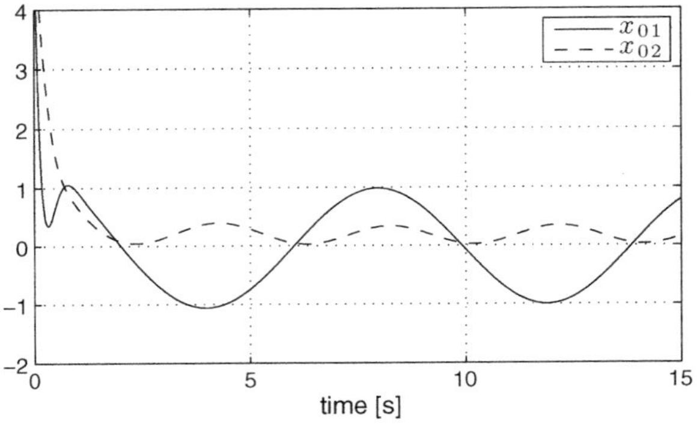

Fig. 8.12 Profile of state trajectories of the controlled Lorenz system with unbounded uncertainties (part 2)

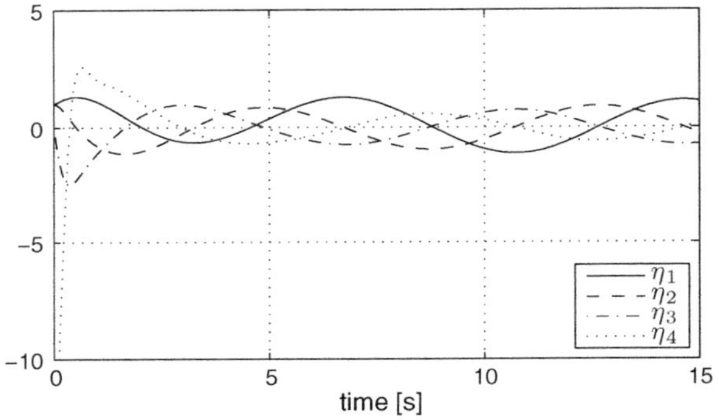

Fig. 8.13 Profile of state trajectories of the controlled Lorenz system with unbounded uncertainties (part 3)

Figs. 8.11–8.14. More specifically, as shown in Fig. 8.11, the tracking error approaches zero asymptotically, and all the state is bounded as shown in Figs. 8.12 and 8.13. In particular, the dynamic gain approaches a finite constant asymptotically as shown in Fig. 8.14.

8.7 Notes and References

The materials in this chapter are adapted from several papers of the authors. When the exosystem is linear, the GRORP was solved for the filter augmented (output feedback) systems in [1] (Sect. 8.2) and for the lower triangular systems in [2] (Sect. 8.3). When the exosystem is nonlinear, the GRORP was first studied for the local case in [3] and then for the global case in [4] (Sect. 8.4). When the boundary of uncertainty

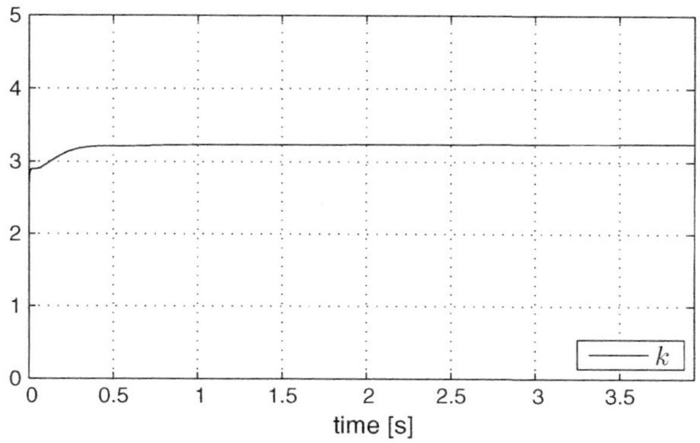

Fig. 8.14 Profile of dynamic gain of the controlled Lorenz system with unbounded uncertainties

is unknown, the GRORP was solved in [5] (Sect. 8.5). The Lorenz system example in Sect. 8.6 first appeared in [6]. The ISS assumption on the inverse dynamics can be relaxed to iISS assumption, see, e.g., [7, 8]. The output regulation problem for nonlinear systems with unknown direction is studied in [9].

8.8 Problems

Problem 8.1 For a linear exosystem $\dot{v} = A_1 v$, define $v^{[i]}$ as in (7.28). Find a constant matrix $A^{[i]}$ such that (see Chapter 4 of [10])

$$\frac{\partial v^{[i]}}{\partial v} A_1 v = A^{[i]} v^{[i]}.$$

Problem 8.2 Solve the GRORP for the following system

$$\begin{aligned} \dot{x}_0 &= -x_0 + 0.2 v_2 x + v_1^2 \\ \dot{x} &= x_0 - 0.1 v_1 x - \sin^2(0.1 w x) + 10 w v_1 + 10 v_2 + 10 u \\ e &= x - 10 v_1 \end{aligned}$$

where the external disturbance $v = \text{col}(v_1,\ v_2)$ is governed by the exosystem

$$\dot{v} = A_1 v,\ A_1 = \begin{bmatrix} 0 & 1 \\ -1 & 0 \end{bmatrix}.$$

It is assumed that $v(t) \in \mathbb{V} = \{v_1^2 + v_2^2 \leq 1\}$ and $-1 \leq w \leq 1$.

Problem 8.3 Solve the GRORP for the following system

$$\begin{aligned}\dot{x}_1 &= w x_1^2 + 2x_2\\ \dot{x}_2 &= u\\ y &= x_1\\ e &= y - v_1\end{aligned}$$

where the external disturbance $v = \text{col}(v_1,\ v_2)$ is governed by the exosystem $\dot{v} = A_1 v$ satisfying $v(t) \in \mathbb{V} = \{v_1^2 + v_2^2 \leq 1\}$ and $-1 \leq w \leq 1$. Consider the following two cases.

(a) $A_1 = \begin{bmatrix} 0 & 1 \\ -1 & 0 \end{bmatrix}$;

(b) $A_1 = \begin{bmatrix} 0 & 2 \\ -2 & 0 \end{bmatrix}$.

Problem 8.4 Solve the GRORP for the following system

$$\begin{aligned}\dot{x}_1 &= w_1 x_1 + 2x_2\\ \dot{x}_2 &= -w_2 x_1 x_2 + u\\ y &= x_1\\ e &= y - v_1\end{aligned}$$

where the external disturbance $v = \text{col}(v_1, v_2)$ is governed by the exosystem $\dot{v} = A_1 v$ satisfying $v(t) \in \mathbb{V} = \{v_1^2 + v_2^2 \leq 1\}$ and $-1 \leq w_1, w_2 \leq 1$. Consider the following two cases.

(a) $A_1 = \begin{bmatrix} 0 & 1 \\ -1 & 0 \end{bmatrix}$,

(b) $A_1 = \begin{bmatrix} 0 & 2 \\ -2 & 0 \end{bmatrix}$.

Problem 8.5 Solve the GRSP of (8.82) for $\sigma = 0.8$, $\alpha = 10 \pm 1$, $\beta = 2.5 \pm 0.5$, $\rho = 30 \pm 2$, and $\|v(0)\| \leq 1$.

Problem 8.6 Repeat Problems 8.2, 8.3, and 8.4 when $\mathbb{V}$ and w (or w_1, w_2) are unknown.

Problem 8.7 Solve the GRORP for the composite system (8.1) with $\mathbb{D}$ an arbitrarily unknown compact set and b a nonzero number with unknown sign.

References

1. Chen Z, Huang J (2005) Global robust output regulation for output feedback systems. IEEE Trans. Autom. Control 50:117–121
2. Huang J, Chen Z (2004) A general framework for tackling the output regulation problem. IEEE Trans. Autom. Control 49:2203–2218
3. Chen Z, Huang J (2005) Robust output regulation with nonlinear exosystems. Automatica 41(8):1447–1454
4. Yang X, Huang J (2012) New results on robust output regulation of nonlinear systems with a nonlinear exosystem. Int. J. Robust Nonlinear Control 22:1703–1719
5. Chen Z, Huang J (2004) Dissipativity, stabilization, and regulation of cascade-connected systems. IEEE Trans. Autom. Control 49:635–650
6. Xu D, Huang J (2010) Robust adaptive control of a class of nonlinear systems and its applications. IEEE Trans. Circuits Syst. I Regul. Pap. 57:691–702
7. Xu D, Huang J (2011) Output regulation for output feedback systems with iISS inverse dynamics. IEEE Trans. Autom. Control 133:044503
8. Xu D, Huang J, Jiang Z-P (2013) Global adaptive output regulation for a class of nonlinear systems using output feedback. Automatica 49:2184–2191
9. Liu L, Huang J (2006) Global robust output regulation of output feedback systems with unknown high-frequency gain sign. IEEE Trans. Autom. Control 51:625–631
10. Huang J (2004) Nonlinear output regulation problem: theory and Applications. SIAM, Philadelphia

Chapter 9
Output Regulation with Uncertain Exosystems

In this chapter, we turn to the problem of the global robust output regulation problem with uncertain exosystems. We will still follow the framework established in Chap. 7 to handle this problem. However, due to the presence of the unknown parameter σ in the exosystem, we cannot perform exactly the same transformation as in (7.60) to obtain the transformed augmented system (7.61). Instead, we need to modify the transformation (7.60) to obtain a transformed augmented system which cannot be globally stabilized by the robust control technique, but can be globally stabilized by the adaptive control technique studied in Chap. 5 under some appropriate assumptions. These assumptions guarantee that the uncertain parameter caused by unknown σ in the augmented system satisfies the linear parameterization property. In Sect. 9.1, we study systems with relative degree one by output feedback control. In Sect. 9.2, we further study systems of the form (8.30) by partial state feedback control. As shown in Sect. 8.2, this class of systems contains the filter extended system of the output feedback system (8.19) as a special case. Thus, the solution of the GRORP for this class of systems via partial state feedback control also lead to the solution of the GRORP for the output feedback systems (8.19) via output feedback control. In Sect. 9.3, we study the disturbance rejection problem for the lower triangular systems. In Sect. 9.4, we apply the result in Sect. 9.3 to the disturbance rejection problem of the Fitzhugh–Nagumo system. Finally, the notes and references are given in Sect. 9.5.

9.1 Systems with Relative Degree One

We first consider the following composite system:

$$\begin{aligned} \dot{x}_0 &= f_0(x_0, x, v(t), w) \\ \dot{x} &= f(x_0, x, v(t), w) + b(w)u \\ \dot{v} &= A_1(\sigma)v \\ y &= x. \end{aligned} \tag{9.1}$$

Z. Chen and J. Huang, *Stabilization and Regulation of Nonlinear Systems*, Advanced Textbooks in Control and Signal Processing,
DOI 10.1007/978-3-319-08834-1_9

Like in (8.1), the system (9.1) is composed of a nonlinear system with relative degree one and a linear exosystem satisfying Assumption 7.2. Thus all notation in (9.1) has the same interpretation as that in (8.1). As in Chap. 8, the tracking error is defined as

$$e = y - y_{\mathrm{d}}, \ y_{\mathrm{d}} = h_{\mathrm{d}}(v, w). \tag{9.2}$$

We recall that the general solution of the current exosystem (7.6) is a sum of finitely many sinusoidal functions with their frequencies depending on the eigenvalues of $A_1(\sigma)$ and their amplitudes and initial phases on the initial condition $v_0 = v(0)$. Let $d(t) = \mathrm{col}(v(t), w, \sigma) \in \mathbb{D}$ with $\mathbb{D} = \mathbb{V} \times \mathbb{W} \times \mathbb{S}$. We assume, for all $t \geq 0$, $d(t) \in \mathbb{D}$.

We are now ready to consider the GRORP for the system (9.1) on a prescribed compact set $\mathbb{D}$. Like in Chap. 8, we still need two assumptions slightly modified from Assumptions 8.1 and 8.2 as follows.

Assumption 9.1 The function $b(w)$ is away from zero, e.g., $b(w) > 0, \forall w \in \mathbb{W}$.

Assumption 9.2 There exists a sufficiently smooth function $\mathbf{x}_0(v, w, \sigma)$ with $\mathbf{x}_0(0, 0, 0) = 0$ satisfying, for all $d \in \mathbb{D}$,

$$\frac{\partial \mathbf{x}_0(v, w, \sigma)}{\partial v} A_1(\sigma) v = f_0(\mathbf{x}_0(v, w, \sigma), h_{\mathrm{d}}(v, w), v, w). \tag{9.3}$$

Under Assumptions 9.1 and 9.2, the solution to the regulator equations is

$$\mathrm{col}(\mathbf{x}_0(v, w, \sigma), \mathbf{x}(v, w), \mathbf{u}(v, w, \sigma))$$

where

$$\begin{aligned}
\mathbf{x}(v, w) &= h_{\mathrm{d}}(v, w) \\
\mathbf{u}(v, w, \sigma) &= b^{-1}(w) \left[\frac{\partial h_{\mathrm{d}}(v, w)}{\partial v} A_1(\sigma) v - f(\mathbf{x}_0(v, w, \sigma), h_{\mathrm{d}}(v, w), v, w) \right].
\end{aligned}$$

Like in Chap. 8, for convenience, we denote $\mathbf{x}_2(v, w, \sigma) = \mathbf{u}(v, w, \sigma)$ in what follows.

We will only consider the case where the system (9.1) admits a linear internal model. For this purpose, we modify Assumption 8.3 to the following:

Assumption 9.3 $\mathbf{x}_2(v, w, \sigma) \in \mathbb{R}$ is a polynomial function in v with coefficients depending possibly on w and σ.

Remark 9.1 By Lemma 7.3, under Assumption 9.3, the composite system (9.1) admits a steady-state generator of the following form

$$\dot{\theta}(v, w, \sigma) = \Phi(\sigma)\theta(v, w, \sigma), \quad \mathbf{x}_2(v, w, \sigma) = \Psi\theta(v, w, \sigma) \tag{9.4}$$

where, for some integer $\ell > 0$, $\theta : \mathbb{R}^l \mapsto \mathbb{R}^\ell$ is a sufficiently smooth function vanishing at the origin, and the pair $(\Psi, \Phi(\sigma))$ is observable.

Under Assumption 9.3, for any controllable pair (M, N) with $M \in \mathbb{R}^{\ell\times\ell}$ Hurwitz and $N \in \mathbb{R}^{\ell\times 1}$, there is a unique nonsingular matrix $T(\sigma)$ satisfying the Sylvester equation $T(\sigma)\Phi(\sigma) - MT(\sigma) = N\Psi$. Then, associated with the steady-state generator (9.4), we can construct an internal model candidate on the mapping $\tau(\theta, \sigma) = T(\sigma)\theta$ as follows:

$$\dot{\eta} = M\eta + Nu. \tag{9.5}$$

Attaching the internal model (9.5) to the given plant yields an augmented system. Performing on the augmented system the following coordinate and input transformation suggested in (8.9) and (8.11):

$$\begin{aligned}
\bar{\eta} &= \eta - T(\sigma)\theta(v, w, \sigma) - Nb^{-1}(w)\bar{x} \\
\bar{x}_0 &= x_0 - \mathbf{x}_0(v, w, \sigma) \\
\bar{x} &= x - \mathbf{x}(v, w) \\
\bar{u} &= u - \Psi T^{-1}(\sigma)\eta
\end{aligned} \tag{9.6}$$

gives the transformed augmented system as follows:

$$\begin{aligned}
\dot{\bar{x}}_0 &= \bar{f}_0(\bar{x}_0, \bar{x}, d) \\
\dot{\bar{\eta}} &= M\bar{\eta} + \rho(\bar{x}_0, \bar{x}, d) \\
\dot{\bar{x}} &= \bar{f}(\bar{x}_0, \bar{\eta}, \bar{x}, d) + b(w)\bar{u}.
\end{aligned} \tag{9.7}$$

In (9.7), various quantities are defined in the same way as those in (8.12) with $\delta = 0$. The only difference is that the matrix T in (8.12) is replaced by $T(\sigma)$.

Nevertheless, note that $\bar{u} = u - \Psi T^{-1}(\sigma)\eta$ which relies on the unknown parameter σ. Even if we can still globally robustly stabilize (9.7) by a static controller $\bar{u}$, this controller will lead to a controller of the form (8.16) which is not implementable since the matrix T depends on σ. Thus, instead of performing the transformation (9.6), we will perform the following transformation:

$$\begin{aligned}
\bar{\eta} &= \eta - T(\sigma)\theta(v, w, \sigma) - Nb^{-1}(w)\bar{x} \\
\bar{x}_0 &= x_0 - \mathbf{x}_0(v, w, \sigma) \\
\bar{x} &= x - \mathbf{x}(v, w)
\end{aligned} \tag{9.8}$$

which leads to the following system:

$$\begin{aligned}
\dot{\bar{x}}_0 &= \bar{f}_0(\bar{x}_0, \bar{x}, d) \\
\dot{\bar{\eta}} &= M\bar{\eta} + \rho(\bar{x}_0, \bar{x}, d) \\
\dot{\bar{x}} &= \bar{f}(\bar{x}_0, \bar{\eta}, \bar{x}, d) - b(w)\eta^{\mathsf{T}}\mu + b(w)u
\end{aligned} \tag{9.9}$$

where $\mu^{\mathsf{T}} = \Psi T^{-1}(\sigma)$.

Let $z = \mathrm{col}(\bar{x}_0, \tilde{\eta})$, and

$$q(z, \bar{x}, d) = \begin{bmatrix} \bar{f}_0(\bar{x}_0, \bar{x}, d) \\ M\tilde{\eta} + \rho(\bar{x}_0, \bar{x}, d) \end{bmatrix}$$

Then the system (9.9) can be put in the following form

$$\begin{aligned} \dot{z} &= q(z, \bar{x}, d) \\ \dot{\bar{x}} &= \bar{f}(z, \bar{x}, d) - b(w)\eta^{\mathsf{T}}\mu + b(w)u \end{aligned} \tag{9.10}$$

The system (9.10) is in the form (5.26) with $f_a(x, t) = -\eta(t)$. Assumption 5.1 is equivalent to Assumption 9.1. To invoke Theorem 5.3, we need one more assumption:

Assumption 9.4 The subsystem $\dot{\bar{x}}_0 = \bar{f}(\bar{x}_0, \bar{x}, d)$ has an ISS Lyapunov function $V(\bar{x}_0)$, i.e.,

$$V(\bar{x}_0) \sim \{\underline{\alpha}, \bar{\alpha}, \alpha, \sigma \mid \dot{\bar{x}}_0 = \bar{f}(\bar{x}_0, \bar{x}, d)\}$$

and

$$\limsup_{s \to 0^+} \frac{s^2}{\alpha(s)} < \infty, \quad \limsup_{s \to 0^+} \frac{\sigma(s)}{s^2} < \infty.$$

By Lemma 8.2, under Assumption 9.4, the system $\dot{z} = q(z, \bar{x}, d)$ has an ISS Lyapunov function $U(z)$, i.e.,

$$U(z) \sim \{\underline{\beta}, \bar{\beta}, \beta, \varsigma \mid \dot{z} = q(z, \bar{x}, d)\}$$

and

$$\limsup_{s \to 0^+} \frac{s^2}{\beta(s)} < \infty, \quad \limsup_{s \to 0^+} \frac{\varsigma(s)}{s^2} < \infty.$$

By Theorem 5.3, the GASP of (9.10) is solved by a controller of the following form

$$\begin{aligned} u &= \eta^{\mathsf{T}}\hat{\mu} - \rho(\bar{x})\bar{x} \\ \dot{\hat{\mu}} &= -\Lambda\bar{x}\eta, \;\; \Lambda = \Lambda^{\mathsf{T}} > 0 \end{aligned} \tag{9.11}$$

Therefore, for any initial condition, the state of the closed-loop system composed of (9.10) and (9.11) is bounded and $\lim_{t\to\infty} \mathrm{col}(z(t), \bar{x}(t)) = 0$. Thus, from (9.8), we can see that, for any initial condition, the state of the closed-loop system composed of (9.1) and the controller (9.1) is also bounded. Since $e = \bar{x}$, we also have $\lim_{t\to\infty} e(t) = 0$. Thus, the GRORP of (9.1) with $\mathbb{D}$ a known compact set is solved by the following controller

$$
\begin{aligned}
u &= \eta^{\mathsf{T}}\hat{\mu} - \rho(e)e \\
\dot{\hat{\mu}} &= -\Lambda e\eta, \ \Lambda = \Lambda^{\mathsf{T}} > 0 \\
\dot{\eta} &= M\eta + Nu.
\end{aligned} \tag{9.12}
$$

In summary, we have the following result.

Theorem 9.1 *Under Assumptions 7.2, 9.1–9.4, the GRORP for the system (9.1) is solved on the prescribed compact set $\mathbb{D}$ by a controller of the form (9.12).*

Example 9.1 In Sect. 8.6, the GRORP for the system (8.76) is studied when the frequency σ of the exosystem (8.77) is known, e.g., $\sigma = 0.8$. However, when σ is unknown, A_1 should be replaced by $A_1(\sigma)$, and hence T by $T(\sigma)$. As a result, the controller (8.87) is not implementable and we need to apply the adaptive technique developed in this section. Using the transformation (9.8) instead of (9.6), the augmented system (8.82) is modified to the form (9.10), i.e.,

$$
\begin{aligned}
\dot{\bar{x}}_0 &= \begin{bmatrix} -\alpha\bar{x}_{o1} + \alpha\bar{x} \\ -\beta\bar{x}_{o2} + (\bar{x}_{o1} + \mathbf{x}_{o1}(v, w))(\bar{x} + v_1) - \mathbf{x}_{o1}(v, w)v_1 \end{bmatrix} \\
\dot{\tilde{\eta}} &= M\tilde{\eta} + MN\bar{x} - Na(\bar{x}_0, \bar{x}, v, w) \\
\dot{\bar{x}} &= a(\bar{x}_0, \bar{x}, v, w) + \Psi T^{-1}(\tilde{\eta} + N\bar{x}) - \eta^{\mathsf{T}}\mu + u
\end{aligned} \tag{9.13}
$$

where $\mu^{\mathsf{T}} = \Psi T(\sigma)^{-1}$. By Theorem 9.1, we can obtain the following adaptive controller:

$$
\begin{aligned}
u &= \hat{\mu}^{\mathsf{T}}\eta - (k_1e^6 + k_2)e \\
\dot{\hat{\mu}} &= -e\eta \\
\dot{\eta} &= M\eta + Nu
\end{aligned} \tag{9.14}
$$

where $\hat{\mu} \in \mathbb{R}^4$ is the estimation for μ. Note that

$$
\begin{aligned}
\Psi T(\sigma)^{-1}\eta &= [4 - 9\sigma^4 \ \ 12 \ \ 13 - 10\sigma^2 \ \ 6]\eta \\
&= [4 \ \ 12 \ \ 13 \ \ 6]\eta + [-9\sigma^4 \ \ -10\sigma^2][\eta_1 \ \eta_3]^{\mathsf{T}}.
\end{aligned}
$$

If we define $\mu^{\mathsf{T}} = [-9\sigma^4 \ \ -10\sigma^2]$, then we can obtain an alternative controller as follows:

$$
\begin{aligned}
u &= [4 \ \ 12 \ \ 13 \ \ 6]\eta + \hat{\mu}^{\mathsf{T}}[\eta_1 \ \eta_3]^{\mathsf{T}} - (k_1e^6 + k_2)e \\
\dot{\hat{\mu}} &= -e[\eta_1 \ \eta_3]^{\mathsf{T}} \\
\dot{\eta} &= M\eta + Nu
\end{aligned} \tag{9.15}
$$

where $\hat{\mu} \in \mathbb{R}^2$. The performance of the controller (9.15) is simulated with $\hat{\mu}(0) = [0, 0]$. All other parameters and the initial conditions are the same as those in Sect. 8.6. It is noted that the parameter $\sigma = 0.8$ is unknown to the controller. The tracking

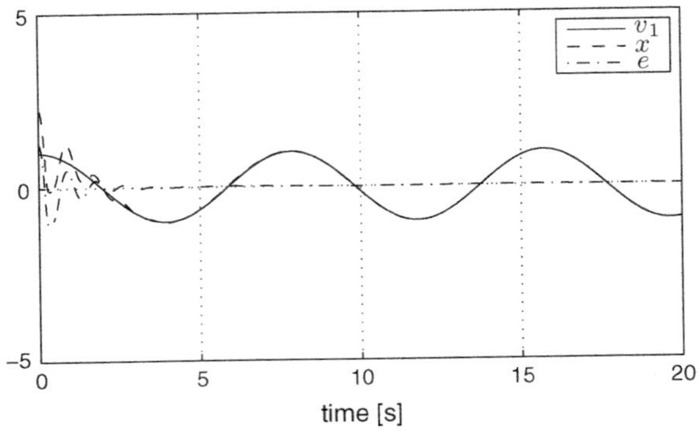

Fig. 9.1 Profile of state trajectories of the closed-loop system in Example 9.1 (part 1)

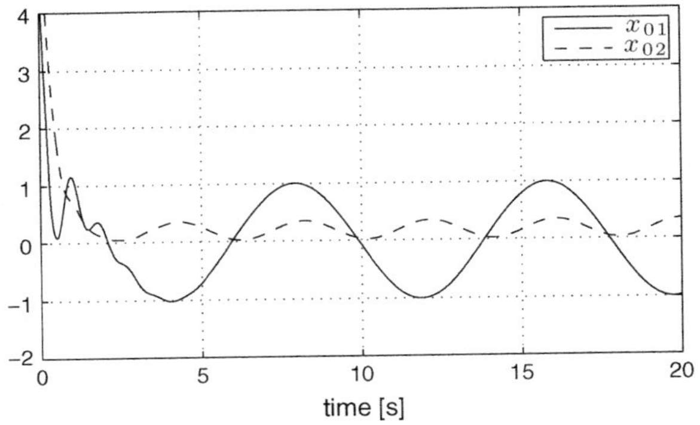

Fig. 9.2 Profile of state trajectories of the closed-loop system in Example 9.1 (part 2)

error approaches zero asymptotically as shown in Fig. 9.1, and all state variables are bounded as shown in Figs. 9.2 and 9.3. The profile of the estimated parameters is illustrated in Fig. 9.4.

Next we will further ascertain conditions under which the unknown parameter vector $\hat{\mu}$ converges to $\mu = (\Psi T^{-1}(\sigma))^{\mathsf{T}}$ as t tends to infinity. For this purpose, we note that Corollary 5.2 gives the following result.

Corollary 9.1 *In Theorem 9.1, the closed-loop system composed of (9.1) and (9.12) has the property:*

$$\lim_{t\to\infty} (\hat{\mu}(t) - \mu)^{\mathsf{T}} T(\sigma)\theta(v(t), w, \sigma) = 0. \tag{9.16}$$

Proof From Corollary 5.2, we have

$$\lim_{t\to\infty} (\hat{\mu}(t) - \mu)^{\mathsf{T}} \eta(t) = 0,$$

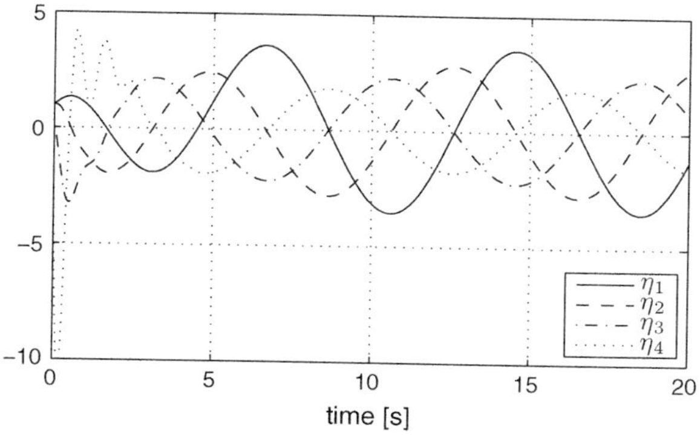

Fig. 9.3 Profile of state trajectories of the closed-loop system in Example 9.1 (part 3)

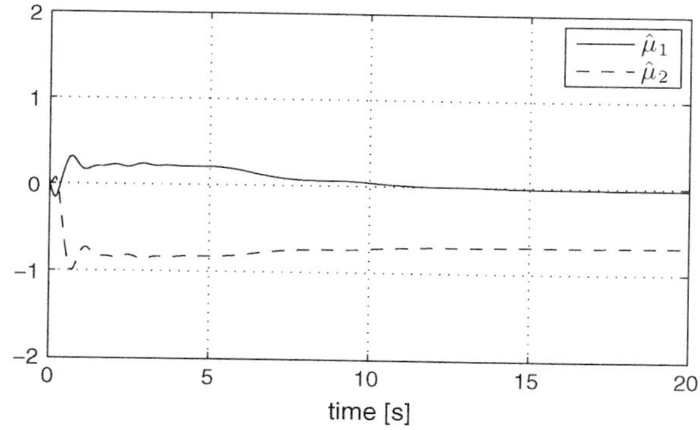

Fig. 9.4 Profile of estimated parameters of the closed-loop system in Example 9.1

which, together with $\lim_{t\to\infty}(\eta(t) - T(\sigma)\theta(v(t), w, \sigma)) = 0$, implies (9.16). □

It is known from Corollary 5.2 that $\lim_{t\to\infty}(\hat{\mu}(t) - \mu) = 0$ if the signal $\theta(v(t), w, \sigma)$ is PE. By Lemma 7.2, if $\mathbf{x}_2(v, w, \sigma)$ is a polynomial in v, then $\mathbf{x}_2(v(t), w, \sigma)$ is a trigonometric polynomial of the form

$$\mathbf{x}_2(v(t), w, \sigma) = \sum_{i=1}^{\ell} C_i(v_0, w, \sigma)e^{j\hat{\omega}_i t}, \ C_i(v_0, w, \sigma) \in \mathbb{C}, \hat{\omega}_i \in \mathbb{R} \tag{9.17}$$

where $\hat{\omega}_i = -\hat{\omega}_{1+\ell-i}$, $\hat{\omega}_i \neq \hat{\omega}_j$ for $i \neq j$, $C_i = C^*_{1+\ell-i}$, and $C_i(v_0, w, \sigma)$ is not identically zero for $v_0 \in \mathbb{R}^{l_1}, w \in \mathbb{R}^{l_2}, \sigma \in \mathbb{R}^{l_3}$. Moreover, $\mathbf{x}_2(v, w, \sigma)$ has a minimal zeroing polynomial with degree ℓ. We call the internal model (9.5) the minimal internal model if $\eta \in \mathbb{R}^{\ell}$.

We are now ready to state the conclusion as follows.

Theorem 9.2 *In Theorem 9.1, suppose the internal model is of the minimal dimension, and* (v_0, w, σ) *is such that none of* $C_i(v_0, w, \sigma)$, $i = 1, \ldots, \ell$, *in (9.17) is zero. Then* $\lim_{t\to\infty}(\hat{\mu}(t) - \mu) = 0$.

Proof Denote $f(t) = \theta(v(t), w, \sigma) \in \mathbb{R}^\ell$. It suffices to show that f is PE. For this purpose, note that

$$\lim_{T_0\to\infty} \frac{1}{T_0} \int_t^{t+T_0} f(s)e^{-j\hat{\omega}_i s} ds = C_i(v_0, w, \sigma)\bar{f}(\hat{\omega}_i), \quad i = 1, \ldots, \ell$$

uniformly in t where

$$\bar{f}(\hat{\omega}_i) = \text{col}(1, j\hat{\omega}_i, (j\hat{\omega}_i)^2, \ldots, (j\hat{\omega}_i)^{\ell-1}).$$

Thus, $f(t) \in \mathbb{R}^\ell$ has spectral lines at ℓ frequencies $\hat{\omega}_i$, $i = 1, \ldots, \ell$.

Since $\hat{\omega}_i$, $i = 1, \ldots, \ell$, are distinct and (v_0, w, σ) is such that none of $C_i(v_0, w, \sigma)$ is zero, the vectors $\hat{f}(\hat{\omega}_i) = C_i(v_0, w, \sigma)\bar{f}(\hat{\omega}_i)$, $i = 1, \ldots, \ell$, are linearly independent. By Lemma 2.3, $f(t)$ is PE. The proof is thus completed. □

Remark 9.2 The employment of the minimal internal model guarantees that the dimension of the unknown vector μ is no more than twice as many as the sinusoids in the signal $\mathbf{x}_2(v, w, \sigma)$. Thus, the employment of a minimal internal model guarantees the satisfaction of the PE condition. This condition is indispensable. In fact, if the dimension of the internal model were $\hat{\ell} > \ell$ for some integer $\hat{\ell}$, then θ would be as follows

$$\theta(v, w, \sigma) = \text{col}(\mathbf{x}_2(v, w, \sigma)), \dot{\mathbf{x}}_2(v, w, \sigma), \ldots, \mathbf{x}_2^{(\hat{\ell}-1)}(v, w, \sigma)).$$

Let $c = [c_1, \ldots, c_{\hat{\ell}}]^\mathsf{T}$ be any constant row vector of dimension $\hat{\ell}$. Then the fact that $c^\mathsf{T} f(t) = 0$ for all t would imply

$$c_1 + c_2(j\hat{\omega}_i) + \cdots + c_{\hat{\ell}}(j\hat{\omega}_i)^{\hat{\ell}-1} = 0 \tag{9.18}$$

for $i = 1, \ldots, \ell$. But (9.18) would not imply $c = 0$ since $\hat{\ell} > \ell$. On the other hand, assume the minimal internal model is employed and (v_0, w, σ) is such that $C_i(v_0, w, \sigma) = 0$ for some i. Then, $\hat{f}(\hat{\omega}_i)$, $i = 1, \ldots, \ell$, are not linearly independent. Thus the assumption that (v_0, w, σ) is such that none of $C_i(v_0, w, \sigma)$ is zero is also indispensable.

9.2 Systems with High Relative Degree

In this section, we will consider the following class of systems:

$$\begin{aligned}\dot{x}_0 &= f_0(x_0, x_1, v(t), w)\\ \dot{x}_1 &= f(x_0, x_1, v(t), w) + b(w)x_2\\ \dot{x}_i &= -\lambda_{i-1}x_i + x_{i+1},\ i = 2, \ldots, r\\ \dot{v} &= A_1(\sigma)\\ y &= x_1,\end{aligned} \tag{9.19}$$

where the tracking error is also defined by (9.2). We assume (9.19) satisfies the same assumptions in Sect. 9.1, i.e., Assumptions 7.2, 9.1–9.4.

Systems of the form (9.19) has been encountered in Sect. 8.2 with $\sigma \in \mathbb{R}^0$. It is known that (9.19) is a special case of the lower triangular systems. We will consider solving the GRORP of (9.19) with $y_{\rm m} = \text{col}(e, x_2, \ldots, x_r)$. As shown in Sect. 8.2, the system (9.19) can also be viewed as the filter extended system of the output feedback system (8.19). Thus, solving the GRORP of (9.19) with $y_{\rm m} = \text{col}(e, x_2, \ldots, x_r)$ will also lead to the solution of the GRORP for the output feedback systems of the form (8.19) by the output feedback control law.

Under Assumption 9.3, the composite system (9.19) admits the steady-state generator with output x_2 as follows

$$\dot{\theta}(v, w, \sigma) = \Phi(\sigma)\theta(v, w, \sigma), \quad \mathbf{x}_2(v, w, \sigma) = \Psi\theta(v, w, \sigma). \tag{9.20}$$

Correspondingly, for any controllable pair (M, N) with $M \in \mathbb{R}^{\ell\times\ell}$ Hurwitz and $N \in \mathbb{R}^{\ell\times 1}$, there is a unique nonsingular matrix T satisfying the Sylvester equation $T(\sigma)\Phi(\sigma) - MT(\sigma) = N\Psi$. Then, associated with the above steady-state generator, we can construct an internal model candidate on the mapping $\tau(\theta, \sigma) = T(\sigma)\theta$ as follows:

$$\dot{\eta} = M\eta + Nx_2. \tag{9.21}$$

Apply the following coordinate transformation

$$\begin{aligned}\bar{\eta} &= \eta - T(\sigma)\theta(v, w, \sigma) - Nb^{-1}(w)\bar{x}_1\\ \bar{x}_0 &= x_0 - \mathbf{x}_0(v, w, \sigma)\\ \bar{x}_1 &= x_1 - \mathbf{x}_1(v, w).\end{aligned} \tag{9.22}$$

Then, the augmented system can be put in the following form

$$\begin{aligned}\dot{z} &= q(z, \bar{x}_1, d)\\ \dot{\bar{x}}_1 &= \bar{f}(z, \bar{x}_1, d) - b(w)\eta^{\mathsf{T}}\mu + b(w)x_2\\ \dot{x}_i &= -\lambda_{i-1}x_i + x_{i+1},\ i = 2, \ldots, r,\end{aligned} \tag{9.23}$$

where the dynamics $\dot{z} = q(z, \bar{x}_1, d)$ are equivalent to

$$\begin{aligned}\dot{\bar{x}}_0 &= \bar{f}_0(\bar{x}_0, \bar{x}_1, d)\\ \dot{\tilde{\eta}} &= M\tilde{\eta} + \rho(\bar{x}_0, \bar{x}_1, d).\end{aligned}$$

The system (9.23) is in the form (5.25) with $f_a(x_1, t) = -\eta(t)$. By Theorem 5.4, there exists a controller of the form

$$\begin{aligned}u &= \varsigma_r(\bar{x}_1, x_2, \ldots, x_r, \hat{\mu}, \hat{b}, t)\\ \dot{\hat{\mu}} &= \tau_r(\bar{x}_1, x_2, \ldots, x_r, \hat{\mu}, \hat{b}, t)\\ \dot{\hat{b}} &= \varpi_r(\bar{x}_1, x_2, \ldots, x_r, \hat{\mu}, \hat{b}, t)\end{aligned} \tag{9.24}$$

that solves the GARP for (9.23) with the performance output $\bar{x}_1$ where various functions in the above control law can be recursively calculated by (5.43) and (5.44).

In (5.25), if $f_a(x_1, t)$ is replaced by $-\eta(t)$ where η is generated by (9.21), then, we have

$$\frac{\partial \varsigma_{i-1}(\overrightarrow{x}_{i-1}, \hat{\mu}, \hat{b}, t)}{\partial t} = \frac{\partial \varsigma_{i-1}(\overrightarrow{x}_{i-1}, \hat{\mu}, \hat{b}, t)}{\partial \eta}(M\eta + Nx_2).$$

To emphasize the reliance of the control law on the variable η, we can modify (9.24) to the following

$$\begin{aligned}u &= \varsigma_r(\bar{x}_1, x_2, \ldots, x_r, \hat{\mu}, \hat{b}, \eta)\\ \dot{\hat{\mu}} &= \tau_r(\bar{x}_1, x_2, \ldots, x_r, \hat{\mu}, \hat{b}, \eta)\\ \dot{\hat{b}} &= \varpi_r(\bar{x}_1, x_2, \ldots, x_r, \hat{\mu}, \hat{b}, \eta).\end{aligned}$$

As a result, the GRORP for the system (9.19) is solved by the following controller

$$\begin{aligned}u &= \varsigma_r(e, x_2, \ldots, x_r, \hat{\mu}, \hat{b}, \eta)\\ \dot{\hat{\mu}} &= \tau_r(e, x_2, \ldots, x_r, \hat{\mu}, \hat{b}, \eta)\\ \dot{\hat{b}} &= \varpi_r(e, x_2, \ldots, x_r, \hat{\mu}, \hat{b}, \eta)\\ \dot{\eta} &= M\eta + Nx_2.\end{aligned} \tag{9.25}$$

In summary, we have the following result.

Theorem 9.3 *Consider the composite system (9.19). Under Assumptions 7.2, 9.1–9.4, the GRORP for the system (9.19) is solved on the prescribed compact set $\mathbb{D}$ by a controller of the form (9.25).*

As in Sect. 9.1, we can also obtain the conditions under which the unknown parameter vector $\hat{\mu}$ converges to $\mu = (\Psi T^{-1}(\sigma))^{\mathsf{T}}$ as t tends to infinity.

Corollary 9.2 *In Theorem 9.3, the closed-loop system composed of (9.19) and (9.25) has the property:*

$$\lim_{t\to\infty} (\hat{\mu}(t) - \mu)^{\top} T(\sigma)\theta(v(t), w, \sigma) = 0. \tag{9.26}$$

Theorem 9.4 *In Theorem 9.3, suppose the internal model is of the minimal dimension ℓ, and (v_0, w, σ) is such that none of $C_i(v_0, w, \sigma)$, $i = 1, \ldots, \ell$, in (9.17) is zero. Then* $\lim_{t\to\infty}(\hat{\mu}(t) - \mu) = 0$.

Example 9.2 Consider the GRORP for the following nonlinear system

$$\begin{aligned} \dot{x}_1 &= w(\sin x_1 - v_1) + x_2 \\ \dot{x}_2 &= u \\ \dot{v} &= A_1(\sigma)v \\ e &= x_1 \end{aligned} \tag{9.27}$$

where

$$v = \begin{bmatrix} v_1 \\ v_2 \end{bmatrix}, \quad A_1(\sigma) = \begin{bmatrix} 0 & \sigma \\ -\sigma & 0 \end{bmatrix}, \quad 0.1 \le \sigma \le 2,$$

and $-1 \le w \le 1$ is an unknown parameter.

It can be verified that all assumptions needed in Theorem 9.3 are satisfied. In particular, with $\mathbf{x}_2(v, w, \sigma) = wv_1$, we have

$$\theta(v, w, \sigma) = \begin{bmatrix} wv_1 & \sigma w v_2 \end{bmatrix}^{\top},$$

and the following relationship:

$$\frac{d^2\mathbf{x}_2(v, w, \sigma)}{dt^2} + \sigma^2 \mathbf{x}_2(v, w, \sigma) = 0.$$

Thus, there is a steady-state generator

$$\dot{\theta}(v, w, \sigma) = \Phi(\sigma)\theta(v, w, \sigma), \;\; \mathbf{x}_2(v, w, \sigma) = \Psi\theta(v, w, \sigma)$$

where

$$\Phi(\sigma) = \begin{bmatrix} 0 & 1 \\ -\sigma^2 & 0 \end{bmatrix}, \quad \Psi = \begin{bmatrix} 1 & 0 \end{bmatrix}.$$

Next, pick a pair of controllable matrices

$$M = \begin{bmatrix} 0 & 1 \\ -1 & -2 \end{bmatrix}, \quad N = \begin{bmatrix} 0 \\ 1 \end{bmatrix}.$$

Then, solving the Sylvester equation $MT(\sigma) + N\Psi = T(\sigma)\Phi(\sigma)$ gives

$$\begin{aligned} T^{-1}(\sigma) &= \begin{bmatrix} 1-\sigma^2 & 2 \\ -2\sigma^2 & 1-\sigma^2 \end{bmatrix} \\ \Psi T^{-1}(\sigma) &= \begin{bmatrix} 1-\sigma^2 & 2 \end{bmatrix}. \end{aligned} \tag{9.28}$$

Thus, we can obtain an internal model on the mapping $\tau(\theta, \sigma) = T(\sigma)\theta$ as follows:

$$\dot{\eta} = M\eta + Nx_2. \tag{9.29}$$

Performing the coordinate transformation $z = \eta - T(\sigma)\theta(v, w, \sigma) - Nx_1$ on (9.27) and (9.29) gives

$$\begin{aligned} \dot{z} &= Mz + MNx_1 - Nw\sin x_1 \\ \dot{x}_1 &= w\sin x_1 + \Psi T^{-1}(\sigma)(z + Nx_1) - \eta^{\mathsf{T}}\mu + x_2 \\ \dot{x}_2 &= u \end{aligned} \tag{9.30}$$

with $\mu^{\mathsf{T}} = \Psi T^{-1}(\sigma)$.

It remains to solve the GASP for the system (9.30) using Theorem 5.4. It is easy to verify that Assumption 5.2 is satisfied. Thus, we can obtain the following controller:

$$\begin{aligned} u &= \mathcal{S}_2(x_1, x_2, \eta, \hat{\mu}) \\ \dot{\hat{\mu}} &= \tau_2(x_1, x_2, \eta, \hat{\mu}) \\ \dot{\eta} &= M\eta + Nx_2 \end{aligned} \tag{9.31}$$

where the functions $\mathcal{S}_2$ and τ_2 are calculated as follows.

First, we note

$$|w\sin x_1 + \Psi T^{-1}(\sigma)(z + Nx_1)| \leq \sqrt{13}\|z\| + 3|x_1|.$$

Pick a matrix $P = \begin{bmatrix} 3 & 1 \\ 1 & 1 \end{bmatrix}$ which satisfies $PM + M^{\mathsf{T}}P = -2I$. Let $\Delta = 1 + 2 \times 13 = 27$. Let $V(z) = 27z^{\mathsf{T}}Pz$ whose derivative along the trajectory of the z-subsystem satisfies

$$\begin{aligned} \dot{V}(z) &= 27\left[-2\|z\|^2 + 2z^{\mathsf{T}}P(MNx_1 - Nw\sin x_1)\right] \\ &\leq -27\|z\|^2 + 27 \times 8x_1^2. \end{aligned}$$

Then, letting $\rho_1 = 230$ gives

$$\mathcal{S}_1(x_1, \eta, \hat{\mu}) = \eta^{\mathsf{T}}\hat{\mu} - 230x_1, \quad \tau_1(x_1, \eta) = -x_1\eta.$$

The remaining functions are given in order by

$$\rho_2 = 230^2/2 + 9/4, \ \varrho_2(\eta) = -230(-\eta^{\mathsf{T}}\hat{\mu} + x_2),$$

and

$$\tau_2(x_1, x_2, \eta, \hat{\mu}) = -x_1\eta - (x_2 - \eta^{\mathsf{T}}\hat{\mu} + 230x_1)(230\eta)$$
$$\nu_2(x_1, x_2, \eta, \hat{\mu}) = \hat{\mu}^{\mathsf{T}}(M\eta + Nx_2) + \eta^{\mathsf{T}}\tau_2(x_1, x_2, \eta, \hat{\mu}).$$

Finally, we obtain

$$\mathcal{S}_2(x_1, x_2, \eta, \hat{\mu}) = -230(-\eta^{\mathsf{T}}\hat{\mu} + x_2) - \rho_2(x_2 - \eta^{\mathsf{T}}\hat{\mu} + 230x_1) + \nu_2(x_1, x_2, \eta, \hat{\mu}).$$

To investigate the convergence issue of the parameter $\hat{\mu}$, note that the minimal zero polynomial of $\mathbf{u}(v, w)$ is $\lambda^2 + \sigma^2$. Thus the internal model (9.29) is the minimal internal model. It can be verified that (9.17) holds for

$$\ell = 2, \ \hat{\omega}_1 = -\hat{\omega}_2 = \sigma$$
$$C_1(v_0, w, \sigma) = C_2^*(v_0, w, \sigma) = w\frac{v_1(0)}{2} + w\frac{v_2(0)}{2j}.$$

Since, for all nonzero v_0 and w, none of C_i is zero for $i = 1, 2$. The condition for the convergence of the unknown parameter $\hat{\mu}^{\mathsf{T}}$ to the true value $\mu^{\mathsf{T}} = \Psi T^{-1}(\sigma)$ is thus satisfied.

The performance of the controller is simulated with $\sigma = 0.5$, $w = 1$, and the initial condition $v(0) = [1, 0]^{\mathsf{T}}$, $x(0) = [1, 20]^{\mathsf{T}}$, $\eta(0) = 0$, and $\hat{\mu}(0) = 0$. As a result, $\mu = [0.75, 2]^{\mathsf{T}}$. Figure 9.5 shows the tracking error, and Fig. 9.6 shows the response of all state variables. Finally, Fig. 9.7 shows the convergence of the estimated parameters to their true value.

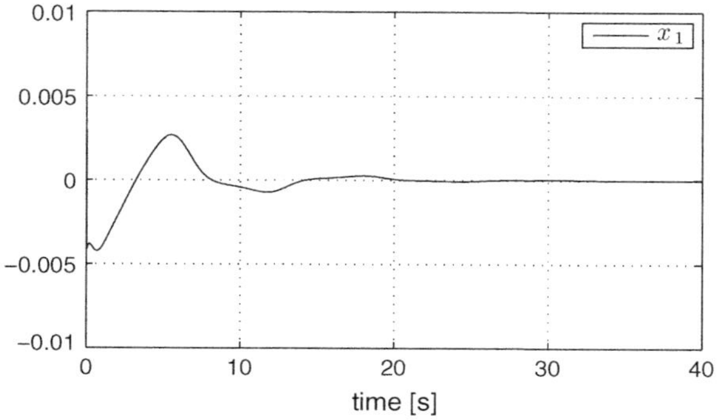

Fig. 9.5 Profile of state trajectories of the closed-loop system in Example 9.2 (part 1)

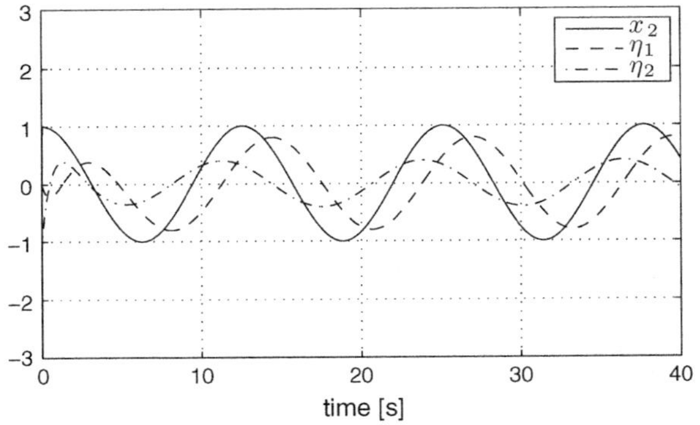

Fig. 9.6 Profile of state trajectories of the closed-loop system in Example 9.2 (part 2)

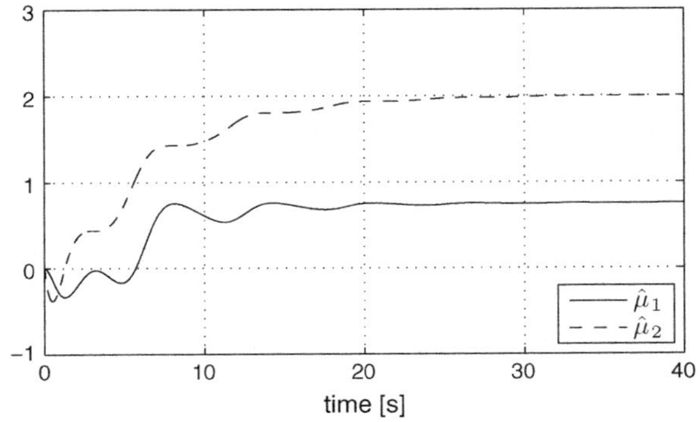

Fig. 9.7 Profile of estimated parameters of the closed-loop system in Example 9.2 (part 3)

9.3 Disturbance Rejection of Lower Triangular Systems

In this section, we turn to a class of more general lower triangular systems as follows:

$$\begin{aligned}
\dot{x}_0 &= f_0(x_0, x_1, w) \\
\dot{x}_i &= f_i(x_0, \overrightarrow{x}_i, w) + b_i(w)x_{i+1}, \ i = 1, \ldots, r-1 \\
\dot{x}_r &= f_r(x_0, \overrightarrow{x}_r, v(t), w) + b(w)u \\
\dot{v} &= A_1(\sigma)v \\
y &= x_1
\end{aligned} \tag{9.32}$$

where the notation in (9.32) is interpreted the same way as (8.41) except that the matrix A_1 is uncertain and satisfies Assumption 7.2. We will consider the disturbance rejection problem of (9.32) by defining $e = y$.

First, we need the following assumption.

Assumption 9.5 For $i = 1, \ldots, r-1$, the function $b_i(w)$ is away from zero, e.g., $b_i(w) > 0, \forall w \in \mathbb{W}$. The function $b(w)$ is away from zero, e.g., $b(w) > 0, \forall w \in \mathbb{W}$.

Under Assumption 9.5, the solution to the regulator equations of the system (9.32) exists globally and can be obtained as $\mathbf{x}_i(v, w) = 0,\ i = 1, \ldots, r$, and

$$\mathbf{u}(v, w) = -b^{-1}(w) f_r(0, v, w).$$

We also assume the following:

Assumption 9.6 The function $\mathbf{u}(v, w) \in \mathbb{R}$ is a polynomial function of v, with coefficients depending on w.

Assumption 9.7 The subsystem $\dot{x}_0 = f(x_0, x_1, w)$ has an ISS Lyapunov function $V(x_0)$, i.e.,

$$V(x_0) \sim \{\underline{\alpha}, \bar{\alpha}, \alpha, \sigma \mid \dot{x}_0 = f(x_0, x_1, w)\}$$

and

$$\limsup_{s \to 0^+} \frac{s^2}{\alpha(s)} < \infty, \quad \limsup_{s \to 0^+} \frac{\sigma(s)}{s^2} < \infty.$$

Under Assumption 9.6, Lemma 7.1 gives a steady-state generator with output u:

$$\dot{\theta}(v, w, \sigma) = \Phi(\sigma)\theta(v, w, \sigma), \quad \mathbf{u}(v, w) = \Psi\theta(v, w, \sigma). \tag{9.33}$$

In particular, the pair $(\Psi, \Phi(\sigma))$ is observable. Correspondingly, for any controllable pair (M, N) with $M \in \mathbb{R}^{\ell \times \ell}$ Hurwitz and $N \in \mathbb{R}^{\ell \times 1}$, there is a unique nonsingular matrix $T(\sigma)$ satisfying the Sylvester equation $T(\sigma)\Phi(\sigma) - MT(\sigma) = N\Psi$. Then, associated with the steady-state generator (9.33), we can construct an internal model candidate on the mapping $\tau(\theta, \sigma) = T(\sigma)\theta$ as follows:

$$\dot{\eta} = M\eta + Nu. \tag{9.34}$$

Next, we modify the coordinate and input transformation (8.48) and (8.51) to the following:

$$\begin{aligned} z_1 &= x_0 \\ \tilde{\eta} &= \eta - T(\sigma)\theta(v, w, \sigma) - Nb^{-1}x_r \end{aligned}$$

Performing the above transformation on (9.32) and (9.34) gives the transformed augmented system of the following form:

$$
\begin{aligned}
\dot{z}_1 &= q_1(z_1, x_1, w) \\
\dot{x}_i &= f_i(z_1, \overrightarrow{x}_i, w) + b_i(w)x_{i+1}, \ i = 1, \ldots, r-1 \\
\dot{\tilde{\eta}} &= M\tilde{\eta} + \rho(z_1, \overrightarrow{x}_{r-1}, x_r, d) \\
\dot{x}_r &= \bar{f}_r(z_1, \overrightarrow{x}_{r-1}, \tilde{\eta}, x_r, d) - b(w)\eta^{\mathsf{T}}\mu + b(w)u
\end{aligned} \tag{9.35}
$$

where $q_1 = f_0$, $\mu^{\mathsf{T}} = \Psi T^{-1}(\sigma)$ and

$$
\begin{aligned}
\rho(z_1, \overrightarrow{x}_{r-1}, x_r, d) &= MNb^{-1}(w)x_r - Nb^{-1}(w) \\
&\quad [f_r(z_1, \overrightarrow{x}_{r-1}, x_r, v, w) - f_r(0, v, w)] \\
\bar{f}_r(z_1, \overrightarrow{x}_{r-1}, \tilde{\eta}, x_r, d) &= f_r(z_1, \overrightarrow{x}_r, v, w) - f_r(0, v, w) \\
&\quad + b(w)\Psi T^{-1}(\sigma)(\tilde{\eta} + Nb^{-1}(w)x_r).
\end{aligned}
$$

Let $\bar{x}_0 = \operatorname{col}(z_1, \overrightarrow{x}_{r-1})$, and

$$
\bar{f}_0(\bar{x}_0, x_r, d) = \begin{bmatrix} q_1(z_1, x_1, w) \\ f_1(x_0, \overrightarrow{x}_1, w) + b_1(w)x_2 \\ \vdots \\ f_{r-1}(z_1, \overrightarrow{x}_{r-1}, w) + b_{r-1}(w)x_r \end{bmatrix}.
$$

Then we have

$$
\dot{\bar{x}}_0 = \bar{f}_0(\bar{x}_0, x_r, d). \tag{9.36}
$$

Thus the system (9.35) can be put in the following form:

$$
\begin{aligned}
\dot{\bar{x}}_0 &= \bar{f}_0(\bar{x}_0, x_r, d) \\
\dot{\tilde{\eta}} &= M\tilde{\eta} + \rho(\bar{x}_0, x_r, d) \\
\dot{x}_r &= \bar{f}_r(\bar{x}_0, \tilde{\eta}, x_r, d) - b(w)\eta^{\mathsf{T}}\mu + b(w)u.
\end{aligned} \tag{9.37}
$$

Now consider the $\bar{x}_0$ subsystem of (9.37). With the following coordinate transformation

$$
\begin{aligned}
\mathcal{X}_1 &= x_1 \\
\mathcal{X}_{i+1} &= x_{i+1} + \rho_i(\mathcal{X}_i)\mathcal{X}_i, \quad i = 1, \ldots, r-1,
\end{aligned}
$$

the system (9.37) takes the following form, with $\mathcal{X}_0 = \operatorname{col}(z_1, \mathcal{X}_1, \ldots, \mathcal{X}_{r-1})$,

$$\begin{aligned}
\dot{\mathcal{X}}_0 &= \varphi(\mathcal{X}_0, \mathcal{X}_r, d) \\
\dot{\tilde{\eta}} &= M\tilde{\eta} + \tilde{\rho}(\mathcal{X}_0, \mathcal{X}_r, d) \\
\dot{\mathcal{X}}_r &= \tilde{f}_r(\mathcal{X}_0, \tilde{\eta}, \mathcal{X}_r, d) - b(w)\eta^{\mathsf{T}}\mu + b(w)u
\end{aligned} \tag{9.38}$$

where

$$\begin{aligned}
\tilde{\rho}(\mathcal{X}_0, \mathcal{X}_r, d) &= \rho(\bar{x}_0, x_r, d) \\
\tilde{f}_r(\mathcal{X}_0, \tilde{\eta}, \mathcal{X}_r, d) &= \bar{f}_r(\bar{x}_0, \tilde{\eta}, x_r, d) + \frac{\partial(\rho_{r-1}(\mathcal{X}_{r-1})\mathcal{X}_{r-1})}{\partial \mathcal{X}_{r-1}}\dot{\mathcal{X}}_{r-1},
\end{aligned}$$

and $\dot{\mathcal{X}}_{r-1}$ is equal to the last element of the vector function $\varphi(\mathcal{X}_0, \mathcal{X}_r, d)$. The vector function φ is defined in Sect. 4.1 [see φ_i in (4.9) for the general procedure]. Under Assumption 9.7, by Theorem 4.2, there exist functions $\rho_1, \ldots, \rho_{r-1}$ such that the system $\dot{\mathcal{X}}_0 = \varphi(\mathcal{X}_0, \mathcal{X}_r, d)$ has an ISS Lyapunov function, i.e.,

$$W(\mathcal{X}_0, \mathcal{X}_r) \sim \{\underline{\beta}, \bar{\beta}, \beta, \varsigma, \mid \dot{\mathcal{X}}_0 = \varphi(\mathcal{X}_0, \mathcal{X}_r, d)\},$$

and

$$\limsup_{s \to 0^+} \frac{s^2}{\beta(s)} < \infty, \quad \limsup_{s \to 0^+} \frac{\varsigma(s)}{s^2} < \infty.$$

It can be seen that (9.38) is in the same form as (9.9), and, under Assumption 9.7, the $\mathcal{X}_0$-subsystem of (9.38) satisfies the same assumption as the $\bar{x}_0$-subsystem of (9.9) does. Thus, by Theorem 9.1, we immediately have the following result.

Theorem 9.5 *Consider the composite system (9.32). Under Assumptions 7.2, 9.5–9.7, the GRORP of (9.32) is solved on the prescribed compact set $\mathbb{D}$ by a controller of the following form:*

$$\begin{aligned}
u &= \eta^{\mathsf{T}}\hat{\mu} - \rho_r(\mathcal{X}_r)\mathcal{X}_r \\
\mathcal{X}_{i+1} &= x_{i+1} + \rho_i(\mathcal{X}_i)\mathcal{X}_i, \qquad i = r-1, \ldots, 2 \\
\mathcal{X}_1 &= x_1 \\
\dot{\hat{\mu}} &= -\Lambda \mathcal{X}_r \eta, \ \ \Lambda = \Lambda^{\mathsf{T}} > 0 \\
\dot{\eta} &= M\eta + Nu.
\end{aligned}$$

Let ℓ be the degree of the minimal zeroing polynomial of $\mathbf{u}(v, w)$. Then, $\mathbf{u}(v(t), w)$ can be written in a trigonometric polynomial form

$$\mathbf{u}(v(t), w) = \sum_{i=1}^{\ell} C_i(v_0, w)e^{j\hat{\omega}_i t}, \ \ C_i(v_0, w) \in \mathbb{C}, \hat{\omega}_i \in \mathbb{R} \tag{9.39}$$

where $\hat{\omega}_i = -\hat{\omega}_{1+\ell-i}$, $\hat{\omega}_i \neq \hat{\omega}_j$ for $i \neq j$, $C_i = C^*_{1+\ell-i}$, and $C_i(v_0, w)$ is not identically zero for $v_0 \in \mathbb{R}^{l_1}$ and $w \in \mathbb{R}^{l_2}$. As a counterpart of Theorem 9.2, we have the following result. The proof is left as an exercise for readers.

Theorem 9.6 *In Theorem 9.5, suppose the internal model is of the minimal dimension, and (v_0, w) is such that none of $C_i(v_0, w)$, $i = 1, \ldots, \ell$, in (9.39) is zero. Then* $\lim_{t\to\infty}(\hat{\mu}(t) - \mu) = 0$.

Example 9.3 Consider the following nonlinear system

$$\begin{aligned}
\dot{z}_1 &= a_{11}z_1 + a_{12}x_1\\
\dot{z}_2 &= a_3 z_2 + z_1 x_1\\
\dot{x}_1 &= a_{21}z_1 + a_{22}x_1 - z_1 z_2 + x_2\\
\dot{x}_2 &= a_{41}z_1 + a_{42}z_2 x_1 + u + d(t)\\
e &= x_1
\end{aligned} \tag{9.40}$$

which is of the form (9.32) with $r = 2$ and $x_0 = [z_1,\ z_2]^{\mathrm{T}}$. Let

$$a = \mathrm{col}(a_{11}, a_{12}, a_{21}, a_{22}, a_3, a_{41}, a_{42})$$

be the unknown parameters in the system with $a_{11} < 0$, $a_3 < 0$. Suppose $a = \bar{a} + w$ where $\bar{a}$ is the nominal value and $w \in \mathbb{W}$ with $\mathbb{W} \subset \mathbb{R}^7$ being a prescribed compact set such that $a_{11} < 0$, $a_3 < 0$. It is noted that the system (9.40) is general enough to cover the classes of hyperchaotic Lorenz systems in Sect. 3.4. Assume $d(t) = A_m \cos(\sigma t + \phi)$ for some unknown constants A_m, σ, ϕ. We introduce the following exosystem

$$\dot{v} = A_1(\sigma)v, \quad v(0) = v_0, \quad t \geq 0 \tag{9.41}$$

where

$$v = \begin{bmatrix} v_1 \\ v_2 \end{bmatrix}, \quad A_1(\sigma) = \begin{bmatrix} 0 & \sigma \\ -\sigma & 0 \end{bmatrix}.$$

It can be seen that $d(t) = v_1(t)$ when $v_0 = [A_m \cos\phi, -A_m \sin\phi]^{\mathrm{T}}$. We will apply Theorem 9.5 to design a state feedback controller to achieve the objective of asymptotic disturbance rejection.

First, Assumption 9.6 is satisfied by noting $\mathbf{u}(v) = -v_1$. Thus, we have

$$\theta(v, w, \sigma) = \begin{bmatrix} -v_1 & -\sigma v_2 \end{bmatrix}^{\mathrm{T}},$$

and

$$\frac{d^2\mathbf{u}(v, w)}{dt^2} + \sigma^2 \mathbf{u}(v, w) = 0,$$

which leads to a steady-state generator

$$\dot{\theta}(v, w, \sigma) = \Phi(\sigma)\theta(v, w, \sigma), \quad \mathbf{u}(v, w) = \Psi\theta(v, w, \sigma)$$

with output u where

$$\Phi(\sigma) = \begin{bmatrix} 0 & 1 \\ -\sigma^2 & 0 \end{bmatrix}, \quad \Psi = \begin{bmatrix} 1 & 0 \end{bmatrix}.$$

Next, pick a pair of controllable matrices

$$M = \begin{bmatrix} 0 & 1 \\ -m_1 & -m_2 \end{bmatrix}, \quad N = \begin{bmatrix} 0 \\ 1 \end{bmatrix}$$

where $m_1, m_2 > 0$. Since the pair $(\Psi, \Phi(\sigma))$ is observable, solving the Sylvester equation $MT(\sigma) + N\Psi = T(\sigma)\Phi(\sigma)$ gives

$$\begin{aligned} T^{-1}(\sigma) &= \begin{bmatrix} m_1 - \sigma^2 & m_2 \\ -m_2\sigma^2 & m_1 - \sigma^2 \end{bmatrix} \\ \Psi T^{-1}(\sigma) &= \begin{bmatrix} m_1 - \sigma^2 & m_2 \end{bmatrix}. \end{aligned} \tag{9.42}$$

Thus, we can obtain the internal model candidate on the mapping $\tau(\theta, \sigma) = T(\sigma)\theta$ as follows:

$$\dot{\eta} = M\eta + Nu. \tag{9.43}$$

Performing the coordinate transformation $\tilde{\eta} = \eta - T(\sigma)\theta(v, w) - Nx_2$ on (9.40) and (9.43) gives

$$\begin{aligned} \dot{z}_1 &= a_{11}z_1 + a_{12}x_1 \\ \dot{z}_2 &= a_3 z_2 + z_1 x_1 \\ \dot{x}_1 &= a_{21}z_1 + a_{22}x_1 - z_1 z_2 + x_2 \\ \dot{\tilde{\eta}} &= M\tilde{\eta} + MNx_2 - N(a_{41}z_1 + a_{42}z_2 x_1) \\ \dot{x}_2 &= a_{41}z_1 + a_{42}z_2 x_1 + \Psi T^{-1}(\sigma)(\tilde{\eta} + Nx_2) - \eta^{\mathsf{T}}\mu + u \\ e &= x_1 \end{aligned} \tag{9.44}$$

where $\mu^{\mathsf{T}} = \Psi T^{-1}(\sigma)$. It can be verified that the system (9.44) satisfies Assumption 9.6. Now applying Theorem 9.5 gives an adaptive state feedback controller

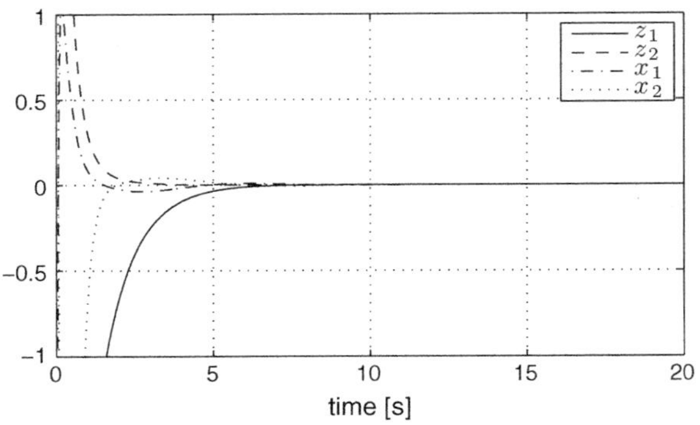

Fig. 9.8 Profile of state trajectories of the closed-loop system in Example 9.3 (part 1)

$$\begin{aligned} u &= -\rho_2(\mathcal{X}_2)\mathcal{X}_2 + \eta^{\mathsf{T}}\hat{\mu} \\ \mathcal{X}_2 &= x_2 + \rho_1(x_1)x_1 \\ \dot{\eta} &= M\eta + Nu \\ \dot{\hat{\mu}} &= -\mathcal{X}_2\eta. \end{aligned}$$

where the functions ρ_1 and ρ_2 depend on the size of $\mathbb{D}$. For example, with $\mathbb{D} = \{v(t) \mid \|v(t)\| \leq 1\} \times \{w \mid w \in \mathbb{R}^7, \| w \| \leq 0.2\} \times \{\sigma \mid 0.5 \leq \sigma \leq 1.5\}$, we can obtain $\rho_1(x_1) = 10 + 4x_1^2$ and $\rho_2(\mathcal{X}_2) = 50 + 8\mathcal{X}_2^8$.

Furthermore, it can be seen that the minimal zeroing polynomial of $\mathbf{u}(v, w)$ is $\lambda^2 + \sigma^2$. Thus the minimal internal model is of dimension two. It can be verified that (9.39) holds for

$$\begin{aligned} \ell = 2, \ \hat{\omega}_1 &= -\hat{\omega}_2 = \sigma \\ C_1(v_0, w) = C_2^*(v_0, w) &= -\frac{v_1(0)}{2} - \frac{v_2(0)}{2j}. \end{aligned}$$

Since the internal model is minimal, and for all nonzero v_0, none of C_i is zero for $i = 1, 2$. The condition for the convergence of the unknown parameter $\hat{\mu}^{\mathsf{T}}$ to the true value $\mu^{\mathsf{T}} = \Psi T^{-1}(\sigma)$ is thus satisfied.

The simulation is conducted with $a_{11} = -1$, $a_{12} = 0.2$, $a_3 = -2$, $a_{21} = 0.1$, $a_{22} = 0.2$, $a_{41} = -0.1$, and $a_{42} = 0.1$. The disturbance is $d(t) = 20\cos(\sigma t)$ with $\sigma = 1$. In the controller, $m_1 = 2$ and $m_2 = 3$ are used. As a result, we have $\mu = [1, 3]^{\mathsf{T}}$. The initial conditions are $z_1(0) = -5$, $z_2(0) = 5$, $x_1(0) = -5$, $x_2(0) = 5$, $\eta(0) = [0, 0]^{\mathsf{T}}$, and $\hat{\mu}(0) = [0, 0]^{\mathsf{T}}$. The performance of the controller is shown in Figs. 9.8, 9.9 and 9.10. In particular, the performance output is shown in Fig. 9.8. All other state variables are bounded as shown in Fig. 9.9. Figure 9.10 further shows that the estimated parameters converge to their true values asymptotically.

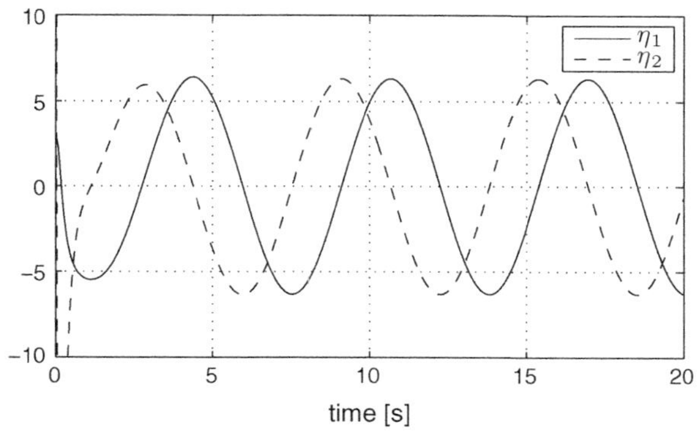

Fig. 9.9 Profile of state trajectories of the closed-loop system in Example 9.3 (part 2)

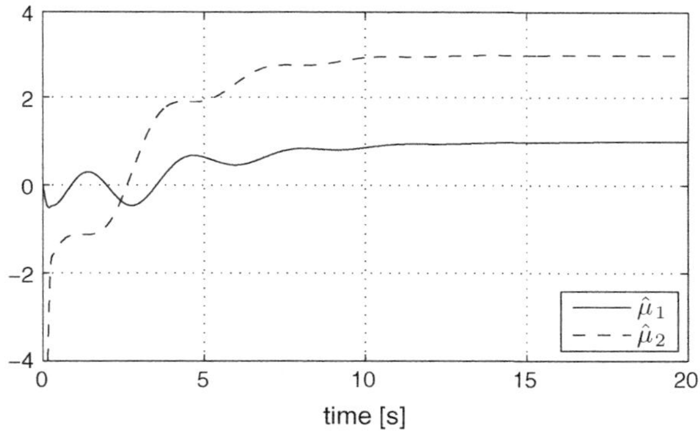

Fig. 9.10 Profile of estimated parameters of the closed-loop system in Example 9.3

9.4 Disturbance Rejection of the FitzHugh–Nagumo Model

Consider the controlled FitzHugh–Nagumo model described by (3.45) with $a = b = c = d = 1$, i.e.,

$$\begin{aligned} \dot{z} &= \begin{bmatrix} -\epsilon & 0 \\ 0 & -\varepsilon \end{bmatrix} z + \begin{bmatrix} \epsilon \\ -\varepsilon \end{bmatrix} y \\ \dot{y} &= y - y^3/3 + [-1 \ 1]z + I(t) + u. \end{aligned} \tag{9.45}$$

Here we consider $I(t) = A_m \sin(\sigma t + \phi)$ as the external disturbance. The objective is to design the control input u for achieving global disturbance rejection in the sense that the trajectory of the closed-loop system starting from any initial condition is bounded and $\lim_{t\to\infty} \mathrm{col}(z(t), y(t)) = 0$ subject to any $I(t)$. Also, we allow the system parameters ϵ and ε to be unknown positive numbers. The problem can be formulated as the GRORP studied in this chapter. In fact, the external disturbance $I(t) = A_m \sin(\sigma t + \phi)$ can be generated by the following exosystem with $I(t) = v_1(t)$:

$$\dot{v} = A_1(\sigma)v, \quad v(0) = v_0 = \begin{bmatrix} A_m \sin\phi \\ A_m \cos\phi \end{bmatrix}, \quad A_1(\sigma) = \begin{bmatrix} 0 & \sigma \\ -\sigma & 0 \end{bmatrix}. \tag{9.46}$$

We assume the frequency $\sigma > 0$ is unknown.

Let $x_0 = z$ and $x = y$. Then the system (9.45) can be rewritten as

$$\begin{aligned} \dot{x}_0 &= \begin{bmatrix} -\epsilon & 0 \\ 0 & -\varepsilon \end{bmatrix} x_0 + \begin{bmatrix} \epsilon \\ -\varepsilon \end{bmatrix} x \\ \dot{x} &= x - x^3/3 + [-1 \ 1]x_0 + v_1 + u. \end{aligned} \tag{9.47}$$

which is of the form (9.32). So, Theorem 9.5 can be applied.

First, we need to verify Assumption 9.6, which is trivially satisfied by noting $\mathbf{u}(v, w) = -v_1$. Thus, we have

$$\theta(v, w, \sigma) = \begin{bmatrix} -v_1 & -\sigma v_2 \end{bmatrix}^{\mathsf{T}},$$

and

$$\frac{d^2\mathbf{u}(v, w)}{dt^2} + \sigma^2\mathbf{u}(v, w) = 0,$$

which leads to the following steady-state generator:

$$\dot{\theta}(v, w, \sigma) = \Phi(\sigma)\theta(v, w, \sigma), \ \mathbf{u}(v, w) = \Psi\theta(v, w, \sigma)$$

with output u where

$$\Phi(\sigma) = \begin{bmatrix} 0 & 1 \\ -\sigma^2 & 0 \end{bmatrix}, \quad \Psi = \begin{bmatrix} 1 & 0 \end{bmatrix}.$$

Next, pick a pair of controllable matrices

$$M = \begin{bmatrix} 0 & 1 \\ -m_1 & -m_2 \end{bmatrix}, \quad N = \begin{bmatrix} 0 \\ 1 \end{bmatrix}$$

where $m_1, m_2 > 0$. Since the pair $(\Psi, \Phi(\sigma)$ is observable, solving the Sylvester equation $MT(\sigma) + N\Psi = T(\sigma)\Phi(\sigma)$ gives

$$T^{-1}(\sigma) = \begin{bmatrix} m_1 - \sigma^2 & m_2 \\ -m_2\sigma^2 & m_1 - \sigma^2 \end{bmatrix}$$
$$\Psi T^{-1}(\sigma) = \begin{bmatrix} m_1 - \sigma^2 & m_2 \end{bmatrix}. \tag{9.48}$$

Thus, we can obtain the internal model candidate on the mapping $\tau(\theta, \sigma) = T(\sigma)\theta$ as follows:

$$\dot{\eta} = M\eta + Nu. \tag{9.49}$$

Performing the coordinate transformation $\tilde{\eta} = \eta - T(\sigma)\theta(v, w) - Nx$ on (9.47) and (9.49) gives

$$\dot{x}_0 = \begin{bmatrix} -\epsilon & 0 \\ 0 & -\varepsilon \end{bmatrix} x_0 + \begin{bmatrix} \epsilon \\ -\varepsilon \end{bmatrix} x$$
$$\dot{\tilde{\eta}} = M\tilde{\eta} + MNx - N(x - x^3/3 + [-1\ 1]x_0)$$
$$\dot{x} = x - x^3/3 + [-1\ 1]x_0 + \Psi T^{-1}(\sigma)(\tilde{\eta} + Nx) - \eta^{\mathsf{T}}\mu + u \tag{9.50}$$

with $\mu^{\mathsf{T}} = \Psi T^{-1}(\sigma)$. It can be seen that Assumption 9.7 is also satisfied. Thus, we can obtain the following controller

$$u = -x\rho(x) + \eta^{\mathsf{T}}\hat{\mu}$$
$$\dot{\eta} = M\eta + Nu$$
$$\dot{\hat{\mu}} = -x\eta. \tag{9.51}$$

The function $\rho(x)$ depends on the size of ϵ, ε and σ. In particular, when $m_1 = 3$, $m_2 = 1$, $0.1 \le \epsilon, \varepsilon \le 2$ and $0.1 \le \sigma \le 10$, we have $\rho(x) = 10x^4 + 50$.

Furthermore, it can be seen that the minimal zeroing polynomial of $\mathbf{u}(v, w)$ is $\lambda^2 + \sigma^2$. Thus the minimal internal model is of dimension two. It can be verified that (9.39) holds for

$$\ell = 2,\ \hat{\omega}_1 = -\hat{\omega}_2 = \sigma$$
$$C_1(v_0, w) = C_2^*(v_0, w) = -\frac{v_1(0)}{2} - \frac{v_2(0)}{2j}.$$

Since the internal model is minimal, and for all nonzero v_0, none of C_i is zero for $i = 1, 2$. The condition for the convergence of the unknown parameter $\hat{\mu}^{\mathsf{T}}$ to the true value $\mu^{\mathsf{T}} = \Psi T^{-1}(\sigma)$ is thus satisfied.

The simulation is conducted with $\epsilon = 2$, $\varepsilon = 2$, and $\sigma = 1$. The disturbance is $I(t) = 10\cos(\sigma t)$. In the controller, $m_1 = 3$ and $m_2 = 1$ are used. As a result, we have $\mu = [2, 1]^{\mathsf{T}}$. The initial conditions are $x_0(0) = [-5, 5]^{\mathsf{T}}$, $x(0) = -5$, $\eta(0) = [0, 0]^{\mathsf{T}}$, and $\hat{\mu}(0) = [0, 0]^{\mathsf{T}}$. The performance of the controller is shown in Figs. 9.11, 9.12 and 9.13. In particular, the performance output is shown in Fig. 9.11.

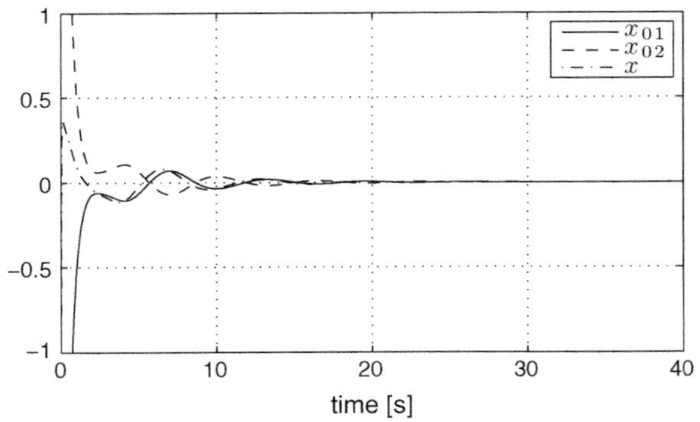

Fig. 9.11 Profile of state trajectories of the controlled FitzHugh–Nagumo model (part 1)

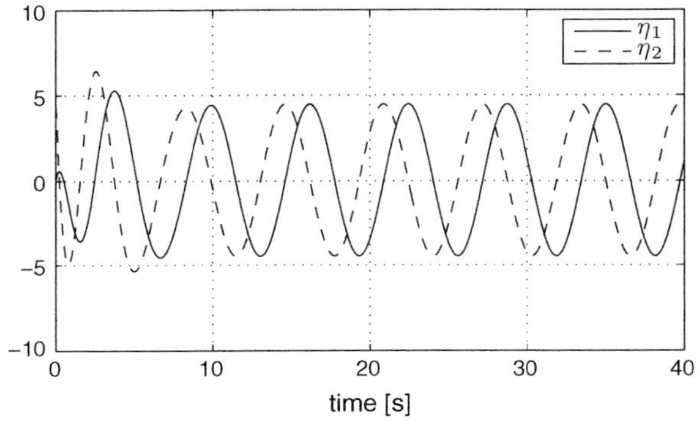

Fig. 9.12 Profile of state trajectories of the controlled FitzHugh–Nagumo model (part 2)

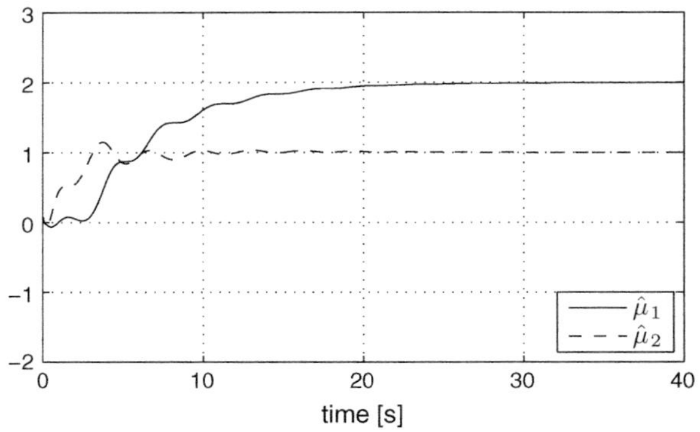

Fig. 9.13 Profile of estimated parameters of the controlled FitzHugh–Nagumo model

All other state variables are bounded as shown in Fig. 9.12. The convergence of the estimated parameters to the true values can be observed from Fig. 9.13.

9.5 Notes and References

The adaptive output regulation problem with uncertain exosystem was first studied in [1] for parametric strict feedback systems. Then, the idea was used for the semi-global robust output regulation of lower triangular systems in [2, 3]. The global case was studied in [4–6] for nonlinear systems in output feedback form. In particular, a class of large-scale systems was studied in [6] which includes the output feedback system as a special case when the number of the subsystems is equal to one. The parameter convergence issue given in this chapter is mainly from [7]. Section 9.1 is adapted from [8]. Section 9.2 is mainly based on [7, 9], and Sect. 9.3 on [10]. The example on FitzHugh–Nagumo model can find its reference in [8]. The more general result on GRORP for lower triangular systems which is not included in this chapter can be found in [11, 12]. The adaptive output regulation problem with uncertain exosystem for an arbitrarily unknown compact set $\mathbb{D}$ can be found in [13].

9.6 Problems

Problem 9.1 Solve the GRORP for the following system

$$\begin{aligned} \dot{x}_0 &= \begin{bmatrix} -1 & 0 \\ x & -2 \end{bmatrix} x_0 + \begin{bmatrix} x \\ 0 \end{bmatrix} \\ \dot{x} &= w x_0 x + u \\ e &= x - v_1 \end{aligned}$$

where the external disturbance $v = \text{col}(v_1, \ v_2)$ is governed by the exosystem

$$\dot{v} = A_1(\sigma)v, \ A_1(\sigma) = \begin{bmatrix} 0 & \sigma \\ -\sigma & 0 \end{bmatrix}$$

It is assumed that $v(t) \in \mathbb{V} = \{v_1^2 + v_2^2 \leq 1\}$, $-1 \leq w \leq 1$ and $1 \leq \sigma \leq 2$.

Problem 9.2 Repeat Problems 8.3 and 8.4 when A_1 is replaced by

$$A_1(\sigma) = \begin{bmatrix} 0 & \sigma \\ -\sigma & 0 \end{bmatrix}, \ \ 1 \leq \sigma \leq 2.$$

Problem 9.3 Extend the results in Sects. 9.1–9.3 to the case where the bound of $\mathbb{D}$ is unknown, or, what is the same, both $v(0)$ and w can be arbitrarily large, by further using the universal adaptive control technique.

Problem 9.4 Consider the system (9.1). Under Assumptions 7.2, 9.1–9.3, suppose the subsystem $\dot{\bar{x}}_0 = \bar{f}_0(\bar{x}_0, \bar{x}, d)$ is iISS with an iISS Lyapunov function $V(\bar{x}_0)$ satisfying $\underline{\alpha}(\|\bar{x}_0\|) \leq V(\bar{x}_0) \leq \bar{\alpha}(\|\bar{x}_0\|)$ for some class $\mathcal{K}_\infty$ functions $\underline{\alpha}$ and $\bar{\alpha}$ such that, along the trajectory of the $\bar{x}_0$-subsystem,

$$\dot{V}(\bar{x}_0) \leq -\alpha(\|\bar{x}_0\|) + \sigma(|\bar{x}|)$$

for some positive definite function α and some class $\mathcal{K}$ function σ. Suppose the function σ satisfies the property $\limsup_{s \to 0^+} \left[\sigma(s)/s^2\right] < \infty$.

Let m_1 and m_2 be some smooth positive functions such that

$$|f(x_0, x, v, w)| \leq m_1(\|x_0\|)\|x_0\| + m_2(|x|)|x|, \ \forall v \in \mathbb{V}, w \in \mathbb{W}.$$

Moreover, $m_1^2(s) = \mathcal{O}[\alpha(s)]$ and $\limsup_{s \to +\infty} \left[m_1^2(s)/\alpha(s)\right] < \infty$ if α is not of class $\mathcal{K}_\infty$. Show that the GRORP for the system (9.1) with $\mathbb{D}$ a known compact set is solved by a controller of the form (9.12).

References

1. Nikiforov VO (1998) Adaptive non-linear tracking with complete compensation of unknown disturbances. Eur J Control 4:132–139
2. Serrani A, Isidori A, Marconi L (2001) Semiglobal nonlinear output regulation with adaptive internal model. IEEE Trans Autom Control 46:1178–1194
3. Delli Priscoli F, Marconi L, Isidori A (2006) A new approach to adaptive nonlinear regulator. SIAM J Control Optim 45:829–855
4. Ding Z (2003) Global stabilization and disturbance suppression of a class of nonlinear systems with uncertain internal model. Automatica 39:471–479
5. Ding Z (2006) Adaptive estimation and rejection of unknown sinusoidal disturbances in a class of non-minimum-phase nonlinear systems. IEE Proc Control Theory Appl 153:379–386
6. Ye XD, Huang J (2003) Decentralized adaptive output regulation for a class of large-scale nonlinear systems. IEEE Trans Autom Control 48:276–281
7. Liu L, Chen Z, Huang J (2009) Parameter convergence and minimal internal model with an adaptive output regulation problem. Automatica 45:1306–1311
8. Xu D, Huang J (2010) Robust adaptive control of a class of nonlinear systems and its applications. IEEE Trans Circuit Syst II Express Briefs 57:691–702
9. Xu D, Huang J (2010) Output regulation design for a class of nonlinear systems with an unknown control direction. ASME J Dyn Syst Meas Contr 132:014503
10. Liu L, Chen Z, Huang J (2011) Global disturbance rejection of lower triangular systems with an unknown linear exosystem. IEEE Trans Autom Control 56(7):1690–1695
11. Chen Z, Huang J (2002) Global tracking of uncertain nonlinear cascaded systems with adaptive internal model. Proceedings of the 41st IEEE conference on decision and control, pp 3855–3862
12. Chen Z, Huang J (2009) Global output regulation with uncertain exosystems. In: Hu X, Jonsson U, Wahlberg B, Ghosh B (eds) Three decades of progress in control science. Springer, Berlin, pp 105–119
13. Xu D, Huang J (2010) Global output regulation for output feedback systems with an uncertain exosystem and its application. Int J Robust Nonlinear Control 20:1678–1691

Chapter 10
Attitude Control of a Rigid Spacecraft

Attitude control of spacecraft systems has been a benchmark control problem and has been extensively studied under various assumptions and scenarios. When a spacecraft system is subject to external disturbances, the attitude control problem poses some specific challenges. In this chapter, we consider the attitude control problem for a spacecraft subject to a class of external disturbances which is a multi-tone sinusoidal function. The techniques introduced in the previous chapters are integrated to solve this problem. This chapter is organized as follows. In Sects. 10.1 and 10.2, we present the model of a rigid spacecraft, and formulate the attitude tracking and disturbance rejection problem, respectively. Then a special case where the model of the spacecraft is known exactly is handled in Sect. 10.3 using the internal model approach studied in Chap. 7. In Sects. 10.4 and 10.5, taking into account the model uncertainty, we detail the approach to dealing with the attitude control and disturbance rejection problem for the case where the frequencies of the disturbance are known and the case where the frequencies of the disturbance are unknown, respectively. The notes and references are given in Sect. 10.6.

10.1 Quaternion Based Rigid Spacecraft Model

In this section, we will present the mathematical model of a rigid spacecraft. The attitude of a three-dimensional rigid body is defined with a set of axes $\{x_b, y_b, z_b\}$ fixed to the body. This set of axes, usually orthogonal coordinates, is named a body reference frame F_b. And another useful reference frame with axes $\{x_i, y_i, z_i\}$ is named an inertial reference frame F_i (see Fig. 10.1). For an earth-orbiting spacecraft, the origin of the inertial coordinate system is at the center of mass of the earth and its direction in space is inertial with respect to the solar system. The coordinate transformation that transforms F_i to F_b is based on the direction cosine matrix C of the form

$$F_b = CF_i.$$

Z. Chen and J. Huang, *Stabilization and Regulation of Nonlinear Systems*,
Advanced Textbooks in Control and Signal Processing,
DOI 10.1007/978-3-319-08834-1_10

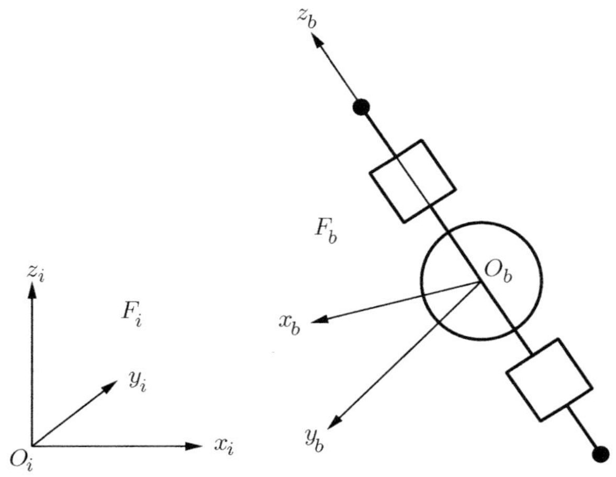

Fig. 10.1 Body reference frame and inertial reference frame

It is easy to check that $\|C\| = 1$ and $C^{\mathrm{T}}C = I_3$ thanks to the orthogonality of the basis vectors.

The matrix C has 9 elements which are subject to 6 constraints. We need to simplify the representation. Usually there are two widely used mathematical representations of spacecraft attitude. One is in terms of Euler angle rotation and the other is based on the unit quaternion. Compared with the first method, the unit quaternion representation has the advantage of being free of singularity, and will thus be adopted here. To introduce the definition of quaternion, we will appeal to the following properties of the direction cosine matrix C. There exist three eigenvalues for C: $\{1, e^{j\phi}, e^{-j\phi}\}$ for $\phi \in \mathbb{R}$. That is, the matrix C has at least one eigenvector with the eigenvalue 1, and this eigenvector is unchanged by the matrix C, i.e.,

$$C\nu = 1\nu.$$

The vector $\nu = [\nu_1, \nu_2, \nu_3]^{\mathrm{T}}$ is called the eigenvector of rotation. This eigenvector has the same components in the body reference frame or the inertial reference frame. In this way, the transform matrix C can be achieved by a single angular rotation about ν over an angle ϕ.

Now, the quaternion is a vector defined by

$$\begin{aligned} q &= [q_1, q_2, q_3, q_4]^{\mathrm{T}} \\ q_i &= \nu_i \sin\left(\frac{\phi}{2}\right) / \|\nu\|, \quad i = 1, 2, 3 \\ q_4 &= \cos\left(\frac{\phi}{2}\right) \end{aligned} \tag{10.1}$$

where $q_v = [q_1, q_2, q_3]^\top$ is called the vector part of q and q_4 is the scalar part of q. Obviously, the quaternion q has the constraint

$$q_v^\top q_v + q_4^2 = 1. \tag{10.2}$$

On the other hand, the direction cosine matrix C can be expressed in term of quaternion as follows:

$$C(q) = (q_4^2 - q_v^\top q_v)I_3 + 2q_v q_v^\top - 2q_4 q_v^\times, \tag{10.3}$$

where, for $q_v \in \mathbb{R}^3$, the notation $q_v^\times$ denotes a skew symmetric matrix as follows:

$$q_v^\times = \begin{bmatrix} 0 & -q_3 & q_2 \\ q_3 & 0 & -q_1 \\ q_2 & q_1 & 0 \end{bmatrix},$$

which satisfies the following properties: for all $a \in \mathbb{R}^3$ and $b \in \mathbb{R}^3$,

$$(a^\times)^\top = -a^\times, \ a^\times b = -b^\times a, \ a^\times a = 0,$$
$$a^\top b^\times a = 0, \ (a+b)^\times = a^\times + b^\times, \ a^\times b^\times = ba^\top - a^\top b I_3,$$
$$(a^\times b)^\times = ba^\top - ab^\top, \ \|a^\times\| = \|a\|.$$

We note that the set of elements of the direction cosine matrix, and the set of elements of the quaternion can be obtained from each other.

To derive the attitude kinematics in terms of the unit quaternion q, we introduce the following notation.

- $\mathbb{Q}$ set of all unit quaternion:
  $\mathbb{Q} = \{q | q = \mathrm{col}(q_v, q_4), q_v \in \mathbb{R}^3, q_4 \in \mathbb{R}, q_v^\top q_v + q_4^2 = 1\}$
- $\odot$ quaternion product: for $q, p \in \mathbb{Q}$,
  $q \odot p = \begin{pmatrix} q_4 p_v + p_4 q_v + q_v^\times p_v \\ q_4 p_4 - q_v^\top p_v \end{pmatrix}$
- q_I quaternion identity: $q_I = \mathrm{col}(0, 0, 0, 1) \in \mathbb{Q}$
- q^{-1} unit quaternion inverse: for $q \in \mathbb{Q}$, $q^{-1} = \mathrm{col}(-q_v, q_4)$
- $q(\cdot)$ for $x \in \mathbb{R}^3$, $q(x) : x \mapsto \mathrm{col}(x, 0)$

In terms of the unit quaternion q, the attitude kinematics of the spacecraft is governed by the following equation:

$$\dot{q} = \frac{1}{2} q \odot q(\Omega) \tag{10.4}$$

where $\Omega \in \mathbb{R}^3$ is the angular velocity of the body frame F_b with respect to the reference frame F_i and is expressed in the frame F_b. The dynamic equation of attitude motion is as follows:

$$J\dot{\Omega} = -\Omega^{\times} J\Omega + u + d \tag{10.5}$$

where $J \in \mathbb{R}^{3\times 3}$ is a constant, positive definite, symmetric, overall inertia matrix of the integrated spacecraft system, $u \in \mathbb{R}^3$ is the control torque, and $d \in \mathbb{R}^3$ is the external disturbance. In this chapter, we assume that the external disturbance $d(t)$ is a combination of a constant and sinusoidal functions, i.e., each component $d_i(t)$, $i = 1, 2, 3$, can be represented as follows:

$$d_i(t) = C_{i0} + \Sigma_{j=1}^{n_i} C_{ij} \sin(b_{ij}t + \Upsilon_{ij}) \tag{10.6}$$

where C_{ij} and Υ_{ij} are arbitrarily unknown amplitudes and phases, and b_{ij}'s are known or unknown frequencies.

Remark 10.1 A simple calculation shows that, along the trajectory of (10.4),

$$\begin{aligned}\frac{d\|q(t)\|^2}{dt} &= 2q_v^{\mathsf{T}}\dot{q}_v + 2q_4\dot{q}_4 \\ &= q_v^{\mathsf{T}}(q_4 I_3 + q_v^{\times})\Omega - q_4 q_v^{\mathsf{T}}\Omega = 0.\end{aligned}$$

Thus, $\|q(0)\| = 1$ implies $\|q(t)\| = 1$ for all $t \geq 0$.

10.2 Problem Formulation

In this section, we will formulate the spacecraft attitude tracking and disturbance rejection problem. Let us suppose the desired attitude motion is generated by the following so-called target system:

$$\dot{q}_d = \frac{1}{2} q_d \odot q(\Omega_d) \tag{10.7}$$

where $q_d = \text{col}(q_{dv}, q_{d4})$ with $q_{dv} = [q_{d1}, q_{d2}, q_{d3}]^{\mathsf{T}}$ is the unit quaternion representing the target attitude and $\Omega_d \in \mathbb{R}^3$ is the target angular velocity. The unit quaternion satisfies $\|q_d\| = 1$. We also assume Ω_d and $\dot{\Omega}_d$ are bounded. Therefore, the reference trajectories $s_d = \text{col}(q_d, \Omega_d, \dot{\Omega}_d)$ are always bounded.

The attitude and angular velocity errors between the spacecraft and the target are defined as follows:

$$\begin{aligned} e &= q_d^{-1} \odot q \\ \omega &= \Omega - C_e \Omega_d \end{aligned} \tag{10.8}$$

where $e = \text{col}(e_v, e_4)$, $e_v = [e_1, e_2, e_3]^\top$, and $C_e = (1 - 2e_v^\top e_v)I_3 + 2e_v e_v^\top - 2e_4 e_v^\times$, and it can be verified that the errors are governed by the following equations

$$\dot{e} = \frac{1}{2} e \odot q(\omega)$$
$$J\dot{\omega} = -(\omega + C_e \Omega_d)^\times J(\omega + C_e\Omega_d) + J(\omega^\times C_e \Omega_d - C_e \dot{\Omega}_d) + u + d. \quad (10.9)$$

By Remark 10.1, $\|e(0)\| = 1$ implies $\|e(t)\| = 1$ for all $t \geq 0$.

Lemma 10.1 *If* $e_v = 0$, *then* $q = \pm q_d$ *and* $C_e = I_3$.

Proof From (10.8), $e_v = 0$ implies $q_{d4}q_v - q_4 q_{dv} = q_{dv}^\times q_v$. But $q_{d4}q_v - q_4 q_{dv}$ and $q_{dv}^\times q_v$ are orthogonal to each other as $[q_{d4}q_v - q_4 q_{dv}]^\top q_{dv}^\times q_v = 0$. As a result, both the vectors are zero, i.e., $q_{d4}q_v - q_4 q_{dv} = q_{dv}^\times q_v = 0$. In other words, $(q_{d4}/q_4)q_v = q_{dv}$. Then, $e_v = 0$ gives $e_4 = \pm 1$ (noting $\|e\| = 1$) and hence

$$\pm 1 = q_{dv}^\top q_v + q_4 q_{d4} = (q_{d4}/q_4)\left[q_v^\top q_v + q_4^2\right] = q_{d4}/q_4.$$

Using the equations $(q_{d4}/q_4)q_v = q_{dv}$ and $q_{d4}/q_4 = \pm 1$, we get $q = \pm q_d$. The proof is thus completed. In particular, $e_v = 0$ gives $C_e = I_3$. □

Remark 10.2 By Lemma 10.1, $e_v = 0$ implies $q = q_d$ or $-q = q_d$. From (10.1), it can be seen that q corresponds to (ν, ϕ) if and only if $-q$ corresponds to $(\nu, \phi + 2\pi)$. Thus, q_d and $-q_d$ are physically the same point.

We are now ready to state the problem as follows.

Attitude Tracking and Disturbance Rejection Problem (ATDRP): *Given a desired target angular velocity* $\Omega_d(t) : [0, \infty) \mapsto \mathbb{R}^3$ *such that* Ω_d *and* $\dot{\Omega}_d$ *are bounded and a desired attitude motion generated by* (10.7), *design a full information* (q, Ω, s_d) *feedback controller, such that, for any disturbances* $d(t)$ *of the form* (10.6) *and any initial state satisfying* $\|q(0)\| = 1$ *and* $\Omega(0) \in \mathbb{R}^3$, *the state of the closed-loop system composed of* (10.4), (10.5) *and the controller is bounded, and*

$$\lim_{t\to\infty} e_v(t) = 0. \quad (10.10)$$

Next, we will convert the above problem into a well formulated global regulation problem. For this purpose, performing the following coordinate transformation

$$x = \omega + K e_v \quad (10.11)$$

gives

$$\dot{e}_v = \frac{1}{2}(e_4 I_3 + e_v^\times)(x - K e_v) \quad (10.12)$$

$$\dot{e}_4 = -\frac{1}{2} e_v^\top (x - K e_v) \quad (10.13)$$

$$\begin{aligned} J\dot{x} = &-(\omega + C\Omega_d)^\times J(\omega + C\Omega_d) + J(\omega^\times C\Omega_d - C\dot{\Omega}_d) \\ &+ JK\left[\frac{1}{2}(e_4 I_3 + e_v^\times)\omega\right] + u + d. \end{aligned} \tag{10.14}$$

We now show that the kinematics subsystem, i.e., (10.12) and (10.13) has some desirable stability property as follows.

Lemma 10.2 *Consider the kinematics subsystem, i.e.,* (10.12) *and* (10.13), *where* K *is some symmetric positive definite matrix. For any piecewise continuous time function* $x(t)$ *defined for* $t \geq 0$ *satisfying* $\lim_{t\to\infty} x(t) = 0$, *and any initial state* $e(0)$ *satisfying* $\|e(0)\| = 1$, *the solution of the subsystem is bounded for all* $t \geq 0$ *and* $\lim_{t\to\infty} e_v(t) = 0$.

Proof First note that since $e_v^\mathrm{T} e_v + e_4^2 = 1$, e_v and e_4 are always bounded. Define a set $\mathbb{B}(\epsilon) = \{x \in \mathbb{R}^3 \mid \|x\| \leq \epsilon\}$. We need to prove that for any $\epsilon > 0$, there exists a finite time $T \geq 0$ such that

$$e_v(t) \in \mathbb{B}(\epsilon),\ \forall t \geq T. \tag{10.15}$$

Let $k_{\min}$ be the minimal eigenvalue of K. Since $\lim_{t\to\infty} x(t) = 0$, there exists a finite time $T_1 \geq 0$ such that

$$\|x(t)\| \leq \frac{1}{2}\epsilon k_{\min},\ \forall t \geq T_1.$$

If there exists no finite time $T_2 \geq T_1$ such that $e_v(T_2) \notin \mathbb{B}(\epsilon)$, then (10.15) holds with $T = T_1$ and the proof is completed. Thus, in what follows, we assume there exists a finite time $T_2 \geq T_1$ such that $e_v(T_2) \notin \mathbb{B}(\epsilon)$. We will show that this assumption leads to the following two claims:

(i) There exists a finite time $T > T_2$ such that $e_v(T) \in \mathbb{B}(\epsilon)$ and $e_v(t) \notin \mathbb{B}(\epsilon),\ \forall T_2 \leq t < T$ (that is, T is the first time at which $e_v(t)$ enters $\mathbb{B}(\epsilon)$);
(ii) $e_v(t) \in \mathbb{B}(\epsilon),\ \forall t \geq T$.

Satisfaction of (10.15) clearly follows from these two claims.

Proof of Claim i: We first prove that

$$\|e_v(t)\| \geq \epsilon \Longrightarrow \dot{e}_4(t) \geq \frac{1}{4}k_{\min}\epsilon^2 > 0,\ \forall t \geq T_1. \tag{10.16}$$

Indeed, we have

$$\frac{1}{2}e_v^\mathrm{T} K e_v \geq \frac{1}{2}k_{\min}\|e_v\|^2 \geq \frac{1}{2}k_{\min}\|e_v\|\epsilon \geq \|e_v\|\|x\| \geq e_v^\mathrm{T} x.$$

Using (10.13), we have,

$$\dot{e}_4 = \frac{1}{2}e_v^{\mathsf{T}} K e_v - \frac{1}{2}e_v^{\mathsf{T}} x \geq \frac{1}{2}k_{\min}\|e_v\|\epsilon - \frac{1}{4}k_{\min}\|e_v\|\epsilon$$
$$= \frac{1}{4}k_{\min}\|e_v\|\epsilon \geq \frac{1}{4}k_{\min}\epsilon^2 > 0, \ \forall\, t \geq T_1.$$

If (i) does not hold, then for all $t \geq T_2$, $e_v(t) \notin \mathbb{B}(\epsilon)$, that is, $\|e_v(t)\| > \epsilon$. From (10.16), $\dot{e}_4(t)$ is positive for all $t \geq T_2$, so there is a finite time $T_3 \geq T_2$ such that $e_4(T_3) > \sqrt{1-\epsilon^2}$. In fact, it suffices to choose

$$T_3 \geq 4\frac{\sqrt{1-\epsilon^2} - e_4(T_2)}{\epsilon^2 k_{\min}} + T_2.$$

As a result, $\|e_v(T_3)\| = \sqrt{1 - e_4^2(T_3)} < \epsilon$, i.e., $e_v(T_3) \in \mathbb{B}(\epsilon)$.

Proof of Claim ii: We first prove $e_4(T) > 0$ as follows:

On one hand, in (i) we know that for $T_2 \leq t < T$, $e_v(t) \notin \mathbb{B}(\epsilon)$, hence $\dot{e}_4(t) > 0$. It implies that $e_4(t)$ is increasing during $T_2 \leq t < T$. Thus $e_4(T) > e_4(T_2)$.

On the other hand, $e_v(T) \in \mathbb{B}(\epsilon)$ gives $\|e_v(T)\| \leq \epsilon$, and $e_v(T_2) \notin \mathbb{B}(\epsilon)$ gives $\|e_v(T_2)\| > \epsilon$. As a result, $\|e_v(T)\| < \|e_v(T_2)\|$, hence $|e_4(T)| > |e_4(T_2)|$.

From $e_4(T) > e_4(T_2)$ and $|e_4(T)| > |e_4(T_2)|$, we have $e_4(T) > 0$.

Suppose $T_4 \geq T$ is the first time the trajectory of e_v is at the edge of $\mathbb{B}(\epsilon)$, that is, $e_v(t) \in \mathbb{B}(\epsilon), \forall T \leq t \leq T_4$. As a result, during $T \leq t \leq T_4$, we have $\|e_v(t)\| \leq \epsilon$, and $|e_4(t)| \geq \sqrt{1-\epsilon^2}$.

From $e_4(T) > 0$, the continuity of $e_4(t)$ and $|e_4(t)| \geq \sqrt{1-\epsilon^2}$, we have $e_4(t) > 0, \forall T \leq t \leq T_4$.

At the edge of $\mathbb{B}(\epsilon)$, we have $\|e_v(T_4)\| = \epsilon$. By (10.16), we have $\dot{e}_4(T_4) > 0$.

Thus, we have $e_4(T_4) > 0$ and $\dot{e}_4(T_4) > 0$, which imply that, at the edge of $\mathbb{B}(\epsilon)$, the time derivative of $|e_4(t)|$ is positive, hence, the time derivative of $\|e_v(t)\|$ is negative. As a result, the trajectories $e_v(t)$ cannot cross the edge of $\mathbb{B}(\epsilon)$ from inside to outside. Claim (ii) is thus proved. □

By Lemma 10.2 and Remark 10.1, we can see that if there exists a dynamic state feedback control law of the form

$$u = \kappa_1(e, x, s_d, v), \quad \dot{v} = \kappa_2(e, x, s_d, v), \tag{10.17}$$

where v is the compensator state, κ_1 and κ_2 are globally defined smooth functions, and e is a bounded piecewise continuous function, such that, for any initial condition $v(0)$, $x(0)$, the solution of the closed-loop system composed of (10.14) and (10.17) is bounded and $\lim_{t\to\infty} x(t) = 0$, then, for any initial condition $v(0)$, $x(0)$, and $e(0)$ satisfying $\|e(0)\| = 1$, the state of the closed-loop system composed of (10.9) and (10.17) is bounded satisfying $\|e(t)\| = 1$, $\lim_{t\to\infty} e_v(t) = 0$, and $\lim_{t\to\infty} \omega(t) = 0$. Hence, the ATDRP is solved by the controller (10.17) under the coordinate transformation (10.8) and (10.11). Thus, we have simplified the ATDRP to a regulation problem of the dynamic subsystem (10.14). The controller (10.17) is a full information feedback controller utilizing the plant state variables q and Ω and the target state s_d.

The system (10.14) contains two types of uncertainties, namely, the parameter uncertainty of the elements of the matrix J, and the unknown disturbance $d(t)$. Without $d(t)$, the stabilization of (10.14) can be achieved by the standard adaptive control approach studied in Chap. 5. On the other hand, without the uncertainty J, the stabilization of (10.14) can be achieved by the internal model based approach described in Chap. 7. With the presence of both types of uncertainties, however, neither the adaptive control approach nor the internal model based approach works alone. To overcome this difficulty, we will combine the adaptive control technique and the internal model design approach. To make the system (10.14) more amenable to adaptive control technique, we need to put system (10.14) in a more standard form. For any vector $x = [x_1, x_2, x_3]^{\mathrm{T}} \in \mathbb{R}^3$, we have

$$Jx = L(x) \begin{bmatrix} J_{11} & J_{22} & J_{33} & J_{23} & J_{13} & J_{12} \end{bmatrix}^{\mathrm{T}}$$

where

$$L(x) = \begin{bmatrix} x_1 & 0 & 0 & 0 & x_3 & x_2 \\ 0 & x_2 & 0 & x_3 & 0 & x_1 \\ 0 & 0 & x_3 & x_2 & x_1 & 0 \end{bmatrix}.$$

When some entries of J are unknown, there exists an unknown vector δ with dimension $0 \leq n_\delta \leq 6$ such that

$$\begin{bmatrix} J_{11} & J_{22} & J_{33} & J_{23} & J_{13} & J_{12} \end{bmatrix}^{\mathrm{T}} = \bar{L}_1 \delta + \bar{L}_0$$

for some known matrices $\bar{L}_1 \in \mathbb{R}^{6 \times n_\delta}$, $\bar{L}_0 \in \mathbb{R}^{6 \times 1}$. Letting $L_1(x) = L(x)\bar{L}_1$ and $L_0(x) = L(x)\bar{L}_0$ gives

$$Jx = L_1(x)\delta + L_0(x). \tag{10.18}$$

Therefore, we have

$$\begin{aligned} -(\omega + C\Omega_d)^{\times} J(\omega + C\Omega_d) + J(\omega^{\times} C\Omega_d - C\dot{\Omega}_d) &= F_1(e, x, s_d)\delta + F_0(e, x, s_d), \\ JK\left(\frac{1}{2}(e_4 I_3 + e_v^{\times})\omega\right) &= G_1(e, x)\delta + G_0(e, x), \end{aligned}$$

where

$$\begin{aligned} F_i(e, x, s_d) &= -(\omega + C\Omega_d)^{\times} L_i(\omega + C\Omega_d) + L_i(\omega^{\times} C\Omega_d - C\dot{\Omega}_d) \\ G_i(e, x) &= \frac{1}{2} L_i \left(K(e_4 I_3 + e_v^{\times})\omega\right), \quad i = 0, 1. \end{aligned} \tag{10.19}$$

In (10.19), we note $\omega = x - Ke_v$. In general, $F_i(e, 0, s_d) \neq 0$ and $G_i(e, 0) \neq 0$.

As a result, the subsystem (10.14) becomes

$$J\dot{x} = (F_1(e, x, s_d) + G_1(e, x))\delta + F_0(e, x, s_d) + G_0(e, x) + u + d. \quad (10.20)$$

The system (10.20) can be further simplified into the form (10.21) upon the input transformation $u = \bar{u} - F_0(e, x, s_d) - G_0(e, x)$.

$$J\dot{x} = (F_1(e, x, s_d) + G_1(e, x))\delta + \bar{u} + d. \quad (10.21)$$

The control problem of the system (10.21) can be handled in two steps. The first step is to design a dynamic compensator based on the internal model principle described in Chap. 7 to account for the unknown disturbance. The dynamic compensator and the system (10.14) constitute the augmented system which will be stabilized using the approach in Chap. 5. For the purpose of motivation, we will first consider a special case where $\delta = 0$.

10.3 A Special Case for Motivation

When $\delta = 0$, the system (10.21) reduces to a linear system. To handle the disturbance rejection problem of the system (10.21) using the internal model approach, we define the following composite system:

$$\begin{aligned} J\dot{x} &= \bar{u} + Ev \\ \dot{v} &= A_1 v \\ e &= y = x \end{aligned} \quad (10.22)$$

where $v \in \mathbb{R}^{(2(n_1+n_2+n_3)+3)}$,

$$A_1 = \begin{bmatrix} A_{11} & 0 & 0 \\ 0 & A_{12} & 0 \\ 0 & 0 & A_{13} \end{bmatrix}, \quad E = \begin{bmatrix} E_1 & 0 & 0 \\ 0 & E_2 & 0 \\ 0 & 0 & E_3 \end{bmatrix} \quad (10.23)$$

with

$$A_{1i} = \text{diag}\left(0, \begin{bmatrix} 0 & b_{i1} \\ -b_{i1} & 0 \end{bmatrix}, \ldots, \begin{bmatrix} 0 & b_{in_i} \\ -b_{in_i} & 0 \end{bmatrix}\right), \quad E_i = [1, 0, \ldots, 0], \quad i = 1, 2, 3.$$

It is clear that the solution of the regulator equations associated with the composite system (10.22) is given by $\mathbf{x}(v) = 0$ and $\mathbf{u}(v) = -Ev$. For $i = 1, 2, 3$, let $\ell_i = 2n_i + 1$, and

$$p_i(\lambda) = \lambda \prod_{j=1}^{n_i} (\lambda^2 + b_{ij}^2) = \lambda^{\ell_i} - \phi_{i,1} - \phi_{i,2}\lambda - \cdots - \phi_{i,\ell_i}\lambda^{(\ell_i - 1)}.$$

Since

$$\mathbf{u}(v(t)) = \mathrm{col}(\mathbf{u}_1(v(t)), \mathbf{u}_2(v(t)), \mathbf{u}_3(v(t))) = -\mathrm{col}(d_1(t), d_2(t), d_3(t)),$$

from the proof of Lemma 7.2, $\mathbf{u}_i(v)$ satisfy the following linear immersion condition:

$$L_{A_1 v}^{\ell_i} \mathbf{u}_i(v) - \phi_{i,1}\mathbf{u}_i(v) - \phi_{i,2} L_{A_1 v}\mathbf{u}_i(v) - \cdots - \phi_{i,\ell_i} L_{A_1 v}^{\ell_i - 1}\mathbf{u}_i(v) = 0. \quad (10.24)$$

Let

$$\Phi_i = \begin{bmatrix} 0 & 1 & 0 & \cdots & 0 \\ 0 & 0 & 1 & \cdots & 0 \\ \vdots & \vdots & \vdots & \cdots & \vdots \\ 0 & 0 & 0 & \cdots & 1 \\ \phi_{i,1} & \phi_{i,2} & \phi_{i,3} & \cdots & \phi_{i,\ell_i} \end{bmatrix}, \quad \Psi_i = \begin{bmatrix} 1 \\ 0 \\ \vdots \\ 0 \end{bmatrix}^{\mathrm{T}}, \quad (10.25)$$

and

$$\theta_i(v) = \begin{bmatrix} \mathbf{u}_i(v) & L_{A_1 v}\mathbf{u}_i(v) & \ldots & L_{A_1 v}^{\ell_i - 1}\mathbf{u}_i(v) \end{bmatrix}^{\mathrm{T}}. \quad (10.26)$$

Then, from Lemmas 7.1 and 7.6, the following linear systems

$$\dot{\theta}_i(v) = \Phi_i \theta_i(v), \;\; \mathbf{u}_i(v) = \Psi_i \theta_i(v), \quad i = 1, 2, 3 \quad (10.27)$$

constitute the steady-state generator for the system (10.22) with output $\bar{u}$.

For simplicity, we assume that the frequencies of the disturbance are all known. Thus, the matrices Φ_i are all known precisely. Corresponding to the steady-state generator (10.27), using Example 7.8 and Lemma 7.6, we can further synthesize an internal model for (10.22) as follows. Let $M_i \in \mathbb{R}^{\ell_i \times \ell_i}$ and $N_i \in \mathbb{R}^{\ell_i \times 1}$ be a pair of controllable matrices with M_i Hurwitz. Let T_i be a nonsingular matrix T_i satisfying the Sylvester equation

$$T_i \Phi_i - M_i T_i = N_i \Psi_i \quad (10.28)$$

which exists since the pair (Φ_i, Ψ_i) is observable. Let

$$\begin{aligned} M &= \mathrm{diag}(M_1, M_2, M_3) \\ N &= \mathrm{diag}(N_1, N_2, N_3) \\ T &= \mathrm{diag}(T_1, T_2, T_3) \\ \Phi &= \mathrm{diag}(\Phi_1, \Phi_2, \Phi_3) \\ \Psi &= \mathrm{diag}(\Psi_1, \Psi_2, \Psi_3). \end{aligned}$$

Then it can be verified that the following dynamic compensator

$$\dot{\eta} = M\eta + N\bar{u} \quad (10.29)$$

where $\eta \in \mathbb{R}^{\ell}$ with $\ell = \ell_1 + \ell_2 + \ell_3$ is an internal model of (10.22) on the mapping $\tau(\theta) = \text{col}(\tau_1(\theta_1), \tau_2(\theta_2), \tau_3(\theta_3))$ with $\tau_i(\theta_i) = T_i\theta_i$, $i = 1, 2, 3$, with output $\bar{u}$.

It is noted that the internal model (10.29) is not unique. Since $\mathbf{x}(v) = 0$, for any sufficiently smooth function $P_0(x)$ with $P_0(0) = 0$, the following dynamic compensator

$$\dot{\eta} = M\eta + N\bar{u} - P_0(x) \tag{10.30}$$

is still an internal model of (10.22) on the mapping τ with output $\bar{u}$. To facilitate the control law design, we will use the internal model (10.30) with $P_0(x) = MNJx$.

Under the following coordinate transformation,

$$\begin{aligned}\bar{\eta} &= \eta - T\theta \\ \tilde{u} &= \bar{u} - \Psi T^{-1}\eta,\end{aligned} \tag{10.31}$$

the augmented system composed of (10.21) and (10.30) takes the following form:

$$\begin{aligned}\dot{\bar{\eta}} &= \left(M + N\Psi T^{-1}\right)\bar{\eta} + N\tilde{u} - MNJx \\ J\dot{x} &= \Psi T^{-1}\bar{\eta} + \tilde{u}.\end{aligned} \tag{10.32}$$

By Proposition 7.1, it suffices to stabilize (10.32) in order to solve the disturbance rejection problem of the composite system (10.22). For this purpose, we perform one more transformation $z = \bar{\eta} - NJx$ on (10.32) leading to the following triangular system:

$$\begin{aligned}\dot{z} &= Mz \\ J\dot{x} &= \Psi T^{-1}NJx + \Psi T^{-1}z + \tilde{u}.\end{aligned} \tag{10.33}$$

Since $\lim_{t\to\infty} z(t) = 0$, letting $\tilde{u} = \left(JK_x - \Psi T^{-1}NJ\right)x$ where $K_x \in \mathbb{R}^{3\times 3}$ is any Hurwitz matrix stabilizes (10.33). Thus, the following controller

$$\begin{aligned}\bar{u} &= \left(JK_x - \Psi T^{-1}NJ\right)x + \Psi T^{-1}\eta \\ \dot{\eta} &= M\eta + N\bar{u} - MNJx\end{aligned}$$

solves the disturbance rejection problem of the composite system (10.22).

10.4 Disturbance with Known Frequencies

When $\delta \neq 0$, the system (10.21) is nonlinear system. The disturbance rejection problem of it is much more difficult. In particular, $x = 0$ is not an equilibrium of (10.21) for all $d \in \mathbb{D}$. Thus, the internal model approach of the previous section cannot be directly applied to the disturbance rejection problem of the system (10.21). Nevertheless, the internal model approach of the previous section can still be used for compensating the unknown disturbance. For simplicity, in this section, we will first consider the case where the frequencies b_{ij} of $d(t)$ is known. We will still use the same internal model of the form (10.30) repeated below

$$\dot{\eta} = M\eta + N\bar{u} - P_0(x) \tag{10.34}$$

where $P_0(x)$ is a sufficiently smooth function vanishing at the origin and its specific form will be determined later.

Consider the augmented system composed of (10.21) and (10.34). Under the following coordinate transformation

$$\bar{\eta} = \eta - T\theta,$$

we have

$$\begin{aligned} \dot{\bar{\eta}} &= (M + N\Psi T^{-1})\bar{\eta} + N\tilde{u} - P_0(x) \\ J\dot{x} &= \{F_1(e, x, s_d) + G_1(e, x)\}\,\delta + \Psi T^{-1}\bar{\eta} - \Psi T^{-1}\eta + \bar{u}. \end{aligned} \tag{10.35}$$

Remark 10.3 The compensator (10.34) is motivated from the robust output regulation theory studied in Chap. 7. However, when $\delta \neq 0$, the disturbance rejection of system (10.21) cannot be directly dealt with by the framework for handling the robust output regulation problem in Chap. 7 for two reasons. First, s_d is not generated by some neutrally stable linear exosystem, and second, $F_i(e, 0, s_d) \neq 0, i = 0, 1$. Thus (10.34) cannot be considered as an internal model of system (10.21). Nevertheless, the idea of the internal model design can still be utilized to handle the unknown disturbance $d(t)$.

In order to solve the ATDRP, it suffices to find a control law to regulate the augmented system (10.35). For this purpose, we need to put the system (10.35) in a more standard form. First, let

$$P_0(x) = MNL_0(x). \tag{10.36}$$

Then a further transformation $z = \bar{\eta} - NJx$ and $\bar{u} = \tilde{u} + \Psi T^{-1}\eta$ gives

$$\begin{aligned} \dot{z} &= Mz + P_1(e, x, s_d)\delta \\ J\dot{x} &= (F_1(e, x, s_d) + G_1(e, x))\delta + \Psi T^{-1}z + \Psi T^{-1}NJx + \tilde{u}, \end{aligned} \tag{10.37}$$

where

$$P_1(e, x, s_d) = MNL_1(x) - N[F_1(e, x, s_d) + G_1(e, x)]. \tag{10.38}$$

Now using (10.18), i.e., $Jx = L_1(x)\delta + L_0(x)$, in (10.37) and then performing the following transformation

$$\tilde{u} = \hat{u} - \Psi T^{-1} N L_0(x) \tag{10.39}$$

converts the system (10.37) to the following form

$$\begin{aligned} \dot{z} &= Mz + P_1(e, x, s_d)\delta \\ J\dot{x} &= g(e, x, s_d)\delta + \Psi T^{-1} z + \hat{u}. \end{aligned} \tag{10.40}$$

where

$$g(e, x, s_d) = F_1(e, x, s_d) + G_1(e, x) + \Psi T^{-1} N L_1(x).$$

We have now completed the conversion of the ATDRP of the spacecraft system into the regulation problem for the system (10.40). This result is summarized as follows.

Lemma 10.3 *Suppose there exists a control law*

$$\hat{u} = \hat{\kappa}_1(e, x, s_d, \eta, \varsigma), \quad \dot{\varsigma} = \hat{\kappa}_2(e, x, s_d, \eta, \varsigma) \tag{10.41}$$

where ς is the compensator state, such that, for any initial condition, the state of the system composed of (10.40) and(10.41) is bounded and $\lim_{t\to\infty} x(t) = 0$. Then the ATDRP of the spacecraft is solved by a control law of the form (10.17) where $v = col(\eta, \varsigma)$, and

$$\begin{aligned} \kappa_1(e, x, s_d, v) = \hat{\kappa}_1(e, x, s_d, \eta, \varsigma) - F_0(e, x, s_d) - G_0(e, x) \\ + \Psi T^{-1}(\eta - N L_0(x)). \end{aligned}$$

Even though (10.40) is nonlinear, it is linear in the uncertain parameter δ. Therefore, it is possible to use the adaptive technique to deal with the regulation problem of (10.40). We have the following result.

Lemma 10.4 *Consider the system (10.40). There exists a control law*

$$\hat{u} = -K_x x - \rho(e, x, s_d, \zeta)\hat{\delta} \tag{10.42}$$

$$\dot{\zeta} = M\zeta + P_1(e, x, s_d) \tag{10.43}$$

$$\dot{\hat{\delta}} = \Lambda \rho^{\mathsf{T}}(e, x, s_d, \zeta) x \tag{10.44}$$

where $\rho(e, x, s_d, \zeta) = \Psi T^{-1}\zeta + g(e, x, s_d)$, K_x is any symmetric positive definite matrix and Λ is any diagonal matrix with positive diagonal entries, such that, for any initial condition, the state of the closed-loop system is bounded and $\lim_{t\to\infty} x(t) = 0$.

Proof Let $\zeta \in \mathbb{R}^{n_1 \times n_\delta}$ be produced by an auxiliary system (10.43) and $\bar{z} = z - \zeta\delta$. Direct calculation shows that

$$\begin{aligned}\dot{\bar{z}} &= Mz + P_1(e, x, s_d)\delta - [M\zeta + P_1(e, x, s_d)]\,\delta = M\bar{z}.\\ J\dot{x} &= g(e, x, s_d)\delta + \Psi T^{-1}z + \hat{u}.\end{aligned} \quad (10.45)$$

It is noted that in system (10.45), the uncertainty term $P_1(e, x, s_d)\delta$ disappears while the newly introduced variable ζ is produced by a known function $P_1(e, x, s_d)$ and matrix M. Therefore, the system (10.43) and (10.45) is in a more benign form and the regulation of this system can be handled as follows. Design a control law (10.42) where the vector $\hat{\delta}$ is used to estimate δ and an update law (10.44). Let Q be the symmetric positive definite matrix satisfying $QM + M^{\mathsf{T}}Q = -I$ and pick a real number $\epsilon \geq \|\Psi T^{-1}\|^2/\|K_x\|$. By choosing

$$V(\bar{z}, x, \tilde{\delta}) = \epsilon\bar{z}^{\mathsf{T}}Q\bar{z} + \frac{1}{2}x^{\mathsf{T}}Jx + \frac{1}{2}\tilde{\delta}^{\mathsf{T}}\Lambda^{-1}\tilde{\delta}$$

with $\tilde{\delta} = \hat{\delta} - \delta$, we have, along (10.43) and (10.45),

$$\begin{aligned}\dot{V}(\bar{z}, x, \tilde{\delta}) &\leq -\epsilon\|\bar{z}\|^2 + \frac{1}{2\epsilon}\|x^{\mathsf{T}}\Psi T^{-1}\|^2 + \frac{1}{2}\epsilon\|\bar{z}\|^2 - x^{\mathsf{T}}K_x x\\ &\leq -a(\bar{z}, x)\end{aligned}$$

with $a(\bar{z}, x) = \epsilon\|\bar{z}\|^2/2 + x^{\mathsf{T}}K_x x/2$. Thus, the state variables $\bar{z}$, x and $\tilde{\delta}$ are bounded. As a result, the state ζ is bounded, and hence z is bounded. By Theorem 2.5, $\lim_{t\to\infty} x(t) = 0$. The proof is thus completed. □

Remark 10.4 The dynamic coordinate transformation technique in the above proof converts the z-subsystem to $\bar{z}$-subsystem which is free of the unknown parameter δ. This manipulation makes it possible to handle the unknown parameter and the dynamic uncertainty separately.

From above, we are ready to reach the main result as follows.

Theorem 10.1 *If the disturbance frequencies are exactly known, then the ATDRP for the system composed of* (10.4), (10.5), (10.6), *and* (10.7) *is solved by a controller of the following form:*

$$\begin{aligned}u &= -K_x x - \rho(e, x, s_d, \zeta)\hat{\delta} - F_0(e, x, s_d) - G_0(e, x)\\ &\quad + \Psi T^{-1}(\eta - NL_0(x))\\ \dot{\zeta} &= M\zeta + P_1(e, x, s_d)\\ \dot{\hat{\delta}} &= \Lambda\rho^{\mathsf{T}}(e, x, s_d, \zeta)x\\ \dot{\eta} &= M\eta + N\bar{u} - P_0(x).\end{aligned} \quad (10.46)$$

where various functions and parameters are explicitly given above.

Example 10.1 We now synthesize a specific controller for the model of the spacecraft taken from [1] where the inertia matrix

$$J = \begin{bmatrix} J_{11} & 1.2 & 0.9 \\ 1.2 & 17 & 1.4 \\ 0.9 & 1.4 & 15 \end{bmatrix} \text{kg} \cdot \text{m}^2$$

contains an unknown parameter $\delta = J_{11}$. We can obtain

$$L_0(x) = \begin{bmatrix} 0 & 1.2 & 0.9 \\ 1.2 & 17 & 1.4 \\ 0.9 & 1.4 & 15 \end{bmatrix} x \text{ kg} \cdot \text{m}^2,$$

$$L_1(x) = \begin{bmatrix} 1 & 0 & 0 \\ 0 & 0 & 0 \\ 0 & 0 & 0 \end{bmatrix} x \text{ kg} \cdot \text{m}^2$$

and hence $F_0(e, x, s_d)$, $F_1(e, x, s_d)$, $G_0(e, x)$, and $G_1(e, x)$ from (10.19).

We consider the disturbance $d_i(t) = C_i \sin(b_i t + \Upsilon_i)$, $i = 1, 2, 3$, with $b_1 = 0.1$, $b_2 = 0.2$ and $b_3 = 0.2$. It is easy to see that

$$\Phi_i = \begin{bmatrix} 0 & 1 \\ -b_i^2 & 0 \end{bmatrix}, \quad \Psi_i = \begin{bmatrix} 1 & 0 \end{bmatrix}.$$

Letting

$$M_i = \begin{bmatrix} 0 & 1 \\ -3 & -2 \end{bmatrix}, \quad N_i = \begin{bmatrix} 0 \\ 1 \end{bmatrix}$$

gives

$$T_i^{-1} = \begin{bmatrix} 3 - b_i^2 & 2 \\ -2b_i^2 & 3 - b_i^2 \end{bmatrix}.$$

Now, we can obtain $P_0(x)$ and $P_1(e, x, s_d)$ from (10.36) and (10.38), respectively. In particular, we note that $P_1(e, x, s_d) \in \mathbb{R}^{6\times 1}$. Then, we can take $\eta \in \mathbb{R}^{6\times 1}$ and $\zeta \in \mathbb{R}^{6\times 1}$. Next, direct calculation shows

$$\rho(e, x, s_d, \zeta) = \Psi T^{-1}\zeta + F_1(e, x, s_d) + G_1(e, x) + \Psi T^{-1} N L_1(x).$$

Now, we are ready to construct the control law (10.46). In particular, we choose the gains of the control law as $K = 0.5I_3$, $K_x = 0.2I_3$, and the update rate matrix $\Lambda = I_3$.

To evaluate the control law, let the target angular velocity be

$$\Omega_d(t) = 0.05\big[\sin(1\pi t/100) \;\; \sin(2\pi t/100) \;\; \sin(3\pi t/100)\big]^{\mathsf{T}} \text{ rad/s}$$

and the initial target unit quaternion be $q_d(0) = [0, 0, 0, 1]^{\mathrm{T}}$. The initial attitude orientation of the spacecraft is $q(0) = [0.3, -0.2, -0.3, 0.8832]^{\mathrm{T}}$, and the initial value of the angular velocity is $\Omega(0) = [0, 0, 0]^{\mathrm{T}}$ rad/s. The disturbance amplitudes are $C_1 = 1$ Nm, $C_2 = 2$ Nm and $C_3 = 0.6$ Nm. The nominal value of the parameter J_{11} is 40. The initial state of the update law is $\hat{\delta}(0) = 10$, and the initial values of all unspecified state variables are 0. The performance of the control law is simulated and shown in Figs. 10.2, 10.3, 10.4, 10.5, 10.6. In particular, the component q_i of the quaternion and the attitude tracking errors $\xi_i = q_i - q_{id}$, $i = 1, 2, 3, 4$, are shown in Figs. 10.2, 10.3, 10.4, 10.5, and the control torque u (Nm) is shown in Fig. 10.6.

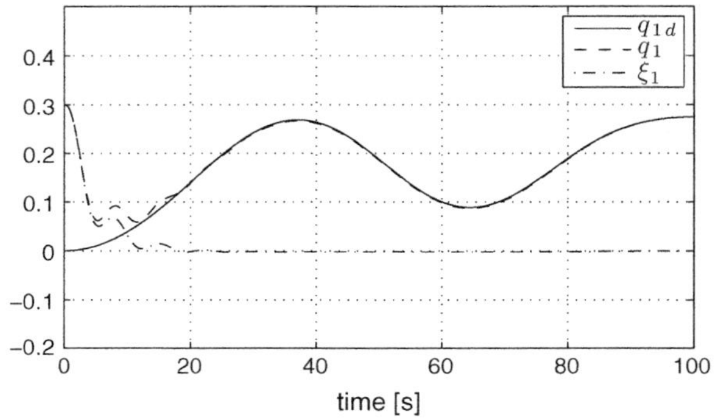

Fig. 10.2 Profile of state trajectories of the controlled spacecraft system (part 1)

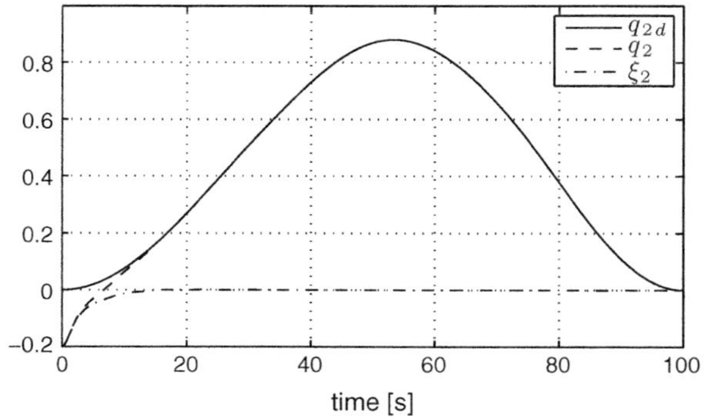

Fig. 10.3 Profile of state trajectories of the controlled spacecraft system (part 2)

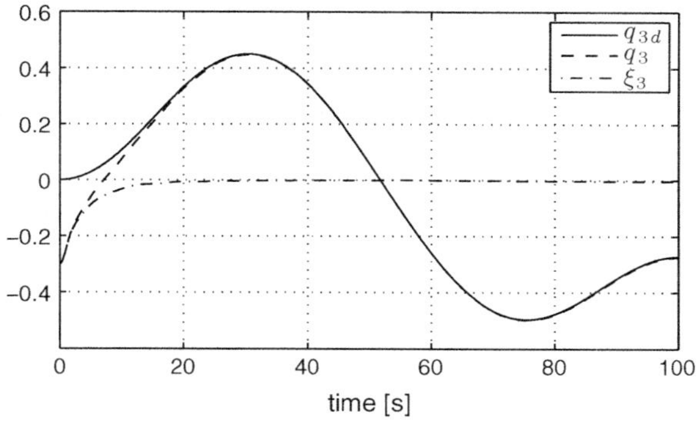

Fig. 10.4 Profile of state trajectories of the controlled spacecraft system (part 3)

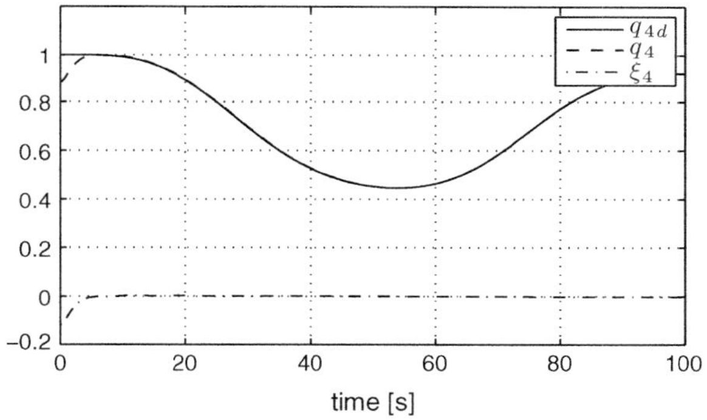

Fig. 10.5 Profile of state trajectories of the controlled spacecraft system (part 4)

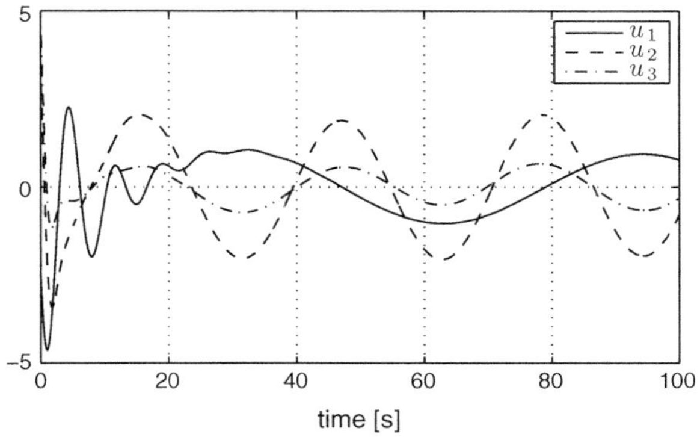

Fig. 10.6 Profile of control trajectories of the controlled spacecraft system

10.5 Disturbance with Unknown Frequencies

When the frequencies b_{ij} of $d(t)$ is known, the regulation problem for the system (10.21) has been studied in the previous section. Next, we will consider a more complicated case where the frequencies b_{ij} are not known. In this case, the coefficients $\phi_{i,1}, \ldots, \phi_{i,\ell_i}$ in (10.24) depend on b_{ij}. Thus, as in Chap. 9, the matrices Φ_i need to be replaced by $\Phi_i(\sigma)$ where σ is an unknown parameter vector depending on b_{ij}. Hence, the solution T to the Sylvester equation depends on σ as well, denoted as $T(\sigma)$. Thus, by replacing T with $T(\sigma)$ in the system (10.37), we obtain the following system:

$$\begin{aligned}\dot{z} &= Mz + P_1(e, x, s_d)\delta \\ J\dot{x} &= [F_1(e, x, s_d) + G_1(e, x)]\delta + \Psi T^{-1}(\sigma)z + \Psi T^{-1}(\sigma)NJx \\ &\quad -\Psi T^{-1}(\sigma)\eta + \bar{u}.\end{aligned} \tag{10.47}$$

Since the term $\Psi T^{-1}(\sigma)\eta$ is not measurable in this case, we cannot perform the transformation $\bar{u} = \tilde{u} + \Psi T^{-1}(\sigma)\eta$ and $\tilde{u} = \hat{u} - \Psi T^{-1}(\sigma)NL_0(x)$ as we did in the previous section. Instead, letting the nominal value of the matrix $T(\sigma)$ be T_0, and performing the transformation $\bar{u} = \tilde{u} + \Psi T_0^{-1}\eta$ and $\tilde{u} = \hat{u} - \Psi T_0^{-1}NL_0(x)$ puts the system (10.47) in the following form:

$$\begin{aligned}\dot{z} &= Mz + P_1(e, x, s_d)\delta \\ J\dot{x} &= \Psi T^{-1}(\sigma)z + g(e, x, s_d)\delta \\ &\quad + [E(\sigma)NL_0(x) + E(\sigma)NL_1(x)\delta - E(\sigma)\eta] + \hat{u}\end{aligned} \tag{10.48}$$

where

$$\begin{aligned}E(\sigma) &= \Psi T^{-1}(\sigma) - \Psi T_0^{-1} \\ P_1(e, x, s_d) &= MNL_1(x) - N[F_1(e, x, s_d) + G_1(e, x)] \\ g(e, x, s_d) &= F_1(e, x, s_d) + G_1(e, x) + \Psi T_0^{-1}NL_1(x).\end{aligned}$$

Now, we have completed the conversion of the ATDRP of the spacecraft into the regulation problem for system (10.48). We can obtain the following result.

Lemma 10.5 *Suppose there exists a control law*

$$\hat{u} = \hat{\kappa}_1(e, x, s_d, \eta, \varsigma), \quad \dot{\varsigma} = \hat{\kappa}_2(e, x, s_d, \eta, \varsigma) \tag{10.49}$$

where ς is the compensator state, such that, for any initial condition, the state of the system composed of (10.48) *and* (10.49) *is bounded and* $\lim_{t\to\infty} x(t) = 0$. *Then the ATDRP is solved by a control law of the form* (10.17) *where $v = col(\eta, \varsigma)$, and*

$$\kappa_1(e, x, s_d, \nu) = \hat{\kappa}_1(e, x, s_d, \eta, \varsigma) - F_0(e, x, s_d) - G_0(e, x) + \Psi T_0^{-1}(\eta - NL_0(x)).$$

If $E(\sigma) = 0$, then the system (10.48) reduces to the one studied in the previous section. But, the term $E(\sigma)$ does not vanish in the current case. Therefore, we need to overcome the difficulty caused by the term $[E(\sigma)NL_0(x)+E(\sigma)NL_1(x)\delta-E(\sigma)\eta]$.

Performing on (10.48) the following dynamic coordinate transformation:

$$\begin{aligned}\bar{z} &= z - \zeta\delta\\ \dot{\zeta} &= M\zeta + P_1(e, x, s_d),\ \ \zeta \in \mathbb{R}^{\ell\times n_\delta}\end{aligned}$$

leads to

$$\begin{aligned}\dot{\bar{z}} &= M\bar{z}\\ J\dot{x} &= \Psi T^{-1}(\sigma)\bar{z} + \Psi T^{-1}(\sigma)\zeta\delta + g(e, x, s_d)\delta\\ &\quad + [E(\sigma)NL_0(x) + E(\sigma)NL_1(x)\delta - E(\sigma)\eta] + \hat{u}.\end{aligned} \tag{10.50}$$

Since M is Hurwitz, we only need to concentrate on the second equation of (10.50). To handle the uncertain term $E(\sigma) \in \mathbb{R}^{3\times\ell}$ by adaptive control technique, we note that the matrix $E(\sigma)$ can be written as follows

$$E(\sigma) = \sum_{j=1}^{\hbar} E^j\omega^j(\sigma) = E[\omega(\sigma)\otimes I_\ell],\ \ E := \left[E^1, \ldots, E^{\hbar}\right],$$
$$\omega(\sigma) := \left[\omega^1(\sigma), \ldots, \omega^{\hbar}(\sigma)\right]^{\mathrm{T}}$$

where $\hbar \geq 1$ is some integer, $E^j \in \mathbb{R}^{3\times\ell}$ is a constant matrix, and $\omega^j(\sigma) \in \mathbb{R}$ a smooth function.

As a result,

$$\begin{aligned}\Psi T^{-1}(\sigma)\zeta\delta &= \Psi T_0^{-1}\zeta\delta + E(\sigma)\zeta\delta = \Psi T_0^{-1}\zeta\delta + E[\omega(\sigma)\otimes I_\ell]\zeta\delta\\ &= \left[\Psi T_0^{-1}\zeta\right]\delta + [E\oplus\zeta]\,[\omega(\sigma)\otimes\delta]\end{aligned}$$

where $\otimes$ is the Kronecker product and $\oplus$ is the Tracy-Singh product, i.e., $E\oplus\zeta = \left[E^1\zeta, \cdots, E^{\hbar}\zeta\right]$. Moreover,

$$\begin{aligned}E(\sigma)[NL_0(x) - \eta] &= E[\omega(\sigma)\otimes I_\ell][NL_0(x) - \eta] = E\oplus[NL_0(x) - \eta]\omega(\sigma)\\ E(\sigma)NL_1(x)\delta &= E[\omega(\sigma)\otimes I_\ell]NL_1(x)\delta = E\oplus[NL_1(x)][\omega(\sigma)\otimes\delta].\end{aligned}$$

Thus, the system (10.50) can be further put in the following form:

$$\begin{aligned}\dot{\bar{z}} &= M\bar{z}\\ J\dot{x} &= \Psi T^{-1}(\sigma)\bar{z} + \rho(e, x, s_d, \eta, \zeta)\mu(\sigma, \delta) + \hat{u}\end{aligned} \tag{10.51}$$

where

$$\rho(e, x, s_d, \eta, \zeta) = \left[\Psi T_0^{-1}\zeta + g(e, x, s_d)\ E \oplus [\zeta + NL_1(x)]\ E \oplus [NL_0(x) - \eta]\right],$$
$$\mu(\sigma, \delta) = \mathrm{col}(\delta, \omega(\sigma) \otimes \delta, \omega(\sigma)).$$

Since the system (10.51) is linear in the uncertain parameters $\mu(\sigma, \delta)$, we can apply the same adaptive control technique as we did in the previous section to construct an adaptive controller for (10.51) as shown in the following lemma.

Lemma 10.6 *For the system* (10.48), *there exists a control law*

$$\begin{aligned}\hat{u} &= -K_x x - \rho(e, x, s_d, \eta, \zeta)\hat{\mu} \\ \dot{\zeta} &= M\zeta + P_1(e, x, s_d) \\ \dot{\hat{\mu}} &= \Lambda\rho^{\mathsf{T}}(e, x, s_d, \eta, \zeta)x\end{aligned} \tag{10.52}$$

where K_x is any symmetric positive definite matrix and Λ any diagonal matrix with positive diagonal entries, such that, for any initial condition, the state of the closed-loop system is bounded and $\lim_{t\to\infty} x(t) = 0$.

Proof Clearly, if the following controller

$$\begin{aligned}\hat{u} &= -K_x x - \rho(e, x, s_d, \eta, \zeta)\hat{\mu}, \\ \dot{\hat{\mu}} &= \Lambda\rho^{\mathsf{T}}(e, x, s_d, \eta, \zeta)x\end{aligned} \tag{10.53}$$

solves the adaptive stabilization problem of the system (10.51), then the controller (10.52) solves the adaptive stabilization problem of the system (10.48). With $\tilde{\mu} = \hat{\mu} - \mu(\sigma, \delta)$, the closed-loop system composed of (10.51) and (10.53) can be put in the following form

$$\begin{aligned}\dot{\bar{z}} &= M\bar{z} \\ J\dot{x} &= \Psi T^{-1}(\sigma)\bar{z} - \rho(e, x, s_d, \eta, \zeta)\tilde{\mu} - K_x x \\ \dot{\tilde{\mu}} &= \Lambda\rho^{\mathsf{T}}(e, x, s_d, \eta, \zeta)x.\end{aligned} \tag{10.54}$$

Now, let Q be the symmetric positive definite matrix satisfying $QM + M^{\mathsf{T}}Q = -I$ and pick a real number $\epsilon \geq \|\Psi T^{-1}(\sigma)\|^2/\|K_x\|$. By choosing

$$V(\bar{z}, x, \tilde{\omega}) = \epsilon\bar{z}^{\mathsf{T}}Q\bar{z} + \frac{1}{2}x^{\mathsf{T}}Jx + \frac{1}{2}\tilde{\mu}^{\mathsf{T}}\Lambda^{-1}\tilde{\mu},$$

we have, along (10.52) and (10.54),

$$\dot{V}(\bar{z}, x, \tilde{\mu}) \leq -\epsilon\|\bar{z}\|^2 + \frac{1}{2\epsilon}\|x^{\mathsf{T}}\Psi T^{-1}(\sigma)\|^2 + \frac{1}{2}\epsilon\|\bar{z}\|^2 - x^{\mathsf{T}}K_x x$$
$$-x^{\mathsf{T}}\rho(e, x, s_d, \eta, \zeta)\tilde{\mu} + x^{\mathsf{T}}\rho(e, x, s_d, \eta, \zeta)\tilde{\mu} \leq -a(\bar{z}, x)$$

where $a(\bar{z}, x) = \epsilon\|\bar{z}\|^2/2 + x^{\mathsf{T}}K_x x/2$. By Theorem 2.5, the state variables $\bar{z}$, x and $\tilde{\mu}$ are bounded and $\lim_{t\to\infty} a(\bar{z}(t), x(t)) = 0$. Thus $\lim_{t\to\infty} x(t) = 0$. □

We now reach the main result as follows.

Theorem 10.2 *The ATDRP for the system composed of* (10.4), (10.5), (10.6), *and* (10.7) *is solved by a controller of the following form:*

$$\begin{aligned}
u &= -K_x x - \rho(e, x, s_d, \eta, \zeta)\hat{\mu} - F_0(e, x, s_d) - G_0(e, x) \\
&\quad + \Psi T_0^{-1}(\eta - NL_0(x)) \\
\dot{\zeta} &= M\zeta + P_1(e, x, s_d) \\
\dot{\hat{\mu}} &= \Lambda\rho^{\mathsf{T}}(e, x, s_d, \eta, \zeta)x \\
\dot{\eta} &= M\eta + N\bar{u} - P_0(x)
\end{aligned} \tag{10.55}$$

where various functions and parameters are explicitly defined above.

Example 10.2 We now synthesize a specific control law with the following numerical data:

$$J = \begin{bmatrix} \delta & 1.2 & 0.9 \\ 1.2 & 17 & 1.4 \\ 0.9 & 1.4 & 15 \end{bmatrix}, \quad d_i(t) = C_i \sin(b_i t + \Upsilon_i), \quad i = 1, 2, 3,$$
$$b_1 = 1,\ b_2 = 0.8,\ b_3 = \sigma$$

with δ and σ unknown. The steady-state generator is defined by the following matrices

$$\Phi_i(\sigma) = \begin{bmatrix} 0 & 1 \\ -b_i^2 & 0 \end{bmatrix}, \quad \Psi_i = \begin{bmatrix} 1 & 0 \end{bmatrix}$$

and the matrices defining the internal model are picked as follows

$$M_i = \begin{bmatrix} 0 & 1 \\ -3 & -2 \end{bmatrix}, \quad N_i = \begin{bmatrix} 0 \\ 1 \end{bmatrix}.$$

Solving the pertinent Sylvester equation gives

$$T_i(\sigma) = \begin{bmatrix} 3 - b_i^2 & -2 \\ 2b_i^2 & 3 - b_i^2 \end{bmatrix} \frac{1}{(3 - b_i^2)^2 + 4b_i^2}, \quad \Psi_i T_i^{-1}(\sigma) = \begin{bmatrix} 3 - b_i^2 & 2 \end{bmatrix}.$$

Assuming the nominal value of $\sigma = 0$, we have $E(\sigma) = L\omega(\sigma)$,

$$\Psi T_0^{-1} = \begin{bmatrix} 2\ 2 & & \\ & 2.36\ 2 & \\ & & 3\ 2 \end{bmatrix}, \quad L = \begin{bmatrix} 0\ 0 & & \\ & 0\ 0 & \\ & & -1\ 0 \end{bmatrix}, \quad \omega(\sigma) = \sigma^2.$$

Direct calculation gives the quantities used in the controller as follows:

$$\rho(e, x, s_d, \eta, \zeta) = \left[\begin{bmatrix} 2\zeta_1 + 2\zeta_2 \\ 2.36\zeta_3 + 2\zeta_4 \\ 3\zeta_5 + 2\zeta_6 \end{bmatrix} + g(e, x, s_d) \begin{bmatrix} 0 \\ 0 \\ -\zeta_5 \end{bmatrix} \begin{bmatrix} 0 \\ 0 \\ \eta_5 \end{bmatrix} \right]$$
$$\mu(\sigma, \delta) = \mathrm{col}(\delta, \sigma^2\delta, \sigma^2).$$

Finally, choose the gains of the control law as $K = 3I_3$, $K_x = 20I_3$, and the update rate parameter $\Lambda = 10I_3$.

To evaluate the performance of the control law by simulation, we let $\sigma = 2$, $\delta = 20$, $\Omega_d = [1, 0, 2]^{\mathsf{T}}$, $C_1 = 1$, $C_2 = 2$, $C_3 = 6$, and the nonzero initial conditions are $\hat{\mu}(0) = [10, 10, 1]^{\mathsf{T}}$, $q_d(0) = [0, 0, 0, 1]^{\mathsf{T}}$, $q(0) = [0.3, -0.2, -0.3, 0.8832]^{\mathsf{T}}$. The simulation results are shown in Figs. 10.7, 10.8, 10.9, 10.10, 10.11, 10.12. In particular, Figs. 10.7, 10.8, 10.9, 10.10 depict the tracking behaviors of q_i and the attitude tracking errors $\xi_i = q_i - q_{id}$, $i = 1, 2, 3, 4$. Figure 10.11 shows that $\hat{\mu}_1$ and $\hat{\mu}_3$ converge to the actual values $\mu_1 = \delta = 20$ and $\mu_3 = \sigma^2 = 4$, respectively. The control torque u (Nm) is illustrated in Fig. 10.12.

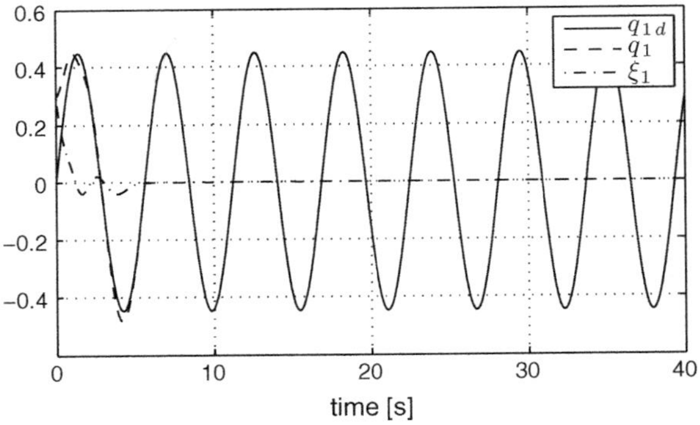

Fig. 10.7 Profile of state trajectories of the controlled spacecraft system with unknown frequencies (part 1)

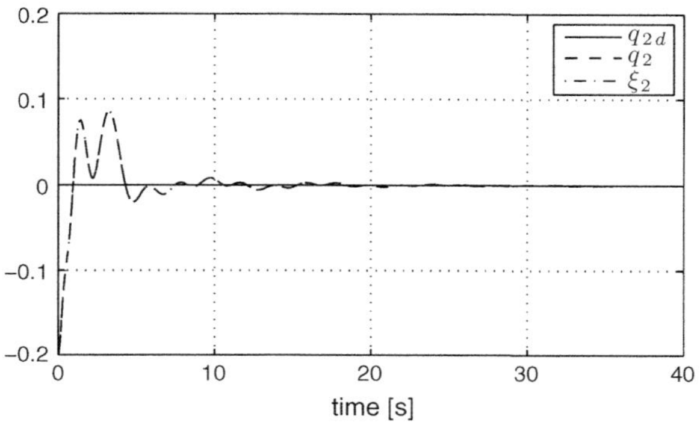

Fig. 10.8 Profile of state trajectories of the controlled spacecraft system with unknown frequencies (part 2)

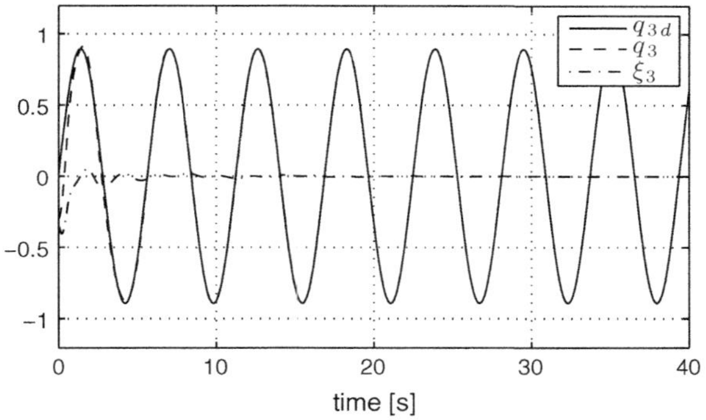

Fig. 10.9 Profile of state trajectories of the controlled spacecraft system with unknown frequencies (part 3)

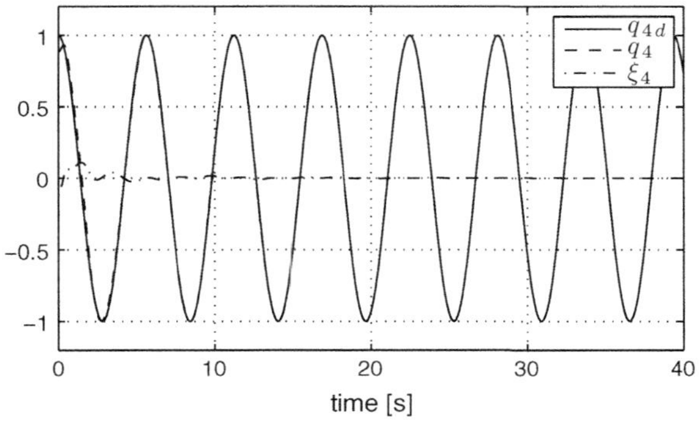

Fig. 10.10 Profile of state trajectories of the controlled spacecraft system with unknown frequencies (part 4)

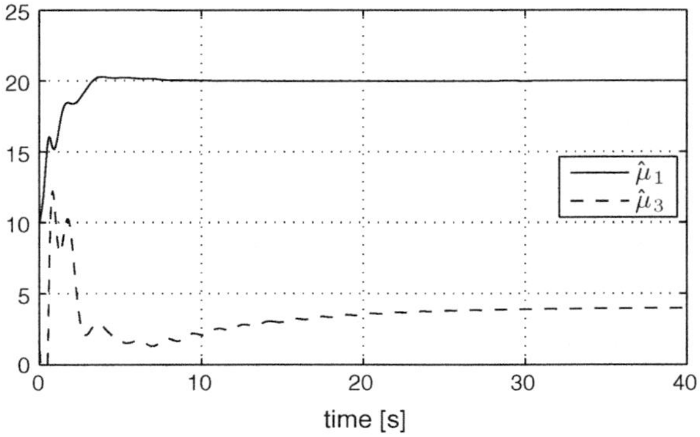

Fig. 10.11 Profile of estimated parameters of the controlled spacecraft system with unknown frequencies

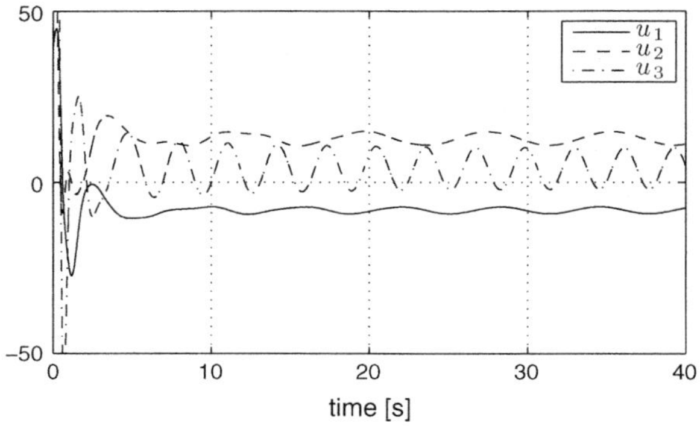

Fig. 10.12 Profile of control trajectories of the controlled spacecraft system with unknown frequencies

10.6 Notes and References

Attitude control has been extensively studied in literature, e.g., [1–4], to name just a few. Attitude control with disturbance attenuation was studied in [5–7], among others. In particular, a so-called inverse optimal adaptive control method was applied in [7] for global attitude tracking, which allows the $\mathcal{L}_2$ gain of the closed-loop system to be chosen arbitrarily small so as to achieve any level of $\mathcal{L}_2$ disturbance attenuation. A global robust adaptive regulation control method was proposed in [8] to solve the attitude tracking and disturbance rejection problem of spacecraft systems for the case where the disturbances are a multi-tone sinusoidal functions with known frequencies. Later, the case where the frequencies of the sinusoidal functions are unknown was further considered in [9]. The mathematical model of a rigid spacecraft in Sect. 10.1 can be found in many books, e.g., [10–12]. Lemmas 10.1 and 10.2 in Sect. 10.2 are from [8, 13], respectively. Sect. 10.4 is mainly based on [8] and Sect. 10.5 on [9].

References

1. Ahmed J, Coppola VT, Bernstein D (1998) Adaptive asymptotic tracking of spacecraft attitude motion with inertia matrix identification. J Guidance Control Dyn 21:684–691
2. Tayebi A (2008) Unit quaternion-based output feedback for the attitude tracking problem. IEEE Trans Autom Control 53(6):1516–1520
3. Yao Y, Yang B, He F, Qiao Y, Cheng D (2008) Attitude control of missile via Fliess expansion. IEEE Trans Control Syst Technol 16(5):959–970
4. Yoon H, Tsiotras P (2002) Spacecraft adaptive attitude and power tracking with variable speed control moment gyroscopes. J Guidance Control Dyn 25(6):1081–1090
5. Kang W (1995) Nonlinear H_∞ control and its applications to rigid spacecraft. IEEE Trans Autom Control 40(7):1281–1285
6. Krstic M, Tsiotras P (1999) Inverse optimal stabilization of a rigid spacecraft. IEEE Trans Autom Control 44:1042–1049
7. Luo W, Chu Y-C, Ling K-V (2005) Inverse optimal adaptive control for attitude tracking of spacecraft. IEEE Trans Autom Control 50:1639–1654
8. Chen Z, Huang J (2009) Attitude tracking and disturbance rejection of rigid spacecraft by adaptive control. IEEE Trans Autom Control 54:600–605
9. Chen Z, Huang J (2013) Attitude tracking of rigid spacecraft subject to disturbances of unknown frequencies. International Journal of Robust and Nonlinear Control. doi:10.1002/rnc.2983
10. Dixon WE, Behal A, Dawson DM, Nagarkatti SP (2003) Nonlinear control of engineering systems: a lyapunov-based approach. Birkhäuser, Boston
11. Klumpp AR (1976) Singularity free extraction of a quaternion from a direction cosine matrix. J Spacecr Rockets 13:754–755
12. Sidi MJ (1997) Spacecraft dynamics and control. Cambridge University Press, Cambridge
13. Yuan JS-C (1988) Closed-loop manipulator control using quaternion feedback. Int J Robust Nonlinear Control 4:434–440

Chapter 11
Appendix

11.1 Some Theorems on Nonlinear Systems

In this section, we summarize some well known results on nonlinear systems without proof. These theorems are cited in Chaps. 2, 3 and 7, respectively.

Theorem 11.1 (Local Existence and Uniqueness) *Assume the function* $f(x,t)$ *in* (1.1) *is piecewise continuous in* t *and locally Lipschitz in* x *around* t_0 *and* x_0, *i.e.*,

$$\|f(x,t) - f(y,t)\| \leq L\|x - y\| \tag{11.1}$$

for all $x, y \in \{x \in \mathbb{R}^n \mid \|x - x_0\| \leq r\}$ *and* $t \in [t_0, t_0+\delta]$ *for some* $r > 0$ *and* $\delta > 0$. *Then, there exist some* $t_1 > t_0$ *and a unique continuous function* $x : [t_0, t_1) \mapsto \mathbb{R}$ *such that* $\dot{x}(t) = f(x(t), t)$, $t_0 \leq t \leq t_1$, *and* $x(t_0) = x_0$.

The function $x(t)$ is called the local solution of (1.1) over $[t_0, t_1]$ satisfying the initial condition $x(t_0) = x_0$. The constant L in (11.1) is called Lipschitz constant.

A global version of Theorem 11.1 is as follows.

Theorem 11.2 (Global Existence and Uniqueness) *Assume the function* $f(x,t)$ *in* (1.1) *is piecewise continuous in* t *and satisfies*

$$\|f(x,t) - f(y,t)\| \leq L\|x - y\| \tag{11.2}$$

for some $L > 0$ *for all* $x, y \in \mathbb{R}^n$ *and all* $t \geq t_0$. *Then, the system* (1.1) *has a unique solution for all* $t \geq t_0$.

Theorem 11.3 (Frobenius Theorem) *Let* $\Delta(x) = span\{f_1(x), \ldots, f_d(x)\}$ *be a* d*-dimensional distribution for a set of smooth vector fields* $f_i(x) : \mathbb{U} \mapsto \mathbb{R}^n$ *where* $\mathbb{U}$ *is an open set of* $\mathbb{R}^n$. *Then, there exist functions* $\eta_i(x)$, $i = 1, \ldots, n-d$, *such that*

Z. Chen and J. Huang, *Stabilization and Regulation of Nonlinear Systems*,
Advanced Textbooks in Control and Signal Processing,
DOI 10.1007/978-3-319-08834-1_11

$$\frac{\partial \eta_i(x)}{\partial x} f(x) = 0, \quad \forall f(x) \in \Delta(x), \ i = 1, \ldots, n - d$$

on $\mathbb{U}$ *if and only if* $\Delta(x)$ *is involutive (i.e., for any two vector fields* $f_1(x), f_2(x) \in \Delta(x)$*, their Lie bracket* $[f_1(x), f_2(x)] \in \Delta(x)$*.)*

Theorem 11.4 (Center Manifold Theorem) *Consider a nonlinear system of the following form*

$$\begin{aligned} \dot{x} &= f(x, v) \\ \dot{v} &= a(v) \end{aligned} \tag{11.3}$$

where $x \in \mathbb{R}^{n_1}$, $v \in \mathbb{R}^{n_2}$*, and the functions* f *and* a *are sufficiently smooth vanishing at the origin. Let* $F = \frac{\partial f}{\partial x}(0, 0)$ *where all the eigenvalues of* F *have nonzero real parts and* $A_1 = \frac{\partial a}{\partial v}(0)$ *where all the eigenvalues of* A_1 *have zero real parts. Then,*

(i) there exist an open neighborhood $\mathbb{V} \subset \mathbb{R}^{n_2}$ *of* $v = 0$ *and a* C^k *function* $\mathbf{x} : \mathbb{R}^{n_2} \mapsto \mathbb{R}^{n_1}$ *with* $\mathbf{x}(0) = 0$*, such that, for all* $v \in \mathbb{V}$*,*

$$\frac{\partial \mathbf{x}(v)}{\partial v} a(v) = f(\mathbf{x}(v), v). \tag{11.4}$$

(ii) the equilibrium point of the system (11.3) *at the origin is Lyapunov stable (asymptotically stable) (unstable) if and only if the equilibrium point* $v = 0$ *of the following system*

$$\dot{v} = a(v) \tag{11.5}$$

is stable (asymptotically stable) (unstable).

(iii) if the equilibrium point $v = 0$ *of the system* (11.5) *is stable, then for any sufficiently small* $\mathrm{col}(x(0), \ v(0))$*, the solution of the system* (11.3) *exists for all* $t > 0$*, and satisfies*

$$\|x(t) - \mathbf{x}(v(t))\| \leq \delta e^{-\lambda t} \|x(0) - \mathbf{x}(v(0))\| \tag{11.6}$$

where δ *and* λ *are positive constants.*

If the equilibrium point of the system (11.5) at $v = 0$ is stable, then the center manifold described in Theorem 11.4 is called the stable center manifold.

11.2 Technical Lemmas

In this section, we present some technical lemmas that are used in the proof of some technical results in the book.

Lemma 11.1 *(i) For any continuous function* $f(x,d) : \mathbb{R}^n \times \mathbb{R}^l \mapsto \mathbb{R}$, *there exist smooth functions* $a(x), b(d) \geq 0$, *such that*

$$|f(x,d)| \leq a(x)b(d). \tag{11.7}$$

(ii) For any continuous function $f(x_1, x_2, d) : \mathbb{R}^{n_1} \times \mathbb{R}^{n_2} \times \mathbb{R}^l \mapsto \mathbb{R}$, *there exist smooth functions* $m_1(x_1, d),\ m_2(x_2, d) \geq 0$, *such that*

$$|f(x_1, x_2, d)| \leq m_1(x_1, d) + m_2(x_2, d). \tag{11.8}$$

(iii) For any continuously differentiable function $f(x,d) : \mathbb{R}^n \times \mathbb{R}^l \mapsto \mathbb{R}$, *satisfying* $f(0,d) = 0$, *there exists a smooth function* $m(x,d) \geq 0$, *such that*

$$|f(x,d)| \leq \|x\| m(x,d). \tag{11.9}$$

(iv) For any continuously differentiable function $f(x_1, \ldots, x_s, d) : \mathbb{R}^{n_1} \times \cdots \times \mathbb{R}^{n_s} \times \mathbb{R}^l \mapsto \mathbb{R}$ *for some* $s \geq 1$, *satisfying* $f(0, \ldots, 0, d) = 0$, *there exist smooth functions* $m_1(x_1, d), \ldots, m_s(x_s, d) \geq 0$, *such that*

$$|f(x_1, \ldots, x_s, d)| \leq \|x_1\| m_1(x_1, d) + \cdots + \|x_s\| m_s(x_s, d). \tag{11.10}$$

Proof (i) Let

$$f_1(x) = \max_{\|d\| \leq \|x\|} |f(x,d)|, \quad \forall x \in \mathbb{R}^n$$

and

$$f_2(d) = \max_{\|x\| \leq \|d\|} |f(x,d)|, \quad \forall d \in \mathbb{R}^l.$$

Then, $|f(x,d)| \leq f_1(x)$ for $\|d\| \leq \|x\|$, and $|f(x,d)| \leq f_2(d)$ for $\|x\| \leq \|d\|$. Thus, for all $(x,d) \in \mathbb{R}^n \times \mathbb{R}^l$,

$$|f(x,d)| \leq f_1(x) + f_2(d).$$

By construction, the functions $f_1(x)$ and $f_2(d)$ are continuous, and hence, can always be dominated by two smooth functions $a(x)$ and $b(d)$ in the sense of

$$f_1(x) + 1 \leq a(x), \quad f_2(d) + 1 \leq b(d).$$

As a result,

$$|f(x,d)| \leq f_1(x) + f_2(d) \leq (f_1(x) + 1)(f_2(d) + 1) \leq a(x)b(d),$$

i.e. (11.7).

(ii) Let

$$f_1(x_1, d) = \max_{\|x_2\| \le \|x_1\|} |f(x_1, x_2, d)|, \quad \forall x_1 \in \mathbb{R}^{n_1},\ d \in \mathbb{R}^l$$

and

$$f_2(x_2, d) = \max_{\|x_1\| \le \|x_2\|} |f(x_1, x_2, d)|, \quad \forall x_2 \in \mathbb{R}^{n_2},\ d \in \mathbb{R}^l.$$

Then, $|f(x_1, x_2, d)| \le f_1(x_1, d)$ for $\|x_2\| \le \|x_1\|$, and $|f(x_1, x_2, d)| \le f_2(x_2, d)$ for $\|x_1\| \le \|x_2\|$. Thus, for all $(x_1, x_2, d) \in \mathbb{R}^{n_1} \times \mathbb{R}^{n_2} \times \mathbb{R}^l$,

$$|f(x_1, x_2, d)| \le f_1(x_1, d) + f_2(x_2, d). \tag{11.11}$$

By construction, the functions $f_1(x_1, d)$ and $f_2(x_2, d)$ are continuous, and hence, can always be dominated by two smooth functions $m_1(x_1, d)$ and $m_2(x_2, d)$ in the sense of

$$f_1(x_1, d) \le m_1(x_1, d), \quad f_2(x_2, d) \le m_2(x_2, d).$$

As a result, (11.8) is true.

(iii) Since $f(x, d)$ is continuously differentiable, the mean value theorem gives

$$|f(x, d) - f(0, d)| \le \|x\| \|f'(cx, d)\|$$

for some c between 0 and 1, where $f'(x, d) = \partial f(x, d)/\partial x$ is the derivative of $f(x, d)$ with respect to x. Let

$$f_1(x, d) = \max_{0 \le c \le 1} \|f'(cx, d)\|,$$

which is continuous by construction, and, hence bounded by a smooth function $m(x, d)$. As $f(0, d) = 0$, one has

$$|f(x, d)| \le \|x\| f_1(x, d) \le \|x\| m(x, d),$$

i.e. (11.9).

(iv) From (iii), one has

$$|f(x_1, \ldots, x_s, d)| \le \|\mathrm{col}(x_1, \ldots, x_s)\| m(x_1, \ldots, x_s, d)$$

for a smooth function $m(x_1, \ldots, x_s, d)$. Let

$$f_i(x_i, d) = \sqrt{s} \max_{\|x_j\| \le \|x_i\|, j \ne i} m(x_1, \ldots, x_s, d), \quad \forall x_i \in \mathbb{R}^{n_i},\ d \in \mathbb{R}^l,\ i = 1, \ldots, s$$

which is continuous by construction, and hence bounded by a smooth function $m_i(x_i, d)$ in the sense of $f_i(x_i, d) \leq m_i(x_i, d)$. Then,

$$|f(x_1, \ldots, x_s, d)| \leq \sqrt{s}\|x_i\| m(x_1, \ldots, x_s, d) \leq \|x_i\| f_i(x_i, d) \leq \|x_i\| m_i(x_i, d)$$

when $\|x_j\| \leq \|x_i\|,\ j \neq i$. As a result, (11.10) is true for all $(x_1, \ldots, x_s) \in \mathbb{R}^{n_1} \times \cdots \times \mathbb{R}^{n_s}$ and $d \in \mathbb{R}^l$. □

Applying (11.7) to (11.8), and (11.10) gives the following corollary.

Corollary 11.1 *(i) For any continuous function $f(x_1, x_2, d) : \mathbb{R}^{n_1} \times \mathbb{R}^{n_2} \times \mathbb{R}^l \mapsto \mathbb{R}$, there exist smooth functions $a_1(d), a_2(d), m_1(x_1), m_2(x_2) \geq 0$, such that*

$$|f(x_1, x_2, d)| \leq a_1(d)m_1(x_1) + a_2(d)m_2(x_2). \tag{11.12}$$

(ii) For any continuously differentiable function $f(x_1, \ldots, x_s, d) : \mathbb{R}^{n_1} \times \cdots \times \mathbb{R}^{n_s} \times \mathbb{R}^l \mapsto \mathbb{R}$ for some $s \geq 1$, satisfying $f(0, \ldots, 0, d) = 0$, there exist smooth functions $a_i(d), m_i(x_i) \geq 0$, $i = 1, \ldots, s$, such that

$$|f(x_1, \ldots, x_s, d)| \leq a_1(d)\|x_1\|m_1(x_1) + \cdots + a_s(d)\|x_s\|m_s(x_s). \tag{11.13}$$

Remark 11.1 If $d \in \mathbb{D}$ where $\mathbb{D}$ is a known compact set, then $a_i(d)m_i(x) \leq a_i m_i(x)$ where $a_i = \max_{d \in \mathbb{D}} a_i(d)$. Since a_i is known, we can always assume $a_i(d) = 1$ for all i in (11.12) and (11.13).

Lemma 11.2 *Let $\alpha(s)$ and $a(s)$ be class $\mathcal{K}_\infty$ functions. For any smooth function $\alpha'(s) : [0, \infty) \mapsto \mathbb{R}$ satisfying $\alpha'(s) = \mathcal{O}[\alpha(s)]$ as $s \to 0^+$, there exists a class $\mathcal{SN}$ function ρ such that $\rho(a(s))\alpha(s) \geq \alpha'(s)$, $s \geq 0$.*

Proof First, we define a function $\rho_1(s) : [0, \infty) \mapsto [0, \infty)$ as follows

$$\rho_1(s) = \sup_{t \in (0, s]} \frac{\alpha'(t)}{\alpha(t)}, \quad s > 0.$$

Because $\alpha'(s) = \mathcal{O}[\alpha(s)]$ as $s \to 0^+$, $\rho_1(s)$ is well defined for all $s > 0$. Also, we define $\rho_1(0) = \lim_{s \to 0^+} \rho_1(s)$. Clearly, $\rho_1(s)$ is a continuous non-decreasing function for $s \geq 0$. By definition, one has $\rho_1(s)\alpha(s) \geq \alpha'(s)$, $\forall s \geq 0$. What is left is to find a class $\mathcal{SN}$ function ρ such that $\rho(a(s)) \geq \rho_1(s)$, equivalently, $\rho(s) \geq \rho_1(a^{-1}(s))$ since $a(s)$ is a class $\mathcal{K}_\infty$ function. Since $\rho_1(a^{-1}(s))$ is a continuous non-decreasing function, we can always find a smooth non-decreasing function (i.e., a class $\mathcal{SN}$ function) $\rho(s)$ such that $\rho(s) \geq \rho_1(a^{-1}(s))$. □

Lemma 11.3 *Let $\alpha_1(s)$ and $\alpha_2(s)$ be class $\mathcal{K}_\infty$ functions. There exist class $\mathcal{K}_\infty$ functions $\beta_1(s)$ and $\beta_2(s)$ such that*

$$\beta_1(\|x\|) \leq \alpha_1(\|x_1\|) + \alpha_2(\|x_2\|) \quad (11.14)$$

$$\beta_2(\|x\|) \geq \alpha_1(\|x_1\|) + \alpha_2(\|x_2\|). \quad (11.15)$$

for any vector $x = col(x_1, x_2)$ *with* $x_1 \in \mathbb{R}^{n_1}$ *and* $x_2 \in \mathbb{R}^{n_2}$.

Proof Pick a class $\mathcal{K}_\infty$ functions β_1 as follows

$$\beta_1(s) \leq \min\{\alpha_1(s/\sqrt{2}), \alpha_2(s/\sqrt{2})\}.$$

By noting

$$\|x\| = \sqrt{\|x_1\|^2 + \|x_2\|^2} \leq \sqrt{2} \max\{\|x_1\|, \|x_2\|\}$$

one has

$$\beta_1(\|x\|) \leq \min\{\alpha_1(\|x_1\|), \alpha_2(\|x_1\|)\} \leq \alpha_1(\|x_1\|),$$

for $\|x_1\| \geq \|x_2\|$; and

$$\beta_1(\|x\|) \leq \min\{\alpha_1(\|x_2\|), \alpha_2(\|x_2\|)\} \leq \alpha_2(\|x_2\|),$$

for $\|x_1\| \leq \|x_2\|$. As a result,

$$\beta_1(\|x\|) \leq \alpha_1(\|x_1\|) + \alpha_2(\|x_2\|).$$

The inequality (11.14) is proved.

Pick the other class $\mathcal{K}_\infty$ functions β_2 as follows

$$\beta_2(s) \geq \alpha_1(s) + \alpha_2(s).$$

One has

$$\beta_2(\|x\|) \geq \alpha_1(\|x_1\|) + \alpha_2(\|x_2\|).$$

The inequality (11.15) is proved. □

11.3 Proof of Theorem 2.12

To prove Theorem 2.12, we need the following lemma.

Lemma 11.4 *Let* β *be a class* $\mathcal{KL}$ *function,* γ *a class* $\mathcal{K}$ *function such that* $\gamma(r) < r$, $\forall r > 0$, *and* $\mu \in (0, 1]$ *a real number. For any nonnegative real numbers* s *and* *d, and any nonnegative real function $z(t) \in L^1_\infty$ satisfying*

$$z(t) \leq \max\left\{\beta(s,t), \gamma\left(\|z_{[\mu t,t]}\|\right), d\right\}, \quad \forall t \geq 0, \tag{11.16}$$

there exists a class $\mathcal{K}_\infty$ function $\hat{\beta}$ such that

$$z(t) \leq \max\left\{\hat{\beta}(s,t), d\right\}, \quad \forall t \geq 0. \tag{11.17}$$

Proof First, we choose a function $\bar{z}(t)$ as follows

$$\bar{z}(t) = \begin{cases} z(t) & \text{if } z(t) > d \\ 0 & \text{otherwise} \end{cases}, \; t \geq 0,$$

clearly, which is real nonnegative function and belongs to L^1_∞. Then we will show that

$$\bar{z}(t) \leq \max\left\{\beta(s,t), \gamma\left(\|\bar{z}_{[\mu t,t]}\|\right)\right\}, \quad \forall t \geq 0. \tag{11.18}$$

To this end, we will consider the following two cases.

(i) $z(t) > d$: On one hand, from (11.16),

$$z(t) \leq \max\left\{\beta(s,t), \gamma\left(\|z_{[\mu t,t]}\|\right)\right\}.$$

On the other hand, $\|\bar{z}_{[\mu t,t]}\| = \|z_{[\mu t,t]}\|$. In fact, at the instant $t_1 \in [\mu t, t]$ when $z(t_1) \geq z(\tau), \ \forall \tau \in [\mu t, t]$, we have $\|z_{[\mu t,t]}\| = z(t_1) \geq z(t) > d$. Thus, $\bar{z}(t_1) = z(t_1) \geq z(\tau) \geq \bar{z}(\tau), \ \forall \tau \in [\mu t, t]$. That is, $\|\bar{z}_{[\mu t,t]}\| = \bar{z}(t_1) = z(t_1) = \|z_{[\mu t,t]}\|$. As a result,

$$\bar{z}(t) = z(t) \leq \max\left\{\beta(s,t), \gamma\left(\|\bar{z}_{[\mu t,t]}\|\right)\right\}.$$

That is, (11.18) holds.

(ii) $z(t) \leq d$: The inequality (11.18) holds since $\bar{z}(t) = 0$.

Now, from (11.18), we have the following claim.

Claim For any $r, \epsilon > 0$, there exists a nonnegative number $T_r(\epsilon)$ such that, if $\bar{z}(t)$ satisfies (11.18) with $s < r$, then $\bar{z}(t) < \epsilon, \ \forall t \geq T_r(\epsilon)$.

Proof Since $\bar{z}(t) \in L^1_\infty$, denote $R = \left\|\bar{z}_{[0,\infty)}\right\|$, which is a finite nonnegative number. If $R = 0$, the proof is trivial. So, we suppose $R > 0$. And let $\delta_1 = R - \gamma(R) > 0$. For any $\delta_2 \in (0, \delta_1)$, there exists a finite $t_1 \geq 0$ such $\bar{z}(t_1) \geq R - \delta_2$. From (11.18), we have

$$R - \delta_2 \leq \bar{z}(t_1) \leq \max\left\{\beta(s,0), \gamma(R)\right\},$$

hence

$$R \leq \max\{\beta(s, 0) + \delta_2, \gamma(R) + \delta_2\}.$$

And $R > \gamma(R) + \delta_2$ gives $R \leq \beta(s, 0) + \delta_2 < \beta(r, 0) + \delta_2$. Since δ_2 can be arbitrarily small, we have $R \leq \beta(r, 0)$. As a result, $\bar{z}(t) \leq \beta(r, 0),\ t \geq 0$.

Next, there exist a real number $0 < \delta_3 < 1$ satisfying

$$\gamma(x) \leq \delta_3 x, \quad \forall x \in [\epsilon/\beta(r, 0)],$$

and a nonnegative integer n satisfying $\delta_3^n < \epsilon/\beta(r, 0)$. Clearly, $\gamma^n(\beta(r, 0)) < \epsilon$. Denote $t_i > 0, i = 1, \ldots, n$ be the first time instant such that

$$\beta(r, t_i) \leq \gamma^i(\beta(r, 0)).$$

And define $\bar{t}_i, i = 0, \ldots, n$ as

$$\bar{t}_0 = 0, \quad \bar{t}_i = \max\left\{t_i, \frac{1}{\mu}\bar{t}_{i-1}\right\}, \quad i = 1, \ldots, n.$$

Now, it can be proved by induction that, for $i = 0, \ldots, n$,

$$\bar{z}(t) \leq \gamma^i(\beta(r, 0)), \forall t \geq \bar{t}_i. \tag{11.19}$$

Indeed, we have shown that (11.19) holds for $i = 0$. Suppose it holds for $n > i > 0$, then for $t \geq \bar{t}_{i+1}$, we have

$$\begin{aligned}
\bar{z}(t) &\leq \max\left\{\beta(s, t), \gamma\left(\|\bar{z}_{[\mu t, t]}\|\right)\right\} \\
&\leq \max\left\{\beta(r, t_{i+1}), \gamma\left(\gamma^i(\beta(r, 0))\right)\right\} \\
&= \gamma^{i+1}(\beta(r, 0)).
\end{aligned}$$

That is, (11.19) holds for $i = 0, \ldots, n$. Now, (11.19) with $i = n$ is

$$\bar{z}(t) \leq \gamma^n(\beta(r, 0)) < \epsilon, \quad \forall t \geq T_r(\epsilon) \tag{11.20}$$

by choosing $T_r(\epsilon) \geq \bar{t}_n$. The proof of the claim is completed.

Using the above claim, we can find a class $\mathcal{KL}$ function $\hat{\beta}$ such that

$$\bar{z}(t) \leq \hat{\beta}(s, t), \quad \forall t \geq 0.$$

By the definition of $\bar{z}(t)$, we note that (11.17) is satisfied.

The existence of $\hat{\beta}$ is given below, which is derived from the result in the proof of Lemma A.1 of [1], and the references within, such as Lemma 2.1.4 and Proposition 2.1.5 of [2], and the proofs of Lemma 3.1 and Proposition 2.5 of [3].

Step 1: From the proof of the claim, it is known that

$$\bar{z}(t) \leq \varphi(s) \tag{11.21}$$

with $\varphi(s) = \beta(s, 0)$.

Step 2: From the proof of the claim, it is clear that $T_r(\epsilon)$ always exists satisfying the claim and the following properties additionally.

(i) For each fixed $r > 0$, $T_r : (0, \infty) \mapsto [0, \infty)$ satisfies $T_r(\epsilon) < \infty$ for any $\epsilon > 0$.
(ii) $T_r(\epsilon_1) \geq T_r(\epsilon_2)$, if $\epsilon_1 \leq \epsilon_2$.

So we can define, for any $r, \epsilon > 0$,

$$\bar{T}_r(\epsilon) = \frac{2}{\epsilon} \int_{\epsilon/2}^{\epsilon} T_r(s) ds.$$

Since T_r is decreasing, $\bar{T}_r$ is well defined and is locally absolutely continuous. Also

$$\bar{T}_r(\epsilon) \geq \frac{2}{\epsilon} T_r(\epsilon) \int_{\epsilon/2}^{\epsilon} ds = T_r(\epsilon).$$

Furthermore,

$$\begin{aligned} \frac{d\bar{T}_r(\epsilon)}{d\epsilon} &= -\frac{2}{\epsilon^2} \int_{\epsilon/2}^{\epsilon} T_r(s) ds + \frac{2}{\epsilon} \left[T_r(\epsilon) - \frac{1}{2} T_r(\epsilon/2) \right] \\ &= \frac{1}{\epsilon} \left[T_r(\epsilon) - \frac{2}{\epsilon^2} \int_{\epsilon/2}^{\epsilon} T_r(s) ds \right] + \frac{1}{\epsilon} \left[T_r(\epsilon) - T_r(\epsilon/2) \right] \\ &= \frac{1}{\epsilon} \left[T_r(\epsilon) - \bar{T}_r(\epsilon) \right] + \frac{1}{\epsilon} \left[T_r(\epsilon) - T_r(\epsilon/2) \right] \\ &\leq 0. \end{aligned}$$

Hence, $\bar{T}_r$ decreases (not necessarily strictly). Finally, define

$$\tilde{T}_r(\epsilon) = \bar{T}_r(\epsilon) + \frac{r}{\epsilon}.$$

Then it follows that, for any fixed r, $\tilde{T}_r : (0, \infty) \mapsto (0, \infty)$ is continuous and strictly decreasing.

Now, for each $r \in (0, \infty)$, denote $\psi_r = \tilde{T}_r^{-1}$. Then

$$\psi_r : (0, \infty) \mapsto (0, \infty)$$

is continuous and strictly decreasing. We also write $\psi_r(0) = \infty$, which is consistent with the fact that

$$\lim_{t \to 0^+} \psi_r(t) = \infty.$$

It follows from the claim and the fact $\tilde{T}_r(\epsilon) \geq T_r(\epsilon)$ that, for any $r, \epsilon > 0$,

$$\bar{z}(t) < \epsilon, \quad \forall t \geq \tilde{T}_r(\epsilon).$$

As $t = \tilde{T}_r(\psi_r(t))$ if $t > 0$, we have

$$\bar{z}(t) < \psi_r(t), \; \forall t > 0.$$

Furthermore, since $\psi_r(0) = \infty$, we obtain

$$\bar{z}(t) \leq \psi_r(t), \quad \forall t \geq 0. \tag{11.22}$$

Step 3: Now for any $s \geq 0$ and $t \geq 0$, let

$$\bar{\psi}(s,t) = \min \left\{ \inf_{r \in (s,\infty)} \psi_r(t), \varphi(s) \right\}.$$

From the above two steps, we have

$$\bar{z}(t) \leq \bar{\psi}(s,t), \quad \forall t \geq 0.$$

By its definition, for any fixed t, $\bar{\psi}(\cdot, t)$ is an increasing function (not necessarily strictly). Also because, for any fixed $r \in (0, \infty)$, $\psi_r(t)$ decreases to 0 (this follows from the fact that $\psi_r : (0,\infty) \to (0,\infty)$ is continuous and strictly decreasing), it follows that,

$$\text{for any fixed } s, \; \bar{\psi}(s,t) \text{ decreases to 0 as } t \to \infty.$$

Pick any function

$$\tilde{\psi} : [0,\infty) \times [0,\infty) \mapsto [0,a)$$

for some $a > 0$ (a can be $+\infty$) with the following properties.

(i) For any fixed $t \geq 0$, $\tilde{\psi}(\cdot, t)$ is continuous and strictly increasing.
(ii) For any fixed $s \geq 0$, $\tilde{\psi}(s,t)$ decreases to 0 as $t \to \infty$.
(iii) $\tilde{\psi}(s,t) \geq \bar{\psi}(s,t)$.

Such a function $\tilde{\psi}$ always exists; for instance, it can be constructed as follows. Define first

$$\hat{\psi}(s,t) = \int_s^{s+1} \bar{\psi}(\varsigma, t)d\varsigma.$$

Then $\hat{\psi}(\cdot, t)$ is an absolutely continuous function on every compact subset of $[0, \infty)$, and it satisfies

$$\hat{\psi}(s,t) \geq \bar{\psi}(s,t) \int_s^{s+1} d\varsigma = \bar{\psi}(s,t).$$

It follows that

$$\frac{\partial \hat{\psi}(s,t)}{\partial s} = \bar{\psi}(s+1,t) - \bar{\psi}(s,t) \geq 0, \ a.e.,$$

and hence $\hat{\psi}(\cdot, t)$ is increasing. Also since for any fixed $s, \bar{\psi}(s, \cdot)$ decreases, so does $\hat{\psi}(s, \cdot)$. Note that

$$\begin{aligned} \bar{\psi}(s,t) \leq \bar{\psi}(s,0) &= \min\left\{ \inf_{r \in (s,\infty)} \psi_r(0), \varphi(s) \right\} \\ &= \varphi(s), \end{aligned}$$

(recall that $\phi_r(0) = \infty$), so by Lebesgue dominated convergence theorem, for any fixed $s \geq 0$,

$$\lim_{t \to \infty} \hat{\psi}(s,t) = \int_s^{s+1} \lim_{t \to \infty} \bar{\psi}(\varsigma, t)d\varsigma = 0.$$

Now we see that the function $\hat{\psi}(s,t)$ satisfies all of the requirements for $\tilde{\psi}(s,t)$ except possibly for the strictly increasing property. We define $\tilde{\psi}$ as follows:

$$\tilde{\psi}(s,t) = \hat{\psi}(s,t) + \frac{s}{(s+1)(t+1)}.$$

Clearly it satisfies all the desired properties.

Finally, define

$$\hat{\beta}(s,t) = \sqrt{\varphi(s)}\sqrt{\tilde{\psi}(s,t)}.$$

Then it follows that $\beta(s,t)$ is a $\mathcal{KL}$ function, and

$$\bar{z}(t) \leq \sqrt{\varphi(s)}\sqrt{\tilde{\psi}(s,t)} \leq \hat{\beta}(s,t), \quad \forall t \geq 0,$$

which concludes the proof of the Lemma. □

Proof of Theorem 2.12 First, we note that the inequality (2.85) and the following one,

$$\gamma_2^x \circ \gamma_1^x(s) < s, \quad \forall s > 0, \tag{11.23}$$

imply each other (see [4]). In fact, denote $r_\infty = \lim_{s\to\infty} \gamma_1^x(s)$. Then (2.85) gives

$$\gamma_2^x(s) < (\gamma_1^x)^{-1}(s), \quad \forall\, 0 < s < r_\infty,$$

hence, (11.23).

Now we will prove that the solution of system (2.73) under the connection (2.74) is bounded for all $t \geq t_0$. Suppose this is not the case, for every number $R > 0$, there exists a time $T > t_0$, such that $\|x_1(T)\| > R$ or $\|x_2(T)\| > R$. Without lose of generality, we only consider the case of $\|x_1(T)\| > R$. Choose R such that, with $u = u_c$,

$$R > \max\Big\{\beta_1(\|x_1(t_0)\|, 0), \gamma_1^x \circ \beta_2(\|x_2(t_0)\|, 0), \gamma_1^x \circ \gamma_2^u\left(\left\|u_{[t_0,\infty)}\right\|\right), \\ \gamma_1^u\left(\left\|u_{[t_0,\infty)}\right\|\right)\Big\}.$$

From (2.84) with $i = 1, 2$, we have

$$\left\|x_{1[t_0,T]}\right\| \leq \max\left\{\beta_1(\|x_1(t_0)\|, 0), \gamma_1^x\left(\left\|x_{2[t_0,T]}\right\|\right), \gamma_1^u\left(\left\|u_{[t_0,T]}\right\|\right)\right\} \tag{11.24}$$

$$\left\|x_{2[t_0,T]}\right\| \leq \max\left\{\beta_2(\|x_2(t_0)\|, 0), \gamma_2^x\left(\left\|x_{1[t_0,T])}\right\|\right), \gamma_2^u\left(\left\|u_{[t_0,T]}\right\|\right)\right\}. \tag{11.25}$$

Substituting (11.25) and (11.24) gives

$$\left\|x_{1[t_0,T]}\right\| \leq \max\Big\{\beta_1(\|x_1(t_0)\|, 0), \gamma_1^x \circ \beta_2(\|x_2(t_0)\|, 0), \gamma_1^x \circ \gamma_2^x\left(\left\|x_{1[t_0,T]}\right\|\right), \\ \gamma_1^x \circ \gamma_2^u\left(\left\|u_{[t_0,T]}\right\|\right), \gamma_1^u\left(\left\|u_{[t_0,T]}\right\|\right)\Big\}.$$

Since

$$\left\|x_{1[t_0,T]}\right\| > \gamma_1^x \circ \gamma_2^x\left(\left\|x_{1[t_0,T]}\right\|\right),$$

we have

$$\left\|x_{1[t_0,T]}\right\| \leq \max\Big\{\beta_1(\|x_1(t_0)\|, 0), \gamma_1^x \circ \beta_2(\|x_2(t_0)\|, 0), \gamma_1^x \circ \gamma_2^u\left(\left\|u_{[t_0,\infty)}\right\|\right), \\ \gamma_1^u\left(\left\|u_{[t_0,\infty)}\right\|\right)\Big\} < R, \tag{11.26}$$

which contradicts $\|x_1(T)\| > R$. Therefore, the solution of system (2.73) under the connection (2.74) is bounded, and unique for all $t \geq t_0$. Let $M = \left\|u_{[t_0,\infty)}\right\|$, $M_1 = \max\left\{\gamma_1^x \circ \gamma_2^u(M), \gamma_1^u(M)\right\}$, and $M_2 = \max\left\{\gamma_2^x \circ \gamma_1^u(M), \gamma_2^u(M)\right\}$. From

(11.26), and by symmetry of x_1 and x_2, we have, for all $t \geq t_0$,

$$\|x_1(t)\| \leq \max\left\{\beta_1(\|x_1(t_0)\|, 0), \gamma_1^x \circ \beta_2(\|x_2(t_0)\|, 0), M_1\right\} \leq \max\{\delta_1(\|x(t_0)\|), M_1\}$$
$$\|x_2(t)\| \leq \max\left\{\beta_2(\|x_2(t_0)\|, 0), \gamma_2^x \circ \beta_1(\|x_1(t_0)\|, 0), M_2\right\} \leq \max\{\delta_2(\|x(t_0)\|), M_2\}$$

with $\delta_1(s) = \max\left\{\beta_1(s, 0), \gamma_1^x \circ \beta_2(s, 0)\right\}$ and $\delta_2(s) = \max\left\{\beta_2(s, 0), \gamma_2^x \circ \beta_1(s, 0)\right\}$. Therefore,

$$\begin{aligned}\|x(t)\| &\leq \|x_1(t)\| + \|x_2(t)\| \leq \max\{2\delta_1(\|x(t_0)\|), 2\delta_2(\|x(t_0)\|), 2M_1, 2M_2\} \\ &= \max\{\delta_3(\|x(t_0)\|), M_3\} = x_\infty \end{aligned} \tag{11.27}$$

with $\delta_3(s) = \max\{2\delta_1(s), 2\delta_2(s)\}$ and $M_3 = \max\{2M_1, 2M_2\}$. Hence, $\left\|x_{[t_0,\infty)}\right\| \leq x_\infty$.

Next, for any time $t_1 \geq 0$, we have

$$\begin{aligned}\|x_1(t_1 + t_0)\| &\leq \max\left\{\beta_1(\|x_1(t_1/2 + t_0)\|, t_1/2), \gamma_1^x\left(\left\|x_{2[t_1/2+t_0, t_1+t_0]}\right\|\right), \gamma_1^u(M)\right\} \\ &\leq \max\left\{\beta_1(x_\infty, t_1/2), \gamma_1^x\left(\left\|x_{2[t_1/2+t_0, t_1+t_0]}\right\|\right), \gamma_1^u(M)\right\}\end{aligned} \tag{11.28}$$

and for $\tau \in [t_1/2, t_1]$,

$$\begin{aligned}\|x_2(\tau + t_0)\| &\leq \max\left\{\beta_2(\|x_2(t_1/4 + t_0)\|, \tau - t_1/4), \gamma_2^x\left(\left\|x_{1[t_1/4+t_0, \tau+t_0]}\right\|\right), \gamma_2^u(M)\right\} \\ &\leq \max\left\{\beta_2(x_\infty, t_1/4), \gamma_2^x\left(\left\|x_{1[t_1/4+t_0, t_1+t_0]}\right\|\right), \gamma_2^u(M)\right\}.\end{aligned} \tag{11.29}$$

Substituting (11.29) into (11.28) gives

$$\begin{aligned}\|x_1(t_1 + t_0)\| &\leq \max\left\{\beta_1(x_\infty, t_1/2), \gamma_1^x \circ \beta_2(x_\infty, t_1/4),\right. \\ &\qquad \left.\gamma_1^x \circ \gamma_2^x\left(\left\|x_{1[t_1/4+t_0, t_1+t_0]}\right\|\right), \gamma_1^x \circ \gamma_2^u(M), \gamma_1^u(M)\right\} \\ &\leq \max\left\{\bar{\beta}_1(x_\infty, t_1), \gamma_1^x \circ \gamma_2^x\left(\left\|x_{1[t_1/4+t_0, t_1+t_0]}\right\|\right), M_1\right\}\end{aligned}$$

for a class $\mathcal{KL}$ function $\bar{\beta}_1$ satisfying

$$\bar{\beta}_1(s, t) \geq \max\{\beta_1(s, t/2), \gamma_1^x \circ \beta_2(s, t/4)\}.$$

Next, denote $z_1(t_1) = \|x_1(t_1 + t_0)\|$ gives

$$z_1(t_1) \leq \max\left\{\bar{\beta}_1(x_\infty, t_1), \gamma_1^x \circ \gamma_2^x\left(\left\|z_{1[t_1/4, t_1]}\right\|\right), M_1\right\}.$$

By Lemma 11.4, there exists a class $\mathcal{KL}$ function $\hat{\beta}_1$ such that

$$z_1(t_1) \leq \max\{\hat{\beta}_1(x_\infty, t_1), M_1\}.$$

Hence,

$$\|x_1(t)\| \le \max\{\hat{\beta}_1(x_\infty, t - t_0), M_1\}, \ \forall t \ge t_0.$$

By symmetry of x_1 and x_2, we have,

$$\|x_2(t)\| \le \max\{\hat{\beta}_2(x_\infty, t - t_0), M_2\}, \ \forall t \ge t_0.$$

As a result,

$$\begin{aligned} \|x(t)\| &\le \|x_1(t)\| + \|x_2(t)\| \\ &\le \max\{2\hat{\beta}_1(x_\infty, t - t_0), 2\hat{\beta}_2(x_\infty, t - t_0), 2M_1, 2M_2\} \\ &\le \max\{\beta_3(x_\infty, t - t_0), M_3\}, \ \forall t \ge t_0 \end{aligned} \tag{11.30}$$

for

$$\beta_3(s, t) = \max\{2\hat{\beta}_1(s, t), 2\hat{\beta}_2(s, t)\}.$$

This is not yet the ISS property since x_∞ depends not only on $x(t_0)$ but also on u. To split this dependence, we consider the following two cases.

(i) $\delta_3(\|x(t_0)\|) \ge M_3$: Since $x_\infty = \delta_3(\|x(t_0)\|)$, (11.30) gives

$$\|x(t)\| \le \max\{\beta_3(\delta_3(\|x(t_0)\|), t - t_0), M_3\}, \ \forall t \ge t_0$$

(ii) $\delta_3(\|x(t_0)\|) \le M_3$: From (11.27), we have $\|x(t)\| \le M_3$.

Now, we have obtained the following inequality

$$\|x(t)\| \le \max\left\{\beta(\|x(t_0)\|, t - t_0), \gamma\left(\left\|u_{[t_0,\infty)}\right\|\right)\right\}, \ \forall t \ge t_0 \tag{11.31}$$

for a class $\mathcal{KL}$ function $\beta(s, t) = \beta_3(\delta_3(s), t)$. Since the solution $x(t)$ depends only on $u(\tau)$ for $t_0 \le \tau \le t$, the supremum on the right-hand side of (11.31) can be taken over $[t_0, t]$ which yields (2.86). As a result, the proof is completed. □

11.4 Notes and References

The existence and uniqueness of the solution of a nonlinear system can be found in many places. A convenient reference is [5]. Theorems 11.1 and 11.2 are from Sect. 3.1 of [5]. Frobenius Theorem is taken from Sect. 1.4 of [6]. Center Manifold Theorem can find the references in [7] and Sect. B.1 of [6]. The inequalities in Lemma 11.1 are extended from similar results used in [8, 9], etc. Lemma 11.4 and the proof of Theorem 2.12 are adapted from [10]. An early version of the small gain theorem and its proof can be found in [11].

References

1. Jiang ZP, Teel AR, Praly L (1994) Small-gain theorem for ISS systems and applications. Math Control Signals Syst 7:95–120
2. Lin Y (1992) Lyapunov function techniques for stabilization. Ph.D. thesis, Rutgers University
3. Lin Y, Sontag ED, Wang Y (1993) Recent results on lyapunov theoretic techniques for nonlinear stability. In: Proceedings of the 1994 American control conference. Report SYCON-93-09
4. Isidori A (1999) Nonlinear control systems. Springer, New York
5. Khalil H (2002) Nonlinear systems. Prentice Hall, New Jersey
6. Isidori A (1995) Nonlinear control systems, 3rd edn. Springer, New York
7. Carr J (1981) Applications of the center manifold theory. Springer, USA
8. Praly L, Jiang ZP (1993) Stabilization by output feedback for systems with ISS inverse dynamics. Syst Control Lett 21:19–33
9. Qian C (2001) A continuous feedback approach to global strong stabilization of nonlinear systems. IEEE Trans Autom Control 46:1061–1079
10. Chen Z, Huang J (2005) A simplified small gain theorem for time-varying nonlinear systems. IEEE Trans Autom Control 50:1904–1908
11. Jiang ZP, Mareels I (1997) A small-gain control method for nonlinear cascaded systems with dynamic uncertainties. IEEE Trans Autom Control 42:292–308

Index

Z. Chen and J. Huang, *Stabilization and Regulation of Nonlinear Systems*,
Advanced Textbooks in Control and Signal Processing,
DOI 10.1007/978-3-319-08834-1

The manufacturer's authorised representative in the EU is Springer Nature Customer Service Centre GmbH, Europaplatz 3, 69115 Heidelberg, Germany. If you have any concerns regarding our products, please contact ProductSafety@springernature.com

Printed and bound by CPI Group (UK) Ltd, Croydon, CR0 4YY

23/03/2026

02076658-0005